T0331114

The Sharpe Ratio

The Sharpe Ratio
Statistics and Applications

Steven E. Pav

CRC Press is an imprint of the
Taylor & Francis Group, an **informa** business
A CHAPMAN & HALL BOOK

First edition published 2022
by CRC Press
6000 Broken Sound Parkway NW, Suite 300, Boca Raton, FL 33487-2742

and by CRC Press
2 Park Square, Milton Park, Abingdon, Oxon, OX14 4RN

CRC Press is an imprint of Taylor & Francis Group, LLC

ISBN: 9781032019307 (hbk)
ISBN: 9781032019314 (pbk)
ISBN: 9781003181057 (ebk)

DOI: 10.1201/9781003181057

Typeset in CMR10
by KnowledgeWorks Global Ltd.

For Oskar and Laszlo.

Contents

Foreword

> The future will soon be a thing of the past.
>
> George Carlin

Quantitative methods are important tool for research in economics and finance. Due to the rapid development of computer power and programming methods it becomes now possible to deal with a large amount of data and to analyse complex systems. On the other side, this requires the derivation of new mathematical and statistical methods needed to solve challenging tasks arising in financial applications, in particular when an optimal portfolio is constructed.

The impact of the estimation error on investments has been recognized in finance. During the last several decades researchers and practitioners in financial sector have dealt with challenging tasks of incorporating the parameter uncertainty in the decision process. Quantification of the estimation error has become an important part of optimal portfolio theory which provides a new way of understanding the stochastic behavior of the portfolio returns.

The book "The Sharpe Ratio: Statistics and Applications" written by Steven E. Pav provides an excellent and important contribution to quantitative finance literature. It develops statistical theory for characterizing the properties of the estimated Sharpe ratio and optimal portfolios. Valuable theoretical results are obtained under several assumptions imposed on the distribution of asset returns. Statistical inference procedures, like point estimation, interval estimation, as well as test theory, are developed for the estimators of the Sharpe ratio and optimal portfolios. The author presents the results derived using methods of both the frequentist statistics and Bayesian statistics. The developed theory is illustrated in several real-life examples. Two **R** packages, **SharpeR** and **MarkowitzR**, developed by the author provide very helpful technical support for the implementation of the theory. The results from matrix algebra, probability theory, and statistics presented at the end of the book give necessary mathematical background for better understanding of its main part.

Taras Bodnar
Department of Mathematics
Stockholm University

Preface

> I am at the moment writing a lengthy indictment against our century. When my brain begins to reel from my literary labors, I make an occasional cheese dip.
>
> JOHN KENNEDY TOOLE
> *A Confederacy of Dunces*

I wrote this book for myself. My career as a quantitative strategist began when I answered a Craigslist Help Wanted ad for a machine learning hedge fund. I was a mathematician and hacker, with no formal training in statistics; I was woefully unprepared for the job. During my tenure I read many papers and some books, always looking for, but never finding, The Book for Quants, some arcane manual that contained the techniques for identifying alpha that all real quants used. When I left that world, I set out to write that book. The result, much narrower in scope, you hold now: a book about the Sharpe ratio.

This is not meant to detract from the many fine books available on the topic of quantitative portfolio management and optimization, including Chincarini and Kim [2010], Qian *et al.* [2007], and Best [2010], among others. More generally, the texts by Kennedy [2008], Campbell *et al.* [2012], Rosenthal [2018], and Dixon *et al.* [2020] provide a good overview of econometrics, statistics, computing, and machine learning within finance. In contrast with existing works, this book is unwavering in its focus on the Sharpe ratio: what it means, how to measure it, and how to optimize it.

My hope is that this book is useful both to the practicing quantitative strategist and to the academic studying investing. The book is organized as a textbook, with plenty of examples and exercises for self-study, but it was not designed for any academic class. The material could likely be covered in one semester for advanced undergraduates with prerequisite knowledge of statistical practice and linear algebra. A fair amount of mathematical maturity is also assumed.

I also hope this book can serve as a reference in the field. Most of the results are known facts about the t-statistic and Hotelling's T^2, recycled as results on the Sharpe ratio and the Markowitz portfolio. It also contains what I thought were novel discoveries I have made in the area. [Pav, 2013, 2014, 2015, 2019b,c, 2020b] Some of the material is rather heterodox. If I had known the orthodoxy when I started, I never would have written this book.

None of the material herein should be considered explicit investment advice. The whole point of the book is to empower you to perform your own analyses.

Organization

The first part of this book examines the Sharpe ratio, roughly under the constraint that one can choose to invest some non-specified amount into a single asset. The second part considers optimization of the Sharpe ratio when one can invest fractional amounts, long or short, in a number of assets.

The first chapter introduces the Sharpe ratio and the signal-noise ratio. Next we consider inference on the signal-noise ratio when returns are Gaussian. In the third chapter we consider inference under deviations from the assumption of Gaussian returns, and in the asymptotic case. The fourth chapter considers "overoptimism", when one has measured the Sharpe ratio of many assets, then selects one.

The second part starts with chapter 5, which covers the theory of maximization of the signal-noise ratio if population moments were known. Chapter 6 considers inference on the portfolio problem under Gaussian returns. Chapter 7 generalizes that analysis to the non-Gaussian case. In chapter 8, we examine the difference between the inferred maximal signal-noise ratio and that achieved by a sample portfolio.

Chapter 9 considers "market timing," and was originally intended to be a segue between the two parts of the book. The material there is now mostly subsumed by chapter 6. Finally in chapter 10, we discuss backtesting.

The appendix contains a chapter with much of the preliminary material required to absorb the book.

Technical Details

This document was compiled into LaTeX using the **knitr** package in **R**. The analysis and presentation relied heavily on the **ggplot2**, **dplyr**, **xts**, and **Rcpp** packages. [Xie, 2018; Wickham, 2016; Wickham *et al.*, 2020; Ryan and Ulrich, 2020; Eddelbuettel and François, 2011] Much of the bespoke analysis of the Sharpe ratio and the Markowitz portfolio is (or ought to be) via the author's own **SharpeR** and **MarkowitzR** packages. [Pav, 2020a, 2018] The data sets examined in this text are distributed in the companion **R** package, **tsrsa**, which may eventually contain supporting code in addition to the data. You can install this package via:

```
remotes::install_github("shabbychef/tsrsa/rpkg")
```

If you have any questions about this text or the Sharpe ratio, you may email me at `steven@sharperat.io`.

Acknowledgments

This book was greatly improved by feedback from Michael Lachanski, Eric Wiewiora, Stephen Markacs, Dale Rosenthal, Stephen Taylor, Abraham D. Flaxman, Scott Locklin, Zachary David, and the anonymous reviewer, for which I am grateful. All errors that remain are mine alone. A thank you to Attaboy and Joy Patterson for cover art direction. I am grateful to Rob Calver at Taylor & Francis Group for taking a calculated risk on this book. Many thanks to Mansi Agarwal for editing and proofreading the book. Thank you to Taras Bodnar for writing a thoughtful foreword.

For good cheer and camraderie over the years, I wish to thank my friends Daniel Seifert, Stephen Markacs, Marius Fromm, Paul Ford, Joy Patterson, Heather Rowe, Ian Rothwell, Mark Bishop, Abie Flaxman, Jessi Berklehammer, Devon Yates, Momoko Ono-Kawasaki, Mike Greer and Kaliel Roberts; my colleagues Bill Carty, Mr. Kinoshita and the good people of Kyocera in Kokubu, Japan, Sugihara Kokichi, Jim Bunch, Mike O'Neil, Radu Mihaescu, Jesse Davis, Vincent Pribish, Jake Freifeld, Geoff Anders, Eric Wiewiora, Sirr Less, and Raj Kamat; and the teachers who showed me the beauty of mathematics, including Daniel Hall, Arlen Jividen, David Kelley, Thomas Hull, Addison Frey, Larry Moss, Douglas Hofstadter, Noel Walkington, and Nancy Pav.

I began to research the Sharpe ratio when I worked at Cerebellum Capital. Many thanks to Astro Teller, David Andre and Conrad Gann for encouraging the pursuit of truth.

This work would have been considerably different without the **R** language and the packages freely available from CRAN. My sincere appreciation goes to those who maintain the language and the repository, and who author and maintain packages. Thanks also to the organizers (and a hearty cheers to the community) of the R in Finance conference.

For firm support and love over the years, much love to Richard, Nancy, Bill, and Dan Pav, and to Linda Shipley, Oskar and Laszlo Pav.

Many years ago I found my way through a crossroads through the effort of William Zeimer, to whom I am ever grateful.

Lastly, I must thank Linda, for teaching me the value of everything.

San Francisco,
August, 2021

List of Figures

List of Tables

Symbols

> Mathematics, a veritable sorcerer in our computerized society, while assisting the trier of fact in the search for truth, must not cast a spell over him.
>
> SUPREME COURT OF CALIFORNIA
> *People v. Collins*

Symbol Description

Symbol	Description
$:=$	"defined as." One would read $a := b$ as "a is defined as b."
x_t	The returns of an asset at time t. Time is treated as a discrete index.
$\boldsymbol{x}$	A vector of the returns of an asset, over time.
n	Typically the sample size, the number of observations of the return of an asset or collection of assets.
μ	The population mean return of a single asset.
σ	The population standard deviation of a single asset.
ζ	The population signal-to-noise ratio (SNR), defined as $\zeta := \mu/\sigma$.
$\hat{\mu}$	The unbiased sample mean return of a single asset.
$\hat{\sigma}$	The sample standard deviation of returns of a single asset.
$\hat{\zeta}$	The sample Sharpe ratio, defined as $\hat{\zeta} := \hat{\mu}/\hat{\sigma}$.
r_0	The risk-free, or disastrous rate of return.
α_i	The ith uncentered moment of a random variable, $\operatorname{E}\left[x^i\right]$.
$\hat{\alpha}_i$	The ith uncentered sample moment of a sample, typically computed as the simple mean of the observations to the ith power.
μ_i	The ith centered moment of a random variable, defined as $\operatorname{E}\left[(x - \operatorname{E}\left[x\right])^i\right]$.
$\hat{\mu}_i$	The ith centered sample moment of a sample.
κ_i	The $i+2$th raw cumulant of a random variable.
γ_i	The $i+2$th standardized cumulant of a random variable, equal to the $i+2$th raw cumulant divided by σ^{i+2}.
γ_1	The skew of a random variable.

γ_2 The excess kurtosis of a random variable.

κ The kurtosis factor of an elliptical distribution. This is one third the (regular, not excess) kurtosis of the marginals, and equals 1 for the Gaussian distribution.

$\boldsymbol{f}_t$ Under the attribution model, a l-vector of factors which "explain" or contextualize the returns of an asset at time t.

l The number of factors in the attribution model.

$\boldsymbol{\beta}$ An l-vector of the linear coefficient under the attribution model. Thus we have $\mathrm{E}\left[x_t \,|\, \boldsymbol{f}_t\right] = \boldsymbol{\beta}^\top \boldsymbol{f}_t$.

ζ_g The "ex-factor signal-noise ratio" under the attribution model, defined as $\zeta_g := \left(\boldsymbol{\beta}^\top \boldsymbol{v} - r_0\right)/\sigma$, for some vector $\boldsymbol{v}$.

F An $n \times l$ matrix of observed factors.

$\hat{\boldsymbol{\beta}}$ An l-vector of the estimated linear coefficient.

$\hat{\zeta}_g$ The sample "ex-factor Sharpe ratio" under the attribution model, defined as $\hat{\zeta}_g := \left(\hat{\boldsymbol{\beta}}^\top \boldsymbol{v} - r_0\right)/\hat{\sigma}$.

k Typically the number of assets in the multiple asset case.

X A $n \times k$ matrix of the returns of k assets in the multiple asset case.

$\boldsymbol{\mu}$ The population mean return vector of k assets.

$\mathsf{\Sigma}$ The population covariance matrix of k assets.

$\boldsymbol{\nu}_*$ The maximal SNR portfolio, constructed using population data: $\boldsymbol{\nu}_* := \mathsf{\Sigma}^{-1}\boldsymbol{\mu}$.

ζ_* The SNR of $\boldsymbol{\nu}_*$, taking value $\zeta_* = \sqrt{\boldsymbol{\mu}^\top \mathsf{\Sigma}^{-1}\boldsymbol{\mu}}$.

f The number of features in the conditional expectation model.

$\boldsymbol{f}_{t-1}$ A f-vector of features for predicting or explaining returns.

F A $n \times f$ matrix, each row of which is an observed feature vector.

B A $k \times f$ matrix of linear coefficients mapping a feature to expected returns under the model: $\mathrm{E}\left[\boldsymbol{x}_t \,\middle|\, \boldsymbol{f}_{t-1}\right] = \mathsf{B}\boldsymbol{f}_{t-1}$.

N A matrix of linear coefficient which one multiplies by a feature vector $\boldsymbol{f}_t$ to construct a portfolio, constructed using population data. Also called a "passthrough matrix."

$\hat{\boldsymbol{\mu}}$ The sample mean return vector of k assets.

$\hat{\mathsf{\Sigma}}$ The sample covariance matrix of k assets.

$\hat{\boldsymbol{\nu}}$ A portfolio, built on sample data.

$\hat{\boldsymbol{\nu}}_*$ The sample maximal Sharpe ratio portfolio, $\hat{\boldsymbol{\nu}}_* := \hat{\mathsf{\Sigma}}^{-1}\hat{\boldsymbol{\mu}}$.

$\hat{\zeta}_*$ The Sharpe ratio of $\hat{\boldsymbol{\nu}}_*$, taking value $\hat{\zeta}_* = \sqrt{\hat{\boldsymbol{\mu}}^\top \hat{\mathsf{\Sigma}}^{-1}\hat{\boldsymbol{\mu}}}$.

$\hat{\mathsf{B}}$ The sample estimate of B, constructed via multivariate multiple linear regression.

$\hat{\mathsf{N}}$ A matrix of the linear coefficient which one multiplies by a feature vector $\boldsymbol{f}_{t-1}$ to construct a portfolio, built on sample data.

$\hat{\mathsf{N}}_*$ The sample optimal matrix

which maximizes the Sharpe ratio under the conditional expectation model, defined as $\hat{\mathsf{N}}_* := c\hat{\Sigma}^{-1}\hat{\mathsf{B}}$.

R A risk budget for optimization problems.

T The population Hotelling-Lawley trace.

$\hat{T}$ The sample Hotelling-Lawley trace.

G Often a matrix of portfolios against which one wants to have no covariance. These are 'hedged out' portfolios.

$\boldsymbol{\nu}_{*,\mathsf{I}\backslash\mathsf{G}}$ The population optimal portfolio that is hedged against G.

$\hat{\boldsymbol{\nu}}_{*,\mathsf{I}\backslash\mathsf{G}}$ The sample optimal portfolio that is hedged against G.

$\Delta_{\mathsf{I}\backslash\mathsf{G}}\,\zeta_*^2$ The squared maximal signal-noise ratio of the portfolio that is hedged against G.

$\Delta_{\mathsf{I}\backslash\mathsf{G}}\,\hat{\zeta}_*^2$ The squared maximal Sharpe ratio of $\hat{\boldsymbol{\nu}}_{*,\mathsf{I}\backslash\mathsf{G}}$.

$q\left(\hat{\boldsymbol{\nu}}\right)$ The achieved signal-noise ratio of portfolio $\hat{\boldsymbol{\nu}}$.

$\mathcal{N}\left(\mu, \sigma^2\right)$ The normal distribution with mean μ and variance σ^2.

$\sim$ "Follows the law." We will write $x \sim \mathcal{N}\left(\mu, \sigma^2\right)$ to mean that x follows a normal distribution with mean μ and variance σ^2.

$\propto$ "Proportional to", but also used in the Bayesian context to denote a prior or posterior belief takes a given distribution.

$f_{\text{tas}}\left(\cdot\right)$ The 'TAS' function, defined as tangent of arcsine.

$\odot$ The Hadamard, or elementwise, multiplication operator.

$\otimes$ The Kronecker multiplication operator.

$\operatorname{Diag}\left(\boldsymbol{a}\right)$ The diagonal matrix whose diagonal is $\boldsymbol{a}$.

$\operatorname{diag}\left(\mathsf{A}\right)$ The vector of the diagonal of matrix A.

$A \mid B$ A conditional on B.

$a \vee b$ The maximum of a and b.

$a \wedge b$ The minimum of a and b.

$\rightsquigarrow$ "Converges in distribution," often in the context of the central limit theorem or multivariate delta method.

Part I

The Sharpe Ratio

1

The Sharpe Ratio and the Signal-Noise Ratio

> A man who seeks advice about his actions will not be grateful for the suggestion that he maximise expected utility.
>
> A. D. ROY
> *Safety First and the Holding of Assets*

1.1 Introduction

The Sharpe ratio is arguably the most commonly used metric of the historical performance of financial assets–mutual funds, hedge funds, stocks, *etc.* It is defined as

$$\hat{\zeta} := \frac{\hat{\mu} - r_0}{\hat{\sigma}}, \tag{1.1}$$

where $\hat{\mu}$ is the historical, or sample, mean return of the mutual fund, $\hat{\sigma}$ is the sample standard deviation of returns, and r_0 is some fixed *risk-free* or *disastrous* rate of return. Under the original definition of Sharpe's "reward-to-variability ratio," r_0 was equal to zero. [Sharpe, 1965] One typically uses the vanilla sample mean and the Bessel-corrected sample standard deviation in computing the ratio, *viz.*

$$\hat{\mu} := \frac{1}{n} \sum_{1 \le t \le n} x_t, \quad \hat{\sigma} := \sqrt{\frac{1}{n-1} \sum_{1 \le t \le n} (x_t - \hat{\mu})^2}. \tag{1.2}$$

The Bessel correction, the use of $n - 1$ in the denominator for $\hat{\sigma}$, makes $\hat{\sigma}^2$ an unbiased estimator for σ^2 when returns are independent and identically distributed ("*i.i.d.*") Gaussians.

In this text, *Sharpe ratio* will refer to this quantity, computed from sample statistics, whereas *signal-noise ratio* will refer to the analogously defined population parameter,

$$\zeta := \frac{\mu - r_0}{\sigma}. \tag{1.3}$$

DOI: 10.1201/9781003181057-1

In general, hats will be placed over quantities to denote population estimates. Sharpe himself notes that ideally population values would be used in the computation of his reward-to-variability ratio, but "since the predictions cannot be obtained in any satisfactory manner, ... ex post values must be used–the average rate of return of a portfolio must be substituted for its expected rate of return, and the actual standard deviation of its rate of return for its predicted risk." [Sharpe, 1965, p. 122]

To demonstrate this computation, we introduce a data set of the *Four Factor returns* that will be used throughout the text.

Example 1.1.1 (Fama-French Four Factor monthly returns). The monthly returns of "the Market" portfolio, the small cap portfolio (known as SMB, for "small minus big"), and the value portfolio (known as HML, for "high minus low"), are published periodically by Kenneth French. [French, 2017] These are the celebrated three factor portfolios of Fama and French [1992] The Market returns are the value-weighted returns of all U.S. incorporated stocks listed on one of the three major exchanges, and which match certain selection criteria for price data. Defined in this way, the Market index broadly reflects the increase in value of the U.S. stock market as a whole. Together with these three, we also consider the returns of the momentum portfolio (know as UMD, for "up minus down"), making a total of four factors. [Carhart, 1997] Throughout this text, we refer to these as "the" four factors.

The monthly returns data were downloaded directly from French's site [French, 2017] and stored in the **R** package **tsrsa**, which is a companion package to this text[1]. [Pav, 2019a] One can access this data in **R** as follows, which puts an `xts` object called `mff4` into the environment:

```
library(tsrsa)
data(mff4)
```

The data are distributed as monthly percent changes to the portfolio values, which we call relative returns (see below). The Market return is quoted as an *excess return*, with the risk-free rate subtracted out. The risk-free rate is also tabulated. For our purposes, the raw market returns are needed, so the risk-free rate is added back to the market excess returns. The risk-free rate, denoted RF, is kept as a kind of inflationary benchmark.

The set consists of $1,128$ mo. of data, from Jan 1927 through Dec 2020. The cumulative values of the factor returns (and of the risk-free rate) are shown in Figure 1.1. By "cumulative value," we mean the value of a one dollar initial investment in the portfolio replicating the factor returns, ignoring all costs and trade frictions. These are shown in nominal dollars, ignoring the changes in the real buying power of a dollar due to inflation, which is presumably related to the cumulative value of RF.

[1] Ken French does a great service to the community by computing and publishing his data. However, the data are published in a somewhat irregular form; to simplify access, the data have been repackaged in **tsrsa**.

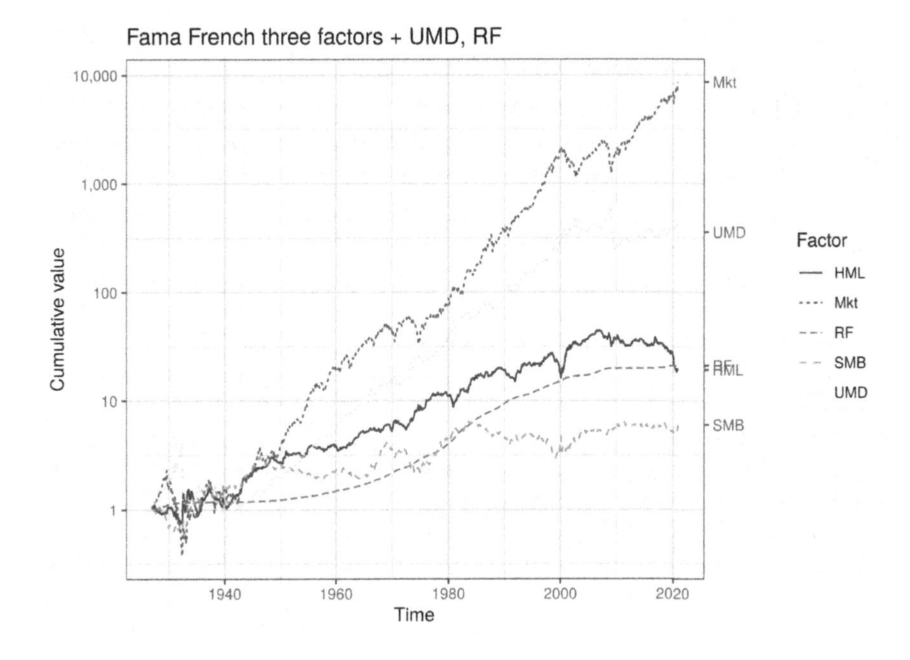

FIGURE 1.1
The cumulative value of the four Fama French factors, plus the risk-free rate, are shown, from Jan 1927 through Dec 2020. Cumulative value is normalized to the initial investment, ignores all trading costs, and is not adjusted for inflation. The y-axis is in log scale.

Typically in this text, the monthly returns of the factors will be used. However, at times the daily returns may be needed. These are available in **tsrsa** as follows:

```
library(tsrsa)
data(dff4)
```

This data set consists of 24,795 days of data from 1926-11-03 through 2020-12-31. As in the monthly returns data set, the daily returns are expressed in percent. ⊣

Example 1.1.2 (Sharpe ratio of the Market). Here we consider the raw returns of "the Market," defined as the Fama French Market factor introduced in Example 1.1.1. The mean monthly return from Jan 1927 to Dec 2020 is 0.95%; the standard deviation is 5.4%. Thus, the Sharpe ratio is $0.18\,\text{mo.}^{-1/2}$. (More on the funny units later.) ⊣

In some situations, there is a natural benchmark, r_t, against which returns are judged. Often this is called a risk-free rate of return, and it will likely vary over the historical period. In this case, the Sharpe ratio is computed over the *excess returns*, $x_t - r_t$. That is $\hat{\zeta}$ is computed as the ratio of the sample mean to the sample standard deviation of $x_t - r_t$.

Example 1.1.3 (Sharpe ratio of the Market, excess returns). Continuing Example 1.1.2, if we compute the *excess* returns of "the Market" by subtracting the risk-free rate from the Market returns, the Sharpe ratio is $0.126\,\text{mo.}^{-1/2}$. ⊣

In practice, the Sharpe ratio is often used to answer the following kinds of questions:

1. "*Should I invest a predetermined amount of money (long) in a given asset?*"
2. "*Should our quant fund replace strategy A with strategy B?*"
3. "*Should we invest in asset B given that we will keep our investment in asset A?*"

Note that each of these questions is binary: we are not asking *how much* should we invest in a given asset, nor whether we should potentially *short* the asset. Rather, the amount and side are predetermined. Later, we will consider portfolio problems, which are less constrained in nature.

We stress again the Sharpe ratio is a *historical* measure, but it is used to guide future investments. But as is often hidden in the fine print, historical performance is no guarantee of future results. In this book, we aim to explain just how much this historical measure can be trusted as an estimate of future results, under certain assumptions.

1.1.1 Which returns?

Let p_t be the price of an asset recorded at time t. We will index prices and returns with t, but will consider time as discrete, rather than continuous, with a fixed period. The *relative return* (also known as "arithmetic," "percent," "discrete," or "simple" return) from t to $t+1$ is defined as

$$\frac{p_{t+1} - p_t}{p_t} = \frac{p_{t+1}}{p_t} - 1. \tag{1.4}$$

The *log return* (also known as "geometric" or "continuous" return) is defined as

$$\log \frac{p_{t+1}}{p_t}. \tag{1.5}$$

The conversion between log and relative returns is simple: if r_t are the relative returns corresponding to log returns l_t, we have

$$l_t = \log\left(1 + r_t\right), \quad r_t = e^{l_t} - 1.$$

There is no clear standard which form of returns should be used in computation of the Sharpe ratio. Log returns are typically recommended because they aggregate over time by summation (*e.g.*, the sum of a week's worth of daily log returns is the weekly log return), and thus taking the mean of them is sensible. For this reason, adjusting the time frame (*e.g.*, annualizing) of log returns is trivial.

However, relative returns have the attractive property that they are additive "laterally": the relative return of a portfolio on a given day is the dollar-weighted mean of the relative returns of each position. This property is important when one considers more general attribution models, or portfolio returns. To make sense of the sums of relative returns one can think of a fund manager who always invests a fixed amount of capital, siphoning off excess returns into cash, or borrowing cash[2] to purchase stock. Under this formulation, the returns aggregate over time by summation just like log returns.

If the value of the underlying asset goes to zero, then the corresponding log return is $-\infty$. That is to say, log returns are unbounded. If there is non-zero probability that such a loss can be observed, then the log returns do not have finite first moments (or any moments), and thus the signal-noise ratio is undefined. This detail is typically swept under the rug[3]. In their favor, relative returns for most (perhaps all) investments are bounded: pick some insanely large number larger than 100, say $10^{10^{10}}$; the absolute value of the return of an investment, in percent, will not exceed this large number, and thus *all* moments of relative returns are finite. On the other hand, relative returns do not average in the way that log returns do: a 100% gain followed by a 100% loss does not average out to "no change in value."

One reason fund managers might use relative returns when reporting Sharpe ratio is because it inflates the results! The "boost" from computing the Sharpe ratio using relative returns is approximately:

$$\frac{\hat{\zeta}_r - \hat{\zeta}}{\hat{\zeta}} \approx \frac{1}{2}\frac{\sum_t {x_t}^2}{\sum_t x_t}, \tag{1.6}$$

where $\hat{\zeta}_r$ is the Sharpe ratio measured using relative returns and $\hat{\zeta}$ uses log returns. This approximation is most accurate for daily returns, and for the modest values of Sharpe ratio one expects to see for real funds. See Exercise 1.17.

It goes without saying that it is generally assumed that prices are measured at equal intervals, and so the returns are "over" comparable time periods. That is, if x_1 is one (market) day's return, then so is x_2, x_3, and so on. (Though see Example 3.1.3 for how to deal with jumbled periodicities.)

Example 1.1.4 (Sharpe ratio of the Market, log returns). Continuing Example 1.1.2, the relative returns of the Market were converted into log returns. The mean monthly log return is 0.008; the standard deviation is 0.054. Thus, the Sharpe ratio is $0.15\,\mathrm{mo.}^{-1/2}$. The boost from computing the Sharpe ratio on relative returns instead of log returns is around 18%. ⊣

Caution. It should be noted, however, that the use of simple returns is considered objectionable in some circles. Because simple returns do not sum as

[2]At no interest.

[3]In Section 3.2.5 we do exactly that.

geometric returns do, summing them to estimate long horizon returns is inaccurate. [Rivin, 2018] Another way to express this is that the change of time units we will introduce in the next section is misleading, and the Sharpe ratio should only be used on the time horizon over which returns are measured. Nonetheless, there should be little in this text that is specific to simple returns, and one may apply most or all of the results to the analysis of log returns.

1.2 Units of Sharpe ratio

The Sharpe ratio is often quoted without units. This is a recipe for disaster, since it is emphatically not a unit-less quantity, and often quoted in units different from those which it is measured.

To compute the units of Sharpe ratio, consider the units of μ and σ. Given two successive returns, x_1, x_2, which are assumed independent, let $x_{1:2} = x_1 + x_2$. Note that $x_{1:2}$ is the return of the asset over the two time periods only if the returns are log returns. The expected value of $x_{1:2}$ is 2μ; the variance of $x_{1:2}$ is $2\sigma^2$. Thus, the units of μ must be "return per time," and the units of σ^2 must be "return squared per time," where the units of return are those in which x are measured. The units of Sharpe ratio are thus "per square root time." Thus, when presented as an "annualized" number, the units of Sharpe ratio are "per square root years," or $\mathrm{yr}^{-1/2}$.

Once the units are explicit, any engineering student armed with sufficient domain knowledge can translate the units, as in the following example[4].

Example 1.2.1 (Annualization). Suppose one computes the sample mean and standard deviation of the (hypothetical) daily log returns of Asset Corp. to be, respectively, 3 $\mathrm{bps}\,\mathrm{day}^{-1}$ and 120 $\mathrm{bps}\,\mathrm{day}^{-1/2}$. The Sharpe ratio is then estimated as $0.025\,\mathrm{day}^{-1/2}$. To translate into "annualized" units, note there are around 252 *market* days per year, thus multiply by the unit quantity $\sqrt{252}\,\mathrm{day}^{1/2}\,\mathrm{yr}^{-1/2}$ to get a Sharpe ratio of around $0.4\,\mathrm{yr}^{-1/2}$. ⊣

Example 1.2.2 (Annualized Sharpe ratio, the Market). Continuing Example 1.1.2, the monthly Sharpe ratio of the Market is $0.18\,\mathrm{mo.}^{-1/2}$. To get annualized terms, multiply by the unit quantity $\sqrt{12}\,\mathrm{mo.}^{1/2}\,\mathrm{yr}^{-1/2}$ to get a Sharpe ratio of around $0.61\,\mathrm{yr}^{-1/2}$. ⊣

This annualization is exact in the case where one computes the Sharpe ratio on log returns, and the returns are independent. That is, if one defines $x_{1:252} = \sum_{1 \le t \le 252} x_t$, where x_t are daily log returns, then $x_{1:252}$ is the actual annual log return, and the signal-noise ratio of $x_{1:252}$ (its expected value divided by its standard deviation) is equal to $\sqrt{252}$ times the signal-noise ratio of x. This

[4]Herein you may see bps (pronounced "*bips*") used to refer to *basis points*, or hundredths of a percent. Sometimes, it is erroneously used to mean 10^{-4} in log return units.

change of units is not exact when dealing with relative returns, unless one imagines investing a constant amount. However, quoting the Sharpe ratio in annualized units for relative returns is often done anyway, because the scale is easier to comprehend, and the computations are approximately correct.

Often the Sharpe ratio is annualized in this manner without checking for potential autocorrelation of returns. Strong positive (negative) autocorrelation tends to cause one to underestimate (overestimate) the volatility of an asset, and thus overestimate (underestimate) the Sharpe ratio. For example, consider a lazy fund manager who provides a historical "daily" mark-to-market constructed by linear interpolation of the quarerly marks of the fund. By removing the daily variation, the standard deviation of returns would be underestimated. On the other extreme, consider a fund with a strong negative autocorrelation of monthly returns, resulting in a sawtooth mark to market: observing this pattern, one could time when to buy and when to sell the fund, in order to get in at a trough and out at a peak, and thus the actual volatility of returns is effectively overstated. See also Exercise 1.30 and Section 3.1.4.

1.2.1 Probability of a loss

Because the units are per square root time, it may make more sense to consider the parameter ζ^{-2}, which has time units. Indeed, if one measures signal-noise ratio on log returns, and assuming $\zeta > 0$, then $\kappa^2\zeta^{-2}$ is the time at which having accumulated a loss (or a loss against the risk-free rate) is a κ standard deviation event. For a fund with a signal-noise ratio of $0.7\,\text{yr}^{-1/2}$, the event of experiencing a loss over $8.163\,\text{yr}$ is a "2-sigma event."

Typically when one states that an outcome is a "2-sigma event," one uses the distribution function of the Gaussian distribution to estimate its probability. The yearly returns of a trading strategy should be very nearly normal, and so the probability so estimated may be accurate.

If the returns are sufficiently well behaved that one can accept the normal approximation, the probability of a loss over period $(\kappa/\zeta)^2$ is bounded from above by $e^{-\kappa^2/2}/\sqrt{2\pi\kappa^2}$, which is to say that doubling the Sharpe ratio of an asset cuts the probability of a loss by much more than a half. To be concrete, under the normal approximation, the probability of a 2-sigma event is around 0.023; the probability of a 4-sigma event is around 3.167×10^{-5}.

Without assuming a distributional shape to returns, however, we can find only a loose bound on the probability of a loss over some period via *Cantelli's inequality.* This inequality states that for a random variable X with mean μ and variance σ^2

$$\Pr\left\{X - \mu \le -\lambda\right\} \le \frac{\sigma^2}{\sigma^2 + \lambda^2}.$$

If μ, σ^2 are measured in units of yr^{-1}, then the probability of a loss over m years is bounded as

$$\Pr\left\{X - m\mu \le -m\mu\right\} \le \frac{m\sigma^2}{m\sigma^2 + m^2\mu^2} = \frac{1}{1 + m\zeta^2}.$$

Thus, over a period of length $(\kappa/\zeta)^2$, the probability of a loss is no greater than $\left(1+\kappa^2\right)^{-1}$.

Example 1.2.3 (Probability of a loss, Market returns). Using the monthly *log* returns of the Market, the Sharpe ratio is computed as $0.52\,\mathrm{yr}^{-1/2}$, which is the value from Example 1.1.4 in annual units. Under Cantelli's inequality, the probability of a down year is no greater than 0.79; under the normal approximation, the probability is around 0.3. For the 94 year sample, the empirical probability of a down year is around 0.24. Thus Cantelli's inequality is far too conservative, and the normal approximation appears reasonable.

Note that the reasoning is circular here: the Sharpe ratio and empirical rate of down years were computed using the same data. ⊣

1.2.2 Interpretation of the Sharpe ratio

The range of sane values one might expect to see of the Sharpe ratio vary depending on the type of asset, the market, relevant risk-free rate, and whether the returns are net or gross[5]. Among mutual funds and hedge funds, loosely, one should consider a signal-noise ratio above $0.5\,\mathrm{yr}^{-1/2}$ to be "good," while values above $1\,\mathrm{yr}^{-1/2}$ to be "very good." Apparently, Warren Buffet's Bershire Hathaway fund achieved a Sharpe ratio of "only" $0.76\,\mathrm{yr}^{-1/2}$ over a 30-year period. [Frazzini *et al.*, 2012] Jim Simons is reported to have achieved a Sharpe ratio of $1.89\,\mathrm{yr}^{-1/2}$ over a 10-year period. [Lux, 2000] Indeed, the Sharpe ratio of Simons' Medallion Fund, net to investors, from 1988 through 2018, is around $1.93\,\mathrm{yr}^{-1/2}$ based on yearly returns. [Zuckerman, 2019]

Those who engage in low-latency, "high-frequency trading" will often demur when asked what their achieved Sharpe ratio is, but may quote numbers as high as $20\,\mathrm{yr}^{-1/2}$ or higher. Often these numbers ignore large fixed costs in maintaining their infrastructure, or keeping up in the technological arms race.

Mkt	SMB	HML	UMD
0.61	0.23	0.32	0.48

TABLE 1.1
The Sharpe ratios of the four factor portfolios are given, in units of $\mathrm{yr}^{-1/2}$, computed on monthly relative returns from Jan 1927 to Dec 2020.

Example 1.2.4 (Sharpe ratio of the Fama French Four Factors). The Sharpe ratios of the Market, SMB, HML and UMD portfolios introduced in Example 1.1.1 were computed on the monthly relative returns, annualized, then tabulated in Table 1.1. The takeaway is that very simple monthly-rebalancing strategies have achieved Sharpe ratios in the ballpark of $0.6\,\mathrm{yr}^{-1/2}$, subject to the survivorship bias of examining the U.S. stock market. ⊣

[5]Of course, the number of samples over which one computes the Sharpe ratio is also relevant; more on this in the sequel.

We introduce another data set which will be used in the remainder of the book, namely the returns of five industry portfolios.

Example 1.2.5 (French 5 Industry monthly returns). Kenneth French tabulates the returns of five portfolios, or baskets, of stocks constructed by their primary industry classification. The five portfolios correspond to "Consumer goods," "Manufacturing," "High-Tech," "Healthcare," and "Other." [French, 2019, 2017] The monthly returns of value-weighted portfolios were downloaded from French's library and available in the **tsrsa** package: [French, 2017; Pav, 2021]

```
library(tsrsa)
data(mind5)
```

The data are distributed as monthly relative returns, quoted in percents. The set consists of 1,128 mo. of data, from Jan 1927 through Dec 2020. The cumulative values of the industry portfolios are shown in Figure 1.2. The portfolios are clearly highly correlated. Monthly returns are scattered against each other in Figure 1.3; the smallest correlation between two portfolios is 0.71. ⊣

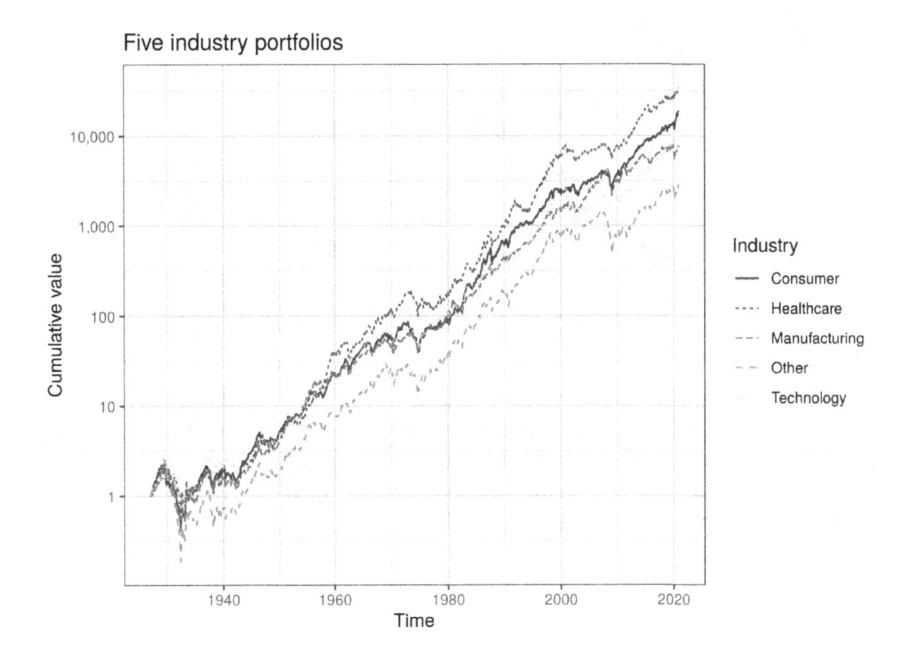

FIGURE 1.2
The cumulative value of the five industry portfolios are shown, from Jan 1927 through Dec 2020. Cumulative value is normalized to the initial investment, ignores all trading costs, and is not adjusted for inflation.

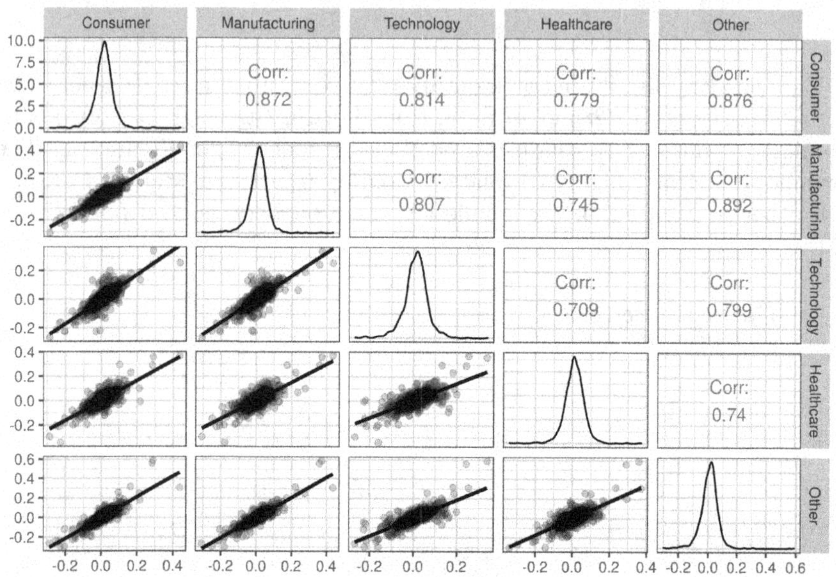

FIGURE 1.3
The monthly returns of the five industry portfolios, from Jan 1927 through Dec 2020 are scattered against each other. In the lower triangle, returns are scattered against each other. Pearson correlations are given in the upper triangle. On the diagonal, the empirical distribution of returns are plotted. Industry portfolio returns are clearly highly correlated.

Example 1.2.6 (Sharpe ratio of the five industry portfolios). The Sharpe ratio of the five industry portfolios introduced in Example 1.2.5 were computed on the monthly relative returns, annualized, then tabulated in Table 1.2. These achieve about the same Sharpe ratio as the Market factor, all around $0.6\,\mathrm{yr}^{-1/2}$. ⊣

Consumer	Manufacturing	Technology	Healthcare	Other
0.67	0.59	0.62	0.68	0.49

TABLE 1.2
The Sharpe ratio of the five industry portfolios are given, in units of $\mathrm{yr}^{-1/2}$, computed on monthly relative returns from Jan 1927 to Dec 2020.

1.3 Historical perspective

That we call this ratio the "Sharpe ratio" is another example of Stigler's Law of Eponymy. [Stigler, 1980] Roy [1952] introduced the "Safety First" criterion as an objective to be maximized in investment decisions. [Sullivan, 2011] Roy suggested one should minimize the probability that returns were below some disastrous rate of return, r_0. Arguing from Chebyshev's inequality, Roy showed that

$$\Pr\left\{x \le r_0\right\} \le \left(\frac{\sigma}{\mu - r_0}\right)^2. \tag{1.7}$$

To minimize the probability of a loss (the left-hand side), one should minimize the quantity on the right, or equivalently maximize $\left(\mu - r_0\right)/\sigma$. Roy then goes on to illustrate what we now call the mean-variance frontier.

As a sample statistic, the Sharpe ratio is fairly similar to the t-statistic. The t-test for the hypothesis $H_0 : \mu \le r_0$ uses the quantity

$$\frac{\hat{\mu} - r_0}{\sqrt{\hat{\sigma}^2/n}} = \sqrt{n}\hat{\zeta}.$$

This modern form of the t-test is *not* the form first considered by Gosset [1908] (writing as "Student"). Gosset originally analyzed the distribution of

$$z = \frac{\hat{\mu}}{s_N} = \frac{\hat{\mu}}{\hat{\sigma}\sqrt{(n-1)/n}} = \hat{\zeta}\sqrt{\frac{n}{n-1}},$$

where s_N is the "standard deviation of the sample," a biased estimate of the population standard deviation that uses n in the denominator instead of $n-1$. Gosset's statistic is asymptotically equivalent to the Sharpe ratio, a fact we will abuse in the sequel.

1.4 Linear attribution models

The Sharpe ratio can be viewed as a specific case of an *attribution model*, or *factor model*. In the general case, one attributes the returns of the asset in question as the linear combination of l factors, one of which is typically the constant one:

$$x_t = \beta_0 1 + \sum_{i}^{l-1} \beta_i f_{i,t} + \epsilon_t, \tag{1.8}$$

where $f_{i,t}$ is the value of some ith "factor" at time t, and the innovations, ϵ, are assumed to be zero mean, and have standard deviation σ. Here we have forced the zeroth factor to be the constant one, $f_{0,t} = 1$.

Given n observations, let F be the $n \times l$ matrix whose rows are the observations of the factors (including a column that is the constant 1), and let $\boldsymbol{x}$ be the n length column vector of returns. One can view F as the "design matrix" of a regression, except it is typically beyond our control, consisting of either random quantities or things like seasonality dummies. The well-known ordinary least squares estimates are then defined as

$$\hat{\boldsymbol{\beta}} := \left(\mathsf{F}^\top \mathsf{F}\right)^{-1} \mathsf{F}^\top \boldsymbol{x}, \qquad \hat{\sigma} := \sqrt{\frac{\left(\boldsymbol{x} - \mathsf{F}\hat{\boldsymbol{\beta}}\right)^\top \left(\boldsymbol{x} - \mathsf{F}\hat{\boldsymbol{\beta}}\right)}{n-l}}. \tag{1.9}$$

The $\hat{\sigma}$ can be expressed as the rescaled Gram matrix of the residuals of the regression, $\boldsymbol{x} - \mathsf{F}\hat{\boldsymbol{\beta}}$.

We can then define an *ex-factor Sharpe ratio* as follows: let $\boldsymbol{v}$ be some non-zero vector, and let r_0 be some risk-free, or disastrous, rate of return. Then define

$$\hat{\zeta}_g := \frac{\hat{\boldsymbol{\beta}}^\top \boldsymbol{v} - r_0}{\hat{\sigma}}. \tag{1.10}$$

In all of the factor models we will consider here, we choose $\boldsymbol{v} = \boldsymbol{e}_0$, the vector of all zeros except a one corresponding to the intercept term. Let *ex-factor signal-noise ratio* be the population analogue.

Remark (Nomenclature). The terms "ex-factor Sharpe ratio" and "ex-factor signal-noise ratio" are idiosyncratic to this text. The author is not aware of any commonly used terms for these concepts, though the term "information ratio" is close. The information ratio refers to the Sharpe ratio of returns from which risky benchmark returns have been subracted. That is, the information ratio is the Sharpe ratio computed on returns in excess of the risky benchmark. Our ex-factor Sharpe ratio involves a regression against the risky benchmark returns.

We note that the intercept term β_0 is often instead written as an α, giving rise to the expression that a strategy "has alpha," meaning returns not explained by those of the benchmark(s). Our notation allows us to simply write all the regression coefficients in one vector as $\boldsymbol{\beta}$.

There are numerous candidates for the factors, and their choice should depend on the return series being modeled. For example, one would choose different factors when modeling the returns of a single company versus those of a broad-market mutual fund versus those of a market-neutral hedge fund, *etc.* Moreover, the choice of factors might depend on the type of analysis being performed. For example, one might be trying to "explain away" the returns of one investment as the returns of another investment (presumably one with smaller fees) plus noise. Alternatively, one might be trying to establish that a given investment does have idiosyncratic alpha (*i.e.*, β_0) without significant exposure to other factors, either because those other factors are some kind of benchmark or because one believes they have zero expectation in the future.

1.4.1 Examples of linear attribution models

- As noted above, the vanilla Sharpe ratio employs a trivial factor model, *viz.* $x_t = \beta_0 1 + \epsilon_t$. This simple model is generally a poor one for describing stock returns; one is more likely to see it applied to the returns of mutual funds, hedge funds, *etc.*
- The simplest refinement to the trivial model would be to include some form of the interest rate, say $x_t = \beta_0 1 + \beta_r f_{r,t} + \epsilon_t$, where $f_{r,t}$ is the properly scaled "risk-free" rate of return. This model can lead to rank-deficiencies unless the interest rate has changed over the observation period. It is actually uncommon to fit the coefficient β_r, however; typically it is assumed to be 1, and the prevailing risk-free rate is subtracted from the return x before any modeling is performed.
- The models above do not take into account the influence of "the market" on the returns of stocks. This suggests a factor model equivalent to the Capital Asset Pricing Model (CAPM)[6], namely $x_t = \beta_0 1 + \beta_M f_{M,t} + \epsilon_t$, where $f_{M,t}$ is the return of "the market" at time t. [Black *et al.*, 1972; Cochrane, 2001; Krause, 2001]

  This is clearly a superior model for stocks and portfolios with a long bias (*e.g.*, typical mutual funds), but may seem inappropriate for, *e.g.*, a long-short balanced hedge fund. It would seem that adding a bunch of irrelevant factors to a factor model could cause one to fail to recognize a significant β_0. That is, a kitchen sink factor model might have low power. It turns out the reduction in power is typically very small, however, while the possibility of reducing type I errors is quite valuable. For example, one might discover that a seemingly long-short balanced fund actually has some market exposure, but no significant idiosyncratic returns (one finds

[6]The term "CAPM" usually encompasses a number of assumptions used to justify the validity of this model for stock returns; here the term is abused to refer only to the factor model.

that $\hat{\beta}_0$ is very small); this is valuable information, since a hedge-fund investor might balk at paying high fees for a return stream that replicates a (much less expensive) ETF plus noise.

- Generalizations of the CAPM model abound. For example, the Fama and French [1992] 3-factor model (I drop the risk-free rate for simplicity):

$$x_t = \beta_0 1 + \beta_{\mathrm{M}} f_{\mathrm{M},t} + \beta_{\mathrm{SMB}} f_{\mathrm{SMB},t} + \beta_{\mathrm{HML}} f_{\mathrm{HML},t} + \epsilon_t, \tag{1.11}$$

where $f_{\mathrm{M},t}$ is the return of "the market," $f_{\mathrm{SMB},t}$ is the return of "small minus big cap" stocks (the difference in returns of these two groups), and $f_{\mathrm{HML},t}$ is the return of "high minus low book value" stocks. Carhart [1997] adds a momentum factor:

$$x_t = \beta_0 1 + \beta_{\mathrm{M}} f_{\mathrm{M},t} + \beta_{\mathrm{SMB}} f_{\mathrm{SMB},t} + \beta_{\mathrm{HML}} f_{\mathrm{HML},t} + \beta_{\mathrm{UMD}} f_{\mathrm{UMD},t} + \epsilon_t, \tag{1.12}$$

where $f_{\mathrm{UMD},t}$ is the return of "ups minus downs," *i.e.*, the returns of the previous period winners minus the returns of previous period losers. Alternative factor models include Elton *et al.* [1995], Chen *et al.* [2011], and Asness *et al.* [2013].
These factor models are designed to explain the returns of stocks, and are typically judged by their parsimony, explanatory power (lack of significant $\hat{\beta}_0$ terms across a wide universe of stocks), orthogonality of the factors, and narrative appeal. However, there is no reason they cannot be used to describe (or explain away, in terms of simpler strategies) the returns of an actively managed portfolio. In fact, there is typically little power lost in a "kitchen sink" approach, if the objective is to eviscerate a proposed trading strategy.

- Henriksson and Merton [1981] describe a technique for detecting market-timing ability in a portfolio. One can cast this model as

$$x_t = \beta_0 1 + \beta_{\mathrm{M}} f_{\mathrm{M},t} + \beta_{\mathrm{HM}} (-f_{\mathrm{M},t})^+ + \epsilon_t,$$

where $f_{\mathrm{M},t}$ are the returns of "the market" the portfolio is putatively timing, and x^+ is the positive part of x.
Actually, one or several factor timing terms could be added to any factor model. Note that unlike the factor returns in models discussed above, one expects $(-f_{\mathrm{M}})^+$ to have significantly non-zero mean. This will cause some decrease in power when testing β_0 for significance. Also note that while Henriksson and Merton intend this model as a positive test for β_{HM}, one could treat the timing component as a factor which one seeks to ignore entirely, or downweight its importance.

- Often the linear factor model is used with a "benchmark" (mutual fund, index, ETF, *etc.*) used as the factor returns. In this case, the process generating x_t may or may not be posited to have zero exposure to the benchmark, but usually one is testing for significant idiosyncratic term. That is, one is hoping to reject the hypothesis that β_0 is zero, and show the returns "have alpha."

- Any of the above models can be augmented by splitting the idiosyncratic term into a constant term and some time-based term. For example, it is often argued that a certain strategy "worked in the past" but no longer does. This implies a splitting of the constant term as

$$x_t = \beta_0 1 + \beta_0{}' f_{0,t} + \sum_i \beta_i f_{i,t} + \epsilon_t,$$

where $f_{0,t} = (n-t)/n$, given n observations. In this case, the idiosyncratic part is an affine function of time, and one can test for β_0 independently of the time-based trend (one can also test whether $\beta_0{}' > 0$ to see if the "alpha" is truly decaying). One can also imagine time-based factors which attempt to address seasonality or "regimes." [Chow, 1960]

Example 1.4.1 (Technology attribution). Consider the monthly simple returns of the Technology industry, one of the five industry portfolios introduced in Example 1.2.5. The returns of this portfolio are attributed to those of the Market portfolio (Example 1.1.1) and an intercept term. The regression fit for the monthly returns against intercept and Market is $\hat{\boldsymbol{\beta}} = [0.095, 0.949]^\top$; the residual volatility is $\hat{\sigma} = 2.322\,\text{mo.}^{-1/2}$. This yields $\hat{\zeta}_g = 0.142\,\text{yr}^{-1/2}$, much smaller than the Sharpe ratio, which we computed as $\hat{\zeta} = 0.617\,\text{yr}^{-1/2}$. Performing an attribution to the returns of the Market has reduced the Sharpe ratio, which is typical. ⊣

Example 1.4.2 (UMD attribution). Consider the monthly simple returns of the Market, SMB, HML, and UMD portfolios, as described in Example 1.1.1. Going whole hog on attribution, the returns of UMD are attributed to the other three factors and an intercept term. The regression fit for the monthly returns, against intercept, Market, SMB, and HML, is $\hat{\boldsymbol{\beta}} = [1.015, -0.217, -0.05, -0.469]^\top$; the residual volatility is $\hat{\sigma} = 4.122\,\text{mo.}^{-1/2}$. This yields $\hat{\zeta}_g = 0.853\,\text{yr}^{-1/2}$. This is somewhat larger than the value of Sharpe ratio computed without the attribution model, namely $\hat{\zeta} = 0.477\,\text{yr}^{-1/2}$, as found in Example 1.2.4. The factor model here acted to increase the expected value slightly and decrease the idiosyncratic volatility slightly. This is somewhat atypical: adding more factors to a model tends to decrease the Sharpe ratio. ⊣

Example 1.4.3 (UMD, Chow test). A strawman argument against momentum strategies is that "they worked in the past," but no longer do. Here, we consider the monthly simple returns of the UMD portfolio, as described in Example 1.1.1. An attribution is made against the factor model where f_0 is the constant 1, and where $f_{1,t}$ is an indicator function for the event that t is prior to 1980-01-01. That is, $f_{1,t}$ is one for all time prior to 1980-01-01 and zero thereafter. This is essentially a *Chow test*, but we are interested in the significance of the intercept term, rather than looking for a structural break. [Chow, 1960; Kennedy, 2008]

The regression fit gives $\hat{\boldsymbol{\beta}} = [0.551, 0.171]^\top$ in units of $\%\,\text{mo.}^{-1}$. That is, the mean returns over the period after 1980-01-01 is $0.551\,\%\,\text{mo.}^{-1}$, while the mean over the period prior to that time is $0.171\,\%\,\text{mo.}^{-1}$ higher. The residual volatility is fit as $\hat{\sigma} = 4.706\,\%\,\text{mo.}^{-1/2}$. This yields $\hat{\zeta}_g = 0.406\,\text{yr}^{-1/2}$, slightly lower than the value of Sharpe ratio computed in the unattributed model, $\hat{\zeta} = 0.477\,\text{yr}^{-1/2}$, as found in Example 1.2.4. Although this is not a proper statistical analysis, one might reasonably believe the signal-noise ratio has decreased since the cutoff date. ⊣

1.4.2 † Heteroskedasticity attribution models

We briefly consider *heteroskedacity* of returns, defined as a difference in the variance across observations. In contrast, when returns have the same variance, the sample exhibits *homoskedasticity*.

We have defined a linear attribution model where some vector of covariates (be they contemporaneous returns of other assets, random predictive "signal" variables, seasonality variables, *etc.*) accounts for the expected returns of an asset, especially how they deviate from the unconditional expected return when conditioning on those covariates. Can we imagine a similar model where some variable(s) explains differences in heteroskedasticity? After all, while forecasting returns is considered difficult, there is a well known conditional heteroskedasticity effect in many markets [Cont, 2001; Nelson, 1991] that ought to be accounted for.

As it is much more difficult to deal with the case of multiple variables that explain heteroskedasticity, we will stick to the scalar case for now. Consider a strictly positive scalar random variable, s_{t-1}, observable at the time the investment decision is required to capture $\boldsymbol{x}_t$. It is more convenient to think of s_{t-1} as a "quietude" variable, or a "weight" for a weighted regression, rather than as a "volatility" feature.

We can then generalize Equation 1.8 to

$$x_t = \beta_0 1 + \sum_{i}^{l-1} \beta_i f_{i,t} + s_{t-1}^{-1} \epsilon_t, \qquad (1.13)$$

where we allow the $f_{i,t}$ to be correlated to, or even equal to, the quietude variable, s_{t-1}, and where the error term ϵ_t is homoskedastic.

We present some examples of conditional heteroskedasticity models:

- The *market clock* or *volatility clock* model posits that returns should be measured on a volatility or volume basis rather than wallclock time. As such, under this theory, if variance of returns is, say, twice the long-term variance, then expected returns ought to be twice the long-term expectation. Another way of putting this is the returns over half a day in this regime should have the same distribution as daily returns over the long term[7]. Supposing that volatility can be measured before the investment

[7] Never mind that this theory is rarely supported by data.

decision, we have

$$x_t = s_{t-1}^{-1}\beta_0 + s_{t-1}^{-1}\epsilon_t,$$

where s_{t-1}^{-1} is the volatility factor. When s_{t-1} has a long-term average of 1, then β_0 is the long-term average return.

Note that for any given time t, the single period return x_t has a signal-noise ratio independent of s_{t-1}, namely β_0 divided by the standard deviation of ϵ_t. When we consider multiple periods, however, the signal-noise ratio will deviate from this value. Note, however, that there is nothing particular to fear in this model about periods of high volatility since the asset holder is perfectly compensated with increased excess returns.

- A more depressing, though perhaps realistic, model of heteroskedasticity is that in periods of higher volatility there is no compensating increase in expected returns, which is to say that "*stuff happens.*" We can express this as

$$x_t = \beta_0 + s_{t-1}^{-1}\epsilon_t,$$

where s_{t-1}^{-1} is the volatility factor. In this case, the single period return x_t has signal-noise ratio proportional to s_{t-1}. That is to say, the risk to reward is higher in periods of lower volatility, and you are not compensated for the extra risk.

- Rather than be forced to choose among these two models of heteroskedasticity, one can consider a mixed model of the form

$$x_t = \beta_0 + s_{t-1}^{-1}\beta_1 + s_{t-1}^{-1}\epsilon_t,$$

where now the coefficients β_0 and β_1 may be zero or non-zero, but can give us either of the two previously mentioned models, volatility clock and stuff happens, with values to be estimated from the data.

The trick for dealing with conditional heteroskedasticity models is to use the quietudes for weighted estimation. That is, rewrite Equation 1.13 as

$$\tilde{x}_t = \beta_0 s_{t-1} + \sum_i^{l-1} \beta_i \tilde{f}_{i,t} + \epsilon_t, \tag{1.14}$$

where we define $\tilde{x}_t := s_{t-1}x_t$, and $\tilde{f}_{i,t} := s_{t-1}f_{i,t}$. We can now fit the data $\tilde{x}_t$ on the new attribution factors $\tilde{f}_{i,t}$ using the machinery of the homoskedastic case.

To illustrate heteroskedasticity modeling, we introduce another data set used throughout the text.

Example 1.4.4 (The VIX). The *VIX index* is a "model-free" estimate of the volatility of the market over the next thirty calendar days, expressed in units of annualized percent. [CBOE, a] Often called "the fear index," the VIX is actually computed from the bid and ask prices of options on the S&P 500 index. The daily value of the VIX index is computed by the CBOE and available from the **tsrsa** package. [CBOE, b; Pav, 2019a] This data set includes the "back-computation" of the VIX using the post-2004 methodology on data back to 1999. The following **R** code gives access to 7,809 days of data, from 1990-01-02 through 2020-12-31:

```
library(tsrsa)
data(dvix)
```

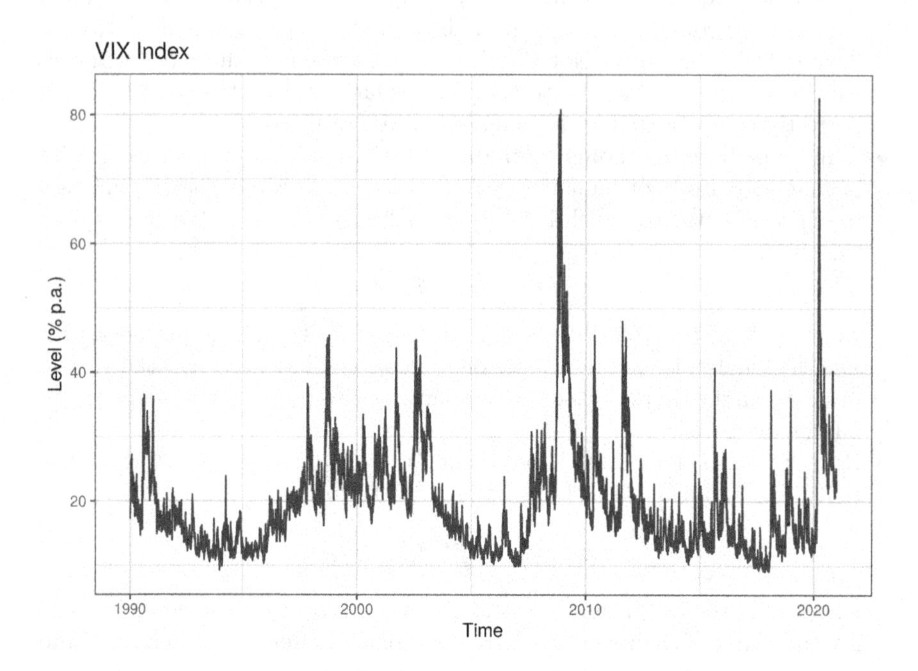

FIGURE 1.4
The level of the VIX index is shown from 1990-01-02 through 2020-12-31.

We plot the VIX index in Figure 1.4. The mean value of the VIX index over this period is approximately 19.5, and the standard deviation is 8.12. The coefficient of variation of the VIX is thus around 0.42. The median level is around 17.5, indicating some amount of positive skew. The empirical skewness is computed as 2.2. We will later need to consider a normalized version of the VIX, with mean value one; we estimate the third centered moment of this normalized VIX to be around 0.159. That is

$$\frac{1}{n}\sum_{1\le i\le n}\left(\frac{v_i}{\frac{1}{n}\sum_{1\le j\le n} v_j} - 1\right)^3 \approx 0.159.$$

The VIX level is clearly autocorrelated. The first differences in the VIX level are mean reverting, as shown in Figure 1.5. Clustering of volatility is one of the "stylized facts" of asset returns. [Cont, 2001] Although the VIX spikes up, then drifts down, one can generally forecast its level by looking at recent historical values. This might be useful, for example, to smooth out returns of a strategy by under-levering when the VIX is high. ⊣

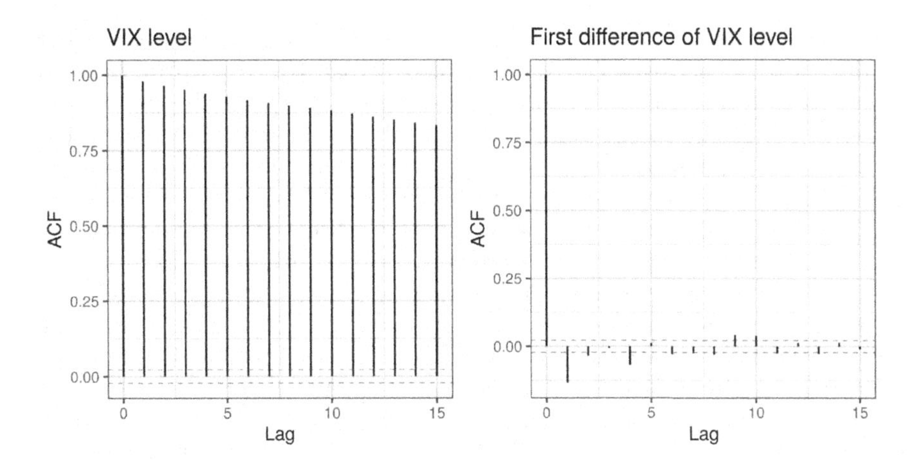

FIGURE 1.5
Autocorrelation plots of the VIX level, and the first difference of the VIX level are shown, along with the standard 95% confidence interval.

Example 1.4.5 (VIX reweighting Market returns). Consider the daily VIX index introduced in Example 1.4.4 as a volatility indicator. We lag it by one trading day, then join to the daily returns of the Market introduced in Example 1.1.1. We rescale the inverse VIX to have unit mean in our sample period, which ranges from 1990-01-03 through 2020-12-31.

Over this period, the regular Sharpe ratio of the Market component is $0.64\,\mathrm{yr}^{-1/2}$, again assuming 252 trading days per year. When we rescale the daily Market returns by the inverse VIX, then compute the Sharpe ratio we compute a value of $0.76\,\mathrm{yr}^{-1/2}$. This is essentially the "pessimistic model" where quietude weighted returns are modeled as having a constant expected return, conditional on the quietude. Under the "vol clock" model, we compute the ex-factor Sharpe ratio as $0.72\,\mathrm{yr}^{-1/2}$. Note that using a classical information criterion-based approach, we slightly prefer the "vol clock" model, but the difference in residual sum of squares between them is so very small that this is hardly a conclusive test. ⊣

1.5 † Should you maximize Sharpe ratio?

In some circles, the question "should you maximize Sharpe ratio?" has an obviously affirmative answer: professional money managers are routinely asked to provide their achieved (or, more dubiously, backtested) Sharpe ratio to prospective clients. To a first-order approximation, it is the *only* quantitative aspect of a trading strategy that clients care about. While volatility, capacity, leverage, fees, lockup terms, *etc.* are also important, the mythical "utility" underpinning most investment theory [Cochrane, 2001; Duffie, 2010; Munk, 2013; Krause, 2001] has zero currency in these discussions: the author has *never* heard (nor heard of) an investor quote their utility function or risk-aversion parameter, while discussion of strategy Sharpe ratio is routine[8].

Pragmatics aside, there are two distinct questions regarding the investment decision and the Sharpe ratio, *viz.* 1. if all the population parameters of investment strategies were somehow known, should one maximize the signal-noise ratio? 2. is the sample Sharpe ratio a good predictor of whatever function of the population parameters one wishes to optimize? While the first question contains a glaring counterfactual, it is important nonetheless, since the second question relies on one's answer to the first.

It is generally assumed that investor's preferences are aligned with the *stochastic dominance* relations. We say that z (first-order) stochastically dominates x if, for all c, $\Pr\{x \geq c\} \leq \Pr\{z \geq c\}$, with inequality holding for some c. First-order stochastic dominance implies second-order stochastic dominance. The definition of second order stochastic dominance is too involved for our purposes here, but it should be noted that if z second-order stochastically dominates x, then z has no smaller mean and no larger variance than x.

While the stochastic dominance relations are typically considered sancrosanct, *i.e.*, all investors would prefer a returns stream to another that it stochastically dominates, stochastic dominance does not form a total ordering.[9] That is, it is possible to construct two returns streams such that neither stochastically dominates the other. For purposes of selecting an investment, one would like to be able to compare any two investments, so a total ordering is required. Here the term "objective" will be used to mean some real-valued function of the population parameters of a returns stream (*i.e.*, μ, σ, *etc.*) which one wishes to maximize. An objective trivially forms a total ordering.

At a bare minimum one expects an objective should be consistent with stochastic dominance. That is, if z stochastically dominates x, it should have a larger objective. Regretably, the signal-noise ratio as an objective function is *not* consistent with stochastic dominance. It is not consistent with second-order stochastic dominance, as most clearly demonstrated by considering the

[8]Often without any acknowledgement of the sample variance of the Sharpe ratio.

[9]The example of "non-transitive dice" illustrates that some forms of probabilistic dominance do not even form partial orders. [Gardner, 2001; Finkelstein and Thorp, 2000]

case $\mu < 0$, where the signal-noise ratio "prefers" higher σ^2. This wrinkle is *not* fixed by assuming away the $\mu < 0$ case[10], since the signal-noise ratio is not even consistent with the stronger first-order stochastic dominance relation, as shown by the counterexample of Hodges [1998], *cf.* Exercise 1.34. [Smetters and Zhang, 2013; Zakamouline and Koekebakker, 2008]

Despite these failings, it is taken as axiomatic in this text that one should optimize the signal-noise ratio, μ/σ. There are numerous reasons to optimize the signal-noise ratio. Among them: 1. the signal-noise ratio bounds the probability of a loss against the disastrous rate, r_0; 2. a large signal-noise ratio approximately bounds the probability of experiencing a large drawdown, as measured in units of volatility; 3. a large signal-noise ratio increases the probability of experiencing a large Sharpe ratio in a fixed period of trading;[11] 4. since the signal-noise ratio is simply defined, it is a relatively easy objective to optimize analytically (again, assuming one knew the population parameters; noisy estimation thereof adds extra complication). Moreover, the deficiencies of the signal-noise ratio are somewhat mitigated for the long-term investor, since the central limit theorem guarantees convergence to normality.

If the goal is maximization of the signal-noise ratio, then the Sharpe ratio is a decent yardstick by which to measure investments, modulo a number of provisos: the Sharpe ratio must be measured over the same length of time, as in, *e.g.*, Sharpe's original paper [Sharpe, 1965], or some correction should be made for different standard errors; if the returns are measured over common time intervals, one must take into account possibly correlated errors; the Sharpe ratio measured on a backtest should be viewed with great suspicion, as backtests are often riddled with methodological and coding errors, and the product of great amounts of data snooping. While, taken together, these make the Sharpe ratio seem like an awful objective on realized returns streams, all other metrics share the same problems. Moreover, many of the proposed alternative metrics defined on achieved returns streams lack any kind of theoretical justification, and many have worse sample variances. See Exercise 1.39.

Example 1.5.1 (Sample variance of the Sortino ratio). The Sortino ratio is defined as the sample mean divided by the downside semivariation. While this sounds like a perfectly reasonable metric, the downside semivariation has slightly higher sample variation than the standard deviation. For the case of symmetric returns, the Sortino ratio should, then, have slightly higher variance than the Sharpe ratio.

To test this, 52 weeks of weekly returns were drawn from a population with $\mu = 0\,\text{wk.}^{-1}, \sigma = 0.02\,\text{wk.}^{-1/2}$, and the Sortino ratio was computed. The signal-noise ratio was also computed. This was repeated 5,000 times. The same computations were then repeated using $\mu = 0.003\,\text{wk.}^{-1}, \sigma = 0.02\,\text{wk.}^{-1/2}$.

---

[10]No investor should consider a returns stream with $\mu < 0$ if they can keep their wealth as cash.

[11]This is a bit (a) circular: a large achieved Sharpe ratio is supposed to instill confidence in prospective investors that $\mu > r_0$, but it is proposed that one achieve a large Sharpe ratio by maximizing $(\mu - r_0)/\sigma$, rather than simply maximizing μ.

Empirical critical values for Frequentist tests based on both statistics were computed by taking the 0.95 quantiles of each statistic under the first population. The empirical powers were then computed by computing the proportion of statistics from the second population exceeding the critical value. The Sortino ratio was found to have a power of approximately 0.255, while the Sharpe ratio had a power of approximately 0.273. ⊣

1.6 † Probability of a loss, revisited

The crux of Roy's justification for the "Safety-First" criterion, which is just the signal-noise ratio, is that it bounds the probability of a loss, defined as a return less than r_0, *cf.* Equation 1.7. [Roy, 1952; Sullivan, 2011] This argument is based on Chebyshev's inequality, and so is a fairly rough upper bound. There are some situations where the signal-noise ratio is exactly monotonic in the probability of a loss. For example, if the returns are drawn from a scale-location family, *i.e.*, if the probability density of x is $\frac{1}{\sigma} f\left(\frac{x-\mu}{\sigma}\right)$, for some function $f(\cdot)$. In this case, the probability that the return is less than r_0 is exactly equal to $F\left(\frac{r_0-\mu}{\sigma}\right)$ for some function $F(\cdot)$. In this case, the signal-noise ratio is consistent with first-order stochastic dominance.

The central limit theorem tells us that, conditional on finite variance, the mean return of x will converge to a Gaussian distribution. Noting that for log returns, by the telescoping property, the mean return is just the total return rescaled, the long-term log return is approximately drawn from a scale-location family. (*cf.* Exercise 1.36.)

One can construct tighter approximations than given by Chebyshev's inequality by considering the classical approximations to the central limit theorem. These will result in objectives which are sensitive to the investor's time horizon and the disastrous rate, r_0. [Smetters and Zhang, 2013; Zakamouline and Koekebakker, 2008; Boudt *et al.*, 2008]

Suppose that one will observe n independent draws from the returns stream, x. The disastrous event is that the observed mean sample return, $\hat{\mu}$, is less than r_0. This is equivalent to

$$\sqrt{n}\frac{\hat{\mu}-\mu}{\sigma} \le \sqrt{n}\frac{r_0-\mu}{\sigma}.$$

The cumulative distribution function of the quantity on the left-hand side can be approximated via some truncation of the *Edgeworth expansion.* [Chernozhukov *et al.*, 2007]

Define[12] $\delta := \sqrt{n}\left(\mu - r_0\right)/\sigma$. The Edgeworth expansion is [Abramowitz

[12] In the sequel, we will see that δ so defined is the non-centrality parameter of the t-distribution associated with the Sharpe ratio.

and Stegun, 1964, 26.2.48]

$$\begin{aligned}\Pr\left\{\sqrt{n}\frac{\hat{\mu}-\mu}{\sigma} \le -\delta\right\} &= \Phi\left(-\delta\right) - \frac{\phi\left(\delta\right)}{\sqrt{n}}\left[\frac{\gamma_1}{6}He_2\left(\delta\right)\right] \\ &+ \frac{\phi\left(\delta\right)}{n}\left[\frac{\gamma_2}{24}He_3\left(\delta\right) + \frac{{\gamma_1}^2}{72}He_5\left(\delta\right)\right] \\ &- \frac{\phi\left(\delta\right)}{n^{3/2}}\left[\frac{\gamma_3}{120}He_4\left(\delta\right) + \frac{\gamma_1\gamma_2}{144}He_6\left(\delta\right) + \frac{{\gamma_1}^3}{1296}He_8\left(\delta\right)\right]\ldots\end{aligned} \tag{1.15}$$

where $\Phi(x)$ and $\phi(x)$ are the cumulative distribution and density functions of the standard unit normal, $He_i(x)$ is the probabilist's Hermite polynomial [Abramowitz and Stegun, 1964, 26.2.31], and γ_i is the standardized $i-2$th cumulant, defined as the $i+2$th cumulant of the distribution divided by σ^{i+2}. It happens to be the case that γ_1 is the skewness, and γ_2 is the excess kurtosis of the distribution.

Truncating beyond the $n^{-1/2}$ term, and applying basic facts of probability, gives

$$\Pr\left\{\hat{\mu} \ge r_0\right\} \approx \Phi\left(\delta\right) + \frac{\phi\left(\delta\right)}{\sqrt{n}}\left[\frac{\gamma_1}{6}\left(\delta^2-1\right)\right]. \tag{1.16}$$

The implication is that the probability that $\hat{\mu}$ exceeds r_0 will be increased if δ is large. Moreover, for a fixed δ, the probability that $\hat{\mu}$ exceeds r_0 is increased for large positive skew if $\delta^2 > 1$, but for large negative skew when $\delta^2 < 1$. The implication is that when δ^2 is "large" (compared to the unit), one would buy lottery tickets, otherwise one would sell lottery tickets[13]. This is asymptotically compatible, as $n \to \infty$, with the commonly held belief that investors universally prefer positive skew.

The Edgeworth expansion suggests a "higher order signal-noise ratio," defined as [Pav, 2015]

$$\zeta_h := -\frac{1}{\sqrt{n}}\Phi^{-1}\left(\Pr\left\{\sqrt{n}\frac{\hat{\mu}-\mu}{\sigma} \le -\delta\right\}\right). \tag{1.17}$$

When all the higher order cumulants are zero, *i.e.*, when returns are normally distributed, the probability of a loss is exactly $\Phi(-\delta)$, so in this case $\zeta_h = \delta/\sqrt{n} = (\mu - r_0)/\sigma$, which is just the signal-noise ratio with a risk-free rate r_0. This definition gives an objective which is consistent with first-order stochastic dominance (Exercise 1.37) and generalizes the signal-noise ratio in an intuitive way that is easy to understand. However, it is essentially investor-dependent in that it is parametrized by n and r_0. [Smetters and Zhang, 2013]

The implicit definition of Equation 1.17 is a bit unwieldy for use as an objective. One would prefer a definition in terms of the cumulants of the

[13]This illustrates the possible mismatch between the objectives of an investor and the money manager: the former is likely concerned about the long term, while the latter may be focused on the short term.

returns stream. Rather than use the Taylor series expansion of $\Phi^{-1}(x)$, one can instead use the *Cornish Fisher expansion* of the sample quantile. [Lee and Lin, 1992; Jaschke, 2001; Ulyanov, 2014]

Let $Y = \sqrt{n}(\hat{\mu} - \mu)/\sigma$. This is a random variable with zero mean and unit standard deviation. Let γ_i be the $i+2$th standardized cumulant of x. The $i+2$th standardized cumulant of Y is $n^{1-i/2}\gamma_i$. The Cornish Fisher expansion is [Abramowitz and Stegun, 1964, 26.2.49]

$$\Pr\{Y \le w\} = \Phi(z),$$

where

$$\begin{aligned} w = z &+ \frac{1}{\sqrt{n}}\left[\frac{\gamma_1}{6} He_2(z)\right] \\ &+ \frac{1}{n}\left[\frac{\gamma_2}{24} He_3(z) - \frac{{\gamma_1}^2}{36}\left[2He_3(z) + He_1(z)\right]\right] \\ &+ \frac{1}{n^{3/2}}\left[\frac{\gamma_3}{120} He_4(z) - \frac{\gamma_1\gamma_2}{24}\left[He_4(z) + He_2(z)\right] + \frac{{\gamma_1}^3}{324}\left[12He_4(z) + 19He_2(z)\right]\right] \\ &+ \ldots \end{aligned} \tag{1.18}$$

To estimate the higher-order signal-noise ratio, plug in $z = -\sqrt{n}\zeta_h$, and $w = -\delta$. This gives

$$\begin{aligned} \zeta_h = \frac{(\mu - r_0)}{\sigma} &+ \frac{1}{n}\left[\frac{\gamma_1}{6} He_2\left(\sqrt{n}\zeta_h\right)\right] \\ &- \frac{1}{n^{3/2}}\left[\frac{\gamma_2}{24} He_3\left(\sqrt{n}\zeta_h\right) - \frac{{\gamma_1}^2}{36}\left[2He_3\left(\sqrt{n}\zeta_h\right) + He_1\left(\sqrt{n}\zeta_h\right)\right]\right] \\ &+ \ldots \end{aligned} \tag{1.19}$$

While this defines ζ_h implicitly, truncation gives polynomial equations, whose roots can be found analytically or numerically. See Exercise 1.38.

Example 1.6.1 (Higher order signal-noise ratio of the Market). Here we consider what might be called the higher order Sharpe ratio of the Market, introduced in Example 1.1.1. The mean simple return of the Market was estimated by sampling, with replacement, 12 months of returns. This process was repeated 5,000 times, and the empirical probability that the mean monthly return was less than $0\%\,\text{mo.}^{-1}$ was computed. The normal quantile was then computed, yielding an approximate higher order Sharpe ratio of $0.469\,\text{yr}^{-1/2}$. The computation was repeated for $r_0 = 0.25\%\,\text{mo.}^{-1}$, yielding a higher order Sharpe ratio of $0.286\,\text{yr}^{-1/2}$. Compare these with the Sharpe ratio computed on monthly returns for these two choices of r_0, which were found to be $0.438\,\text{yr}^{-1/2}$ and $0.276\,\text{yr}^{-1/2}$, respectively. ⊣

1.7 † Drawdowns and the signal-noise ratio

Drawdowns are the quant's bugbear. Though a fund may have a reasonably high signal-noise ratio, it will likely face redemptions and widespread managerial panic if it experiences a large or prolonged drawdown. Moreover, drawdowns are statistically nebulous: the sample maximum drawdown does not correspond in an obvious way to some population parameter; the variance of sample maximum drawdown is usually large and depends on sample size in a subtle way; traded strategies are typically cherry-picked to not have a large maximum drawdown in backtests; the distribution of maximum drawdowns is affected by skew and kurtosis, heteroskedasticity, omitted variable bias, and autocorrelation. Even assuming *i.i.d.* Gaussian returns, modeling drawdowns is non-trivial. [Magdon-Ismail *et al.*, 2002; Becker, 2010]

Given n observations of the mark to market of a single asset, p_t, the drawdown from the high water mark, as a time series, is defined as

$$D_t := \max_{1 \le s \le t} \log\left(\frac{p_s}{p_t}\right). \tag{1.20}$$

As so defined, the drawdown is negative the most extreme peak to point log return, and so is always non-negative[14]. The drawdown can be expressed as a a percent loss from the high watermark as $100\left(1 - e^{-D_t}\right)\%$.

Example 1.7.1 (drawdowns of the Market). Consider the Market factor returns introduced in Example 1.1.1. Its drawdown from high watermark is shown in Figure 1.6. The maximum drawdown over the period is around 80%. ⊣

The *maximum drawdown* is a commonly computed statistic of backtested and live strategy returns. It is literally the maximum of the drawdown series, or

$$M_n := \max_{1 \le t \le n} D_t = \max_{1 \le s < t \le n} \log\left(\frac{p_s}{p_t}\right). \tag{1.21}$$

There is a connection between drawdowns and the signal-noise ratio, which is made obvious when one uses volatility as the units in which drawdown is measured. Let x_t be the *log* returns: $x_t = \log \frac{p_t}{p_{t-1}}$, assumed to be *i.i.d.* Let μ and σ be the population mean and standard deviation of the log returns x_t. Now note that we can express the log returns over the period s to t as

$$\begin{aligned}\log\left(\frac{p_s}{p_t}\right) &= -\sum_{s<q\le t} x_q = -\left([t-s-1]\,\mu + \sigma \sum_{s<q\le t} y_q\right),\\ &= -\sigma\left([t-s-1]\,\zeta + \sum_{s<q\le t} y_q\right),\end{aligned}$$

[14]Under this definition a large drawdown is a bad event, which matches common informal usage of the term.

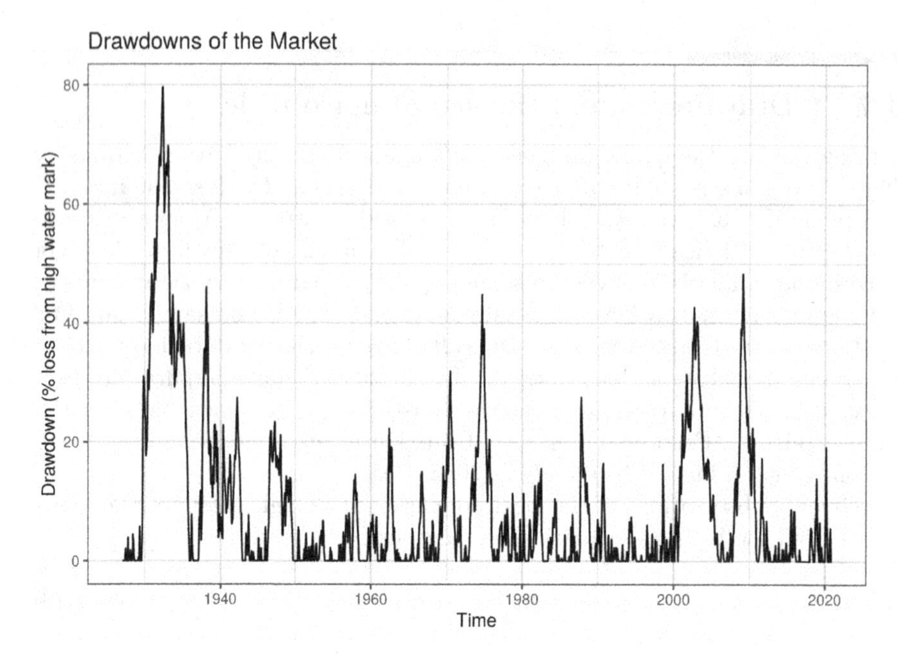

FIGURE 1.6
The drawdown from high watermark of the Market is shown, as a percent loss.

where y_t is some zero-mean, unit-variance random variable that is a shifted, rescaled version of x_t.

Now re-express the maximum drawdown in units of the volatility of log returns at the sampling frequency:

$$\frac{M_n}{\sigma} = -\min_{1 \leq s < t \leq n} \left([t - s - 1]\,\zeta + \sum_{s < q \leq t} y_q \right). \tag{1.22}$$

The volatility is a natural numeraire: one expects an asset with a larger volatility to have larger drawdowns. Moreover, the quantity on the right-hand side is a random variable drawn from a one parameter (the signal-noise ratio) family, rather than a two parameter (location and scale) family. It should be clear, moreover, that within this one parameter family, stochastic dominance of the drawdown distribution is monotonic in the signal-noise ratio. That is, a higher signal-noise ratio leads to a lower probability of a drawdown of a fixed size, *ceterus paribus*.

Worst statistic ever?

Equation 1.22 illustrates one reason why the sample maximum drawdown is a terrible statistic for estimating future performance of a strategy: it depends on volatility and signal-noise ratio in ways that are hard to disambiguate. That is, if one performed a hypothesis test based solely on the sample maximum drawdown, one would reject the null[15] if either the signal-noise ratio were high or the volatility were low.

Second, it is not at all clear that the variance of the sample maximum drawdown statistic is actually decreasing with sample size. That is, if one considers a longer time history, the sample maximum drawdown for any backtest can only increase, but the variance of that statistic may increase as well. For an arbitrarily long backtest history (longer than one could sensibly ever produce) one suspects that all sample maximum drawdowns should be nearly 100%. Depending on the signal-noise ratio, there is likely to be an "unsweet spot" of sample length where the variance of the maximum drawdown is actually *maximized.*

Another problem with sample maximum drawdown is that the statistic typically has a higher variance than other statistics. To illustrate this, 8, 192 Monte Carlo simulations of 5 years of weekly returns data were drawn from two populations. Returns are normally distributed with an equivalent daily volatility of around $0.013\,\text{day}^{-1/2}$. The two populations correspond to the null, with $\zeta = 0\,\text{yr}^{-1/2}$ and the alternative, with $\zeta = 0.75\,\text{yr}^{-1/2}$, where signal-noise ratio is measured on log returns. For each draw of historical data, the sample maximum drawdown and Sharpe ratio were computed. The 0.95 empirical quantile for sample maximum drawdown under the null was around 22%. However, it is not the case that a large number of simulations under the alternative have smaller maximum drawdown. Only 2, 524 do, thus the empirical power is around 0.31. In comparison, the empirical power for the Sharpe ratio statistic in this experiment is around 0.53. (The theoretical power in this case is closer to 0.51. See Section 2.5.3.)

The empirical cumulative distributions of the two statistics for the two populations are shown in Figure 1.7, with vertical lines plotted at the 0.95 empirical quantile for the null population. The lack of power for the sample drawdown is apparent in this plot. See also Exercises 1.39 and 1.40.

1.7.1 Controlling drawdowns via the signal-noise ratio

While the sample maximum drawdown is an inherently flawed statistic, real portfolio drawdowns are nevertheless a serious occupational hazard. How can drawdowns be controlled? Ignoring the contributions to drawdown caused by autocorrelation or skewed distributions, the decomposition in Equation 1.22 indicates that, on a per-vol basis, drawdowns are stochastically monotonic in the signal-noise ratio when choosing from a one parameter family.

One reasonable way a portfolio manager might approach drawdowns is to define a "knockout" drawdown from which she will certainly not recover[16] and

[15]Presumably the null is that the strategy is "not good," but this would have to be made precise in terms of the population parameters to form an actual hypothesis test.

[16]This is certainly a function of the fund's clients, or the PM's boss.

a maximum probability of hitting that knockout in a given epoch (*i.e.*, n). For example, the desired property might be *"the probability of a 40% drawdown in one year is less than 0.1%."* These constrain the acceptable signal-noise ratio and volatility of the fund.[17]

As a risk constraint, this condition shares the hallmark limitation of the value-at-risk (VaR) measure, namely that it may limit the probability of a certain sized drawdown, but not its expected magnitude. For example, underwriting catastrophe insurance may satisfy this drawdown constraint, but may suffer from enormous losses when a drawdown does occur. Nevertheless, this VaR-like constraint is simple to model and may be more useful than harmful.

Fix the one parameter family of distributions on y. Then, for given ϵ, δ, and n, the acceptable funds are defined by the set

$$\mathcal{C}\left(\epsilon, \delta, n\right) := \left\{ (\zeta, \sigma) \mid \sigma > 0, \, \Pr\left\{ M_n \geq \sigma\epsilon \right\} \leq \delta \right\}. \tag{1.23}$$

These are the points in the two-dimensional signal-noise ratio and volatility space that have strictly positive volatility and for which the probability of a drawdown exceeding $\sigma\epsilon$ is less than δ. When the x are daily returns, the range of signal-noise ratio one may reasonably expect for portfolios of equities is fairly modest. In this case, the lower boundary of $\mathcal{C}\left(\epsilon, \delta, n\right)$ can be approximated by a half space:

$$\left\{ (\zeta, \sigma) \in \mathcal{C}\left(\epsilon, \delta, n\right) \mid |\zeta| \leq \zeta_{big} \right\} \approx \left\{ (\zeta, \sigma) \mid \sigma \leq \sigma_0 + b\zeta, \, |\zeta| \leq \zeta_{big} \right\},$$

where σ_0 and b are functions of ϵ, δ, n, and the family of distributions on y. The minimum acceptable signal-noise ratio is $-\sigma_0/b$. It may be the case that σ_0 is negative. The contants σ_0 and b have to be approximated numerically, as in Example 1.7.2. It should be noted that a linear approximation of this form is easy to encode as a constraint in a portfolio optimizer.

Example 1.7.2 (halfspace for drawdown control). 5 years of weekly returns are drawn from a Gaussian distribution with a fixed signal-noise ratio and volatility. The maximum drawdown is computed in volatility units. This is replicated 50,000 times for a fixed value of signal-noise ratio, and the $1 - 0.005$ quantile is computed. Then the volatility such that this quantile is equal to a 40% loss is computed, abusing the rescaling relationship of Equation 1.22. This is repeated for signal-noise ratio ranging from $0\,\mathrm{yr}^{-1/2}$ to $2.5\,\mathrm{yr}^{-1/2}$. This estimated cutoff volatility of log returns, in annualized units (*i.e.*, log return $\mathrm{yr}^{-1/2}$) is plotted versus the signal-noise ratio (in units of $\mathrm{yr}^{-1/2}$), along with a linear fit, in Figure 1.8. The linear fit is $\sigma_0 \approx 0.086$ logret $\mathrm{yr}^{-1/2}$ and $b \approx 0.091$ logret. As indicated in the figure, a quadratic model is likely to give a better fit, a finding supported by an F test. See also Exercise 1.41. ⊣

[17] Another problem with this formulation is that humans typically have poor intuition for rare events.

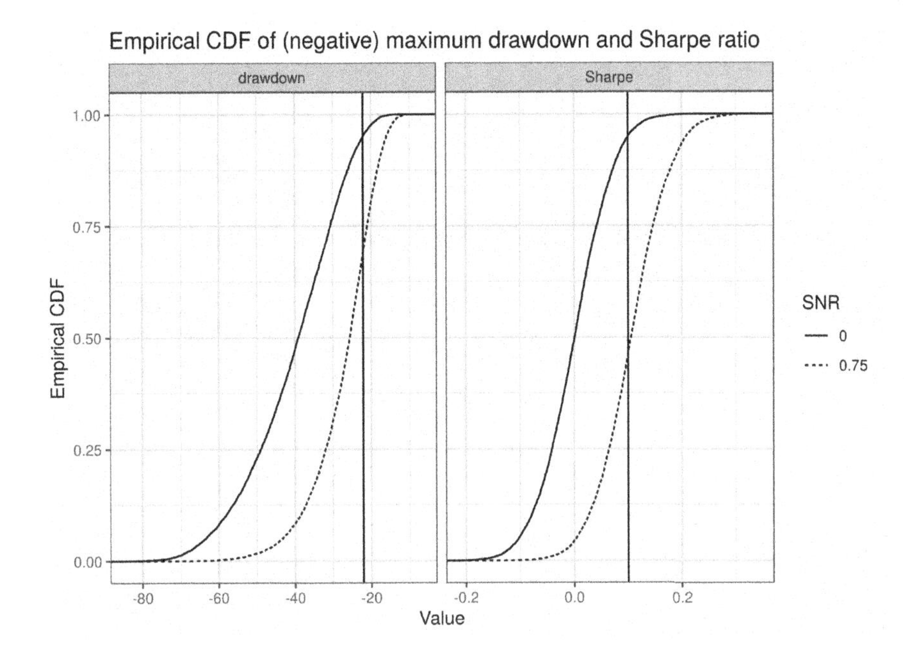

FIGURE 1.7

8192 Monte Carlo simulations of 5 years of weekly returns data were drawn from two populations. The first has a signal-noise ratio of $0\,\mathrm{yr}^{-1/2}$, the other $0.75\,\mathrm{yr}^{-1/2}$, where signal-noise ratio is measured on *log* returns. Returns are normally distributed with a daily volatility of around 0.013. The sample maximum drawdown and the Sharpe ratio were computed for each series. On the left, empirical CDFs of the *negative* maximum drawdown (in percent) are shown for the two populations; on the right, the empirical CDFs of the Sharpe ratio are plotted. For both these statistics, larger values is considered better. Vertical lines are plotted at the 0.95 empirical quantile for the $\zeta = 0\,\mathrm{yr}^{-1/2}$ case. The power of the implicit Sharpe ratio test is much higher than for the sample maximum drawdown.

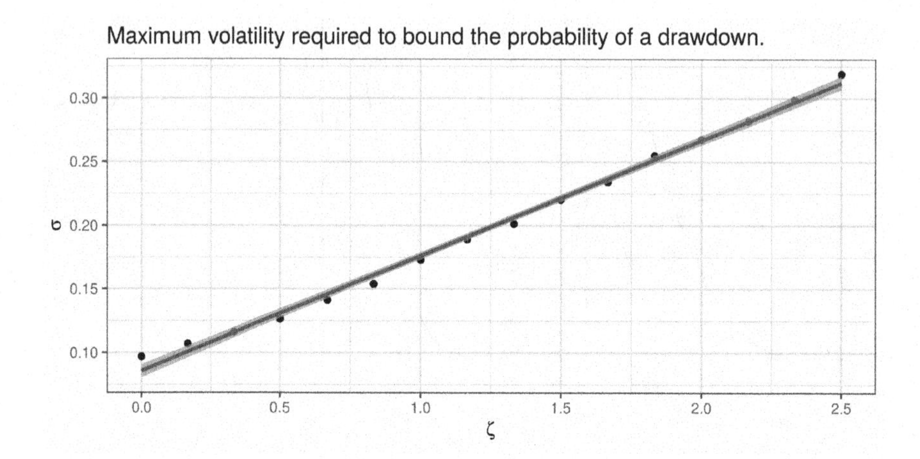

FIGURE 1.8
The maximum annualized volatility of log returns such that the probability of a drawdown of 40% is less than 0.005, as estimated by 50,000 Monte Carlo simulations of weekly Gaussian returns is plotted versus the annualized signal-noise ratio. The fit is given by $\sigma \approx 0.086\ \text{logret}\,\text{yr}^{-1/2} + 0.091\ \text{logret}\zeta$.

1.8 Exercises

* **Ex. 1.1 Note on exercises** If you use this book for self-study, the exercises are essential to check your understanding. They should stimulate your thinking and curiosity.

1. Exercises which are more difficult than others will typically be marked with a star, as above.
2. Some questions may be marked with the symbol "‽" to denote that the author is not sure that there is a nice answer. Some exercises are untested and may not have easy answers. Some exercises are marked as research problems: these might be worthy of a paper.
3. Some exercises might be labeled as "boring." The point of these is not to bore you, rather to ensure that you have understood some computational recipe well enough to reproduce on your own.

Ex. 1.2 Units conversion 1. Convert a Sharpe ratio of $0.2\,\text{mo.}^{-1/2}$ computed on monthly returns to annual units.
2. Convert an annual Sharpe ratio of $0.8\,\text{yr}^{-1/2}$ to daily units.

Ex. 1.3 Compute the Sharpe ratio Select some publicly traded company you consider successful. Compute the Sharpe ratio of its log returns over the past five years.

Ex. 1.4 Returns aggregation How do returns aggregate?

1. Let x_t be the relative return of an asset over some period. Show that the relative return of the asset over n periods is
$$\prod_{1 \le t \le n} (1 + x_t) - 1.$$
2. Now let x_t be log returns of an asset over some period. Show that the total log return of the asset over n periods is the sum $\sum_{1 \le t \le n} x_t$.

Ex. 1.5 Leverage By borrowing money, one can purchase an asset *with leverage*: One borrows l proportion of one's wealth, purchases an asset worth $1 + l$ proportion of one's wealth, waits a single period, sells the asset, and returns the original loan, with interest paid. (In the simple formulation, we ignore the borrow costs.)

1. What is the relative return of a leveraged long holder of an asset?
2. What is the log return of a leveraged long holder?

Ex. 1.6 Shorting Suppose that the relative return of an asset is x. Suppose that you *short* the asset: you borrow (say, without interest) from a long holder, sell the borrowed asset, wait one period, buy the asset (preferrably at a lower price), and return it to the lender.

1. What is the one-period relative return of the short holder?
2. Suppose $-1 \leq x \leq M$, where $M > 1$ is some large maximum return. What are the bounds of the returns to the short holder?
3. What is the one-period log return of shorting? Is it necessarily well-defined?
4. Suppose one holds the asset short for multiple periods, without any rebalancing. The relative returns of the asset for each period are $x_1, x_2, \ldots, x_n$. What is the relative return of the short holder?

Ex. 1.7 Returns of a portfolio Show that relative returns are laterally additive. That is, suppose you hold $a_i > 0$ proportion of your wealth in asset i, which experiences return x_i in a given period. Show that the relative return of your wealth is $\sum_i a_i x_i$. Does this still hold when one uses $a_i < 0$ to represent holding an asset short? Does this still hold when allowing $a_i > 1$ to represent buying with leverage?

Ex. 1.8 Concavity of log returns Use calculus to show that log returns are always less than the equivalent relative returns. That is, show that $\log(1 + x) \leq x$, where x is the relative return.

Ex. 1.9 The fundamental law of active management Suppose you could invest an equal proportion of your wealth in n independent assets, each with identical mean and variance, μ and σ^2, respectively. Show that the signal-noise ratio of your portfolio, measured on relative returns, scales as $\sqrt{n}$.
This is an example of the *fundamental* law of active *management,* which vaguely states that a constant "edge" aggregates as the square root of the number of independent opportunities to apply the edge. [Grinold and Kahn, 1999] Note that the same scaling applies in the translation of time units of the signal-noise ratio.

Ex. 1.10 The opposite of a bad strategy Suppose that Alice and Bob have antithetical views of the market. If Alice holds a_i proportion of her wealth in an asset at a given time, Bob shorts a_i proportion of his wealth in that asset, and vice versa. (That is, imagine that at each day's close, Alice and Bob instantaneously trade so that their portfolios have this property, incur no trade costs, get the same trade price, *etc.*)

1. If Alice's relative return over some period is x, what is Bob's relative return over the same period?
2. Can it be the case that *both* Alice and Bob lose money over a single rebalance period?
3. Can it be the case that *both* Alice and Bob lose money over the span of multiple rebalance periods?

4. Suppose you traded at random, *i.e.*, used a random number generator to allocate your wealth among some fixed set of assets. Ignoring trading costs, why should it be the case that your expected *log* returns are negative?

Ex. 1.11 Leveraged ETFs Certain *leveraged ETFs* purport to have daily relative returns which are some multiple of the daily relative returns of some index (or some other ETF). [Zhang, 2010] It is noted that these products tend to underperform in "flat" markets.
Let x be the relative return of an ETF tracking an index, and let y be the relative returns of a $3\times$ levered ETF on the same index. Suppose that the ETF makes a two period "round trip": $x_1 = \epsilon, x_2 = -\epsilon/\left(1+\epsilon\right)$.

1. Show that the total relative return of x is zero.
2. What is the total relative return of y? Plot this as a function of ϵ. Include negative values of ϵ.
3. Is there an arbitrage opportunity if one can trade an ETF on an index and a leveraged ETF on the same index?

Ex. 1.12 Leverage and log return signal-noise ratio Let μ, σ^2 be the mean and variance of the relative returns of a strategy. By levering up the strategy by a factor of k, the mean becomes $k\mu$ and the variance $k^2\sigma^2$. Let ζ_g be the signal-noise ratio measured on the *log* returns. Using the approximation

$$\zeta_g \approx \frac{\mu - \frac{1}{2}\sigma^2}{\sigma},$$

what happens to ζ_g as k becomes large? How do you interpret these results?

Ex. 1.13 Translation to log returns The transformation from relative returns to log returns is given by $f_{\log}\left(x\right) = \log\left(1+x\right)$.

1. Compute the Taylor series expansion of $f_{\log}\left(\cdot\right)$ around zero.
2. Let x be the relative returns and let $\alpha_i = \mathrm{E}\left[x^i\right]$. Find the series representation of $\mathrm{E}\left[f_{\log}\left(x\right)\right]$ in terms of α_i. What is the two term approximation of $\mathrm{E}\left[f_{\log}\left(x\right)\right]$?
3. Compute the first three terms of the expansion of $\left(f_{\log}\left(\cdot\right)\right)^2$. What is the two term approximation of $\mathrm{E}\left[\left(f_{\log}\left(x\right)\right)^2\right]$?
4. What is the two term approximation of
$\mathrm{E}\left[\left(f_{\log}\left(x\right)\right)^2\right] - \left(\mathrm{E}\left[f_{\log}\left(x\right)\right]\right)^2$?
5. Using the fact that

$$\frac{1}{\sqrt{\sigma^2+\epsilon}} \approx \frac{1}{\sigma}\left(1 - \frac{1}{2}\frac{\epsilon}{\sigma^2}\right),$$

express the signal-noise ratio computed on log returns in terms of that computed on relative returns and some of the moments α_i.

6. Suppose the relative returns are normally distributed with mean $\mu \neq 0$ and variance σ^2. Then we have $\alpha_1 = \mu$, $\alpha_2 = \mu + \sigma^2$, $\alpha_3 = \mu^3 + \mu\sigma^2$, $\alpha_4 = \mu^4 + 6\mu^2\sigma^2 + 3\sigma^4$. Write an approximation to the signal-noise ratio measured on log returns in terms of μ and σ.

Ex. 1.14 Translation to relative returns Repeat the previous exercise, but make the reverse translation. That is, let the transformation from log returns to relative returns be given by $f\mathrm{exp}(x) = e^x - 1$. Let x be the log returns and let $\alpha_i = \mathrm{E}\left[x^i\right]$. Express the signal-noise ratio measured on relative returns in terms of the α_i using Taylor's series expansions.

Ex. 1.15 Fractal returns Suppose that daily *log* returns, x, are normally distributed with mean μ and variance σ^2.

1. Show that the weekly returns are normally distributed.
2. Assume hourly returns are independent and identically distributed. Argue that they are normally distributed. (*Hint:* use Cramér's theorem.)

* 3. If x are normally distributed relative returns, are weekly returns normally distributed? Hourly returns?

Ex. 1.16 Fractal returns II Compare the daily relative returns of the Market component of the Fama French factors to "fortnightly" returns. That is, first scale up the daily returns by multiplying them by 10, calling this sample A. Then convert relative returns to log returns, sum them in groups of 10, and convert back to relative returns to get sample B.

1. Perform a two-sample empirical Q-Q plot of the two.
2. Perform a two-sample non-parametric test of equality of the two distributions, like the Kolmogorov-Smirnov test, or the Baumgartner-Weiß-Schindler test. [Baumgartner *et al.*, 1998; Pav, 2016a]
3. Which returns "look more normal"? That is, Q-Q plot both in turn against a normal law.
4. Perform the same comparisions on random data: let the fake daily returns be drawn from a fat-tailed distribution, say a $t(4)$ distribution, but independent of each other. Draw a sample of the same size as the daily market returns and with the same standard deviation. Do you see similar changes to the fortnightly returns? Can we blame the apparent "smoothing" of real fortnightly returns to mean reversion, or is it purely an effect of the central limit theorem?

* **Ex. 1.17 Boost from relative returns** Suppose ζ_r is the signal-noise ratio measured on relative returns, and ζ_g is the signal-noise ratio measured using log returns.

1. Show that
$$\frac{\zeta_r - \zeta_g}{\zeta_g} \approx \frac{1}{2}\frac{\alpha_{2,g}}{\mu_g},$$
where $\mu_g = \mathrm{E}\left[x\right]$ and $\alpha_{2,g} = \mathrm{E}\left[x^2\right]$ and x are the log returns.
2. Suppose $\mu_g = 3\times 10^{-4}\,\mathrm{day}^{-1}$ and $\sigma_g = 1.3\times 10^{-2}\,\mathrm{day}^{-1/2}$ are the mean and standard deviation of daily log returns. Compute the boost from computing signal-noise ratio on relative returns instead of log returns, expressed as a percent, the left-hand side of the above approximation. Also evaluate the right-hand side.
3. Write an approximation for ζ_g, the signal-noise ratio on log returns, in terms of ζ_r and σ_r, the signal-noise ratio and volatility of relative returns. What is the order of the error? Does the approximation improve in the case where log returns have zero skew?

Ex. 1.18 Cantelli's inequality Describe a random variable for which Cantelli's inequality is an equality. Is this a reasonable model for the returns of *e.g.*, a mutual fund?

* **Ex. 1.19 Higher order moments** It is often said that investors have positive appetite for odd order moments of returns (mean and skew), and negative appetite for even order moments (volatility, kurtosis). Here you will expand Roy's argument to include higher order moments. [Sullivan, 2011] Let $r_0 \leq \mu$ be the "disastrous rate" of return. Let $p > 2$.

1. Show that
$$\Pr\left\{x \leq r_0\right\} \leq \frac{\mu_p}{\left|\mu - r_0\right|^p},$$
where $\mu_p := \mathrm{E}\left[(x-\mu)^p\right]$. (*Hint:* Express the probability as an integral of a dummy indicator and multiply the numerator and denominator by $\left|x-\mu\right|^p$, then bound the integral. This should be similar to the proof of Chebyshev's inequality.)
2. For even p, rewrite this as
$$\Pr\left\{x \leq r_0\right\} \leq \left(\frac{{\gamma_p}^{1/p}}{\zeta}\right)^p,$$
where ζ includes the "risk-free" rate r_0, and
$$\gamma_p := \frac{\mathrm{E}\left[(x-\mu)^p\right]}{\sigma^p}$$
is the standardized pth moment. The conclusion is that one should maximize $\zeta/{\gamma_p}^{1/p}$.
3. Besides the fact that this ignores odd-order moments, do you think this line of argument will improve upon Roy's original argument? (*Hint:*
$$\Pr\left\{x \leq r_0\right\} \leq \frac{1}{\zeta} \wedge \bigwedge_{p=2,4,\ldots} \left(\frac{{\gamma_p}^{1/p}}{\zeta}\right)^p,$$

and thus only the smallest bound matters. (We write $a \wedge b$ to mean the minimum of a and b, and $\bigwedge$ to mean the minimum over a number of terms.) Estimate the bounds for $p = 2, 4, 6$ using empirical data from returns of, *e.g.*, a broad market ETF.)

Ex. 1.20 Contours of Sharpe ratio In the (μ, σ)-space, plot contours of constant signal-noise ratio, assuming $r_0 = 0$. Plot them again assuming $r_0 = 0.01$.

Ex. 1.21 The Sharpe ratio as yardstick The Sharpe ratio is used as a metric to compare funds. However, it is often lamented that when the sample mean is negative, use of this metric "prefers" higher volatility funds.

1. Suppose you observe, for two different funds, $\hat{\zeta}_1 < \hat{\zeta}_2 < 0$. Is there justification to prefer the second fund to the first?
2. Suppose you are omniscient and know that $\zeta_1 < \zeta_2 < 0$. Is there any justified preference?

Ex. 1.22 Generalized Sharpe ratio Confirm that the ex-factor Sharpe ratio defined in Equation 1.10, for the case where the factors are only the constant one, corresponds to the traditional Sharpe ratio defined in Equation 1.1

* **Ex. 1.23 Approximate Sharpe ratio** For a given return, x, let $\alpha_2 = \mathrm{E}\left[x^2\right]$ be the uncentered second moment. Define

$$f_{\text{tas}}(x) := \tan(\arcsin(x)) = \frac{x}{\sqrt{1 - x^2}}. \tag{1.24}$$

1. What is the domain of $f_{\text{tas}}(\cdot)$? Show that $|\mu| \le \sqrt{\alpha_2}$
2. Show that $\zeta = f_{\text{tas}}\left(\mu/\sqrt{\alpha_2}\right)$. Černý proposed that $\mu/\sqrt{\alpha_2}$ be named the *Hansen ratio.* [Černý, 2020]
3. Compute the first derivative of $f_{\text{tas}}(\cdot)$. Via Taylor's theorem, derive the linear approximation to $f_{\text{tas}}(\cdot)$; call it $\bar{f}_{\text{tas}}(\cdot)$.
4. Show that $f_{\text{tas}}(\cdot)$ is an increasing function.
5. Derive the Lagrange form of the remainder to the linear approximation.
6. Assuming $|x| \le 0.1$, find an upper bound on the absolute error, $\left|f_{\text{tas}}(x) - \bar{f}_{\text{tas}}(x)\right|$. Find an upper bound on the absolute relative error $\left|1 - \bar{f}_{\text{tas}}(x) / f_{\text{tas}}(x)\right|$.
7. Having observed n returns $x_1, \ldots, x_n$, let the first two sample moments be

$$\hat{\mu} = \frac{1}{n} \sum_{1 \le t \le n} x_t, \quad \hat{\alpha}_2 = \frac{1}{n} \sum_{1 \le t \le n} {x_t}^2.$$

Show that $\hat{\zeta} = \sqrt{\frac{n-1}{n}} f_{\text{tas}}\left(\hat{\mu}/\sqrt{\hat{\alpha}_2}\right)$.

* **Ex. 1.24 Market timing** Suppose that a fund engages in *market timing* on an underlying asset. That is, for some ϵ with $|\epsilon| \leq 1$, the fund correctly guesses the sign of the next month's return on the market with probability $\frac{1+\epsilon}{2}$ and takes long or short positions accordingly. (See also Chapter 9.)

1. What is the signal-noise ratio of the fund, computed on relative returns? Your answer should be expressed in terms of ϵ and

$$\kappa = \frac{\mathrm{E}\left[|x|\right]}{\sqrt{\mathrm{E}\left[x^2\right]}}.$$

2. Is the signal-noise ratio infinite for $\epsilon = 1$? Why or why not?
3. Compute the value of κ for the case where x follows a normal distribution with mean μ and standard deviation σ. For this value of κ plot ζ as a function of ϵ.
4. For monthly relative returns of the Market factor, $\kappa = 0.719$. For this value of κ, plot ζ as a function of ϵ. Express ζ in annualized terms. What value of ϵ is required to achieve $\zeta = 1\,\mathrm{yr}^{-1/2}$?
5. Compute the value of κ for the case where x follows a normal distribution with mean μ and standard deviation σ. For this value of κ plot ζ as a function of ϵ.

* **Ex. 1.25 Correlation and market timing** Suppose you observe a random variable f_{t-1}, called "a signal," prior to the time needed to make an investment decision to capture the relative return x_t. Suppose that x_t is zero mean, as is f_{t-1}, and the correlation between f_{t-1} and x_t is ρ.

1. Suppose that f_{t-1} takes only values ± 1. If you allocate f_{t-1} of your capital (long or short, depending on the sign of f_{t-1}) to the asset, show that the signal-noise ratio of your trading strategy, measured on relative returns, is $f_{\mathrm{tas}}\left(\rho\right)$, where $f_{\mathrm{tas}}\left(\cdot\right)$ is defined in Equation 1.24.
* 2. Suppose instead that f_{t-1} and x_t are jointly normal, zero mean variables with correlation ρ. Suppose that you allocate cf_{t-1}, long or short, in the asset, where c is chosen to maintain risk constraints. Show that the signal-noise ratio of your trading strategy, measured on relative returns, is $\rho/\sqrt{1+\rho^2}$. (*Hint:* If $\boldsymbol{y} \sim \mathcal{N}\left(\boldsymbol{\mu}, \Sigma\right)$, then $\mathrm{E}\left[\boldsymbol{y}^\top \mathsf{A}\boldsymbol{y}\right] = \mathrm{tr}\left(\mathsf{A}\Sigma\right) + \boldsymbol{\mu}^\top \mathsf{A}\boldsymbol{\mu}$, and $\mathrm{Var}\left(\boldsymbol{y}^\top \mathsf{A}\boldsymbol{y}\right) = 2\,\mathrm{tr}\left(\mathsf{A}\Sigma\mathsf{A}\Sigma\right) + 4\boldsymbol{\mu}^\top \mathsf{A}\Sigma\mathsf{A}\boldsymbol{\mu}$ for symmetric A.)
* 3. Suppose instead that f_{t-1} and x_t are general zero mean variables with finite fourth moments, and with correlation ρ, but make no further assumptions. Suppose that you allocate cf_{t-1}, long or short, in the asset, where c is chosen to maintain risk constraints. Give a lower bound on the signal-noise ratio of your trading strategy, measured on relative returns, in terms of the first four moments of f_{t-1} and x_t. How small can your bound on signal-noise ratio be for a given $\rho > 0$? (*Hint:* apply

the Cauchy-Schwarz inequality to bound the second moment of your returns from above.)

* 4. Can you construct zero mean, finite variance random variables f_{t-1} and x_t with correlation $\rho > 0$, but such that the signal-noise ratio of the market timing portfolio (*i.e.*, the one with returns $cf_{t-1}x_t$, for some positive c) is arbitrarily small?
* 5. Suppose that f_{t-1} and x_t are jointly normal, zero mean variables with correlation ρ, and the variance of f_{t-1} is known. Can you construct a trading strategy with higher signal-noise ratio than the one suggested in the previous question? You may assume the variance of both f_{t-1} and x_t are known. Your strategy should only depend on f_{t-1}, and not on historically achieved returns.
* 6. ‽ Suppose that f_{t-1} and x_t are zero mean random variables with correlation ρ and known variances. Can you find the "maximin" trading strategy and its signal-noise ratio? That is, find the trading strategy conditional on f_{t-1} that has maximal worst-case signal-noise ratio over all random variables with the given properties.

* **Ex. 1.26 Bid/ask bounce** One commonly rediscovered, putative trading strategy is the phantom bid/ask bounce trade. This supposed anomaly is found by assuming that one can simultaneously observe and trade in, *e.g.*, an auction. This odd assumption typically follows from using low-frequency data for backtest simulations.

Assume that the price quoted in the data feed is either at the bid or the ask, that is, from a market participant crossing the spread to sell to, or buy the asset from, a liquidity provider. Let the price p_t be given by $\log p_t = \log m_t + b_t$, where the log of midpoint m_t is a zero mean random walk, and b_t, the (log) bid/ask bounce, takes values of $\pm\epsilon_b$. The presence of the bid/ask bounce causes negative autocorrelation of the log returns $x_{t+1} = \log p_{t+1}/p_t$. If it was assumed one could observe and trade on price p_t, a mean reverting strategy would appear profitable, when in reality the effect cannot be captured.

1. Let the midpoint log returns, $\log m_{t+1} - \log m_t$ be *i.i.d.* and have zero mean and variance σ_m^2. Let the b_t be *i.i.d.*, and let b_t be independent of $m_1, \ldots, m_{t+1}$. Show that the total log returns $x_{t+1} = \log p_{t+1} - \log p_t$ have autocorrelation of
$$\rho = \frac{-\epsilon_b^2}{2\epsilon_b^2 + \sigma_m^2}$$
2. Although it is only an approximation, and not completely correct (see Exercise 1.25, and also note this question deals with log returns, not relative), assume that the trading strategy which allocates $-cx_t$ capital to the asset achieves a signal-noise ratio of $-\rho$. How does the annualized signal-noise ratio scale with the time period Δt between trades? You may assume that the bid/ask bounce are *i.i.d.* at all scales.

3. Estimate σ_m^2 and ϵ_b from bid/ask data for a real asset. (For example, use the data sample for OEX options provided by CBOE DataShop, `https://datashop.cboe.com/mdr-quotes-trades-data`.) At what time scale would the mean reverting strategy have a signal-noise ratio of $4\,\text{yr}^{-1/2}$?
4. Compute ρ in the case that the bid/ask variable is correlated to the midpoint log returns. Let η be the correlation of $\log m_t - \log m_{t-1}$ and b_t. For more on microstructure models, see *e.g.*, Hasbrouck [2007].

* **Ex. 1.27 The signal-noise ratio of a sum** Let x and y be independent random variables. Let μ_x, σ_x, ζ_x be the mean, standard deviation, and signal-noise ratio of x, and so on.

1. For some constants a, b, let $z = ax + by$. Derive an expression for ζ_z in terms of μ_x, σ_x, μ_y, σ_y, a and b.
2. Show that ζ_z is scale invariant with respect to a and b. That is, for any $c > 0$ $ax + by$ and $cax + cby$ have the same signal-noise ratio.
3. Find the a, b which maximize ζ_z. Because of the scale invariance, these can be determined only up to scale, so choose a and b such that $a^2+b^2 = 1$. (This is a basic *portfolio optimization* problem, of which more in the sequel.)
4. Find ζ_z for the optimal a, b.
5. In the above, a and b were denominated in dollar-proportional units. Instead, we could denominate them in units of risk. That is, define $z = (a/\sigma_x)\,x + (b/\sigma_y)\,y$. Find the a and b which maximize ζ_z. To bypass the scale invariance, express your answer as the ratio a/b. What is the takeaway rule of thumb for portfolio optimization?

* **Ex. 1.28 The signal-noise ratio of a product** Let $z = xy$, where x and y are independent random variables. Let ζ_i be the signal-noise ratio of variable i.

1. Derive an expression for ζ_z entirely in terms of ζ_x and ζ_y.
2. Suppose that y is the level of the VIX index. [CBOE, a,b] Then, $\zeta_y \approx 2.399\,\text{day}^{-1/2}$, based on data from 1990-01-02 to 2020-12-31. Suppose $\zeta_x = 0.1\,\text{day}^{-1/2}$. Compute ζ_z.
3. Compute the derivative
$$\frac{\mathrm{d}\zeta_z}{\mathrm{d}\zeta_y}.$$
Does ζ_z have a local maximum with respect to ζ_y?
4. Find
$$\lim_{\zeta_y \to \infty} \zeta_z$$

5. Viewing y as some kind of random "leverage" on the strategy with returns x, what is the best case for an investor in z?

* **Ex. 1.29 Higher order moments, autocorrelated returns** Assume that daily log returns, x, follow an AR(1) process: $x_{t+1} - \mu = \rho(x_t - \mu) + \epsilon_{t+1}$, where ϵ_t are independent zero-mean error terms with standard deviation σ. Let y be the annualized (or weekly, or monthly, *etc.*) log returns of the same process. So, for some fixed m, the number of days in an epoch, let $y_t = \sum_{(t-1)m<i\le tm} x_i$.

1. What is the mean of y in terms of the mean of x, m, and ρ?
2. What is the standard deviation of y in terms of the standard deviation of x, m, and ρ?
3. * What is the third moment of y in terms of the first three moments of x, m and ρ? What is the skew of y?
4. ** What is the fourth moment of y in terms of the first four moments of x, m and ρ? What is the excess kurtosis of y?

* **Ex. 1.30 Autocorrelated returns and signal-noise ratio** In Section 1.2, the method for annualizing the Sharpe ratio was given *for independent returns.* As in Exercise 1.29, assume daily log returns, x, follow an AR(1) process: $x_{t+1} - \mu = \rho(x_t - \mu) + \epsilon_{t+1}$, where ϵ_t are independent zero-mean error terms with standard deviation σ. So, for some fixed n, the number of days in an epoch, let $y_t = \sum_{(t-1)n<i\le tn} x_i$.

1. Compute the signal-noise ratio of x, *cf.* Lo [2002].
2. Compute the signal-noise ratio of y.
3. Compute the ratio of the signal-noise ratio of y to $\sqrt{n}$ times the signal-noise ratio of x. Plot this ratio versus the signal-noise ratio of x for $n = 252$, using $\rho = 0.1$. Plot it against ρ assuming the signal-noise ratio of x is $0.05\,\text{day}^{-1/2}$.

Ex. 1.31 Attribution fits Consider the monthly returns of the Fama-French factor data from **tsrsa**, using code as given in Example 1.1.1. Each of the following asks you to perform a factor attribution. For each of them, compute the ex-factor Sharpe ratio.

1. The "Market" returns have the risk-free rate subtracted from them, but there may be residual exposure of these *excess returns* to the risk-free rate. Perform an attribution of the excess returns of the market to the risk free rate, computing the ex-factor Sharpe ratio.
2. Perform a CAPM attribution on the SMB portfolio returns. For simplicity, use the excess returns of the Market portfolio as the market.
3. Perform a CAPM attribution on the HML portfolio returns.

4. Perform a Chow-test attribution on the excess returns of the Market portfolio, with an indicator factor that is one prior to 1960.
5. Perhaps it is the case that the small cap premium is disappearing and does not currently exist. Test this by attributing the monthly returns of the SMB portfolio against a factor which is linearly decreasing over time, taking value of 1 at the start of the data, and value 0 at the end.
6. Can all the excess returns of the Market be attributed to the "January Effect"? Perform an attribution against an indicator variable that is 1 exactly in January and 0 elsewhere.

Ex. 1.32 Geometric utility Suppose μ and σ^2 are the mean and variance of the simple returns of a strategy, x. Show that $\operatorname{E}\left[\log\left(1+x\right)\right] \approx \mu - \frac{1}{2}\sigma^2$, the mean-variance objective function.

Ex. 1.33 Other objectives Construct several objective functions on the population parameters of returns distributions which sound like plausible objectives for an investor. Which of them are consistent with first-order stochastic dominance?

* **Ex. 1.34 Moment preferences and signal-noise ratio** Suppose x is a returns stream with mean $\mu > 0$ and standard deviation σ. Let y be a contingent claim (a 0/1 Bernoulli random variable) which pays out with probability p, independent of x. Suppose $z = x + y$.

1. Show that z (first-order) stochastically dominates x. That is, show that, for all c, $\Pr\left\{x \geq c\right\} \leq \Pr\left\{z \geq c\right\}$.
2. Find the signal-noise ratio of z.
3. Find conditions on μ, σ^2, p such that the signal-noise ratio of z is *less* than that of x.

* **Ex. 1.35 Is the signal-noise ratio monotonic?** Let x and y be paired random variables.

1. Suppose that $\Pr\left\{x_i \leq y_i\right\} = 1$ for all i. Let ζ_x, and ζ_y be the signal-noise ratios of these variables. Can you prove that $\zeta_x \leq \zeta_y$? If not, can you find a counterexample?
2. Suppose you observe n paired observations of x and y and notice that $x_i \leq y_i$ for all i. You measure their Sharpe ratios, $\hat{\zeta}_x$ and $\hat{\zeta}_y$. Must it be the case that $\hat{\zeta}_x \leq \hat{\zeta}_y$? If not, can you find a counterexample?

Ex. 1.36 Approximate normality Load the daily returns of "the Market" from **tsrsa**. Resample, with replacement, 250 days of market returns, and compute the mean. Repeat this 5,000 times, and Q-Q plot the means against normality.

1. What implications does this have for the argument that signal-noise

ratio is an inappropriate objective because it assumes returns are normal?

Ex. 1.37 Higher order signal-noise ratio Consider the higher order signal-noise ratio, defined in Equation 1.17.

1. Show that the higher order signal-noise ratio is consistent with first-order stochastic dominance.
2. Is it necessarily consistent with second-order stochastic dominance?

Ex. 1.38 Solving for higher order signal-noise ratio Consider the approximate higher order signal-noise ratio given by Equation 1.19.

1. Find the roots of the equation

$$\zeta_h = \frac{(\mu - r_0)}{\sigma} + \frac{\gamma_1}{6}\left(\zeta_h^2 - \frac{1}{n}\right).$$

2. Which root do you suspect gives a closer approximation to the higher order signal-noise ratio defined in Equation 1.17?
3. Suppose $(\mu - r_0)/\sigma = 0.1\,\text{day}^{-1/2}$, $\gamma_1 = 1.5$ and $n = 252\,\text{days}$. Find the approximate higher order signal-noise ratio.

* **Ex. 1.39 Sample variance of alternative metrics** Explore the sample variation of some alternative metrics, as compared to the Sharpe ratio. For example, the Calmar ratio, defined as the three year average return divided by the maximum drawdown over three years. Draw 3 years of daily returns from a population with Gaussian log returns, using $\mu = 0\,\text{day}^{-1}, \sigma = 0.01\,\text{day}^{-1/2}$, then compute the Calmar ratio on the returns. Also compute the Sharpe ratio. Repeat this 1,000 times. Then do the same using $\mu = 0.001\,\text{day}^{-1}, \sigma = 0.01\,\text{day}^{-1/2}$. Estimate, empirically, the power of the test with 0.05 type I rate via the samples of the statistics on these two populations. The Calmar ratio is provided in **R** by the **PerformanceAnalytics** package, as are all the alternative metrics mentioned below, while the Sharpe ratio can be computed via the **SharpeR** package. [Peterson and Carl, 2018; Pav, 2020a]

1. Repeat this exercise for the Sterling ratio.
2. Repeat this exercise for the "upside potential ratio."
3. Repeat this exercise for the maximum drawdown. (You will have to turn the log returns series into a price series to compute drawdown.)
4. Repeat this exercise for the so-called "Fano ratio," defined as $\hat{\mu}/\hat{\sigma}^2$. [Kakushadze and Yu, 2017] (See also Exercise 1.43.)

* **Ex. 1.40 Drawdown and distribution** Draw 2 years of daily log returns from the Gaussian distribution with $\mu = 0\,\text{day}^{-1}, \sigma = 0.01\,\text{day}^{-1/2}$. Turn the returns series into a price series, then compute the maximum drawdown. Repeat this 10,000 times. Compute the 0.95 empirical quantile.

1. Repeat this exercise but using draws from a shifted, rescaled t-distribution with 4 degrees of freedom. You need to make sure you have achieved the proper mean and variance.
2. Repeat this exercise but using draws from a shifted, rescaled, "Lambert W × Gaussian" distribution using $\beta = 0.1$. [Goerg, 2009] (The **LambertW** package is recommended for this task. [Goerg, 2016])
3. How sensitive is the critical maximum drawdown to the distribution from which returns are drawn?

* **Ex. 1.41 Drawdown and distribution II** Repeat the experiment of Example 1.7.2, but draw returns from a shifted, rescaled t-distribution with 4 degrees of freedom. How do the half space constraint regression coefficients change?

Ex. 1.42 Drawdown and distribution III Rej *et al.* [2017] note a connection between the length and depth of the current drawdown and the signal-noise ratio. Perform simulations to confirm their findings. Here the "current drawdown" is not the same as the maximum drawdown, rather it is the difference between the maximum fund value over a history and the recent value. So let x_t be *i.i.d.* Gaussian log returns and let

$$p_t = \exp\left(\sum_{1 \le s \le t} x_s\right).$$

The current drawdown at time t is defined as

$$\max_{1 \le s \le t} p_s - p_t.$$

Simulate 2 years of weekly data using volatility of $\sigma = 0.04\,\mathrm{wk.}^{-1/2}$ and compute the current drawdown at the end of the 2-year period. Repeat this 1,000 times for each choice of $\mu = 0, 0.001\,\mathrm{wk.}^{-1}, \ldots, 0.010\,\mathrm{wk.}^{-1}$.

1. Compute the median drawdown for each choice of μ and scatter plot against ζ^{-1}. Do you see a linear fit?

* **Ex. 1.43 A Drawdown estimator of signal-noise ratio** Challet [2015] introduced an estimator of the signal-noise ratio based on the number of drawdowns (and draw ups) of an asset's price level. This estimator is purported to have lower standard error (and by implication, lower mean square error) than the Sharpe ratio when returns are drawn from leptokurtic distributions.

Test this assertion empirically. The code to compute the drawdown-based estimator is available in the **R** package, **sharpeRratio**. Draw 1,000 days of returns from a t distribution, and compute the Sharpe ratio and the drawdown estimator, then compute the error of each. Repeat this 10,000 times. Perform this experiment for the case where the signal-noise ratio is $0.5\,\mathrm{yr}^{-1/2}$ and again when it is $2\,\mathrm{yr}^{-1/2}$. For the returns distribution vary the degrees of freedom of the t distribution, taking it to be 8, or 4.25. You will compute estimates of the mean square error of each of these two estimators for 4 different parameter settings.

1. Does the drawdown-estimator consistently have lower mean square error? If not, why not?
2. What do you think should happen when returns are normally distributed? Repeat the experiment with normal returns to find out.

Ex. 1.44 **Mr. Buffett's bet** In 2007, Warren Buffett made a million dollar bet with Ted Seides that the Vanguard S&P 500 index fund would outperform the average return of five funds of hedge funds returns over a ten year period. Mr. Buffett prevailed in this bet. [Buffett, 2017; Buffett and Seides, 2007] Let us consider Mr. Seides' odds.
We downloaded the BarclayHedge Equity Market Neutral Index data, then joined to the monthly Market returns data from the Fama French factor returns. In total, we consider 240 months of data from 1997-01-01 to 2016-12-01. Supposing that the Hedge Fund returns were similar to those of the Market Neutral Index (they weren't), and that the S&P fund from Vanguard tracked the Market (a decent approximation), consider the relative returns of one dollar long on Market and one dollar short on the Market neutral index.
Over the sample period, this combined strategy had an empirical mean return of $0.3\%\,\mathrm{mo.}^{-1}$ and a standard deviation of $4.5\%^{-1/2}\,\mathrm{mo.}^{-1/2}$.

1. What is the probability that a normally distributed random variable with mean 0.3 and standard deviation 4.5 will be positive?
2. What is the probability that the sum of 120 independent normally distributed random variables with mean 0.3 and standard deviation 4.5 will be positive?
3. We have taken the sample standard deviation of the difference in returns of the two series, which exhibit sizeable correlation over the sample period (around 0.21). Instead consider the influence of correlation. The two returns series have volatilities of $0.88\%^{-1/2}\,\mathrm{mo.}^{-1/2}$ and $4.6\%^{-1/2}\,\mathrm{mo.}^{-1/2}$. Suppose the correlation between them is 0. What is the volatility of their difference? How would this affect the probability that their difference, assumed normal, would have positive sum over 120 months?
4. Suppose instead that the Fund of Funds has zero expected mean and that the Market has mean return of $0.74\%\,\mathrm{mo.}^{-1}$, and their difference has standard deviation $4.5\%^{-1/2}\,\mathrm{mo.}^{-1/2}$. What is the probability that their difference will be positive over 120 months assuming normal returns?

2

The Sharpe Ratio for Gaussian Returns

Hypotheses non fingo.

Isaac Newton
General Scholium

While normality of returns is a *terrible* model for most market instruments [Cont, 2001], it is a terribly convenient model. For the rest of this chapter, then, unless stated otherwise, we will adhere to this terrible model and assume returns are unconditionally *i.i.d.* Gaussian, *i.e.*, $x_t \sim \mathcal{N}(\mu, \sigma^2)$. In the sequel, we will examine the consequences of assuming normality, independence, homoskedasticity, *etc.*, and correct for them when possible.

The Sharpe ratio is, up to a scaling, the Student t-statistic for testing the mean of a (typically Gaussian) random variable. Sharpe [1965] does not mention this relationship. This connection to t-statistics, perhaps first noted by Miller and Gehr [1978], allows us to translate known statistical results from testing of the mean to testing of the signal-noise ratio.

2.1 The non-central t-distribution

We consider now a classical probability distribution connected to that of the Sharpe ratio. Let Z be a random variable following a normal distribution with mean δ and variance 1, which we write as $Z \sim \mathcal{N}(\delta, 1)$. Let X, independent of Z take a chi-square distribution with ν degrees of freedom, written $X \sim \chi^2(\nu)$. Z is like our sample mean and X is like our sample standard deviation. The variable

$$t = \frac{Z}{\sqrt{X/\nu}}$$

follows a *non-central t distribution* with non-centrality parameter δ and ν degrees of freedom. [Johnson and Welch, 1940; Scholz, 2007; Owen, 1968] We write this as $t \sim t(\nu, \delta)$. As a special case, when $\delta = 0$, t follows a (central) t distribution. (The non-central t is a special case of the more general *doubly* non-central t distribution, wherein X follows a non-central chi-square distribution.)

To capture the statistics of the Sharpe ratio, we need to shift and scale

DOI: 10.1201/9781003181057-2

the numerator and denominator. Suppose now that $Z \sim \mathcal{N}\left(\mu_0, \sigma^2/\nu\right)$ independently of $(\nu - 1) X/\sigma^2 \sim \chi^2(\nu - 1)$. Then

$$\sqrt{\nu}\frac{Z - \mu_1}{\sqrt{X}} \sim t\left(\delta = \sqrt{\nu}\frac{\mu_0 - \mu_1}{\sigma}, \nu - 1\right). \tag{2.1}$$

The non-centrality parameter δ is $\sqrt{\nu}$ times the difference in signal-noise ratio values, μ_0/σ and μ_1/σ.

2.1.1 Distribution of the Sharpe ratio

Now let $x_1, x_2, \ldots, x_n$ be *i.i.d.* draws from a normal distribution $\mathcal{N}(\mu, \sigma)$. Let $\hat{\mu} := \sum_t x_t/n$ and $\hat{\sigma}^2 := \sum_t (x_t - \hat{\mu})^2/(n-1)$ be the unbiased sample mean and variance statistics. It happens to be the case that $\hat{\mu}$ and $\hat{\sigma}$ are independent with $\hat{\mu} \sim \mathcal{N}\left(\mu, \frac{\sigma^2}{n}\right)$, and $(n-1)\hat{\sigma}^2/\sigma^2 \sim \chi^2(n-1)$. [Bain and Engelhardt, 1992; Spiegel and Stephens, 2007] This gives the distribution of the Sharpe ratio under the assumption of Gaussian returns. That is

$$\sqrt{n}\hat{\zeta} = \sqrt{n}\frac{\hat{\mu} - r_0}{\hat{\sigma}} \sim t\left(\sqrt{n}\frac{\mu - r_0}{\sigma} = \sqrt{n}\zeta, n - 1\right). \tag{2.2}$$

Note the non-centrality parameter, $\delta = \sqrt{n}\left(\mu - r_0\right)/\sigma$, looks like the sample statistic $\sqrt{n}\left(\hat{\mu} - r_0\right)/\hat{\sigma}$, but defined with population quantities. Informally, it is the "population analogue" of the sample statistic.

2.1.2 Distribution of the ex-factor Sharpe ratio

Now consider the ex-factor Sharpe ratio introduced in Equation 1.10 of Section 1.4. Suppose that $x_1, x_2, \ldots, x_n$ are the observed returns, and that $\boldsymbol{f}_1, \boldsymbol{f}_2, \ldots, \boldsymbol{f}_n$ are the corresponding factors in a factor model. Let F be the $n \times l$ matrix whose rows are the vectors $\boldsymbol{f}_t^{\top}$. Assume the model is properly specified: conditional on observing $\boldsymbol{f}_t$,

$$x_t = \boldsymbol{f}_t^{\top}\beta + \epsilon_t,$$

where the errors are *i.i.d.* normal, $\epsilon_t \sim \mathcal{N}(0, \sigma)$. The multiple linear regression estimates are

$$\hat{\boldsymbol{\beta}} := \left(\mathsf{F}^{\top}\mathsf{F}\right)^{-1}\mathsf{F}^{\top}\boldsymbol{x}, \qquad \hat{\sigma} := \sqrt{\frac{\left(\boldsymbol{x} - \mathsf{F}\hat{\boldsymbol{\beta}}\right)^{\top}\left(\boldsymbol{x} - \mathsf{F}\hat{\boldsymbol{\beta}}\right)}{n - l}}. \tag{2.3}$$

Then, conditional on observing F,

$$\hat{\boldsymbol{\beta}} \sim \mathcal{N}\left(\boldsymbol{\beta}, \sigma^2\left(\mathsf{F}^{\top}\mathsf{F}\right)^{-1}\right), \quad (n - l)\frac{\hat{\sigma}^2}{\sigma^2} \sim \chi^2(n - l),$$

and the estimators are independent. [Bain and Engelhardt, 1992; Rao *et al.*, 2010]

The ex-factor Sharpe ratio is then distributed as

$$\hat{\zeta}_g := \frac{\hat{\boldsymbol{\beta}}^\top \boldsymbol{v} - r_0}{\hat{\sigma}}$$
$$\sim \frac{\mathcal{N}\left(\boldsymbol{\beta}^\top \boldsymbol{v} - r_0, \sigma^2 \boldsymbol{v}^\top \left(\mathsf{F}^\top \mathsf{F}\right)^{-1} \boldsymbol{v}\right)}{\sqrt{\sigma^2 \chi^2 \left(n-l\right) / \left(n-l\right)}},$$
$$\sim c^{-1} \frac{\mathcal{N}\left(c\frac{\boldsymbol{\beta}^\top \boldsymbol{v} - r_0}{\sigma}, 1\right)}{\sqrt{\chi^2 \left(n-l\right) / \left(n-l\right)}},$$

for $c = \left(\boldsymbol{v}^\top \left(\mathsf{F}^\top \mathsf{F}\right)^{-1} \boldsymbol{v}\right)^{-1/2}$. Thus we have

$$c\hat{\zeta}_g \sim t\left(c\zeta_g, n-l\right), \tag{2.4}$$

with $\zeta_g := \frac{\boldsymbol{\beta}^\top \boldsymbol{v} - r_0}{\sigma}$. Compare this with Equation 2.2. Most of the results we will explore concerning the Sharpe ratio have equivalent results for the ex-factor Sharpe ratio. However, keeping track of the scaling parameter c would be unwieldy, so generalizing the results will be left as an exercise for the reader. See also Exercise 2.21.

While many of the results in this chapter have analogous forms for the ex-factor Sharpe ratio, strictly speaking they only apply to the case where the factors $\boldsymbol{f}_t$ are *deterministic*, and controlled by the experimenter. This is decidedly *not* the case for many models in which attribution will be performed, *e.g.*, CAPM, the Fama-French models, Henriksson-Merton, *etc.* Indeed, few, if any, of the models listed in Section 1.4.1 feature deterministic $\boldsymbol{f}_t$. The issue is that the randomness of the $\boldsymbol{f}_t$ contributes to estimation error. Methods for dealing with attribution for random covariates will be considered in Chapter 3.

2.2 † Density and distribution of the Sharpe ratio

Since the Sharpe ratio is distributed as a non-central t up to scaling, its density, cumulative distribution, and quantile functions can be defined in terms of their non-central t counterparts, as follows:

$$\begin{aligned} f_{SR}\left(\hat{\zeta}; \zeta, n\right) &= n^{-1/2} f_t\left(\sqrt{n}\hat{\zeta}; \sqrt{n}\zeta, n-1\right), \\ F_{SR}\left(\hat{\zeta}; \zeta, n\right) &= F_t\left(\sqrt{n}\hat{\zeta}; \sqrt{n}\zeta, n-1\right), \\ SR_p\left(\zeta, n\right) &= n^{-1/2} t_p\left(\sqrt{n}\zeta, n-1\right). \end{aligned} \tag{2.5}$$

Here $f_{SR}\left(\hat{\zeta};\zeta,n\right)$ is the density (or PDF) of the Sharpe ratio distribution with signal-noise ratio ζ on n samples; $F_{SR}\left(\hat{\zeta};\zeta,n\right)$ is the cumulative distribution (or CDF); and $SR_p\left(\zeta,n\right)$ is the pth quantile. The functions on the right hand side of the equations are the PDF, CDF, and quantile of the non-central t distribution.

The PDF, CDF, and quantile functions of the non-central t have stock implementations. For example, in **R**, these are available as `dt`, `pt`, and `qt`. As a practical matter, these should be used instead of writing your own. *If your interests are practical, rather than theoretical, you can safely skip the rest of this section.*

2.2.1 † The density of the Sharpe ratio

First, let us compute the density of the Sharpe ratio under *i.i.d.* normal returns. Following Walck [1996], let $\hat{\zeta} = z/\sqrt{x}$, where $z \sim \mathcal{N}\left(\zeta, \frac{1}{n}\right)$ independently of $x \sim \chi^2\left(n-1\right)$. The joint density of independent x, z is the product of their densities,

$$f\left(x,z;\zeta,n\right) = \left[\frac{1}{2^{\frac{n-1}{2}}\Gamma\left(\frac{n-1}{2}\right)} x^{\frac{n-1}{2}-1} e^{-\frac{x}{2}}\right]\left[\sqrt{\frac{n}{2\pi}} e^{-\frac{n(z-\zeta)^2}{2}}\right]. \tag{2.6}$$

Now we make the transformation $\left[x,z\right]^{\top} \to \left[x, z/\sqrt{x}\right]^{\top} = \left[x,\hat{\zeta}\right]^{\top}$. To get the density of the transformed variables, we note that the determinant of the Jacobian of the inverse transform is $\sqrt{x}$. Thus

$$f\left(x,\hat{\zeta};\zeta,n\right) = \sqrt{x} f\left(x,z;\zeta,n\right) = \sqrt{x} f\left(x,\hat{\zeta}\sqrt{x};\zeta,n\right).$$

To get the density of $\hat{\zeta}$, integrate out x. This gives the integral form of the density as

$$\begin{aligned} f_{SR}\left(\hat{\zeta};\zeta,n\right) &= \int_0^\infty \sqrt{x}\left[\frac{x^{\frac{n-1}{2}-1} e^{-\frac{x}{2}}}{2^{\frac{n-1}{2}}\Gamma\left(\frac{n-1}{2}\right)}\right]\left[\sqrt{\frac{n}{2\pi}}\exp\left(-\frac{n\left(\hat{\zeta}\sqrt{x}-\zeta\right)^2}{2}\right)\right] dx, \\ &= \frac{\sqrt{n/2}}{\sqrt{2\pi}\Gamma\left(\frac{n-1}{2}\right)} \int_0^\infty \left(\frac{x}{2}\right)^{\frac{n-2}{2}} \exp\left(-\frac{x}{2} - \frac{n\left(\hat{\zeta}\sqrt{x}-\zeta\right)^2}{2}\right) dx. \end{aligned} \tag{2.7}$$

2.2.2 The distribution and quantile of the Sharpe ratio

Finding the CDF of the Sharpe ratio consists of "simply" integrating the PDF given in Equation 2.7. Witkovsky [2013] gives a few different forms for the CDF of the non-central t distribution. The equivalent formulations for the

Sharpe ratio distribution are as follows, for $\hat{\zeta} > 0$,

$$\begin{aligned} F_{SR}\left(\hat{\zeta};\zeta,n\right) &= \frac{1}{\sqrt{n-1}}\int_0^{\infty}\Phi\left(\hat{\zeta}\sqrt{\frac{n}{n-1}}z-\sqrt{n}\zeta\right)f_{\chi^2}\left(z;n-1\right)\,\mathrm{d}z,\\ &= \Phi\left(-\sqrt{n}\zeta\right)+\int_{-\sqrt{n}\zeta}^{\infty}\left(1-F_{\chi^2}\left(\frac{(n-1)\left(z+\sqrt{n}\zeta\right)^2}{n\hat{\zeta}^2};n-1\right)\right)\phi\left(z\right)\,\mathrm{d}z,\\ &= \Phi\left(-\sqrt{n}\zeta\right)+\frac{1}{2}\sum_{i=0}^{\infty}\left[\left(\frac{n\zeta^2}{2}\right)^{i}\frac{e^{-n\zeta^2/2}}{i!}\mathrm{I}_{\frac{n\hat{\zeta}^2}{n-1+n\hat{\zeta}^2}}\left(i+\frac{1}{2},\frac{n-1}{2}\right)\right]\\ &+\frac{1}{2}\sum_{i=0}^{\infty}\left[\left(\frac{n\zeta^2}{2}\right)^{i+1/2}\frac{e^{-n\zeta^2/2}}{\Gamma\left(i+3/2\right)}\mathrm{I}_{\frac{n\hat{\zeta}^2}{n-1+n\hat{\zeta}^2}}\left(i+1,\frac{n-1}{2}\right)\right]. \end{aligned} \tag{2.8}$$

Here $\mathrm{I}_x\left(a,b\right)$ is the regularized incomplete beta function [Abramowitz and Stegun, 1964, 6.6.2], and $F_{\chi^2}\left(x;\nu\right)$ and $f_{\chi^2}\left(x;\nu\right)$ are, respectively, the CDF and PDF of the chi-square distribution with ν degrees of freedom. The connection between the non-central t distribution and the lambda prime distribution (*cf.* Section 2.4) are evident in some of these forms. The last form is essentially that used in the standard computation of the CDF of the t distribution via "AS 243." [Lenth, 1989; Guenther, 1978]

The inverse CDF, or quantile function, of the non-central t-distribution is not easily expressed in compact notation. The "exact" computation, provided by `qt` in **R**, is due to Hill [1970, 1981]. Akahira [1995] present an approximation to the quantile function, first described by Akahira *et al.* [1995], based on the Cornish Fisher expansion[1]. The Akahira approximation generalizes the quantile approximation of Johnson and Welch [1940].

2.3 Moments of the Sharpe ratio

The moments of the non-central t-distribution are known, and can easily be translated into those of the Sharpe ratio. [Hogben *et al.*, 1961; Walck, 1996; Wikipedia contributors, 2011] Suppose that $t \sim t\left(\delta, n-1\right)$. Then, for $0 \le i < n-1$,

$$\mathrm{E}\left[t^i\right] = \left(\frac{n-1}{2}\right)^{i/2}\frac{\Gamma\left(\frac{n-1-i}{2}\right)}{\Gamma\left(\frac{n-1}{2}\right)}\left(e^{-\delta^2/2}\frac{\mathrm{d}^i e^{\delta^2/2}}{\mathrm{d}\delta^i}\right). \tag{2.9}$$

The quantity in parenthesis is a variant of the "probabilist's Hermite polynomial," but has all positive coefficients in δ. [Abramowitz and Stegun, 1964] Since $\sqrt{n}\hat{\zeta} \sim t\left(\sqrt{n}\zeta, n-1\right)$ under our assumption of normal returns, the moments of the Sharpe ratio are

$$\alpha_i\left(\hat{\zeta}\right) := \mathrm{E}\left[\hat{\zeta}^i\right] = \left(\frac{n-1}{2}\right)^{i/2}\frac{\Gamma\left(\frac{n-1-i}{2}\right)}{\Gamma\left(\frac{n-1}{2}\right)}\frac{1}{n^{i/2}}e^{-n\zeta^2/2}\frac{\mathrm{d}^i e^{n\zeta^2/2}}{\mathrm{d}\left(\sqrt{n}\zeta\right)^i}, \tag{2.10}$$

[1] Akahira use the inversion that connects the t-distribution to the lambda prime distribution, see Section 2.4.

and so

$$\begin{aligned}
\alpha_1\left(\hat{\zeta}\right) &= \left(\frac{n-1}{2}\right)^{1/2} \frac{\Gamma\left(\frac{n-2}{2}\right)}{\Gamma\left(\frac{n-1}{2}\right)} \left(\frac{n^{\frac{1}{2}}\zeta}{n^{\frac{1}{2}}}\right) = d_n \zeta,\\
\alpha_2\left(\hat{\zeta}\right) &= \left(\frac{n-1}{2}\right) \frac{\Gamma\left(\frac{n-3}{2}\right)}{\Gamma\left(\frac{n-1}{2}\right)} \left(\frac{1+n\zeta^2}{n}\right) = \frac{n-1}{n-3}\left(\frac{1+n\zeta^2}{n}\right),\\
\alpha_3\left(\hat{\zeta}\right) &= \left(\frac{n-1}{2}\right)^{3/2} \frac{\Gamma\left(\frac{n-4}{2}\right)}{\Gamma\left(\frac{n-1}{2}\right)} \left(\frac{3n^{\frac{1}{2}}\zeta + n^{\frac{3}{2}}\zeta^3}{n^{\frac{3}{2}}}\right) = \frac{n-1}{n-4} d_n \frac{\zeta}{n}\left(3+n\zeta^2\right),\\
\alpha_4\left(\hat{\zeta}\right) &= \left(\frac{n-1}{2}\right)^{2} \frac{\Gamma\left(\frac{n-5}{2}\right)}{\Gamma\left(\frac{n-1}{2}\right)} \left(\frac{3+6n\zeta^2+n^2\zeta^4}{n^2}\right)\\
&= \frac{(n-1)^2}{(n-3)(n-5)}\left(\frac{3+6n\zeta^2+n^2\zeta^4}{n^2}\right),\\
\alpha_5\left(\hat{\zeta}\right) &= \left(\frac{n-1}{2}\right)^{5/2} \frac{\Gamma\left(\frac{n-6}{2}\right)}{\Gamma\left(\frac{n-1}{2}\right)} \left(\frac{15n^{\frac{1}{2}}\zeta + 10n^{\frac{3}{2}}\zeta^3 + n^{\frac{5}{2}}\zeta^5}{n^{\frac{5}{2}}}\right)\\
&= \frac{(n-1)^2}{(n-4)(n-6)} d_n \frac{\zeta}{n^2}\left(15+10n\zeta^2+n^2\zeta^4\right),
\end{aligned} \tag{2.11}$$

where we have defined

$$d_n := \sqrt{\frac{n-1}{2}}\frac{\Gamma\left(\frac{n-2}{2}\right)}{\Gamma\left(\frac{n-1}{2}\right)}, \tag{2.12}$$

as the "bias term." The even moments enjoy cancellation due to definition of the Gamma function, while the odd moments have this irreducible constant, d_n. Thus we have

$$\begin{aligned}
\operatorname{E}\left[\hat{\zeta}\right] &= d_n\zeta,\\
\operatorname{Var}\left(\hat{\zeta}\right) &= \frac{(1+n\zeta^2)(n-1)}{n(n-3)} - \left(d_n\zeta\right)^2,\\
\operatorname{skew}\left(\hat{\zeta}\right) &= \frac{\alpha_3\left(\hat{\zeta}\right) - 3\alpha_2\left(\hat{\zeta}\right)\alpha_1\left(\hat{\zeta}\right) + 2\alpha_1\left(\hat{\zeta}\right)^3}{\left(\alpha_2\left(\hat{\zeta}\right) - \alpha_1\left(\hat{\zeta}\right)^2\right)^{3/2}},\\
\operatorname{ex\,kurtosis}\left(\hat{\zeta}\right) &= \frac{\alpha_4\left(\hat{\zeta}\right) - 4\alpha_3\left(\hat{\zeta}\right)\alpha_1\left(\hat{\zeta}\right) + 6\alpha_2\left(\hat{\zeta}\right)\alpha_1\left(\hat{\zeta}\right)^2 - 3\alpha_1\left(\hat{\zeta}\right)^4}{\left(\alpha_2\left(\hat{\zeta}\right) - \alpha_1\left(\hat{\zeta}\right)^2\right)^2} - 3.
\end{aligned} \tag{2.13}$$

Example 2.3.1 (Cumulants of the Sharpe ratio). Suppose you observe 60 months of Gaussian log returns with $\zeta = 0.3\,\text{mo.}^{-1/2}$. Then $\operatorname{E}\left[\hat{\zeta}\right] = 0.304\,\text{mo.}^{-1/2}$, $\operatorname{Var}\left(\hat{\zeta}\right) = 0.018\,\text{mo.}^{-1}$, $\operatorname{skew}\left(\hat{\zeta}\right) = 0.12$, $\operatorname{ex\,kurtosis}\left(\hat{\zeta}\right) = 0.129$. The skewness and excess kurtosis are unitless quantities by definition.

To check these calculations, 10^7 samples from the Sharpe ratio distribution with $\zeta = 0.3\,\text{mo.}^{-1/2}$ and $n = 60$ months were drawn. The empirical mean of the Sharpe ratio was $0.304\,\text{mo.}^{-1/2}$, the empirical variance was $0.018\,\text{mo.}^{-1}$, and the empirical skew and excess kurtosis were measured to be 0.121, and 0.129. ⊣

2.3.1 The Sharpe ratio is biased

The geometric bias term, d_n, is related to the constant c_4 from the statistical control literature via

$$d_n = \frac{n-1}{n-2} c_4 \left(n\right).$$

The bias term does not equal one, thus the Sharpe ratio is a biased estimator of the signal-noise ratio (when it is nonzero), as first described by Miller and Gehr. [Miller and Gehr, 1978; Jobson and Korkie, 1981; Bao and Ullah, 2006]

The bias is multiplicative and larger than one, so the Sharpe ratio will overestimate the signal-noise ratio when the latter is positive, and underestimate it when it is negative. The bias term is a function of sample size only, and approaches one fairly quickly. A decent asymptotic approximation [NIST, 2011] to d_n is given by

$$d_{n+1} = 1 + \frac{3}{4n} + \frac{25}{32n^2} + \mathcal{O}\left(n^{-3}\right). \tag{2.14}$$

In Figure 2.1, $1 - d_n$ is plotted versus n, along with this approximation. Higher order formulae for the bias of the Sharpe ratio for non-Gaussian returns are given in Section 3.2.3, and suggest the approximation

$$d_{n+1} \approx 1 + \frac{3}{4n} + \frac{49}{32n^2}.$$

In Section 3.2.3, we will give the bias of the Sharpe ratio for general, non-Gaussian returns.

Example 2.3.2 (Bias in Sharpe ratio). Looking at one year's worth of returns with monthly marks, the bias is fairly large: $d_{12} = 1.075$, *i.e.*, almost 8%. When looking at one year's worth of weekly marks, the bias is more modest: $d_{52} = 1.015$; for a year of daily marks $d_{252} = 1.003$. When $n > 100$, say, this bias is negligible. ⊣

2.3.2 Moments under up-sampling

Suppose, as a prospective investor in a fund, you are given one year's worth of daily log returns of the fund, say $n = 252$. You are offered, for a small fee, the option of viewing the log returns of every minute the market is open, a 390-fold increase in n. Should you accept this offer? Again, in this chapter we are assuming returns are *i.i.d.* normally distributed. How will this up-sampling affect the moments of the sample Sharpe ratio? Your intution should tell

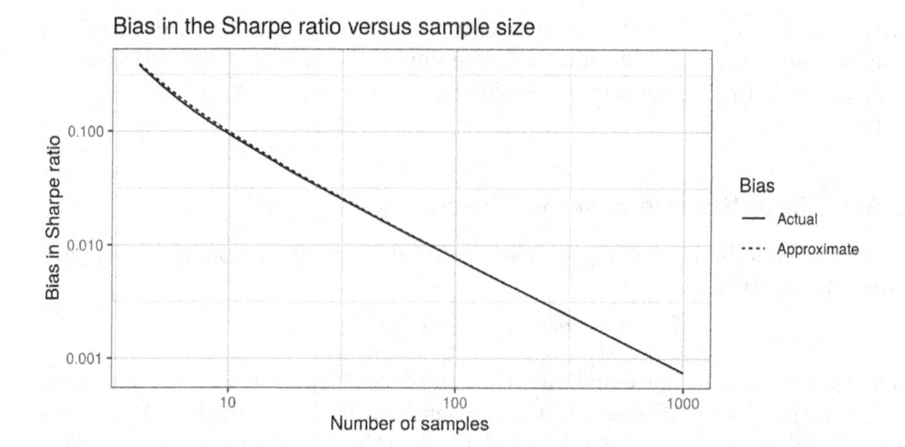

FIGURE 2.1
The relative bias of the Sharpe ratio, $d_n - 1$ is plotted versus sample size. The approximation given by $d_{n+1} \approx 1 + \frac{3}{4n} + \frac{25}{32n^2}$ is also plotted.

you that under the *i.i.d.* assumption, not much can be gleaned from higher resolution data, even though it appears to be a huge increase in the sample size. Indeed, because the daily returns and minutely log returns have the same sum, the mean over minute marks is exactly the daily mean divided by 390; only a change in the sample variance could cause the annualized Sharpe ratio to change.

To be concrete, let $\hat{\zeta}$ be the Sharpe ratio computed based on n marks, and let k be some positive integer. Let $\hat{\zeta}_k$ be the Sharpe ratio computed over the same time period, but with each log return divided into k pieces. Under the assumption of normality,

$$\sqrt{kn}\hat{\zeta}_k \sim t\left(\sqrt{kn}\zeta_k, kn - 1\right),$$

where $\zeta_k = \zeta/\sqrt{k}$ to make the units match. (That is, we are measuring returns over a higher frequency time scale, so the signal-noise ratio decreases in the usual way.) Similarly, because of how the units are defined, we should compare $\hat{\zeta}$ to $\sqrt{k}\hat{\zeta}_k$, since they have the same units. The moments of this estimate, with limits as $k \to \infty$, are

$$
\begin{aligned}
\mathrm{E}\left[\left(\sqrt{k}\hat{\zeta}_k\right)\right] &= \sqrt{k} d_{kn} \zeta_k = d_{kn}\zeta \to \zeta, \\
\mathrm{E}\left[\left(\sqrt{k}\hat{\zeta}_k\right)^2\right] &= k\frac{kn-1}{kn-3}\left(\frac{1+kn\zeta_k^2}{kn}\right) \to \left(\frac{1+n\zeta^2}{n}\right), \\
\mathrm{E}\left[\left(\sqrt{k}\hat{\zeta}_k\right)^3\right] &= k^{3/2}\frac{kn-1}{kn-4} d_{kn}\frac{\zeta_k}{kn}\left(3+kn\zeta_k^2\right) \to \frac{\zeta}{n}\left(3+n\zeta^2\right), \\
\mathrm{E}\left[\left(\sqrt{k}\hat{\zeta}_k\right)^4\right] &= k^2\frac{(kn-1)^2}{(kn-3)(kn-5)}\left(\frac{3+6kn\zeta_k^2+k^2n^2\zeta_k^4}{k^2n^2}\right), \\
&\to \left(\frac{3+6n\zeta^2+n^2\zeta^4}{n^2}\right),
\end{aligned}
\tag{2.15}
$$

and so on. All terms in Equation 2.11 containing $\sqrt{n}\hat{\zeta}$ remain unchanged (they are unitless), while the ratios of terms in n converge to 1. The decrease in bias and variance in using $\sqrt{k}\hat{\zeta}_k$ instead of $\hat{\zeta}$ will be small, indeed, even when n is modest, say bigger than only 100. As $k \to \infty$, in fact, we have $\sqrt{k}\hat{\zeta}_k \xrightarrow{d} \mathcal{N}\left(\zeta, \frac{1}{n}\right)$. Thus, the variance in the Sharpe ratio statistic is limited by the *length of time* over which we measure returns, and only weakly dependent on sample size (when sample size is reasonably large). It is not clear that changes in the higher order moments will have any meaningful impact on inference.

Example 2.3.3 (Up-sampling). Suppose you observe 252 days of Gaussian log returns with $\zeta = 0.05\,\mathrm{day}^{-1/2} = 0.794\,\mathrm{yr}^{-1/2}$. You also observe the up-sampled data with $k = 390$. Then the percent change in the mean is

$$
100\% \frac{\mathrm{E}\left[\sqrt{k}\hat{\zeta}_k\right] - \mathrm{E}\left[\hat{\zeta}\right]}{\mathrm{E}\left[\hat{\zeta}\right]} = -0.298\%,
$$

which is very small. The similarly defined percent change in the variance is -0.92%; in the skewness -99.7%; in the excess kurtosis -99.7%. ⊣

Example 2.3.4 (Down-sampling Market returns). The Sharpe ratio of the Market portfolio, introduced in Example 1.1.1, was computed on relative returns from Jan 1927 to Dec 2020. Returns were resampled at frequencies up to yearly, then the Sharpe ratios were computed, and tabulated in Table 2.1. The signal-noise ratios are all around $0.61\,\mathrm{yr}^{-1/2}$. We do not know which estimate is more accurate, but we also do not see radically different Sharpe ratios from different resampling frequencies. ⊣

Daily	Weekly	Monthly	Quarterly	Yearly
0.64	0.63	0.61	0.54	0.60

TABLE 2.1
The Sharpe ratio of the Market factor is shown, in units of $\text{yr}^{-1/2}$, computed at different sampling frequencies, based on relative returns from Jan 1927 to Dec 2020.

2.3.3 Unbiased estimation and efficiency

While $\hat{\zeta}$ is a biased estimator for ζ, it is *asymptotically* unbiased. Moreover, we can easily construct an unbiased estimator

$$\tilde{\zeta} := \frac{\hat{\zeta}}{d_n}. \tag{2.16}$$

This estimator is only of academic interest, as any fund manager reporting this statistic would be putting themselves at a relative disadvantage, since it is smaller than the biased estimator[2].

It is well known that $\left[\hat{\mu}, \hat{\sigma}^2\right]^\top$ is a sufficient statistic for the parameters $\left[\mu, \sigma^2\right]^\top$ of a normal distribution. [Bain and Engelhardt, 1992] Because $\left[\tilde{\zeta}, \hat{\sigma}\right]^\top$ is a one-to-one transformation of this sufficient statistic, it is sufficient as well.

Suppose that $x_1, x_2, \ldots, x_n$ are drawn *i.i.d.* from a normal distribution with unknown signal-noise ratio and variance. Suppose one has an (vector) estimator of the signal-noise ratio and the variance. The Fisher information matrix can easily be shown to be:

$$\mathcal{I}\left(\zeta, \sigma^2\right) = n \begin{bmatrix} 1 & \frac{\zeta}{2\sigma^2} \\ \frac{\zeta}{2\sigma^2} & \frac{2+\zeta^2}{4\sigma^4} \end{bmatrix} \tag{2.17}$$

Inverting the Fisher information matrix gives the Cramer-Rao lower bound for an unbiased vector estimator of signal-noise ratio and variance:

$$\mathcal{I}^{-1}\left(\zeta, \sigma^2\right) = \frac{1}{n} \begin{bmatrix} 1 + \zeta^2/2 & -\zeta\sigma^2 \\ -\zeta\sigma^2 & 2\sigma^4 \end{bmatrix} \tag{2.18}$$

Now consider the estimator $\left[\tilde{\zeta}, \hat{\sigma}^2\right]^\top$. This is an unbiased estimator for $\left[\zeta, \sigma^2\right]^\top$. One can show that the variance of this estimator is

$$\operatorname{Var}\left(\left[\tilde{\zeta}, \hat{\sigma}^2\right]^\top\right) = \begin{bmatrix} \frac{(1+n\zeta^2)(n-1)}{{d_n}^2 n(n-3)} - \zeta^2 & \zeta\sigma^2\left(\frac{1}{d_n} - 1\right) \\ \zeta\sigma^2\left(\frac{1}{d_n} - 1\right) & \frac{2\sigma^4}{n-1} \end{bmatrix} \tag{2.19}$$

[2] Assuming their Sharpe ratio is positive!

The variance of $\tilde{\zeta}$ follows from Equation 2.13. The cross terms follow from the independence of the sample mean and variance and from the unbiasedness of the two estimators. The variance of $\hat{\sigma}^2$ is well known.

Since $d_n = 1 + \frac{3}{4(n-1)} + \mathcal{O}\left(n^{-2}\right)$, the asymptotic variance of $\tilde{\zeta}$ is $\frac{(n-1)+\frac{n}{2}\zeta^2}{(n+(3/2))(n-3)} + \mathcal{O}\left(n^{-2}\right)$, and the covariance of $\tilde{\zeta}$ and $\hat{\sigma}^2$ is $-\zeta\hat{\sigma}^2\frac{3}{4n} + \mathcal{O}\left(n^{-2}\right)$. Thus the estimator $\left[\tilde{\zeta}, \hat{\sigma}^2\right]^\top$ is asymptotically *efficient*, *i.e.*, it achieves the Cramer-Rao lower bound asymptotically.

2.4 † The lambda prime distribution

Now we turn our attention to Lecoutre's *lambda prime* distribution, which is, in some sense, "dual" to the t-distribution. The connection between the lambda prime and the non-central t distribution shows up in the construction of frequentist confidence intervals (see Section 2.5.1), hypothesis tests on Sharpe ratio for independent samples (Section 2.5.3), and the Bayesian analysis (Section 2.7). It is also the "confidence distribution" of the non-central t, and appears in Fisher's work on Fiducial inference.

The lambda prime is defined as follows: let $Z \sim \mathcal{N}(0,1)$ independently of $\chi^2 \sim \chi^2(\nu)$. Then $\lambda = Z + t\sqrt{\chi^2/\nu}$ follows a lambda prime distribution with parameter t and degrees of freedom ν. [Lecoutre, 1999, 2007; Poitevineau and Lecoutre, 2010] Let us write this as $\lambda \sim \lambda'(t,\nu)$, and we write the density, cumulative distribution, and quantile functions of the lambda prime distribution as, respectively, $f_{\lambda'}(x;t,\nu), F_{\lambda'}(x;t,\nu)$, and $\lambda'_q(t,\nu)$.

The connection between the two distributions is apparent. Suppose that $t \sim t(\delta,\nu)$. This means that there are independent random variables $Z \sim \mathcal{N}(0,1)$ and $\chi^2 \sim \chi^2(\nu)$ such that

$$t = \frac{\delta + Z}{\sqrt{\chi^2/\nu}}.$$

This can be rearranged as

$$\delta = t\sqrt{\chi^2/\nu} - Z.$$

Because the normal distribution is symmetric, $-Z$ has the same distribution as Z, so conditional on observing t, we have $\delta \sim \lambda'(t,\nu)$.

This also connects the cumulative distribution functions (equivalently, the quantile functions) of the two distributions. For example, we have

$$F_t(x;\delta,\nu) = 1 - F_{\lambda'}(\delta;x,\nu), \tag{2.20}$$

and thus

$$F_{SR}(x;\zeta,n) = F_t\left(\sqrt{n}x;\sqrt{n}\zeta,n-1\right) = 1 - F_{\lambda'}\left(\sqrt{n}\zeta;\sqrt{n}x,n-1\right). \tag{2.21}$$

See Exercise 2.10. We can also define confidence interval on δ in terms of the quantile function of the lambda prime distribution. An α confidence bound for the non-centrality parameter, δ, conditional on observing t, is ${\lambda'}_{1-\alpha}\left(t,\nu\right)$. Thus the confidence interval given in Equation 2.27 (see later) can be expressed as

$$\frac{1}{\sqrt{n}}\left[{\lambda'}_{\alpha/2}\left(\sqrt{n}\hat{\zeta},n-1\right),{\lambda'}_{(2-\alpha)/2}\left(\sqrt{n}\hat{\zeta},n-1\right)\right). \tag{2.22}$$

The distribution and quantile functions of the lambda prime can be evaluated directly [Poitevineau and Lecoutre, 2010], but they can be more easily implemented via off-the-shelf implementations of the distribution and quantile of the t distribution. For our purposes, we will at times need the distribution and quantile of a more general distribution, described below.

2.5 Frequentist inference on the signal-noise ratio

2.5.1 Confidence intervals

The variance of $\hat{\zeta}$ is given in Equation 2.13. Using the asymptotic expansion $d_n = 1 + \frac{3}{4(n-1)} + \mathcal{O}\left(n^{-2}\right)$, the variance of $\hat{\zeta}$ is approximated by

$$\operatorname{Var}\left(\hat{\zeta}\right) = \frac{(1+n\zeta^2)(n-1)}{n(n-3)} - \left(d_n\zeta\right)^2 \approx \frac{n-1}{n\left(n-3\right)} + \frac{\zeta^2}{2\left(n-3\right)} + \zeta^2\mathcal{O}\left(n^{-2}\right). \tag{2.23}$$

This is often quoted in the following more convenient form, which is asymptotically equivalent,

$$se\left(\hat{\zeta}\right) := \sqrt{\operatorname{Var}\left(\hat{\zeta}\right)} \approx \sqrt{\frac{1+\frac{\zeta^2}{2}}{n}}. \tag{2.24}$$

This is the most widely used form of the standard error of the Sharpe ratio. This standard error was described by Jobson and Korkie [1981], and later by Lo [2002]. The equivalent result concerning the non-central t-distribution (which is the Sharpe ratio up to scaling by $\sqrt{n}$) was published in 1940 by Johnson and Welch [1940]. We will refer to Equation 2.24 informally as the "vanilla" standard error, or the "Johnson and Welch" formula.

Since the signal-noise ratio, $\hat{\zeta}$, is unknown, it is typically approximated with the Sharpe ratio (the so-called "plug-in" method), giving the following approximate $1-\alpha$ confidence interval on the signal-noise ratio:

$$\hat{\zeta} \pm z_{\alpha/2}\sqrt{\frac{1+\frac{\hat{\zeta}^2}{2}}{n}},$$

where $z_{\alpha/2}$ is the $\alpha/2$ quantile of the normal distribution. In practice, the asymptotically equivalent form

$$\hat{\zeta} \pm z_{\alpha/2}\sqrt{\frac{1 + \frac{\hat{\zeta}^2}{2}}{n-1}} \tag{2.25}$$

has better small sample coverage for normal returns.

We can take this one step further for the small sample case by adjusting for the bias in the Sharpe ratio. Using the Taylor approximation ${d_n}^{-1} \approx 1 - \frac{3}{4(n-1)}$ gives the approximate $1-\alpha$ confidence interval

$$\hat{\zeta}\left(1 - \frac{3}{4\left(n-1\right)}\right) \pm z_{\alpha/2}\sqrt{\frac{1 + \frac{\hat{\zeta}^2}{2}}{n-1}} \tag{2.26}$$

We can find confidence intervals on ζ assuming only normality of x (or large n and an appeal to the Central Limit Theorem), by inversion of the cumulative distribution of the non-central t-distribution. A $1-\alpha$ confidence interval on ζ has endpoints $[\zeta_l, \zeta_u]$ defined implicitly by

$$1 - \alpha/2 = F_t\left(\sqrt{n}\hat{\zeta}; \sqrt{n}\zeta_l, n-1\right), \quad \alpha/2 = F_t\left(\sqrt{n}\hat{\zeta}; \sqrt{n}\zeta_u, n-1\right), \tag{2.27}$$

where $F_t\left(x; \delta, n-1\right)$ is the CDF of the non-central t-distribution with non-centrality parameter δ and $n-1$ degrees of freedom. Computationally, this method requires one to invert the CDF (*e.g.*, by Brent's method [Brent, 1973]), which is slower than approximations based on asymptotic normality. The endpoints of this confidence interval, ζ_l and ζ_u are actually quantiles of the lambda prime distribution. (*cf.* Section 2.4)

Mertens [2002] gives the form of standard error

$$se\left(\hat{\zeta}\right) \approx \sqrt{\frac{1 + \frac{2+\gamma_2}{4}\zeta^2 - \gamma_1\zeta}{n}}, \tag{2.28}$$

where γ_1 is the skew, and γ_2 is the *excess* kurtosis of the returns distribution. [Opdyke, 2007; Bao, 2009] These are both zero for normally distributed returns, and so Mertens' form reduces to the Johnson & Welch form of the standard error. In practice when returns are not assumed to be Gaussian, the cumulants are unknown and have to be estimated from the data, which results in some mis-estimation of the standard error when skew is extreme. We will consider Mertens' form further in the sequel, see Chapter 3.

Example 2.5.1 (Confidence interval, Market returns). Consider the monthly relative returns of the Market, introduced in Example 1.1.1, which span from Jan 1927 to Dec 2020. The Sharpe ratio was measured to be $0.614\,\text{yr}^{-1/2}$. A 95% confidence interval, based on the "exact" method was computed as $[0.41, 0.817)\ \text{yr}^{-1/2}$. ⊣

Example 2.5.2 (Confidence interval, UMD returns, attribution model). Consider the monthly relative returns of UMD, under an attribution against intercept, Market, SMB and HML, using the data introduced in Example 1.1.1, 1,128 months of data from Jan 1927 to Dec 2020. The ex-factor Sharpe ratio was measured to be $0.246\,\text{mo.}^{-1/2}$. A 95% confidence interval, based on the "exact" method was computed as $[0.186, 0.306)\,\text{mo.}^{-1/2}$. ⊣

2.5.2 † Symmetric confidence intervals

The implicit confidence intervals on ζ given above are "symmetric" in the sense that they are typically computed with equal type I error rates on both sides. That is, a $1-\alpha$ confidence interval $[\zeta_l, \zeta_u]$ is usually computed such that

$$\Pr\left\{\zeta < \zeta_l\right\} = \alpha/2 = \Pr\left\{\zeta > \zeta_u\right\}.$$

However, these are typically not symmetric around the observed Sharpe ratio, $\hat{\zeta}$, rather they are usually slightly imbalanced. However, symmetric confidence intervals can easily be constructed numerically by finding δ such that

$$F_t\left(\sqrt{n}\hat{\zeta}; \sqrt{n}\left(\hat{\zeta}-\delta\right), n-1\right) - F_t\left(\sqrt{n}\hat{\zeta}; \sqrt{n}\left(\hat{\zeta}+\delta\right), n-1\right) = 1-\alpha.$$

The advantage of a symmetric interval is that we can express it as

$$\Pr\left\{\left|\zeta-\hat{\zeta}\right| \le \delta\right\} = 1-\alpha,$$

a fact which we will use later.

Example 2.5.3 (Confidence intervals, Symmetric). Consider the case of 504 daily observations of some hypothetical asset's returns, which result in a measured Sharpe ratio of exactly $0.6\,\text{yr}^{-1/2}$, assuming 252 days per year. The "exact" 95% confidence interval is computed as $[-0.787, 1.986)\,\text{yr}^{-1/2}$, which we can write as $\left[\hat{\zeta}-1.387, \hat{\zeta}+1.386\right)\,\text{yr}^{-1/2}$. These are *almost* symmetric.

Numerically we compute the symmetric interval as approximately

$$\hat{\zeta} \pm 1.386\,\text{yr}^{-1/2}.$$

This is very close to the exact interval. It is also close to, but narrower than, the confidence interval computed by the standard error approximation, Equation 2.25, which we compute as

$$\hat{\zeta} \pm 1.388\,\text{yr}^{-1/2}.$$

⊣

The symmetry condition is useful in the following situation. Suppose you observe the Sharpe ratio of an asset; if it is positive, you will hold the asset long, otherwise you will hold the asset short. This simple "opportunistic" strategy seems perfectly natural, though perhaps somewhat naïve. It is not obvious how you would compute a confidence interval on your achieved returns, since it would depend not just on the population parameter, ζ, but also on the statistic $\hat{\zeta}$, which you will also use to compute your confidence interval.

We can easily compute such intervals. Starting from the symmetric confidence interval condition, $\Pr\left\{\left|\zeta - \hat{\zeta}\right| \le \delta\right\} = 1 - \alpha$, multiply the inside of the absolute value by a ± 1 in the form of $\operatorname{sign}\left(\hat{\zeta}\right)$ to arrive at

$$\Pr\left\{\left|\operatorname{sign}\left(\hat{\zeta}\right)\zeta - \left|\hat{\zeta}\right|\right| \le \delta\right\} = 1 - \alpha.$$

And thus

$$\left|\hat{\zeta}\right| \pm \delta \tag{2.29}$$

forms a $1 - \alpha$ confidence interval on the returns of this simple opportunistic strategy.

While the confidence interval is symmetric about $\left|\hat{\zeta}\right|$, the rates of type I errors are typically not balanced on the two sides. Moreover, the imbalance depends on the unknown signal-noise ratio: when $|\zeta|$ is near 0, the type I errors will mostly be found below the lower bound; conversely when $|\zeta|$ is "large," the type I errors will be more evenly balanced.

To illustrate this effect, we consider the rate of type I errors separately for "lower" and "upper" violations of the symmetric confidence bound. These can be defined, respectively, as

$$\Pr\left\{\operatorname{sign}\left(\hat{\zeta}\right)\zeta \le \left|\hat{\zeta}\right| - \delta\right\}, \quad \text{and} \quad \Pr\left\{\operatorname{sign}\left(\hat{\zeta}\right)\zeta \ge \left|\hat{\zeta}\right| + \delta\right\}.$$

We construct the symmetric confidence intervals using the standard error approximation of Equation 2.25, with the actual signal-noise ratio, *i.e.*, we are setting $\delta = \left|z_{\alpha/2}\right|\sqrt{\frac{1+\frac{\hat{\zeta}^2}{2}}{n-1}}$. We consider the case of 3 years of daily data, at 252 days per year. We compute the lower and upper type I rates, and plot them against ζ ranging from $0\,\mathrm{yr}^{-1/2}$ to $2.5\,\mathrm{yr}^{-1/2}$ in Figure 2.2. When $\zeta \le \left|z_{\alpha/2}\right|\sqrt{\frac{1+\frac{\hat{\zeta}^2}{2}}{n-1}}$ we see that there are no upper type I errors. (See Exercise 2.25.)

Note this imbalance in type I rates is a function of the unknown signal-noise ratio, and not of the observed Sharpe ratio. It is not possible to say much about the balance of type I errors conditional on the Sharpe ratio without leaning on a prior distribution for ζ, which would take us out of the Frequentist framework.

Note that one can turn the symmetric confidence interval $\left|\hat{\zeta}\right| \pm \delta$ into a one-sided "symmetric" confidence interval

$$\left[\left|\hat{\zeta}\right| - \delta, \infty\right) \tag{2.30}$$

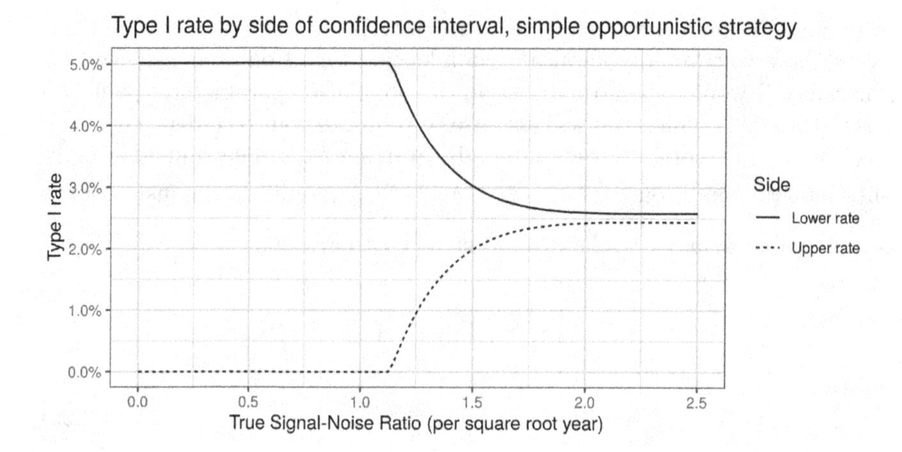

FIGURE 2.2
Rates of upper and lower type I errors are plotted versus ζ for the case of 3 years of daily data, using the approximate standard error from Equation 2.25. The symmetric confidence interval has a total type I rate of approximately 0.05.

by extending one side. This confidence interval should have coverage between $1 - \alpha$ and $1 - \alpha/2$ depending on the unknown signal-noise ratio.

One can also analyze the opportunistic strategy via conditional inference, *cf.* Section 4.1.5.

2.5.3 Hypothesis tests

There are a few statistical tests for hypotheses involving the signal-noise ratio. The classical t-test for the mean can be considered a hypothesis test on the signal-noise ratio with a disastrous rate of return. In each of these, the sample Sharpe ratio is used.

1 sample mean test

The classical one-sample test for mean involves a t-statistic which is like a Sharpe ratio with constant benchmark. Thus to test the null hypothesis:

$$H_0 : \mu = \mu_0 \quad \text{versus} \quad H_1 : \mu > \mu_0,$$

we reject if the statistic

$$t_0 = \sqrt{n}\frac{\hat{\mu} - \mu_0}{\hat{\sigma}}$$

is greater than $t_{1-\alpha}(n-1)$, the $1 - \alpha$ quantile of the (central) t-distribution with $n - 1$ degrees of freedom.

If $\mu = \mu_1 > \mu_0$, then the power of this test is

$$1 - F_t\left(t_{1-\alpha}\left(n-1\right); \delta_1, n-1\right),$$

where $\delta_1 = \sqrt{n}\left(\mu_1 - \mu_0\right)/\sigma$ and $F_t\left(x; \delta, n-1\right)$ is the cumulative distribution function of the non-central t-distribution with non-centrality parameter δ and $n-1$ degrees of freedom. [Scholz, 2007; Owen, 1968]

1 sample signal-noise ratio test

A one-sample test for signal-noise ratio can also be interpreted via the t-statistic. To test:

$$H_0 : \zeta = \zeta_0 \quad \text{versus} \quad H_1 : \zeta > \zeta_0,$$

we reject if the statistic $t = \sqrt{n}\hat{\zeta}$ is greater than $t_{1-\alpha}\left(\delta_0, n-1\right)$, the $1-\alpha$ quantile of the non-central t-distribution with $n-1$ degrees of freedom and non-centrality parameter $\delta_0 = \sqrt{n}\zeta_0$. Equivalently we reject if $\hat{\zeta} > SR_{1-\alpha}\left(\zeta_0, n\right)$.

If $\zeta = \zeta_1 > \zeta_0$, then the power of this test is

$$1 - F_t\left(t_{1-\alpha}\left(\delta_0, n-1\right); \delta_1, n-1\right),$$

where $\delta_1 = \sqrt{n}\zeta_1$ and $F_t\left(x; \delta, n-1\right)$ is the cumulative distribution function of the non-central t-distribution with non-centrality parameter δ and $n-1$ degrees of freedom. [Scholz, 2007; Owen, 1968] Equivalently the power can be expressed as

$$1 - F_{SR}\left(SR_{1-\alpha}\left(\zeta_0, n\right); \zeta_1, n\right).$$

Hypothesis tests of the signal-noise ratio involving dependent returns will be considered later, see Section 3.3 and Equation 3.48.

Example 2.5.4 (Hypothesis testing, the Market). Consider the monthly relative returns of the Market, introduced in Example 1.1.1. To test the null hypothesis, $H_0 : \mu = 0.5\%\,\text{mo.}^{-1}$, we essentially compute the Sharpe ratio with benchmark rate of $0.5\%\,\text{mo.}^{-1}$, which is computed as $0.29\,\text{yr}^{-1/2}$. Multiplying this by the square root of $n = 1{,}128\,\text{mo.}$ gives the t-stat of 9.75. We reject the null at the 0.05 level.

To test the null hypothesis, $H_0 : \zeta = 0.3\,\text{yr}^{-1/2}$, against the alternative $H_1 : \zeta > 0.3\,\text{yr}^{-1/2}$, we compute $\hat{\zeta} = 0.614\,\text{yr}^{-1/2}$. The 0.975 quantile of the Sharpe ratio distribution under the null for $n = 1{,}128\,\text{mo.}$ is $SR_{0.975}\left(0.3\,\text{yr}^{-1/2}, 1{,}128\right) = 0.503\,\text{yr}^{-1/2}$, and thus we reject the null hypothesis at the 0.025 level. This is essentially the same conclusion as could be drawn from the confidence interval given in Example 2.5.1. ⊣

2.5.4 † Two one-sided and other intersection union tests

It is sometimes lamented that hypothesis tests only allow you to cast doubt on the null hypothesis, rather than somehow positively assert some condition on the population parameters. The use of "Two One-Sided Tests" is sometimes prescribed[3] for this situation, as they potentially allow one to reject "towards" some relationship, rather than away from one. [Lakens, 2016]

Suppose the goal is to establish that the signal-noise ratio of an asset is equal to some fixed value, say ζ_0. Since strict equality is unlikely to be true, one instead seeks to establish equality within some range, or up to some uncertainty. So let us say that ζ is equivalent to ζ_0 if and only if $\zeta_l \le \zeta \le \zeta_h$, for some suitably defined ζ_l, ζ_h selected to give the proper precision to the notion of equivalence.

To try to show equivalence, one performs a test under the null hypothesis of *unequivalence*, with the hope of rejecting in favor of equivalence. That is, one tests

$$H_0 : \zeta < \zeta_l \text{ or } \zeta > \zeta_h \quad \text{versus} \quad H_1 : \zeta_l \le \zeta \le \zeta_h.$$

To perform this test, one conducts two hypothesis tests, namely

$$\begin{aligned} H_{0a} &: \zeta < \zeta_l \quad \text{versus} \quad H_{1a} : \zeta \ge \zeta_l, \text{ and} \\ H_{0b} &: \zeta > \zeta_h \quad \text{versus} \quad H_{1b} : \zeta \le \zeta_h. \end{aligned}$$

If one rejects *both* H_{0a} and H_{0b}, then effectively one is rejecting H_0 "in favor of" the alternative hypothesis of equality, H_1. To conduct these tests, as outlined in Section 2.5.3, we reject at the α level if

$$t_{1-\alpha}\left(\sqrt{n}\zeta_l, n-1\right) \le \sqrt{n}\hat{\zeta} \le t_{\alpha}\left(\sqrt{n}\zeta_h, n-1\right).$$

Here, as above, $t_{\alpha}\left(\delta, n-1\right)$ is the α quantile of the non-central t-distribution with $n-1$ degrees of freedom and non-centrality parameter δ.

If we truly have equivalence, that is if $\zeta_l \le \zeta \le \zeta_h$, then the power of this test is

$$\left[1 - F_t\left(t_{1-\alpha}\left(\sqrt{n}\zeta_l, n-1\right); \sqrt{n}\zeta, n-1\right)\right] \bigwedge \left[F_t\left(t_{\alpha}\left(\sqrt{n}\zeta_h, n-1\right); \sqrt{n}\zeta, n-1\right)\right].$$

That is, the power of the test is the *minimum* of the powers of the two sub-hypothesis tests.

This combined test is an example of an *intersection union test*, wherein one is testing a null hypothesis that can be expressed as the *union* of multiple testable hypotheses. The critical region, wherein one rejects based on the sample statistic (the Sharpe ratio in this case) is the *intersection* of critical regions of the separate tests[4] [Sen, 2007; Berger, 1996]

In a similar manner, any of the equality tests of Section 2.5.3 and Section 2.5.5 can be reformulated as intersection union tests, see Exercise 2.24.

---

[3]To be fair, Bayesian methods are far more popular for this task.

[4]In contrast, in a union intersection test, one is testing the null which is the intersection of hypotheses, and the critical region is the union of critical regions. The order of words in the nomenclature here seems somewhat arbitrary.

Example 2.5.5 (TOST for signal-noise ratio equivalent to zero). Consider the case of testing equivalence of the signal-noise ratio to zero by means of a symmetric interval, with $\zeta_l = -0.2\,\mathrm{yr}^{-1/2}$, $\zeta_h = 0.2\,\mathrm{yr}^{-1/2}$ at the type I rate of 0.05. Suppose one observes $n = 1,008$ days of data. Then, the critical region for TOST in this case is $[-0.148, 0.148)\ \mathrm{yr}^{-1/2}$. We reject the null hypothesis precisely when the Sharpe ratio falls in this interval. The power of this test, the probability of correctly rejecting the null when the signal-noise ratio falls within $[-0.2, 0.2)\ \mathrm{yr}^{-1/2}$ for a type I rate of 0.05, is shown in Figure 2.3. We see that this test has very high power in this case when the signal-noise ratio is between about $-0.1\,\mathrm{yr}^{-1/2}$ and $0.1\,\mathrm{yr}^{-1/2}$. ⊣

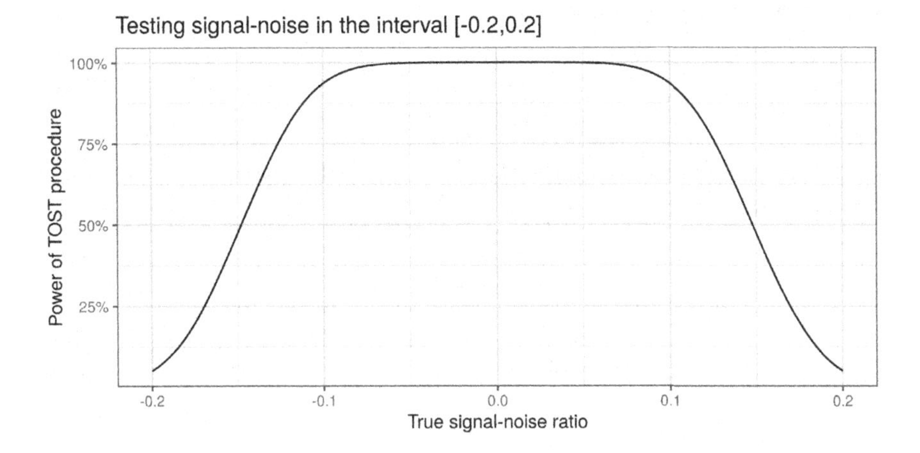

FIGURE 2.3
Power of the two-sided test for $-0.2\,\mathrm{yr}^{-1/2} \leq \zeta \leq 0.2\,\mathrm{yr}^{-1/2}$ for the case of 1,008 daily observations is plotted versus the true signal-noise ratio, for a type I rate of 0.05.

2.5.5 Hypothesis tests for the linear attribution model

Given n observations of the returns and the factors, let $\boldsymbol{x}$ be the vector of returns and let F be the $n \times l$ matrix consisting of the returns of the l factors and a column of all ones. Again we should stress that the factors must be deterministic, as otherwise their variability would contribute to extra uncertainty in the test statistics. This requirement is relaxed in Section 3.4, where we consider random Gaussian returns and factors. The ordinary least squares estimator for the regression coefficients is expressed by the "normal equations":

$$\hat{\boldsymbol{\beta}} = \left(\mathsf{F}^{\top}\mathsf{F}\right)^{-1}\mathsf{F}^{\top}\boldsymbol{x}.$$

The estimated variance of the error term is $\hat{\sigma}^2 = \left(\boldsymbol{x} - \mathsf{F}\hat{\beta}\right)^\top \left(\boldsymbol{x} - \mathsf{F}\hat{\beta}\right)/(n - l)$.

1 sample test for regression coefficients

The classical t-test for regression coefficients tests the null hypothesis:

$$H_0 : \beta^\top \boldsymbol{v} = c \quad \text{versus} \quad H_1 : \beta^\top \boldsymbol{v} > c,$$

for some conformable vector $\boldsymbol{v}$ and constant c. To perform this test, we construct the regression t-statistic

$$t = \frac{\hat{\beta}^\top \boldsymbol{v} - c}{\hat{\sigma}\sqrt{\boldsymbol{v}^\top \left(\mathsf{F}^\top \mathsf{F}\right)^{-1} \boldsymbol{v}}}. \tag{2.31}$$

This statistic should be distributed as a non-central t-distribution with non-centrality parameter

$$\delta = \frac{\beta^\top \boldsymbol{v} - c}{\sigma\sqrt{\boldsymbol{v}^\top \left(\mathsf{F}^\top \mathsf{F}\right)^{-1} \boldsymbol{v}}},$$

and $n - l$ degrees of freedom. Thus, we reject the null if t is greater than $t_{1-\alpha}\left(n - l\right)$, the $1-\alpha$ quantile of the (central) t-distribution with $n-l$ degrees of freedom.

1 sample test for ex-factor signal-noise ratio

To test the null hypothesis:

$$H_0 : \beta^\top \boldsymbol{v} = \sigma c \quad \text{versus} \quad H_1 : \beta^\top \boldsymbol{v} > \sigma c,$$

for given $\boldsymbol{v}$ and c, one constructs the t-statistic

$$t = \frac{\hat{\beta}^\top \boldsymbol{v}}{\hat{\sigma}\sqrt{\boldsymbol{v}^\top \left(\mathsf{F}^\top \mathsf{F}\right)^{-1} \boldsymbol{v}}}. \tag{2.32}$$

Under the null this statistic should be distributed as a non-central t-distribution with non-centrality parameter

$$\delta = \frac{c}{\sqrt{\boldsymbol{v}^\top \left(\mathsf{F}^\top \mathsf{F}\right)^{-1} \boldsymbol{v}}},$$

and $n - l$ degrees of freedom. Thus we reject the null if t is greater than $t_{1-\alpha}\left(\delta, n - l\right)$, the $1 - \alpha$ quantile of the non-central t-distribution with $n - l$ degrees of freedom and non-centrality parameter δ.

Equivalently, by definition,

$$\hat{\zeta}_g = \frac{\hat{\boldsymbol{\beta}}^\top \boldsymbol{v}}{\hat{\sigma}},$$

thus we should compare $\hat{\zeta}_g$ to the cutoff value of $rt_{1-\alpha}\left(c/r, n-l\right)$, where $r = \sqrt{\boldsymbol{v}^\top \left(\mathsf{F}^\top \mathsf{F}\right)^{-1} \boldsymbol{v}}$ is the rescaling parameter.

Example 2.5.6 (Hypothesis testing, UMD, attribution model). Consider the monthly relative returns of the Market, SMB, HML, and UMD portfolios, introduced in Example 1.1.1. Ignoring the error in the factors, we perform an attribution of the returns of UMD as some exposure to Market, SMB and HML. The ex-factor Sharpe ratio of the residual term under this attribution model is computed to be approximately $0.246\,\text{mo.}^{-1/2}$ on $n = 1,128\,\text{mo.}$ This actually corresponds to a rather large t statistic.

To test the null hypothesis that the *residual* mean of UMD is $0.5\%\,\text{mo.}^{-1}$, we compute the t-statistic to have value of 4.128, and reject the null at the 0.05 level.

To test the null hypothesis that the ex-factor signal-noise ratio of UMD is equal to $0.1\,\text{mo.}^{-1/2}$, we compute the 0.95 quantile of the non-central t distribution with non-centrality parameter $\delta = 3.303$ and $n - l = 1124$ degrees of freedom. This has value 4.958, equivalent to comparing the computed value of the ex-factor Sharpe ratio to $0.15\,\text{mo.}^{-1/2}$. Since we compute the ex-factor Sharpe ratio to be around $0.246\,\text{mo.}^{-1/2}$, we reject the null at the 0.05 level. ⊣

Example 2.5.7 (Hypothesis testing, Technology ex-factor signal-noise ratio). Consider the monthly relative returns of the Technology industry portfolio, introduced in Example 1.2.5. We align these with the corresponding monthly Market returns. Joining the two series, we have 1,128 months of data, ranging from Jan 1927 through Dec 2020. We compute the "beta" of Technology to the market to be close to one, taking value 0.949. We compute the ex-factor Sharpe ratio to be around $0.041\,\text{mo.}^{-1/2}$.

To test the null hypothesis that the ex-factor signal-noise ratio of Technology is equal to $0.01\,\text{mo.}^{-1/2}$, we compute the 0.95 quantile of the non-central t distribution with non-centrality parameter $\delta = 0.331$ and $n - l = 1126$ degrees of freedom. This has value 1.977, leading to cutoff value $0.06\,\text{mo.}^{-1/2} > 0.041\,\text{mo.}^{-1/2}$, and we fail to reject the null at the 0.05 level. ⊣

Example 2.5.8 (Hypothesis testing, VIX reweighting Market returns). Consider the daily Market returns scaled by inverse previous day VIX weights, as introduced in Example 1.4.5. Joining the two series, we have 7,808 days of data, ranging from 1990-01-03 through 2020-12-31. The inverse VIX weights have been rescaled to have unit mean in the sample period. We perform the attribution

$$\tilde{x_t} = x_t s_{t-1} = \beta_0 s_{t-1} + \beta_1 + \epsilon_t,$$

and test $H_0 : \beta^\top \boldsymbol{v} = \sigma c$ with $c = 0.317\,\text{yr}^{-1/2}$ and $\boldsymbol{v}^\top = \left[1, 1\right]^\top$.

We compute $\hat{\zeta}_g = 0.76\,\mathrm{yr}^{-1/2}$. To test at the $\alpha = 0.05$ rate, we compare this to $rt_{1-\alpha}\left(c/r, n-l\right) = 0.613\,\mathrm{yr}^{-1/2}$, and narrowly reject the null hypothesis. ⊣

2.5.6 † Confidence intervals for the linear attribution model

The hypothesis tests given above can be converted into confidence intervals on the ex-factor signal-noise ratio. Such confidence intervals can be constructed by inverting the non-central t distribution, in analogue to the confidence intervals on the signal-noise ratio discussed in Section 2.5.1.

Again, we are assuming we observe n-vector of returns $\boldsymbol{x}$ and $n \times l$ matrix of associated factors F. The returns are Gaussian and independent, but are related to the factors which we assume here are deterministic. Let $\boldsymbol{v}$ be a given l vector. Then a $1-\alpha$ confidence interval on the quantity $\zeta_g = \beta^\top \boldsymbol{v}/\sigma$ has endpoints $[\zeta_{g,l}, \zeta_{g,u}]$ defined implicitly by

$$1-\alpha/2 = F_t\left(\hat{\zeta}_g/r; \zeta_{g,l}/r, n-l\right), \quad \alpha/2 = F_t\left(\hat{\zeta}_g/r; \zeta_{g,u}/r, n-l\right), \tag{2.33}$$

where $\hat{\zeta}_g = \hat{\boldsymbol{\beta}}^\top \boldsymbol{v}/\hat{\sigma}$ is the ex-factor Sharpe ratio, $r = \sqrt{\boldsymbol{v}^\top \left(\mathsf{F}^\top \mathsf{F}\right)^{-1} \boldsymbol{v}}$ is a rescaling parameter, and $F_t\left(x; \delta, n-l\right)$ is the CDF of the non-central t-distribution with non-centrality parameter δ and $n-l$ degrees of freedom. (*cf.* Equation 2.33.)

Additionally by using the connection to the t-distribution as we did in Section 2.5.1, we can approximate the standard error of the ex-factor Sharpe ratio as

$$se\left(\hat{\zeta}_g\right) := \sqrt{\mathrm{Var}\left(\hat{\zeta}_g\right)} \approx \sqrt{\frac{1 + \frac{\zeta_g^2}{2r^2(n-l+1)}}{n-l+1}}, \tag{2.34}$$

where again $r = \sqrt{\boldsymbol{v}^\top \left(\mathsf{F}^\top \mathsf{F}\right)^{-1} \boldsymbol{v}}$ is the rescaling parameter.

Example 2.5.9 (Confidence intervals, UMD, attribution model). Consider the monthly relative returns of the Market, SMB, HML, and UMD portfolios, introduced in Example 1.1.1. We perform an attribution of the returns of UMD as some exposure to Market, SMB and HML, treating the latter returns as deterministic. The ex-factor Sharpe ratio of the residual term under this attribution model is computed to be approximately $0.246\,\mathrm{mo.}^{-1/2}$ on $n = 1{,}128\,\mathrm{mo.}$ We compute an "exact" 95% confidence interval on the ex-factor Sharpe ratio as $[0.186, 0.306)\,\mathrm{mo.}^{-1/2}$.

By plugging in the sample value, we estimate the standard error of the ex-factor Sharpe ratio via Equation 2.34 to be approximately $0.03\,\mathrm{mo.}^{-1/2}$. Then we can compute an approximate 95% confidence interval on the ex-factor Sharpe ratio as $[0.187, 0.306)\,\mathrm{mo.}^{-1/2}$. *cf.* Example 3.4.1. ⊣

Example 2.5.10 (Confidence interval, Technology ex-factor signal-noise ratio). Consider the attribution of monthly Technology industry returns to those of the Market, as described in Example 2.5.7. The ex-factor Sharpe ratio of the residual term under this attribution model is computed to be approximately $0.041\,\mathrm{mo.}^{-1/2}$ on $n = 1,128\,\mathrm{mo}$. We compute an "exact" 90% confidence interval on the ex-factor Sharpe ratio as $[-0.009, 0.091)\ \mathrm{mo.}^{-1/2}$.

By plugging in the sample value, we estimate the standard error of the ex-factor Sharpe ratio via Equation 2.34 to be approximately $0.03\,\mathrm{mo.}^{-1/2}$. Then we can compute an approximate 90% confidence interval on the ex-factor Sharpe ratio as $[-0.008, 0.09)\ \mathrm{mo.}^{-1/2}$, which contains zero. ⊣

2.5.7 Type I errors, true incidence rate, and false discovery

Caution (On "Significance"). Often, in the social sciences and elsewhere, a p-value smaller than 0.05 is deemed "significant," a word which has come to have no meaning other than "exhibiting a p-value smaller than 0.05."

One should recognize, moreover, that the incidence rate of profitable trading strategies is certainly very low, likely lower than 0.05. In fact, some forms of the Efficient Markets Hypothesis would posit that the incidence rate is actually zero, and all trading strategies are type I errors. [Malkiel, 2003] One should exercise extreme caution, and use Bayes rule and some amount of guesstimation to avoid false discoveries. We outline how to deal with multiple testing in Chapter 4.

Example 2.5.11 (Testing, incidence, and false discovery rate). Suppose that you sample randomly from trading strategies where the incidence rate of "good" strategies is 0.001. If one employs a test for strategies with a 0.001 type I rate, and a power of 0.8, the false discovery rate will be as high as 0.555. Which is to say around half of strategies which pass the significance test will not actually be "good." If, however, the incidence rate is as low as 10^{-6}, using the same test with the same type I and type II rates, the false discovery rate jumps to 0.999; the vast majority of strategies passing the test are type I errors. ⊣

Unfortunately, the true incidence rate is unknown. Moreover, there is often no real gold standard for determining whether a trading strategy is in fact "good": even stellar performance of a strategy in real trading for some fixed time after the analysis was performed may be the result of a type I error, and not an indication of true goodness.

It should also be noted that this simple model of randomly selecting trading strategies for significance testing is often an inaccurate description of strategy development. Typically strategies are developed by building off of well known ideas and theories, sometimes from published studies, and often involves sequential refinement of code and ideas, typically only accepting changes which improve some metric (*e.g.*, Sharpe ratio) of backtested returns. Dealing with the false discovery rate under this model of strategy development is addressed in Chapter 4. See also Exercise 2.29 to Exercise 2.30.

2.5.8 Power and sample size

Consider the test of the hypothesis $H_0 : \zeta = 0$, against the alternative $H_1 : \zeta > 0$. Note that this is equivalent to the traditional t-test for zero mean, *i.e.*, for testing the hypothesis $H_0 : \mu = 0$. A *power rule* ties together the (unknown) true effect size (ζ), sample size (n), and the type I and type II rates implicitly into a single equation. Typically, starting from three of these quantities, one infers the fourth, as illustrated by the following use cases:

1. Suppose you wanted to analyze a pairs trade on a pair of stocks which have only existed for two years. Is this enough data assuming the signal-noise ratio is $2.0\ \mathrm{yr}^{-1/2}$?
2. Suppose investors in a fund you manage want to "see some returns" within a year otherwise they will withdraw their investment. What signal-noise ratio should you be hunting for so that, with probability one half, the actual returns will "look good" over the next year?
3. Suppose you observe three months of a fund's returns, and you fail to reject the null under the one sample t-test. Assuming the signal-noise ratio of the process is $1.5\ \mathrm{yr}^{-1/2}$, what is the probability of a type II error?

The power equation can be derived simply. The hypothesis test for $H_0 : \zeta = \zeta_0$ against the alternative $H_1 : \zeta > \zeta_0$ rejects the null precisely when

$$\hat{\zeta} \geq SR_{1-\alpha}\left(\zeta_0, n\right),$$

where the quantity on the right hand side is the $1 - \alpha$ quantile of the Sharpe ratio distribution with signal-noise ratio ζ_0 on sample size n. Now suppose that we wish the test to have power $1 - \beta$ when the true signal-noise ratio is ζ_e. This requires the cutoff value to be the β quantile when the signal-noise ratio is ζ_e. Thus the power equation is

$$SR_{1-\alpha}\left(\zeta_0, n\right) = SR_{\beta}\left(\zeta_e, n\right). \tag{2.35}$$

Because the quantile function is the inverse CDF, this can be expressed in two other, equivalent, ways, *viz.*

$$1 - \alpha = F_{SR}\left(SR_{\beta}\left(\zeta_e, n\right); \zeta_0, n\right), \quad \text{or,} \quad F_{SR}\left(SR_{1-\alpha}\left(\zeta_0, n\right); \zeta_e, n\right) = \beta. \tag{2.36}$$

From one of these two equations, one can infer either α or β as needed. If the goal is to infer the requisite ζ_e or n, numerical search using Equation 2.35 is indicated.

Example 2.5.12 (Basic power computations). Suppose you wish to test $H_0 : \zeta = 0\,\mathrm{yr}^{-1/2}$ against $H_1 : \zeta > 0\,\mathrm{yr}^{-1/2}$, with a 0.05 type I rate. Given 4 years of daily observations, the significance test rejects when $\hat{\zeta}$ exceeds $0.823\,\mathrm{yr}^{-1/2}$. Supposing $\zeta = 1.5\,\mathrm{yr}^{-1/2}$, the power of the test is 0.912. ⊣

For sufficiently large sample size (say $n \geq 30$), the power law for the t-test of $H_0 : \zeta = 0$ is well approximated by

$$n \approx \frac{c}{\zeta^2}, \tag{2.37}$$

where the constant c depends on the type I rate and the type II rates, and whether one is performing a one- or two-sided test. This relationship was first noted by Johnson and Welch [1940]. Unlike the type I rate, which is traditionally set at 0.05, there is no widely accepted value of power.

Values of the coefficient c are given for one and two-sided t-tests at different power levels in Table 2.2. The case of $\alpha = 0.05, 1-\beta = 0.80$ is known as "Lehr's rule." [van Belle, 2002; Lehr, 1992]

	One sided	Two sided
Power = 0.25	0.96	1.68
Power = 0.50	2.72	3.86
Power = 0.80	6.20	7.87

TABLE 2.2
Scaling of sample size with respect to ζ^2 required to achieve a fixed power in the t-test, at a fixed $\alpha = 0.05$ rate.

Consider now the scaling in the rule $n \approx c\zeta^{-2}$. If the signal-noise ratio ζ is given in daily units, the sample size will be in days. One annualizes ζ by multiplying by the square root of the number of days per year, which downscales n appropriately. That is, if ζ is quoted in annualized terms, this rule of thumb gives the number of *years* of observations required. This is very convenient since we usually think of ζ and $\hat{\zeta}$ in annualized terms.

The following rule of thumb may prove useful:

Power rule

The number of years required to reject non-zero mean with power of one half is around $2.7/\zeta^2$.

The mnemonic form of this is "$e = nz^2$." Note that Euler's number appears here coincidentally, as it is nearly equal to $\left[\Phi^{-1}(0.95)\right]^2$. The relative error in this approximation for determining the sample size is shown in Figure 2.4, as a function of ζ; the error is smaller than one percent in the tested range.

The power rules are sobering indeed. Suppose you were a hedge fund manager whose investors threatened to perform a one-sided t-test after one year. If your strategy's signal-to-noise ratio is less than $1.649\,\mathrm{yr}^{-1/2}$ (a value which should be considered "very good"), your chances of "passing" the t-test are less than fifty percent.

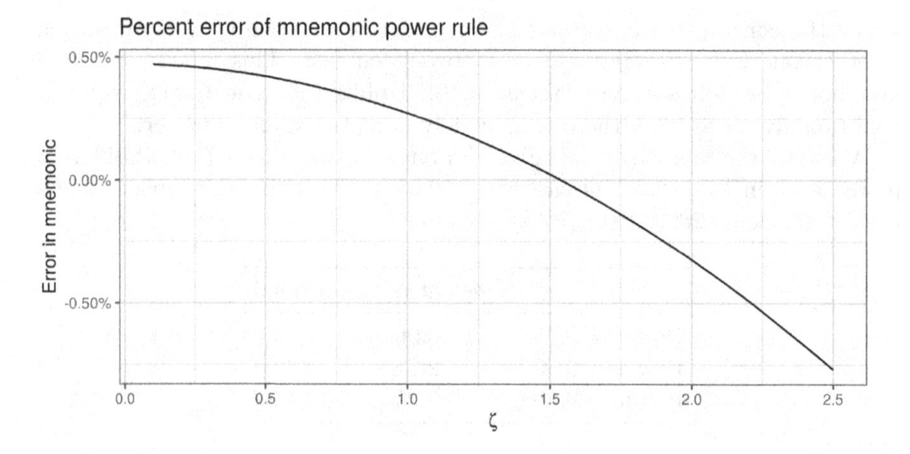

FIGURE 2.4
The percent error of the power mnemonic $e \approx n\zeta^2$ is plotted versus ζ.

2.5.9 † Frequentist prediction intervals

Suppose, based on a sample of size n_1, you observed $\hat{\zeta}_1$ for some asset stream. What can you expect of the Sharpe ratio for n future observations? Though this is a question similar to that answered by the power rules given in Section 2.5.8, the power rules are deficient because 1. they rely on the unknown population parameter, ζ, when only a noisy estimate is available, and 2. they make statements within the dichotomy of type I and type II errors. To correct this, one may use a Frequentist *prediction interval*, which is an interval which, conditional on $\hat{\zeta}_1$ and n_1, contains the Sharpe ratio of those future observations with some specified probability, under replication.[5] The "under replication" clause here means that if you repeated the full experiment of generating $\hat{\zeta}_1$, n_1, constructing the prediction interval, then observing $\hat{\zeta}_2$, you should find that $\hat{\zeta}_2$ is within the interval with the given probability.

Suppose you observe $\hat{\zeta}_1$ on n_1 observations of normally distributed *i.i.d.* returns, then observe $\hat{\zeta}_2$ on n_2 observations from the same returns stream. We can write

$$\hat{\zeta}_1\sqrt{\chi_1^2/\left(n_1-1\right)} + Z_1/\sqrt{n_1} = \zeta = \hat{\zeta}_2\sqrt{\chi_2^2/\left(n_2-1\right)} + Z_2/\sqrt{n_2}, \qquad (2.38)$$

where the $Z_i \sim \mathcal{N}(0,1)$, and the $\chi_i^2 \sim \chi^2\left(n_i-1\right)$ are independent. If there were an exact two-sample test for equal signal-noise ratios on independent samples, one would use that for this problem. Instead an approximation must be used.

[5]Typically "prediction interval" is reserved for an interval around a single future observation, while "tolerance interval" is used for multiple future observations. Our application is somewhat between these two.

Let $\hat{\zeta}_i = \zeta + s_i\epsilon_i$, for $i = 1, 2$ where s_i is the standard error of the Sharpe ratio, $\hat{\zeta}_i$ based on n_i observations, and ϵ_i is a zero mean, unit variance random variable. Usually we have $s_i = s/\sqrt{n_i}$. Then if the ϵ_i are independent, we have

$$\hat{\zeta}_2 = \hat{\zeta}_1 + \sqrt{1 + \frac{n_1}{n_2}} \frac{s}{\sqrt{n_1}} \epsilon. \tag{2.39}$$

Thus the prediction interval around $\hat{\zeta}_1$ is inflated by a factor of

$$c = \sqrt{1 + \frac{n_1}{n_2}} \tag{2.40}$$

compared to the equivalent confidence interval.

Example 2.5.13 (Prediction intervals, the Market). Consider the monthly relative returns of the Market, introduced in Example 1.1.1. The Sharpe ratio on $n_1 = 1,128$ mo. was computed to be $\hat{\zeta}_1 = 0.177\,\text{mo.}^{-1/2}$. For $n_2 = 12$ mo., the 95% prediction interval on $\hat{\zeta}_2$ was computed, via the normal approximation and standard error of Equation 2.24 to be approximately $[-0.4, 0.75)\,\text{mo.}^{-1/2}$.

For comparison, if one simply assumes that $\zeta = \hat{\zeta}_1 = 0.177\,\text{mo.}^{-1/2}$, then the 95% prediction interval for the Sharpe ratio on $n_2 = 12$ mo. is $[-0.42, 0.86)\,\text{mo.}^{-1/2}$. ⊣

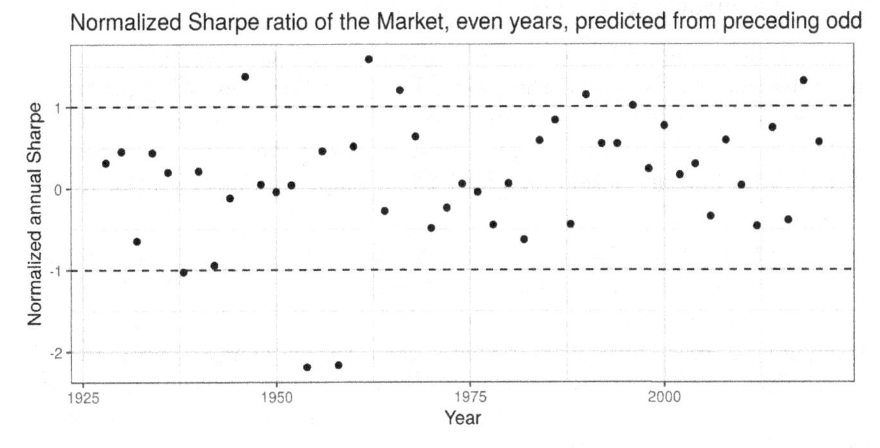

FIGURE 2.5
The normalized Sharpe ratio of daily returns of the Market in even-numbered years are plotted, normalized to the 0.9 prediction interval computed using the previous year's Sharpe ratio. The normalization is such that a value of -1 corresponds to the lower limit of the prediction interval, 1 to the upper limit. The empirical coverage is approximately 0.81. See Example 2.5.14.

Example 2.5.14 (Prediction interval coverage, the Market). Consider the daily relative returns of the Market, introduced in Example 1.1.1. For each odd-numbered year from 1927 through 2019, the Sharpe ratio is computed using daily returns in the odd numbered year. Then a 0.9 prediction interval for Sharpe ratio is constructed based on the number of days in the following (even-numbered) year. The realized Sharpe ratio is then computed on the following even-numbered year, and compared to the prediction interval. The empirical coverage is approximately 0.81, much smaller than the nominal coverage. Thus the prediction intervals are too conservative, likely due to omitted variable bias or volatility clustering.

In Figure 2.5, the realized Sharpe ratio is plotted, for even-numbered years, relativized to the prediction intervals, so that a value of 1 corresponds to the realized Sharpe ratio exactly equaling the upper limit of the prediction interval, and a -1 corresponds to the lower limit. There are 9 values falling outside the prediction intervals.

To diagnose the conservative prediction intervals, the daily returns were permuted, and the experiment repeated: prediction intervals on Sharpe ratio computed using odd "years," then compared to "realized" Sharpe ratio in even "years." Repeating this process 100 times, the empirical coverage is approximately 0.89, equal to the nominal value. Thus it appears that non-normality and the standard error approximation are *not* to blame for the poor coverage, rather autocorrelation of returns or volatility in actual market returns. ⊣

ex-factor Sharpe ratio prediction intervals

We note that this approximate form only requires the computation and inflation of a standard error, and thus could be applied to computing prediction intervals on the ex-factor Sharpe ratio. However, the standard error of Equation 2.34 does not take into account randomness of the factors F_i, and may not deliver faithful prediction intervals. We attempt to correct for this in Section 3.5.2.

Example 2.5.15 (Prediction intervals, Technology ex-factor signal-noise ratio). Consider the attribution of monthly Technology industry returns to those of the Market, as described in Example 2.5.7. As in Example 2.5.14, for each odd-numbered year from 1927 through 2017, the ex-factor Sharpe ratio is computed using 12 monthly returns in the odd numbered year. Then a 0.9 prediction interval for ex-factor Sharpe ratio is constructed for the following (even-numbered) year. Prediction intervals are constructed by plugging in the observed ex-factor Sharpe ratio into Equation 2.34 to compute a standard error, then multiplying by $\sqrt{2}$.

The realized ex-factor Sharpe ratio is then computed on the following even-numbered year, and compared to the prediction interval. The empirical coverage is approximately 0.7, much smaller than the nominal coverage. Again the prediction intervals are too conservative, although in this case one suspects the poor coverage may be due to the rough approximation of Equation 2.34, the small sample size (12 monthly returns in-sample and out-of-sample), or due to autocorrelation, or correlated heteroskedasticity, *etc.*

To test this, as in Example 2.5.14 we permute the monthly paired returns of Technology and the Market and repeat the experiment, computing the ex-factor Sharpe ratio on the odd (shuffled) "years" and comparing them to realized ex-factor Sharpe ratio on even "years." Repeating this process 100 times with different shufflings of the data, the empirical coverage is approximately 0.88, much closer to the nominal value. We blame autocorrelation of returns or volatility for the poor performance of the prediction intervals, and not non-normality of returns, small sample size, or the asymptotic approximation. ⊣

2.6 † Likelihoodist inference on the signal-noise ratio

We will now consider the likelihoodist's view of the Sharpe ratio. While philosophically distinct from the frequentist's view, the likelihoodist will arrive at similar conclusions to the frequentist.

We start with the likelihood of ζ given $\hat{\zeta}$. This is merely the density, given in Equation 2.7, but expressed as a function of ζ:

$$\mathcal{L}_{SR}\left(\zeta \,\middle|\, \hat{\zeta}, n\right) := f_{SR}\left(\hat{\zeta}; \zeta, n\right).$$

The expression for the density of the Sharpe ratio given in Equation 2.7 suffices to point out there is unlikely to be a simple closed form for the maximum likelihood estimate, MLE. To find the MLE, take the derivative of the likelihood:

$$\frac{\partial \mathcal{L}_{SR}\left(\zeta \,\middle|\, \hat{\zeta}, n\right)}{\partial \zeta} = \\ \frac{\sqrt{n/2}}{\sqrt{2\pi}\Gamma\left(\frac{n-1}{2}\right)} \int_0^\infty n\left(\hat{\zeta}\sqrt{x} - \zeta\right)\left(\frac{x}{2}\right)^{\frac{n-2}{2}} \exp\left(-\frac{x + n\left(\hat{\zeta}\sqrt{x} - \zeta\right)^2}{2}\right) \mathrm{d}x.$$

Then the MLE satisfies

$$\zeta_{\mathrm{MLE}} = \frac{\hat{\zeta} \int_0^\infty \sqrt{x}\left(\frac{x}{2}\right)^{\frac{n-2}{2}} \exp\left(-\frac{x+n\left(\hat{\zeta}\sqrt{x}-\zeta_{\mathrm{MLE}}\right)^2}{2}\right) \mathrm{d}x}{\int_0^\infty \left(\frac{x}{2}\right)^{\frac{n-2}{2}} \exp\left(-\frac{x+n\left(\hat{\zeta}\sqrt{x}-\zeta_{\mathrm{MLE}}\right)^2}{2}\right) \mathrm{d}x}. \tag{2.41}$$

Thus $\zeta_{\mathrm{MLE}}/\hat{\zeta}$ is the ratio of two integrals of positive functions, and is a positive number. Therefore the MLE estimate has the same *sign* as $\hat{\zeta}$. Unlike the unbiased estimator, which is smaller than $\hat{\zeta}$ when the latter is positive, it appears that the MLE is often *larger* than $\hat{\zeta}$ in this case; see Example 2.6.1.

For the purposes of estimating the likelihood, or the MLE, however, we are likely better off using off-the-shelf implementations of the density of the non-central t-distribution, and then finding the maximum using *e.g.*, a golden section search in the neighborhood of $\hat{\zeta}$.

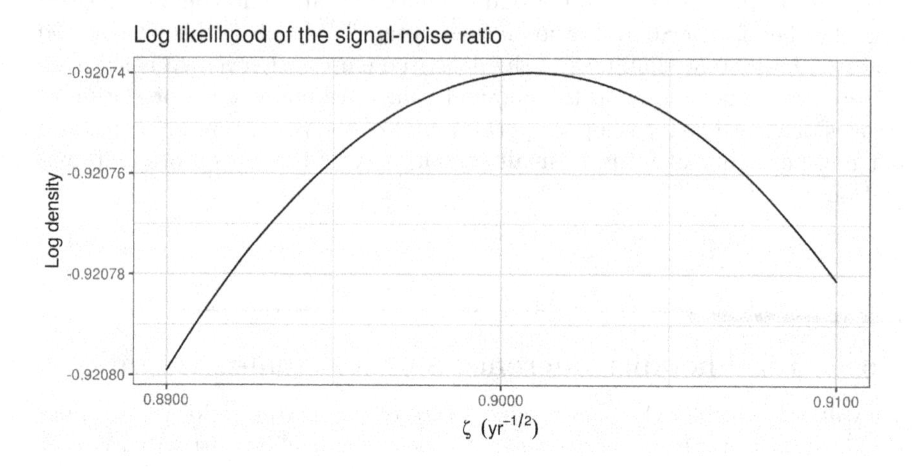

FIGURE 2.6
The log density of $\hat{\zeta} = 0.9\,\mathrm{yr}^{-1/2}$ is plotted as a function of the signal-noise ratio, for $n = 252$ daily observations.

Example 2.6.1 (MLE). Suppose, based on 252 days of Gaussian log returns, you observe $\hat{\zeta} = 0.9\,\mathrm{yr}^{-1/2}$. The likelihood of $\hat{\zeta}$ as a function of ζ is plotted in Figure 2.6. The MLE is *not* equal to $\hat{\zeta}$, rather it takes a slightly larger value. We have $100\% \left(\frac{\zeta_{\mathrm{MLE}}}{\hat{\zeta}} - 1\right) \approx 0.099\%$. ⊣

Example 2.6.2 (MLE, Market returns). Consider the monthly relative returns of the Market, introduced in Example 1.1.1. We compute $\hat{\zeta} = 0.614\,\mathrm{yr}^{-1/2}$ for $n = 1,128\,\mathrm{mo}$. The MLE for ζ is found to be approximately $\hat{\zeta} = 0.614\,\mathrm{yr}^{-1/2}$. We have $100\% \left(\frac{\hat{\zeta}_{\mathrm{MLE}}}{\hat{\zeta}} - 1\right) \approx 0.022\%$. ⊣

2.6.1 Likelihood ratio test

To test $H_0 : \zeta = \zeta_0$ versus an unrestricted alternative, find the MLE, then compute the test statistic

$$D = -2 \log \frac{\mathcal{L}_{SR}\left(\zeta_0 \middle| \hat{\zeta}, n\right)}{\mathcal{L}_{SR}\left(\zeta_{\mathrm{MLE}} \middle| \hat{\zeta}, n\right)} = -2 \log \frac{f_{SR}\left(\hat{\zeta}; \zeta_0, n\right)}{f_{SR}\left(\hat{\zeta}; \zeta_{\mathrm{MLE}}, n\right)}. \tag{2.42}$$

The unrestricted model has one extra degree of freedom. For large n, under Wilk's theorem, we expect D to converge to a chi-square random variable with one degree of freedom. See Example 2.6.3.

Example 2.6.3 (LRT, simple example). Suppose, based on 1,518 days of hypothetical Gaussian log returns, you observe $\hat{\zeta} = 1.5\,\mathrm{yr}^{-1/2}$. To test $H_0 : \zeta = 0.5\,\mathrm{yr}^{-1/2}$ against the unrestricted alternative, the log likelihood under H_0 is computed as -3.014; the log likelihood for the MLE is -0.025. The test statistic is then $D = 5.976$. The probability that a chi-square with one degree of freedom takes a value this large is around 0.014, so we may reject H_0 with high confidence. ⊣

To test $H_0 : \zeta = \zeta_0$ versus $H_1 : \zeta = \zeta_1$, compute the test statistic

$$\Lambda = \frac{\mathcal{L}_{SR}\left(\zeta_0 \left| \hat{\zeta}, n\right.\right)}{\mathcal{L}_{SR}\left(\zeta_1 \left| \hat{\zeta}, n\right.\right)}. \tag{2.43}$$

When Λ is small, smaller than some cutoff, we prefer H_1; when it is large, we prefer H_0. To use this test statistic in a null hypothesis test, one must find the cutoff value to reject H_0 in favor of H_1, with the cutoff value chosen to achieve the desired type I rate.

Note that the density of the Sharpe ratio is that of the non-central t distribution, up to scaling: $\mathcal{L}_{SR}\left(\zeta \left| \hat{\zeta}, n\right.\right) = \sqrt{n}\mathcal{L}_t\left(\sqrt{n}\zeta \left| \sqrt{n}\hat{\zeta}, n-1\right.\right)$. Thus

$$\Lambda = \frac{\mathcal{L}_t\left(\sqrt{n}\zeta_0 \left| \sqrt{n}\hat{\zeta}, n-1\right.\right)}{\mathcal{L}_t\left(\sqrt{n}\zeta_1 \left| \sqrt{n}\hat{\zeta}, n-1\right.\right)}.$$

Kruskal [1954] showed that this ratio is monotonic in $\sqrt{n}\hat{\zeta}$, and thus it is monotonic in $\hat{\zeta}$. That is, supposing without loss of generality, that $\zeta_0 < \zeta_1$, the statistic Λ is decreasing in $\hat{\zeta}$. Thus to find a cutoff which achieves the desired type I rate under H_0, we take the $1 - \alpha$ quantile under H_0, and plug it in as the Sharpe ratio. That is, the cutoff value is

$$\Lambda_c = \frac{f_{SR}\left(SR_{1-\alpha}\left(\zeta_0, n\right); \zeta_0, n\right)}{f_{SR}\left(SR_{1-\alpha}\left(\zeta_0, n\right); \zeta_1, n\right)}, \tag{2.44}$$

where $SR_q\left(\zeta_0, n\right)$ is the q quantile of the Sharpe ratio for the given signal-noise ratio and number of observations. This we can find by using off-the-shelf implementations of the quantile function of the non-central t distribution. See Example 2.6.4.

The upshot of all this is that the identity of H_1 is largely irrelevant (other than that it posits a larger signal-noise ratio than the null hypothesis, rather than a smaller one). There is, in fact, no reason to compute this likelihood ratio, and we reject H_0 in favor of *any* larger signal-noise ratio whenever $\hat{\zeta} > SR_{1-\alpha}\left(\zeta_0, n\right)$. This is merely the frequentist test for H_0 presented in Section 2.5.3. However, by the Neyman-Pearson lemma, the likelihood ratio test is the *uniformly most powerful* test of size α for H_0 against the set of alternatives $\zeta > \zeta_0$. [Bain and Engelhardt, 1992] Since the frequentist test rejects exactly when the likelihood ratio test does, it, too, is the UMP for H_0.

Example 2.6.4 (LRT and Hypothesis testing). Suppose you wish to test $H_0 : \zeta = 0.25\,\mathrm{yr}^{-1/2}$ versus $H_1 : \zeta = 1.75\,\mathrm{yr}^{-1/2}$ for a Sharpe ratio observed on 1518 days of Gaussian log returns. To achieve a type I rate of $\alpha = 0.005$, the $1 - \alpha$ quantile under H_0 is computed as $SR_{1-\alpha}(\zeta_0, n) = 1.304\,\mathrm{yr}^{-1/2}$. The cutoff value for the likelihood ratio is $\Lambda_c = 0.066$. ⊣

Example 2.6.5 (LRT, Market returns). Consider the monthly relative returns of the Market, introduced in Example 1.1.1. Suppose you wish to test $H_0 : \zeta = 0.25\,\mathrm{yr}^{-1/2}$ versus $H_1 : \zeta = 0.75\,\mathrm{yr}^{-1/2}$ for the Market, assuming normality of returns. To achieve a type I rate of $\alpha = 0.01$, the $1 - \alpha$ quantile under H_0 is computed as $SR_{1-\alpha}(\zeta_0, n) = 0.491\,\mathrm{yr}^{-1/2}$. Since we observe $\hat{\zeta} = 0.614\,\mathrm{yr}^{-1/2}$, we reject H_0 at the 0.01 level. As noted above, the identity of H_1 is irrelevant, other than that it posits a larger signal-noise ratio than H_0. ⊣

2.7 † Bayesian inference on the signal-noise ratio

We now consider how a Bayesian might perform inference on the Sharpe ratio. In contrast to the Frequentist paradigm, wherein one aims to achieve a fixed frequency of type I errors, Bayesian inference focuses on updating the beliefs of the practitioner.

Bayesian inference is usually described in terms of conjugate priors because it simplifies the exposition: the prior and posterior belief follow the same distributional form, and the data acts to update the hyperparameters of that distributional form. It is, however, a bit odd to presume that the practitioner's beliefs must follow some form we prescribe.

That said, Bayesian analysis of observations drawn from a Gaussian distribution with unknown mean and variance usually proceeds via one of two families of conjugate priors. [Murphy, 2007; Gelman *et al.*, 2013] These are either priors on the mean and variance, μ and σ^2, or priors on the mean and *precision*, defined as σ^{-2}. We will focus on the latter, applying some transformation to arrive at a prior on the signal-noise ratio, ζ, and some transform of the variance.

One commonly used conjugate prior for mean and precision is the "Normal-Inverse-Gamma," under which one has an unconditional inverse gamma prior distribution on σ^2 (this is, up to scaling, one over a chi-square), and, conditional on σ, a normal prior on μ. [Gelman *et al.*, 2013, 3.3] These can be stated as

$$\begin{aligned} \sigma^2 &\propto \Gamma^{-1}\left(m_0/2, m_0\sigma_0^2/2\right), \\ \mu \left| \sigma^2 \right. &\propto \mathcal{N}\left(\mu_0, \sigma^2/n_0\right), \end{aligned} \tag{2.45}$$

where σ_0^2, m_0, μ_0 and n_0 are the *hyper-parameters.* The density function for the inverse gamma law is given by

$$F_{\Gamma^{-1}}\left(x;a,b\right) = \frac{b^a}{\Gamma\left(a\right)} x^{-(a+1)} e^{-\frac{b}{x}}, \quad \text{for } x > 0. \tag{2.46}$$

Together the hyper-parameters parametrize the prior belief.

While the normal distribution is familiar, the inverse gamma may be less so. To give some intuition regarding this distribution, suppose that $x \sim \Gamma^{-1}\left(m_0/2, m_0\sigma_0^2/2\right)$. The probability density of x is unimodal with mode at $m_0\sigma_0^2/\left(m_0+2\right)$, and the expected value of x is $m_0\sigma_0^2/\left(m_0-2\right)$, assuming $m_0 > 2$, otherwise it is infinite. If $m_0 > 4$, then the variance of x is

$$\left(\frac{m_0\sigma_0^2}{m_0-2}\right)^2 \frac{2}{m_0-4},$$

otherwise it is infinite. In general if $m_0 > 2k$ then the kth moment of x exists. The density of x is a somewhat non-descript unimodal distribution on the positive reals.

After observing n *i.i.d.* draws from a normal distribution, $\mathcal{N}\left(\mu,\sigma\right)$, say $x_1, x_2, \ldots, x_n$, let $\hat{\mu}$ and $\hat{\sigma}$ be the sample estimates from Equation 1.2. The posterior is then

$$\begin{aligned} \sigma^2 &\propto \Gamma^{-1}\left(m_1/2, m_1\sigma_1^2/2\right), \\ \mu \left|\sigma^2\right. &\propto \mathcal{N}\left(\mu_1, \sigma^2/n_1\right), \end{aligned} \tag{2.47}$$

where

$$n_1 = n_0 + n, \qquad \mu_1 = \frac{n_0\mu_0 + n\hat{\mu}}{n_1}, \tag{2.48}$$

$$m_1 = m_0 + n, \qquad \sigma_1^2 = \frac{m_0\sigma_0^2 + \left(n-1\right)\hat{\sigma}^2 + \frac{n_0 n}{n_1}\left(\mu_0 - \hat{\mu}\right)^2}{m_1}. \tag{2.49}$$

Given the update rules, one could interpret μ_0 as the sample mean over n_0 *pseudo-observations* of x_t. By pseudo-observation, we mean some fictional sample of the returns x_t that one has observed with the mind's eye and that inform one's beliefs. Similarly, σ_0^2 is the sample variance over m_0 pseudo-observations. If the pseudo-observations were real, then the update rules above would describe exactly how one would combine the mean and variance estimates from two distinct samples[6]. Under this interpretation the hyperparameters $n_0 = 0 = m_0$ correspond to having no pseudo-observations, and thus should be viewed as "objective" in some sense. Indeed, these correspond to a noniformative *Jeffreys' prior*, which is in invariant to reparametrizations. It should be noted that the selection of an "objective" prior is a thorny philosophical issue. [Kass and Wasserman, 1996; Yang and Berger, 1996]

This prior on the mean and variance can be transformed to one on the

[6]Up to the use of a Bessel-corrected denominator for the sample variance.

variance and the signal-noise ratio, where the former is a nuisance parameter. Transforming Equation 2.45, we arrive at

$$\begin{aligned} \sigma^2 &\propto \Gamma^{-1}\left(m_0/2, m_0\sigma_0^2/2\right), \\ \zeta\left|\sigma^2\right. &\propto \mathcal{N}\left(\frac{\mu_0}{\sigma}, 1/n_0\right), \end{aligned} \tag{2.50}$$

Marginalizing out σ^2, we arrive at a lambda prime prior (*cf.* Section 2.4)

$$\sqrt{n_0}\zeta \propto \lambda'\left(\sqrt{n_0}\zeta_0, m_0\right), \tag{2.51}$$

where $\zeta_0 = \mu_0/\sigma_0$. The marginal posterior can be written as

$$\sqrt{n_1}\zeta \propto \lambda'\left(\sqrt{n_1}\zeta_1, m_1\right), \tag{2.52}$$

where

$$n_1 = n_0 + n, \qquad \zeta_1 = \frac{n_0\zeta_0\sigma_0 + n\hat{\zeta}\hat{\sigma}}{n_1\sigma_1}, \tag{2.53}$$

$$m_1 = m_0 + n, \quad \sigma_1^2 = \frac{m_0\sigma_0^2 + (n-1)\,\hat{\sigma}^2 + \frac{n_0 n}{n_1}\left(\zeta_0\sigma_0 - \hat{\zeta}\hat{\sigma}\right)^2}{m_1}, \tag{2.54}$$

where $\hat{\zeta} = \hat{\mu}/\hat{\sigma}$.

One is tempted to rearrange the equations to try to achieve an update formula for belief on ζ that relies on $\hat{\zeta}$ alone. However, the $\hat{\zeta}$ follows a non-central t distribution up to scaling. Conjugate priors can only be found for random variables whose density comes from a certain class of functions, the so-called exponential family. The non-central t distribution is not in this family, thus there is no hope of finding a conjugate prior for it. [Gelman *et al.*, 2013]

Example 2.7.1 (Basic Bayesian update). Suppose your prior corresponds to the hyperparameters $\zeta_0 = 0.015\,\mathrm{day}^{-1/2}$, $n_0 = 10\,\mathrm{days}$, $\sigma_0 = 0.013\,\mathrm{day}^{-1/2}$, $m_0 = 100\,\mathrm{days}$. (These prior hyperparameters were chosen merely to illustrate the mechanics of computation.) One then observes $n = 252$ days of returns with $\hat{\zeta} = 0.009\,\mathrm{day}^{-1/2}$ and $\hat{\sigma} = 0.017\,\mathrm{day}^{-1/2}$. The posterior is then $\zeta_1 = 0.01\,\mathrm{day}^{-1/2}$, $n_1 = 262\,\mathrm{days}$, $\sigma_1 = 0.016\,\mathrm{day}^{-1/2}$, $m_1 = 352\,\mathrm{days}$. 10^5 samples were drawn from the prior and posterior distributions; their empirical densities are plotted in Figure 2.7. The signal-noise ratio is plotted in units of $\mathrm{day}^{-1/2}$. We observe that 48.1% of the draws from the prior distribution are less than zero, while only 43.8% of the posterior draws are negative. Thus the evidence has reduced somewhat the uncertainty in profitability of the strategy, but considerable doubt remains. ⊣

Example 2.7.2 (Bayesian update, Market returns). Consider the monthly relative returns of the Market, introduced in Example 1.1.1. Suppose your prior (constructed prior to Jan 1927!) was $\zeta_0 = 0.125\,\mathrm{mo.}^{-1/2}$, $n_0 = 12\,\mathrm{mo.}$, $\sigma_0 = 4\%\,\mathrm{mo.}^{-1/2}$, $m_0 = 72\,\mathrm{mo.}$

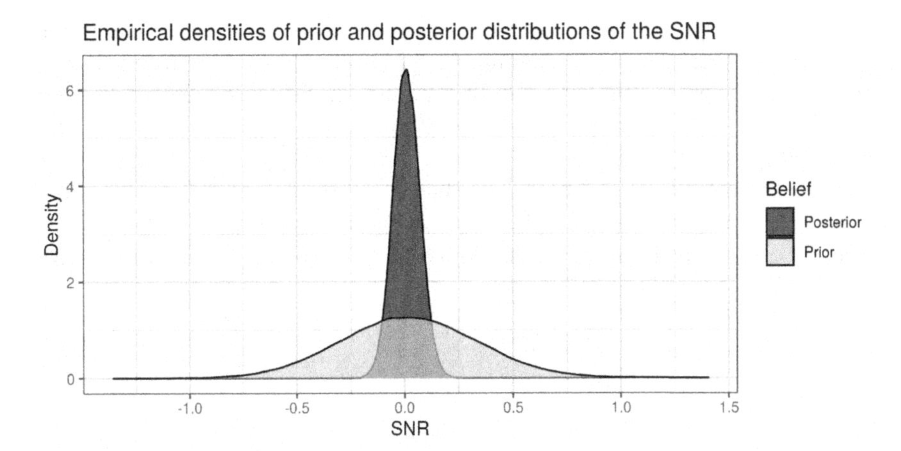

FIGURE 2.7
100,000 samples were drawn from both the prior and posterior distributions for Example 2.7.1. The empirical densities are then plotted.

One then observes $n = 1,128$ months of returns with $\hat{\zeta} = 0.177\,\text{mo.}^{-1/2}$ and $\hat{\sigma} = 5.354\%\,\text{mo.}^{-1/2}$. The posterior is then $\zeta_1 = 0.179\,\text{mo.}^{-1/2}$, $n_1 = 1,140\,\text{mo.}$, $\sigma_1 = 5.281\%\,\text{mo.}^{-1/2}$, $m_1 = 1,200\,\text{mo.}$ ⊣

Example 2.7.2 illustrates some of the problems with selecting a prior. In this case it is hard to imagine how one would have informative beliefs regarding the returns of the Market which were not somehow influenced by the past 90 years of actually observed returns. For this reason, a non-informative prior might be recommended here. However, the prior used in the example is relatively uninformative, corresponding to just a few years of pseudo-observations, when compared to the sample size of actual observations available.

2.7.1 Bayesian inference on the ex-factor signal-noise ratio

Now consider the attribution model for returns, Equation 1.8, which we can write as

$$x_t = \boldsymbol{\beta}^\top \boldsymbol{f}_t + \epsilon_t.$$

The error term ϵ_t is zero mean and has variance σ^2. We assume the $\boldsymbol{f}_t$ is deterministic.

The analogue of the Normal-Inverse-Gamma prior for this regression problem is stated as

$$\begin{aligned} \sigma^2 &\propto \Gamma^{-1}\left(m_0/2, m_0\sigma_0^2/2\right), \\ \boldsymbol{\beta}\left|\sigma^2\right. &\propto \mathcal{N}\left(\boldsymbol{\beta}_0, \sigma^2{\Lambda_0}^{-1}\right), \end{aligned} \tag{2.55}$$

where σ_0^2, m_0 are the Bayesian hyperparameters for the scale and degrees of freedom of the error term, while $\boldsymbol{\beta}_0$ is the location hyperparameter for the regression coefficients, and the matrix Λ_0 parametrizes uncertainty in the regression coefficients. When Λ_0 is diagonal, one can view the elements on the diagonal as the number of pseudo-observations in one's prior on each of the elements of $\boldsymbol{\beta}$.

As in Section 2.1.2, assume one has n observations of $\boldsymbol{f}_t$, stacked row-wise in the $n \times l$ matrix, F, and the corresponding returns stacked in vector $\boldsymbol{x}$. As in Equation 2.3, define $\hat{\boldsymbol{\beta}} := \left(\mathsf{F}^\top\mathsf{F}\right)^{-1}\mathsf{F}^\top\boldsymbol{x}$ and $\hat{\sigma} := \sqrt{\left(\boldsymbol{x} - \mathsf{F}\hat{\boldsymbol{\beta}}\right)^\top\left(\boldsymbol{x} - \mathsf{F}\hat{\boldsymbol{\beta}}\right)(n-l)^{-1}}$. The posterior distribution is

$$\begin{aligned} \sigma^2 &\propto \Gamma^{-1}\left(m_1/2, m_1\sigma_1^2/2\right), \\ \boldsymbol{\beta}\left|\sigma^2\right. &\propto \mathcal{N}\left(\boldsymbol{\beta}_1, \sigma^2{\Lambda_1}^{-1}\right), \end{aligned} \tag{2.56}$$

where

$$\Lambda_1 = \Lambda_0 + \mathsf{F}^\top\mathsf{F}, \tag{2.57}$$

$$\boldsymbol{\beta}_1 = {\Lambda_1}^{-1}\left(\Lambda_0\boldsymbol{\beta}_0 + \mathsf{F}^\top\mathsf{F}\hat{\boldsymbol{\beta}}\right), \tag{2.58}$$

$$m_1 = m_0 + n, \tag{2.59}$$

$$\sigma_1^2 = \frac{m_0\sigma_0^2 + (n-l)\,\hat{\sigma}^2 + \hat{\boldsymbol{\beta}}^\top\mathsf{F}^\top\mathsf{F}\hat{\boldsymbol{\beta}} + {\boldsymbol{\beta}_0}^\top\Lambda_0\boldsymbol{\beta}_0 - {\boldsymbol{\beta}_1}^\top\Lambda_1\boldsymbol{\beta}_1}{m_1}. \tag{2.60}$$

A non-informative prior corresponds to $\Lambda_0 = 0, \boldsymbol{\beta}_0 = \boldsymbol{0}, \sigma_0^2 = 0, m_0 = 0$.

Recall we defined the ex-factor signal-noise ratio as (ignoring the risk-free rate) $\boldsymbol{\beta}^\top\boldsymbol{v}/\sigma^2$, for some pre-specified direction $\boldsymbol{v}$. To aid our analysis, we can collapse the prior or posterior "along" the direction $\boldsymbol{v}$:

$$\begin{aligned} \sigma^2 &\propto \Gamma^{-1}\left(m_i/2, m_i\sigma_i^2/2\right), \\ \boldsymbol{v}^\top\boldsymbol{\beta}\left|\sigma^2\right. &\propto \mathcal{N}\left(\boldsymbol{v}^\top\boldsymbol{\beta}_i, \sigma^2\boldsymbol{v}^\top{\Lambda_i}^{-1}\boldsymbol{v}\right), \end{aligned} \tag{2.61}$$

where $i = 0$ for the prior and $i = 1$ for the posterior. As in the unattributed model, marginalizing out σ^2, we have a lambda prime prior and posterior:

$$\left(\boldsymbol{v}^\top{\Lambda_i}^{-1}\boldsymbol{v}\right)^{-1/2}\frac{\boldsymbol{\beta}^\top\boldsymbol{v}}{\sigma} = \left(\boldsymbol{v}^\top{\Lambda_i}^{-1}\boldsymbol{v}\right)^{-1/2}\zeta_g \propto \lambda'\left(\left(\boldsymbol{v}^\top{\Lambda_i}^{-1}\boldsymbol{v}\right)^{-1/2}\zeta_{g,i}, m_i\right), \tag{2.62}$$

where $\zeta_{g,i} := {\boldsymbol{\beta}_i}^\top\boldsymbol{v}/\sigma_i$.

Example 2.7.3 (Bayesian update, UMD returns, attribution model). Consider the monthly relative returns of UMD, under an attribution against intercept, Market, SMB and HML, using the data introduced in Example 1.1.1. Let $\boldsymbol{\beta}$ (and thus the columns of F) be ordered by intercept (the idiosyncratic term), then SMB, then HML. To keep track of the units, let $\mathsf{U} = \mathrm{Diag}\left(\left[\text{mo.}, \%, \%, \%\right]^\top\right)$ be the matrix of units.

Suppose your prior (constructed prior to Jan 1927!) was

$$\Lambda_0 = \mathsf{U}^\top \begin{bmatrix} 60.00 & 0.00 & 0.00 & 0.00 \\ 0.00 & 12.00 & 0.00 & 0.00 \\ 0.00 & 0.00 & 12.00 & 0.00 \\ 0.00 & 0.00 & 0.00 & 12.00 \end{bmatrix} \mathsf{U},$$

$$\boldsymbol{\beta}_0 = \mathsf{U}^{-1}[0.5, 0, 0, 0]^\top\%, \quad m_0 = 12\,\text{mo.}, \quad \sigma_0 = 3\%\,\text{mo.}^{-1/2}.$$

Again, these are just made up numbers, but the m_0 and diagonal of Λ_0 can be interpreted as the number of months of pseudo-observations in the prior.

One then observes $n = 1,128$ months of returns with

$$\mathsf{F}^\top\mathsf{F} = \mathsf{U}^\top \begin{bmatrix} 1128.00 & 1070.17 & 234.60 & 363.32 \\ 1070.17 & 33325.49 & 6435.78 & 5419.69 \\ 234.60 & 6435.78 & 11492.63 & 1689.78 \\ 363.32 & 5419.69 & 1689.78 & 13984.86 \end{bmatrix} \mathsf{U},$$

$$\hat{\boldsymbol{\beta}} = \mathsf{U}^{-1}[1.015, -0.217, -0.05, -0.469]^\top\%,$$

$$n = 1,128\,\text{mo.}, \quad \hat{\sigma} = 4.122\%\,\text{mo.}^{-1/2}.$$

The posterior is then

$$\Lambda_1 = \mathsf{U}^\top \begin{bmatrix} 1188.00 & 1070.17 & 234.60 & 363.32 \\ 1070.17 & 33337.49 & 6435.78 & 5419.69 \\ 234.60 & 6435.78 & 11504.63 & 1689.78 \\ 363.32 & 5419.69 & 1689.78 & 13996.86 \end{bmatrix} \mathsf{U},$$

$$\boldsymbol{\beta}_1 = \mathsf{U}^{-1}[0.988, -0.217, -0.05, -0.468]^\top\%,$$

$$m_1 = 1,140\,\text{mo.}, \quad \sigma_1 = 4.106\%\,\text{mo.}^{-1/2}.$$

The ex-factor signal-noise ratio for idiosyncratic returns under the attribution model corresponds to $\boldsymbol{v} = \mathsf{U}[1, 0, 0, 0]^\top$. Thus the posterior belief on ζ_g, marginalizing out σ, can be expressed as

$$33.925\,\text{mo.}^{1/2}\%^{-1}\zeta_g \propto \lambda'\left(8.164, 1140\right).$$

⊣

2.7.2 Credible intervals on the signal-noise ratio

One can construct credible intervals on the signal-noise ratio based on the posterior, so-called *posterior intervals*, via quantiles of the lambda prime distribution. [Gelman *et al.*, 2013] For example, a $(1-\alpha)$ credible interval on ζ is given by

$$\frac{1}{\sqrt{n_1}}\left[\lambda'_{\alpha/2}\left(\sqrt{n_1}\zeta_1, m_1\right), \lambda'_{1-\alpha/2}\left(\sqrt{n_1}\zeta_1, m_1\right)\right). \tag{2.63}$$

For the case of noninformative priors (corresponding to $n_0 = m_0 = 0$), this is equivalent to

$$\frac{1}{\sqrt{n}}\left[{\lambda'}_{\alpha/2}\left(\sqrt{n}\sqrt{\frac{n}{n-1}}\hat{\zeta},n\right),{\lambda'}_{1-\alpha/2}\left(\sqrt{n}\sqrt{\frac{n}{n-1}}\hat{\zeta},n\right)\right) \tag{2.64}$$

This is *mathematically* equivalent to the Frequentist confidence interval given in Equation 2.22, but replacing n for $n-1$ in the degrees of freedom for σ. Asymptotically, the Bayesian credible interval for an noninformed prior is the same as the Frequentist confidence interval. Alternatively, a Bayesian quant could argue that her Frequentist cousin is a confused Bayesian with prior $n_0 = 0, m_0 = -1, \sigma_0^2 = 0$.

For the ex-factor signal-noise ratio, the $(1-\alpha)$ credible interval on ζ_g is

$$\frac{1}{\sqrt{\boldsymbol{v}^\top \Lambda_1 \boldsymbol{v}}}\left[{\lambda'}_{\alpha/2}\left(\sqrt{\boldsymbol{v}^\top \Lambda_1 \boldsymbol{v}}\zeta_{g,1},m_1\right),{\lambda'}_{1-\alpha/2}\left(\sqrt{\boldsymbol{v}^\top \Lambda_1 \boldsymbol{v}}\zeta_{g,1},m_1\right)\right). \tag{2.65}$$

Example 2.7.4 (Noninformative prior, Market returns). Consider the monthly relative returns of the Market, introduced in Example 1.1.1. Starting from a non-informative prior ($n_0 = m_0 = 0\,\text{mo.}$), one computes $n = 1,128\,\text{mo.}$, $\hat{\zeta} = 0.177\,\text{mo.}^{-1/2}$ and $\hat{\sigma} = 5.354\%\,\text{mo.}^{-1/2}$. The posterior is then $\zeta_1 = 0.177\,\text{mo.}^{-1/2}$, $n_1 = 1,128\,\text{mo.}$, $\sigma_1 = 5.352\%\,\text{mo.}^{-1/2}$, $m_1 = 1,128\,\text{mo.}$ From the posterior, a 95% credible interval on ζ is $[0.41, 0.818)\ \text{yr}^{-1/2}$. The Frequentist 95% confidence interval is $[0.41, 0.817)\ \text{yr}^{-1/2}$. The two agree to two decimal places. ⊣

Example 2.7.5 (Noninformative prior, UMD returns, attribution model). Consider the monthly relative returns of UMD, under an attribution against intercept, Market, SMB and HML, using the data introduced in Example 1.1.1. The posterior belief on ζ_g, marginalizing out σ, can be expressed as

$$33.028\,\text{mo.}^{1/2}\%^{-1}\zeta_g \propto \lambda'\left(8.149, 1128\right).$$

From the posterior, a 95% credible interval on ζ_g is $[0.186, 0.307)\ \text{mo.}^{-1/2}$. Compare this to the frequentist confidence interval, found in Example 2.5.2 to be $[0.186, 0.306)\ \text{mo.}^{-1/2}$. ⊣

2.7.3 Posterior prediction intervals on the Sharpe ratio

A Bayesian prediction interval is an interval which contains some fixed proportion of our posterior *belief* about the Sharpe ratio of some future observations. This is very similar to the Frequentist prediction interval introduced in Section 2.5.9, but dances around the issues of frequency and belief that separate the Frequentist and Bayesian.

As in the Frequentist case, our belief is that $\hat{\zeta}_2$, based on n_2 future observations, will be drawn from a compound non-central Sharpe ratio distribution with non-centrality parameter drawn from the posterior distribution.

Effectively this is a "t of lambda prime" distribution, as was the case in the Frequentist setting. We can summarize this as

$$\begin{aligned}\sqrt{n_1}\zeta &\propto \lambda'\left(\sqrt{n_1}\zeta_1, m_1\right),\\ \sqrt{n_2}\hat{\zeta}_2 \,|\zeta &\propto t\left(\sqrt{n_2}\zeta, n_2 - 1\right),\end{aligned} \tag{2.66}$$

although this jumbles up the usual notation, since $\hat{\zeta}_2$ is a quantity one can eventually observe, not a population parameter. Nevertheless, the intent of these equations should be clear. The quantity $\hat{\zeta}_2$ does not take a well known distributional form, but the prediction interval may be easily found via direct Monte Carlo simulations: first draw a value of ζ from the posterior, then simulate a value of $\hat{\zeta}_2$ from a Sharpe ratio distribution with that signal-noise ratio. That is, simulate ζ, then $\hat{\zeta}_2$ using the generative forms from Equation 2.66.

Example 2.7.6 (Predictive intervals, Market returns). Consider the monthly relative returns of the Market, introduced in Example 1.1.1. Based on a prior of $\zeta_0 = 0.125\,\text{mo.}^{-1/2}$, $n_0 = 12\,\text{mo.}$, $\sigma_0 = 4\%\,\text{mo.}^{-1/2}$, $m_0 = 72\,\text{mo.}$, one observes $n = 1,128$ months of returns with $\hat{\zeta} = 0.177\,\text{mo.}^{-1/2}$ and $\hat{\sigma} = 5.354\%\,\text{mo.}^{-1/2}$. The posterior is then $\zeta_1 = 0.179\,\text{mo.}^{-1/2}$, $n_1 = 1,140\,\text{mo.}$, $\sigma_1 = 5.281\%\,\text{mo.}^{-1/2}$, $m_1 = 1,200\,\text{mo.}$

For $n_2 = 12\,\text{mo.}$, the 95% prediction interval on $\hat{\zeta}_2$ was computed, by 10^6 Monte Carlo simulations, to be approximately $[-0.43, 0.87)\,\text{mo.}^{-1/2}$. Compare this to the Frequentist prediction interval found in Example 2.5.13. ⊣

2.8 Exercises

Ex. 2.1 **Units** What are the units of the t-statistic?

Ex. 2.2 ***t* form for ex-factor Sharpe ratio** Derive the equivalents of Equation 2.5 for the ex-factor Sharpe ratio.

Ex. 2.3 **Form of the *t* statistic** Suppose that $Z \sim \mathcal{N}\left(\mu_0, \sigma^2/\nu\right)$ independently of $(\nu - 1) X/\sigma^2 \sim \chi^2(\nu - 1)$. Show that

$$\sqrt{\nu}\frac{Z - \mu_1}{\sqrt{X}} \sim t\left(\sqrt{\nu}\frac{\mu_0 - \mu_1}{\sigma}, \nu - 1\right).$$

Ex. 2.4 **Regarding c_4** The statistical quality control literature defines

$$c_4(n) := \sqrt{\frac{2}{n-1}}\frac{\Gamma\left(\frac{n}{2}\right)}{\Gamma\left(\frac{n-1}{2}\right)}.$$

Prove that $d_n = \frac{n-1}{n-2}c_4(n)$, for d_n defined as in Equation 2.12.

Ex. 2.5 "Robust" tests The t test tests the null hypothesis of zero *mean* return. In social sciences it is often argued that the *median* is a more representative measure of some effect. Why would it be a bad idea to rely on tests of median return?

Ex. 2.6 Probability of a loss Suppose you have n observations of *i.i.d.* Gaussian log returns $x \sim \mathcal{N}(\mu, \sigma)$.

1. Show that the probability of a loss (*i.e.*, $\sum_t x_t \leq 0$) is $\Phi\left(-\sqrt{n}\zeta\right)$, where $\Phi(x)$ is the probability distribution function of the normal distribution.

2. Show that, as a consequence, $f_{SR}(0; \zeta, n) = \Phi\left(-\sqrt{n}\zeta\right)$.
* 3. Is the function $f_{SR}(x; \zeta, n) - \Phi\left(x - \sqrt{n}\zeta\right)$ monotonic in x?
4. Plot the probability of a loss over one year ($n = 252$ days) when ζ ranges from $0.5\,\text{yr}^{-1/2}$ to $2.5\,\text{yr}^{-1/2}$, in logscale. Estimate a power law for the probability of a loss.

Ex. 2.7 Boring Frequentist computations Suppose you observe n daily observations of *i.i.d.* Gaussian log returns $x \sim \mathcal{N}(\mu, \sigma)$.

1. Suppose that $\zeta = 0.1\,\text{day}^{-1/2}$. Compute the probability of observing $\hat{\zeta} \geq 0.02\,\text{day}^{-1/2}$ given $n = 500$.
2. Suppose that $\zeta = 0.1\,\text{day}^{-1/2}$. Compute the expected value of $\hat{\zeta}$ given $n = 500$.
3. Supposing that $\hat{\zeta} = 0.02\,\text{day}^{-1/2}$ based on $n = 500$ days of observations. Compute 95% confidence interval on ζ based on Equation 2.25. Compute them again based on Equation 2.27.

Ex. 2.8 More boring Frequentist computations

1. Compute the ex-factor Sharpe ratio of the Consumer industry returns using an attribution against the Fama French Market, SMB, HML, and UMD returns. Test the hypothesis that the ex-factor signal-noise ratio is zero against the alternative that it is larger than zero.
2. Repeat that test using the Manufacturing industry returns.

Ex. 2.9 Approximate test for equal signal-noise ratio Derive the normal approximation to the hypothesis test of equality of signal-noise ratios for two independent samples.

1. Apply your procedure to test the null hypothesis that the returns of the Market have the same signal-noise ratio before and since January 1970.

Ex. 2.10 Lambda prime Consider the lambda prime distribution defined in Section 2.4.

1. Show that the probability distribution of the non-central t distribution is the survival function of the lambda prime, and vice versa, *i.e.*, Equation 2.20.
2. Show that the lambda prime is the *confidence distribution* associated with the non-central t distribution. That is, show that if $t \sim t(\delta, \nu)$, then $F_{\lambda'}(\delta; t, \nu) \sim \mathcal{U}(0, 1)$. [Singh *et al.*, 2007; Xie and Singh, 2013]
3. Derive the probability density function of the lambda prime distribution. (*Hint:* it is a convolution of a Nakagami distribution density with a normal density.)
4. How does the PDF of the lambda prime relate to the PDF of the non-central t?
5. Suppose $t \sim t(\delta, \nu)$. Is it the case that the MLE of δ conditional on observed t is equal to the *mode* of the lambda prime distribution with parameter t and ν degrees of freedom?

Ex. 2.11 CDF of Sharpe ratio Write code to compute the cumulative distribution function of the Sharpe ratio. It should take observed Sharpe ratio, $\hat{\zeta}$, the signal-noise ratio, ζ, and sample size n, and return the CDF. A simple test of your code is to randomly generate $\hat{\zeta}$ values for a fixed value of ζ and n, apply your CDF function to those $\hat{\zeta}$ values, and Q-Q plot them against the uniform.

Ex. 2.12 CDF of the lambda prime Write code to compute the CDF of the lambda prime distribution. Test it via randomly generating variates.

Ex. 2.13 Moments of Sharpe ratio Given 1,000 daily observations of *i.i.d.* normal log returns with $\zeta = 0.06\,\text{day}^{-1/2}$, compute the first four non-central moments of $\hat{\zeta}$ computed in daily units. Also compute the variance, skewness, and excess kurtosis of $\hat{\zeta}$.

Ex. 2.14 Stabilizing variance of the Sharpe ratio Find the *variance stabilizing transform* for the Sharpe ratio.

1. Express the variance of $\hat{\zeta}$ as $a^2 + b^2 \left(\mathrm{E}\left[\hat{\zeta}\right]\right)^2$, with $a^2 = \frac{n-1}{n(n-3)}$.
2. Following Laubscher [1960], the variance stabilizing transform for $\hat{\zeta}$ is

$$f\left(\hat{\zeta}\right) = \frac{1}{b} \sinh^{-1}\left(\frac{b}{a}\hat{\zeta}\right).$$

   Confirm by simulation that $f\left(\hat{\zeta}\right) - f\left(\mathrm{E}\left[\hat{\zeta}\right]\right)$ is approximately zero mean and unit variance. Is it normal?
3. Consider the hypothesis test

$$H_0 : \zeta_1 = \zeta_2 \quad \text{versus} \quad H_1 : \zeta_1 > \zeta_2,$$

given independent $\hat{\zeta}_1$ and $\hat{\zeta}_2$, on identical sized samples. How might you approximately test this hypothesis using the variance stabilizing transform?

Ex. 2.15 **Up-sampling** Suppose you are given 260 weekly observations of *i.i.d.* normal log returns with $\zeta = 0.10\,\text{wk.}^{-1/2}$.

1. Compute the first four non-central moments of $\hat{\zeta}$, computed in weekly units.
2. Compute the variance, skewness, and excess kurtosis of $\hat{\zeta}$, computed in weekly units where appropriate.
3. Suppose you upsample to daily units, thus $k = 5$. Compute the percent changes in mean, variance, skewness, and excess kurtosis of $\sqrt{k}\hat{\zeta}_k$ versus $\hat{\zeta}$.

Ex. 2.16 **Up-sampling FF3** Consider the monthly and daily returns of the Fama-French factors, available in **tsrsa**. [Pav, 2019a] Compute the Sharpe ratio for the three portfolios, Market, SMB, and HML, for the monthly and the daily returns.

Ex. 2.17 **HFT: not about sample size** It is often clumsily argued that high frequency trading is to be preferred to daily frequency trading "because the sample size is larger." Critique this argument.

Ex. 2.18 **Standard error of the Sharpe ratio** Derive the approximation of Equation 2.23.

Ex. 2.19 **Simulating hypothesis tests under the null** Via Monte Carlo simulation verify the frequentist tests of Section 2.5.3 by drawing samples under the null and computing p-values. Perform 10^5 simulations for each of the following:

1. Confirm the one sample test for signal-noise ratio. Draw $n = 1,024$ days of daily returns from a normal random generator with $\zeta = 0.1\,\text{day}^{-1/2}$. Confirm that $\sqrt{n}\hat{\zeta} \sim t\left(\sqrt{n}\zeta, n-1\right)$. One easy way to perform this task is to compute p-values and confirm they are uniformly distributed by Q-Q plotting them against a uniform law, or by plotting their empirical CDF.
2. Confirm the independent two sample test for equality of signal-noise ratio. For $i = 1, 2$, draw $n_i = 512i$ days daily returns from two independent normal asset streams, both with $\zeta = 0.075\,\text{day}^{-1/2}$. Let the population mean of the ith population be $\mu_i = i \times 10^{-4}\,\text{day}^{-1}$.
3. Confirm the independent k sample test for an equation on signal-noise ratio. For $i = 1, \ldots, 7$, draw $n_i = 512i$ days daily returns from seven independent normal asset streams, with the signal-noise ratio of the ith population $\zeta_i = (0.01 + 0.01i)\,\text{day}^{-1/2}$. Let the population mean of

the ith population be $\mu_i = i \times 10^{-4}\,\text{day}^{-1}$. The null hypothesis being tested is that $\sum_i \zeta_i = 0.35\,\text{day}^{-1/2}$.

Ex. 2.20 **Simulating under the alternative** Via Monte Carlo simulation, check that the frequentist tests of Section 2.5.3 can reject the null, by drawing samples under an alternative hypothesis, computing p-values, and confirming that the p-values are not uniform. (*cf.* Exercise 2.19.) Perform 10^5 simulations for each of the following:

1. Check the one sample test for signal-noise ratio. Draw $n = 1,024$ days of daily returns from a normal random generator with $\zeta = 0.25\,\text{day}^{-1/2}$. Check the hypothesis $H_0 : \zeta = 0$ against the alternative $H_1 : \zeta > 0$. Again, compute p-values and Q-Q plot them against a uniform law.
2. Check the independent two sample test for equality of signal-noise ratio. For $i = 1, 2$, simulate $n_i = 512\,i$ days independent normal daily returns, with $\zeta_i = 0.05\,i\,\text{day}^{-1/2}$. Let the population mean of the ith population be $\mu_i = \sqrt{i} \times 10^{-4}\,\text{day}^{-1}$. Check the hypothesis $H_0 : \zeta_1 = \zeta_2$ against the alternative $H_1 : \zeta_1 < \zeta_2$.
3. Check the independent k sample test for an equation on signal-noise ratio. For $i = 1, \ldots, 7$, draw $n_i = 512\,i$ days daily returns from seven independent normal asset streams, with $\zeta_i = (0.01 + 0.01\,i)\,\text{day}^{-1/2}$. Let the population mean of the ith population be $\mu_i = i \times 10^{-4}\,\text{day}^{-1}$. Check the hypothesis $H_0 : \sum_i \zeta_i = 0\,\text{day}^{-1/2}$ against the alternative $H_1 : \sum_i \zeta_i > 0\,\text{day}^{-1/2}$.

Ex. 2.21 **Rescaling parameter in linear attribution model** The scaling factor $c = \sqrt{\boldsymbol{v}^\top \left(\mathsf{F}^\top \mathsf{F}\right)^{-1} \boldsymbol{v}}$ appears often in testing of the ex-factor Sharpe ratio, in Section 2.5.6. The factor c appears in the denominator, while in the analogous results for the Sharpe ratio, we often instead see $\sqrt{n}$ in the numerator. We wish to show that c is "morally equivalent" to $n^{-1/2}$.

1. For the case where the linear attribution model expresses Sharpe's model, we take F to be the $n \times 1$ matrix of all ones, and $\boldsymbol{v}$ is a scalar one. In this case show that $c = n^{-1/2}$.
2. Although in this chapter we usually consider the matrix F to be deterministic, assume here that the rows of F are independent draws of a zero-mean l-dimensional Gaussian with identity covariance. Let $\boldsymbol{v}$ be the vector consisting of a single one, and $l - 1$ zeros. Then $\boldsymbol{v}^\top \left(\mathsf{F}^\top \mathsf{F}\right)^{-1} \boldsymbol{v}$ follows an inverse Gamma distribution with shape parameter $\alpha = (n - l + 1)/2$ and scale parameter $\beta = 1/2$, and so $\operatorname{E}\left[1/r^2\right] = n - l + 1$. Confirm this relationship via simulation.

Ex. 2.22 **No TOST for you** Consider the two one-sided test of Section 2.5.4.

1. Is it possible that the critical region is empty? What implications does this have?
2. Under what conditions might the critical region be empty?
3. For the equivalence test and type I rate considered in Example 2.5.5, find sample size n such that the critical region is empty.

Ex. 2.23 TOST the market Use the TOST approach of Section 2.5.4 to test the hypothesis

$$\left|\zeta - 0.5\,\mathrm{yr}^{-1/2}\right| > 0.1\,\mathrm{yr}^{-1/2}$$

on the Market returns.

Ex. 2.24 Intersection union tests Construct some intersection union tests as described in Section 2.5.4.

1. For $i = 1, 2$, given n_i *i.i.d.* draws from Gaussian returns from two assets with signal-noise ratios ζ_i, consider the test for

$$H_0 : |\zeta_2 - m\zeta_1 - b| \geq \epsilon \quad \text{versus} \quad H_1 : |\zeta_2 - m\zeta_1 - b| < \epsilon,$$

for given m, b. The hope is that one will reject "towards" the alternative which puts the signal-noise ratios within a band of the line $y = mx + b$. Under what conditions will you reject the null?

2. For the case of ex-factor signal-noise ratio, consider the test for

$$H_0 : \left|\beta^\top \boldsymbol{v} - \sigma c\right| \geq \epsilon \quad \text{versus} \quad H_1 : \left|\beta^\top \boldsymbol{v} - \sigma c\right| < \epsilon,$$

for given $\boldsymbol{v}$ and c. Under what conditions will one reject the null hypothesis?

Ex. 2.25 One-sided violations of symmetric CI Prove that when $\zeta \leq \delta$,

$$\Pr\left\{\operatorname{sign}\left(\hat{\zeta}\right)\zeta \geq \left|\hat{\zeta}\right| + \delta\right\} = 0.$$

Ex. 2.26 Narrowest CI on signal-noise ratio Just as the confidence interval given in Equation 2.27 may not be symmetric about $\hat{\zeta}$, it might not be the narrowest confidence interval for ζ based on the exact method. By "width," we mean $\zeta_u - \zeta_l$. This question is hard to study theoretically, but easy to consider empirically. Assume $\alpha = 0.05$. Take x from 0.001 to 0.0499. Compute ζ_l and ζ_u via

$$0.95 + x = F_t\left(\sqrt{n}\hat{\zeta}; \sqrt{n}\zeta_l, n - 1\right), \quad x = F_t\left(\sqrt{n}\hat{\zeta}; \sqrt{n}\zeta_u, n - 1\right),$$

then plot $\zeta_u - \zeta_l$ versus x. Assume you have observed $\hat{\zeta} = 0.75\,\mathrm{yr}^{-1/2}$ based on 4 years of daily data at a rate of 252 days per year. Where do you observe the narrowest confidence interval?

Ex. 2.27 **Power I** Assume log returns follow an *i.i.d.* Gaussian distribution. Fix the type I rate, α at 0.01, and set the "combined effect size," $\sqrt{n}\zeta_e = 2.0$. Compute the power of the hypothesis test for $H_0 : \zeta = 0$ against the one sided alternative, $H_0 : \zeta > 0$ via Equation 2.36, for various values of n. Does the power vary much?

Ex. 2.28 **Boring power computations** Suppose you observe n daily observations of *i.i.d.* Gaussian log returns $x \sim \mathcal{N}(\mu, \sigma)$, assuming 252 days per year. Consider the power of the hypothesis test for $H_0 : \zeta = \zeta_0$ against the one sided alternative, $H_0 : \zeta > \zeta_0$.

1. For $n = 756$, and taking $\alpha = 0.01$, compute the power for testing $\zeta_0 = 1.0\,\mathrm{yr}^{-1/2}$ assuming $\zeta = 1.5\,\mathrm{yr}^{-1/2}$.
2. For $n = 756$, and taking $\alpha = 0.01$, how large must ζ be to achieve a power of 90% when testing for $\zeta_0 = 0.75\,\mathrm{yr}^{-1/2}$?
3. Suppose you wish to test at a type I rate of $\alpha = 0.001$, and achieve a power of 50%, when testing for $\zeta_0 = 0.5\,\mathrm{yr}^{-1/2}$. Assuming $\zeta = 1.0\,\mathrm{yr}^{-1/2}$, how large must n be?

* **Ex. 2.29** **Toy hedge fund** Suppose you are starting a hedge fund. After one year, your investors will pull all their money if your fund has not achieved a Sharpe ratio of $1.0\,\mathrm{yr}^{-1/2}$ over that year, which we will call "failure." There are two possible causes of failure: bad luck of an actually good strategy, and the "normal" poor performance from trading a strategy which had been selected due to a type I error.

1. What signal-noise ratio do you need to bound the probability of failure due to bad luck to be no greater than ϵ? What value do you get for $\epsilon = 0.10$? Let this number be ζ_0.
2. Suppose that you randomly select a trading strategy, backtest it over a 5 year period, then perform a hypothesis test for $H_0 : \zeta = 0$ against $H_1 : \zeta > 0$. If the strategy "passes" the test, you trade it for a year, otherwise you randomly select another until you find one which does. What is the cutoff for the hypothesis test to achieve a type I rate of $\alpha = 0.05$?
3. To estimate the total probability of failure, you must compute the false discovery rate of the hypothesis test. This requires knowledge of the distribution of signal-noise ratio in the population from which you randomly sample. For simplicity, assume the population is bimodal: with probability 10^{-6}, you select a strategy with $\zeta = \zeta_0$, otherwise you select a strategy with $\zeta = 0\,\mathrm{yr}^{-1/2}$. Assume that the strategies' returns are independent in the backtest period (this is very unrealistic). What is the false discovery rate of your process? (Here, the false discovery rate is the probability that a strategy which passes the hypothesis test is a type I error, rather than a truly good strategy.)
4. What is the total probability of failure? Note that this should include

the possibility that a strategy which is a type I error has actually performed well when traded live.

5. Suppose your fund does not fail, rather your achieved Sharpe ratio passes the hurdle rate. What is the probability that this was due to a type I error strategy?

* **Ex. 2.30** **Toy hedge fund II** Suppose you are starting a hedge fund. You have at your disposal a random stream of strategies which have independent returns.

1. Suppose your sampling process draws strategies with signal-noise ratio $\zeta \sim \mathcal{N}\left(-0.25\,\mathrm{yr}^{-1/2}, 0.5\,\mathrm{yr}^{-1/2}\right)$. You draw a strategy, observe 4 years of backtest, perform a hypothesis test for $H_0 : \zeta = 1.0\,\mathrm{yr}^{-1/2}$ against the alternative $H_1 : \zeta > 1.0\,\mathrm{yr}^{-1/2}$. Those strategies for which you reject the null, with $\alpha = 0.05$, "pass" to live trading. The distribution of signal-noise ratio of strategies which pass does not have a closed form. Via simulation, estimate the mean and standard deviation of signal-noise ratio of this distribution. Is it nearly normal? What is the approximate probability that a randomly drawn strategy will pass the significance test filter? Why is the mean signal-noise ratio so much lower than the hypothesis test cutoff?
2. To improve the quality of passed strategies, one is tempted to use a more stringent test. Perform your simulations again, but testing for $H_0 : \zeta = 1.5\,\mathrm{yr}^{-1/2}$ against the alternative $H_1 : \zeta > 1.5\,\mathrm{yr}^{-1/2}$. Does the mean of passed strategies increase? How is the probability of passing the filter affected?

* 3. Can you imagine a case where increasing the cutoff of the hypothesis test (either via requiring a smaller type I rate, or via testing for a more stringent null hypothesis) could lead to a *decrease* in the expected signal-noise ratio of strategies which pass the test?

* **Ex. 2.31** **Distribution of optimum Sharpe ratio**
Let $p_1, p_2, \ldots, p_k$ be independent random variables each uniform on $[0, 1)$. Consider the jth *order statistic* of the p_i, which is the jth largest of the p_i, call it $p_{(j)}$. It takes a beta distribution:

$$p_{(j)} \sim \mathcal{B}\left(j, k + 1 - j\right).$$

(See also Section 4.1.1.)

1. Suppose you observe k series of returns, each of length n, and each independent of each other. Suppose all returns are normally distributed. Furthermore suppose that the signal-noise ratios are all zero. Compute their Sharpe ratios and consider their order statistics: $\hat{\zeta}_{(1)} \leq \hat{\zeta}_{(2)} \leq \ldots \leq \hat{\zeta}_{(k)}$. What is the distribution of $\hat{\zeta}_{(j)}$?
2. Write code to compute the CDF of $\hat{\zeta}_{(j)}$ under these assumptions.

3. Perform a simulation: generate 1,000 days of independent normal returns with zero mean, and compute the Sharpe ratio; repeat this 100 times and record the largest Sharpe ratio observed; repeat that 2,500 times and feed these 2,500 maximal Sharpe ratio values into your CDF code. The resultant values should be uniform. Q-Q plot them against a uniform law.
4. Write code to compute the quantile function of the jth largest of k independent realizations of the Sharpe ratio for normal returns and a common (possibly nonzero) signal-noise ratio.
5. Suppose you have an army of 100 quants, each of whom generates 10 strategies, all of which are independent and have normally distributed returns. You will pick the strategy which demonstrates the highest Sharpe ratio based on a three year backtest of daily returns, with 252 trading days per year. Your goal for live trading is a signal-noise ratio of $0.9\,\mathrm{yr}^{-1/2}$. How large should the largest Sharpe ratio be to trade it assuming a type I rate of 0.10?
6. Consider the test of the hypothesis on the maximal signal-noise ratio of a set of strategies with independent returns. That is, consider testing

$$H_0 : \bigvee_{1 \le i \le k} \zeta_i = 0$$

against the alternative

$$H_1 : \bigvee_{1 \le i \le k} \zeta_i > 0.$$

This can be tested by computing the CDF of $\zeta_{(k)}$, having observed independent series of normal returns each of size n, and rejecting the null if the CDF exceeds $1 - \alpha$. Does this achieve the nominal type I rate of α? What is the power of this test?

Ex. 2.32 **Freakout intervals** Losses during the early days of a quantitative strategy's live trading often leads portfolio managers and quants to freak out. Thus one can think of prediction intervals as "freakout intervals." Suppose you observe Sharpe ratio of $\hat{\zeta}_1$ based on $n = 1,250$ days of returns of a strategy. Assume there are 252 trading days per year.

1. Compute approximate 98% prediction interval for the Sharpe ratio of a further 21 days of returns if $\hat{\zeta}_1 = 1\,\mathrm{yr}^{-1/2}$.
2. Write a function that computes approximate lower 1% prediction bounds for the Sharpe ratio for 21 days of future returns, and plot this for values of $\hat{\zeta}_1$ ranging from 0 to $2\,\mathrm{yr}^{-1/2}$
3. You observed $\hat{\zeta}_1 = 1.5\,\mathrm{yr}^{-1/2}$ based on $n = 1,250$ days of returns in backtests. In the first 15 days of trading live, the strategy has achieved a Sharpe ratio of $\hat{\zeta}_2 = -2.0\,\mathrm{yr}^{-1/2}$. Do you freak out and pull the plug?

* 4. Suppose you observe the past returns of 200 strategies with independent returns over a period of $n = 1,250$ days, and pick the one with the maximal Sharpe ratio over that period. It achieved a Sharpe ratio of $\hat{\zeta}_1 = 1.75\,\mathrm{yr}^{-1/2}$. Write a function that computes approximate lower 1% prediction bounds for the Sharpe ratio of this strategy for 21 days of future returns.

Ex. 2.33 Prediction interval on the market Download the *daily* returns of the Market factor, as described in Example 1.1.1. Using the daily data from January 01, 1930 through November 31, 1930, construct a 0.90 prediction interval for the Sharpe ratio for the period December 01, 1930 through December 31, 1930. Compute the actual Sharpe ratio for that period and record whether it was within the 0.90 prediction interval. Repeat this for each year from 1930 through 2020. What proportion of December Sharpes fell within the prediction interval?

* **Ex. 2.34 Prediction interval, attribution model**
Repeat Example 2.5.14, but compute a prediction interval on the ex-factor Sharpe ratio of the UMD factor attributed against the returns of the Market, SMB and HML. Perform computations on daily returns, and find the empirical coverage.

Ex. 2.35 Fisher information Derive Equation 2.17. As a starting point, use the well known Fisher information of the mean and variance for Gaussian random variables [Pawitan, 2001]

$$\mathcal{I}\left(\mu, \sigma^2\right) = n \begin{bmatrix} \frac{1}{\sigma^2} & 0 \\ 0 & \frac{1}{2\sigma^4} \end{bmatrix}. \tag{2.67}$$

Then reparametrize the problem in terms of ζ and σ^2. This is by viewing

$$\begin{bmatrix} \mu \\ \sigma^2 \end{bmatrix} = f\left(\begin{bmatrix} \zeta \\ \sigma^2 \end{bmatrix}\right)$$

Then compute

$$\mathcal{I}\left(\zeta, \sigma^2\right) = \mathsf{J}^{\top} \mathcal{I}\left(f\left(\begin{bmatrix} \zeta \\ \sigma^2 \end{bmatrix}\right)\right) \mathsf{J},$$

where J is the matrix derivative of the function $f(\cdot)$ in the numerator layout convention.

Ex. 2.36 LRT Consider the likelihood ratio test of $H_0 : \zeta = 0.8\,\mathrm{yr}^{-1/2}$ against the unrestricted alternative. Suppose, given 1,000 days of Gaussian log returns, you observe $\hat{\zeta} = 1.4\,\mathrm{yr}^{-1/2}$.

1. Compute the MLE, ζ_{MLE}.
2. Compute the LRT statistic, D.
3. Interpret D as a chi-square random variable with 1 degree of freedom. Would you reject H_0 at the $\alpha = 0.01$ level?

Ex. 2.37 LRT II Consider the likelihood ratio test of $H_0 : \zeta = 0.5\,\mathrm{yr}^{-1/2}$ versus $H_1 : \zeta = 0.8\,\mathrm{yr}^{-1/2}$ for a Sharpe ratio observed on 2,000 days of Gaussian log returns. What is the cutoff Λ_c needed to achieve a type I rate of $\alpha = 0.01$? What is the smallest Sharpe ratio for which this test rejects H_0?

Ex. 2.38 Likelihood intervals A p *likelihood interval* for ζ is defined as a set

$$\left\{ \zeta \left| \frac{\mathcal{L}_{SR}\left(\zeta \left| \hat{\zeta}, n\right.\right)}{\mathcal{L}_{SR}\left(\zeta_{\mathrm{MLE}} \left| \hat{\zeta}, n\right.\right)} \geq p \right. \right\}$$

It is said that a 14.7% likelihood interval will be equivalent to the 95% confidence interval in certain cases. Continue Example 2.7.4 by finding the 14.7% likelihood interval for the signal-noise ratio, given $\hat{\zeta} = 0.177\,\mathrm{mo.}^{-1/2}$ for $n = 1,128$ months of observations.

Ex. 2.39 Boring Bayesian computations Assume a prior of $\zeta_0 = 0.2\,\mathrm{mo.}^{-1/2}$, $n_0 = 24\,\mathrm{mo.}$, $\sigma_0 = 1.5\%\,\mathrm{mo.}^{-1/2}$, $m_0 = 72\,\mathrm{mo.}$

1. Suppose you observe $n = 60\,\mathrm{mo.}$ of returns with $\hat{\zeta} = 0.31\,\mathrm{mo.}^{-1/2}$ and $\hat{\sigma} = 2\%\,\mathrm{mo.}^{-1/2}$. Compute the posterior hyperparameters.
2. Construct a 95% credible interval on ζ based on that posterior.
3. Compute an approximate 99% posterior prediction interval on $\hat{\zeta}_2$ drawn from a future $n_2 = 12\,\mathrm{mo.}$ of returns.

Ex. 2.40 Jeffreys' prior One variant of the Jeffreys' prior is [Kass and Wasserman, 1996]

$$f\left(\left[\begin{array}{c}\zeta\\ \sigma^2\end{array}\right]\right) \propto \sqrt{\left|\mathcal{I}\left(\zeta, \sigma^2\right)\right|}.$$

Derive the Jeffrey's prior from Equation 2.17.

Ex. 2.41 Approximate conjugate prior Using the normal approximation for Sharpe ratio given in Equation 2.24, construct a approximate conjugate prior for signal-noise ratio. The prior and posterior densities should be normal. That is, the prior should take the form

$$\zeta \propto \mathcal{N}\left(\zeta_0, \gamma_0^2\right), \tag{2.68}$$

where ζ_0 and γ_0^2 are the prior hyperparameters.

1. Assuming normal likelihood for ζ given an observed $\hat{\zeta}$ on n observations, what is the posterior belief?
2. What settings of the prior hyperparameters correspond to a non-informative prior?

3. Continue Example 2.7.4 by finding the 95% credible interval on the Market returns under this approximate posterior, using a non-informative prior.
4. Continue Example 2.7.6 by finding the 95% posterior prediction interval for $\hat{\zeta}_2$ over $n_2 = 12\,\text{mo.}$, assuming a non-informative prior. (An "informative" prior was used in that example, so one does not expect the results to match exactly, unless the data overwhelm the prior.)

Ex. 2.42 Bayesian inference, sum of signal-noise ratios
In Section 2.5.3, a Frequentist test is quoted for the null hypothesis

$$H_0 : \sum_i a_i \zeta_i = b \quad \text{versus} \quad H_1 : \sum_i a_i \zeta_i > b,$$

given n_i independent draws from Gaussian returns from k assets with signal-noise ratios ζ_i, and fixed $a_1, a_2, \ldots, a_k, b$. Construct a conjugate Bayesian prior and posterior for the sum $\sum_i a_i \zeta_i$.

Ex. 2.43 Bayesian sum of ex-factor signal-noise ratios
In Section 2.5.5, a Frequentist test is quoted for the null hypothesis

$$H_0 : \sum_i \frac{{\beta_i}^{\top} \boldsymbol{v_i}}{\sigma_i} = c \quad \text{versus} \quad H_1 : \sum_i \frac{{\beta_i}^{\top} \boldsymbol{v_i}}{\sigma_i} > c,$$

given n_i independent draws from factor models with Gaussian errors on k assets, and fixed $\boldsymbol{v_1}, \boldsymbol{v_2}, \ldots, \boldsymbol{v_k}, c$. Construct a conjugate Bayesian prior and posterior for the sum $\sum_i {\beta_i}^{\top} \boldsymbol{v_i} / \sigma_i$.

Ex. 2.44 Kalman filter on signal-noise ratio
Consider Bayesian Recursive Estimation on ζ, which can be viewed as the *Kalman Filter.*

1. Suppose that $x \sim \lambda'(t, \nu)$ and $z \sim \mathcal{N}(0, \sigma_z^2)$ are independent. Show that
$$\frac{x + z}{\sqrt{1 + \sigma_z^2}} \sim \lambda'\left(\frac{t}{\sqrt{1 + \sigma_z^2}}, \nu\right).$$
2. Suppose that your prediction for a small change in ζ at a given point in time is
$$\zeta \mapsto a\zeta + z,$$
with $z \sim \mathcal{N}(0, \sigma_z^2)$. Given the prior of Equation 2.45, show that the prior after the prediction phase is
$$\begin{aligned} \sigma^2 &\propto \Gamma^{-1}\left(m_0/2, m_0\sigma_0^2/2\right), \\ \mu^- \left|\sigma^2\right. &\propto \mathcal{N}\left(a\mu_0, \sigma^2\left(\sigma_z^2 + a^2/n_0\right)\right). \end{aligned} \tag{2.69}$$

3. Show that the update step has the form

$$\begin{aligned} \sigma^2 &\propto \Gamma^{-1}\left(m_1/2, m_1\sigma_1^2/2\right), \\ \mu\left|\sigma^2\right. &\propto \mathcal{N}\left(\mu_1, \sigma^2/n_1\right), \end{aligned} \tag{2.70}$$

for

$$n_1 = \frac{n_0}{n_0\sigma_z^2 + a^2} + n, \quad \mu_1 = \frac{\frac{n_0}{n_0\sigma_z^2+a^2}a\mu_0 + n\hat{\mu}}{n_1}, \tag{2.71}$$

$$m_1 = m_0 + n, \quad \sigma_1^2 = \frac{m_0\sigma_0^2 + (n-1)\,\hat{\sigma}^2 + \frac{n_0 n}{(n_0\sigma_z^2+a^2)n_1}\left(a\mu_0 - \hat{\mu}\right)^2}{m_1}. \tag{2.72}$$

4. Marginalize out σ^2 to arrive at a lambda prime posterior for ζ.

Ex. 2.45 Approximate Kalman filter on signal-noise ratio
Derive the Kalman filter under the approximate conjugate prior for ζ of Exercise 2.41, assuming the prediction step of

$$\zeta \mapsto a\zeta + b + z,$$

with $z \sim \mathcal{N}\left(0, \sigma_z^2\right)$.

Ex. 2.46 Fama French, Kalman filter Implement the Kalman filter of Exercise 2.44. Apply it to the Market series of the Fama French monthly factor data. Let $a = 1$ and let σ_z for the monthly change in signal-noise ratio be $0.05\,\text{mo.}^{-1/2}$.

1. Run the code with different values of σ_z, and plot the running estimates of ζ. How might you select a good value of σ_z? How would you select a σ_z that responds to market events?
2. Does the ζ in your predict step forecast $\hat{\zeta}$ well?

Ex. 2.47 Market winter/summer Consider the hypothesis that the signal-noise ratio of the Market is higher in the summer than the winter. Define the summer as the returns of the months of May through August, inclusive, and the winter as November through February. Use the monthly returns of the Fama-French factor data from **tsrsa**, using code as given in Example 1.1.1.

1. Compute the Sharpe ratio of the Market for the Summer and Winter, along with a confidence interval.
2. Perform the frequentist hypothesis test for equality of signal-noise ratio of the two periods.

Ex. 2.48 Market takes a weekend Consider the hypothesis that the signal-noise ratio of the Market is higher from Friday close to Monday close than from Wednesday close to Thursday close. Use the daily returns of the Fama-French factor data from **tsrsa**.

1. Compute the Sharpe ratio of the Market for EOD Mondays and EOD Thursdays, along with a confidence interval.
2. Perform the frequentist hypothesis test for equality of signal-noise ratio of the two periods.

* **Ex. 2.49** **Test for bounded non-stationarity** One objection to estimating the signal-noise ratio using long returns histories is that "returns are not stationary." (Indeed, this seems to be the lesson of Example 2.5.14, wherein prediction intervals failed to achieve nominal coverage for returns of the Market.) One response would be to give up and embrace your own mortality and powerlessness.
An optimist, however, might allow for a small amount of non-stationarity in returns. Suppose you measure returns over k years. Let the signal-noise ratio in the ith year be denoted by ζ_i. Suppose for a given $V, \epsilon > 0$, you wanted to test the null hypothesis

$$H_0 : \zeta_k = V, \ |\zeta_{i+1} - \zeta_i| \leq \epsilon, \text{ for } i = 1, 2, \ldots, k - 1.$$

1. Using the normal approximation for Sharpe ratio, how would you test for this null hypothesis?
2. How would you construct confidence interval for ζ_k under the assumption that $|\zeta_{i+1} - \zeta_i| \leq \epsilon$ for $i = 1, 2, \ldots, k - 1$?
3. Taking $\epsilon = 0.05\,\text{yr}^{-1/2}$, construct a 95% confidence interval for the most recent year's signal-noise ratio of the Market.

Ex. 2.50 **Prediction intervals, vol-weighted Market** Perform the analysis of Example 2.5.13, but rescaling Market returns by an inverse volatility, as outlined in Section 1.4.2 and Equation 1.14. Compute the rolling twelve month mean absolute return of the Market, delay it by one month, invert it, then normalize to mean 1.
Based on the entire sample available to you, compute a 95% prediction interval on the Sharpe ratio of inverse volatility weighted returns for a further $n_2 = 12\,\text{mo.}$ of out-of-sample data.

Ex. 2.51 **Sign Sharpe ratio** Suppose that $x \sim \mathcal{N}\left(\mu, \sigma^2\right)$.

1. Show that
$$\operatorname{E}\left[\operatorname{sign}\left(x\right)\right] = 2\Phi\left(\frac{\mu}{\sigma}\right) - 1.$$
2. Via the Maclaurin series expansion of the error function, show that
$$\operatorname{E}\left[\operatorname{sign}\left(x\right)\right] \approx \sqrt{\frac{2}{\pi}}\left(\frac{\mu}{\sigma} - \frac{1}{6}\left(\frac{\mu}{\sigma}\right)^3 + \frac{1}{40}\left(\frac{\mu}{\sigma}\right)^5 - \ldots\right).$$

3. Confirm this experimentally with ζ ranging from 0 to 0.3. For daily returns, would you be comfortable dropping higher order terms in ζ?
4. This relationship suggests the use of a *sign Sharpe ratio*, defined as

$$\check{\zeta} := \sqrt{\frac{\pi}{2}\frac{1}{n}} \sum_{1 \le t \le n} \operatorname{sign}\left(x_t\right). \tag{2.73}$$

   This statistic is nearly unbiased for ζ for Gaussian returns. Show, however, that the standard error of $\check{\zeta}$ is approximately equal to $\sqrt{\frac{\pi}{2n}}$, quite a bit larger than that of $\hat{\zeta}$.
5. Suppose an asset has Gaussian returns with an observed proportion of positive daily returns (or "win rate") of $(50 + e)\,\%$. Show that this would lead to an estimated signal-noise ratio of $e\sqrt{2\pi}\sqrt{252}\,\mathrm{yr}^{-1/2} \approx 0.4e\,\mathrm{yr}^{-1/2}$. Thus a 55% win rate translates to $\check{\zeta} = 2\,\mathrm{yr}^{-1/2}$.
6. Show that $\check{\zeta}$ is nearly unbiased for heteroskedastic Gaussian returns. That is, suppose that $x_t \sim \mathcal{N}\left(\zeta\sigma_t, \sigma_t^2\right)$; show that $\check{\zeta}$ is nearly unbiased for ζ.
7. While $\check{\zeta}$ is a nearly unbiased estimator of ζ for Gaussian returns, it will not be for skewed returns. Show that $\mathrm{E}\left[\check{\zeta}\right] \approx \zeta - \frac{\gamma_1}{6}$, where γ_1 is the skewness of returns. (*Hint:* use the Edgeworth expansion of Equation 1.15.) Check this via simulation.
8. Is $\check{\zeta}$ nearly unbiased for unskewed, leptokurtic returns? Check via simulations.

* 9. ‽ Can you find some other function $f(\cdot)$ such that $\mathrm{E}\left[f(x)\right] = \zeta$?

3

The Sharpe Ratio for Other Returns

> It's such a fine line between
> stupid and clever.
>
> DAVID ST. HUBBINS
> *This is Spinal Tap*

In the previous chapter, we considered the distribution of the Sharpe ratio for the case where returns are 1. independent, 2. identically distributed, 3. drawn from a normal distribution, 4. independently of previously observable state variables. In fact, probably each of these conditions is violated by real returns series. In this chapter, we will consider the effect of these assumptions, and try to correct for them where necessary. For the most part, we will find that the Sharpe ratio is robust to these assumptions, and Mertens' formula for the standard error does a reasonable job of correcting for non-normality of returns.

Here we are tempted to catalog the tests for detecting deviances from assumptions, but for a few issues:

1. For some of these tests, it is almost a foregone conclusion that they will reject the null of *i.i.d.* Gaussian on real returns. For example, normality tests, dozens of which exist each of varying power under different alternatives, often will reject the null for large samples from ostensibly very good Gaussian pseudo-random number generators!
2. Some of these tests provide little indication of the effect size, thus the magnitude of the problem (and whether it matters for inference on the signal-noise ratio) is hard to determine.
3. There is a huge body of literature on each of these topics, far too large for us to summarize.

Instead, we illustrate a few deviances for one data set, the returns of the Market.

Example 3.0.1 (Is the Market *i.i.d.* normal?). Consider the monthly relative returns of the Market, introduced in Example 1.1.1. First, we apply the **R** function, `shapiro.test` to the monthly returns. As expected, the test rejects the null of normal returns, reporting a p-value of 5.219×10^{-24}.

We calculate the autocorrelation of raw monthly returns, and the autocorrelation of the absolute value of monthly returns. These are plotted for up to 36 lags in Figure 3.1. The first autocorrelation of the raw returns is computed

DOI: 10.1201/9781003181057-3

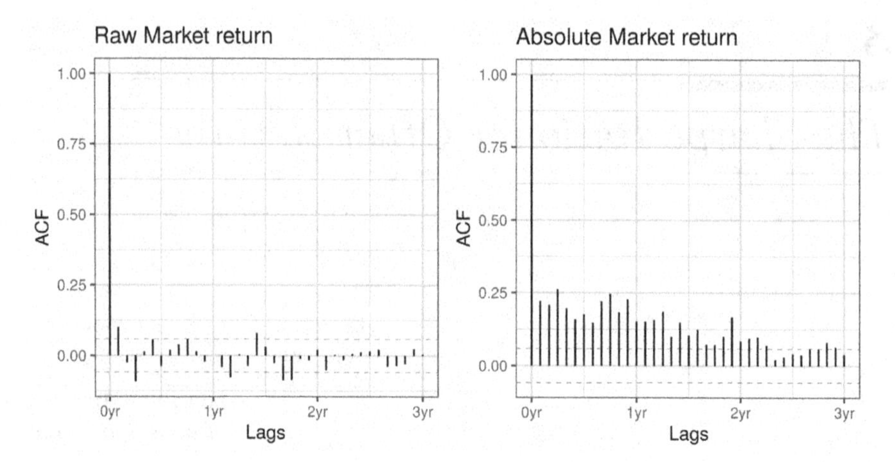

FIGURE 3.1
Autocorrelation plots of the raw and absolute monthly relative returns of the Market over the period from Jan 1927 to Dec 2020 are shown.

to be approximately 0.1, which seems rather high. One suspects that such a high autocorrelation would have been "arb'ed" out by market participants. While "bid-ask bounce" (*cf.* Exercise 1.26) would contribute to an apparent *negative* autocorrelation of returns, it is unlikely to have much of any effect at this time scale. Rather, it turns out that the trade to capture this autocorrelation effect has a very small implied signal-noise ratio. *cf.* Exercise 1.25.

The first autocorrelation of absolute returns is computed to be 0.22. This indicates autoregressive conditional heteroskedasticity ("ARCH"). An interesting question is whether periods of higher volatility are paired with a concomitant increase in mean return, as one theory of heteroskedasticity is that it is caused by a difference in "market time" and "wallclock time." (*cf.* Section 1.4.2.) Under this theory, price discovery occurs faster during periods of higher volatility, thus one should see mean return scale as the *square* of volatility[1]. This does *not* appear to occur for the Market. In Figure 3.2, the mean of daily log returns is plotted against the standard deviation of daily log returns for years from 1926 through 2020. The slope of the regression line is negative, and statistically significantly so at the 0.05 level, but there are many caveats with such a statement: one suspects that significance result is highly sensitive to a few "outliers[2]," that the results might have been different if one had considered quarterly aggregation instead of annual, that one ought to regress against the log of volatility, and so on. Nevertheless, there is hardly

[1]Another way of viewing this is that the signal-noise ratio should be "reannualized" to match market time, and thus should increase as the square root of volatility.

[2]It seems a shame to call them outliers, since they correspond to huge macroeconomic shifts, *e.g.*, in the years 1931, 2008, and so on.

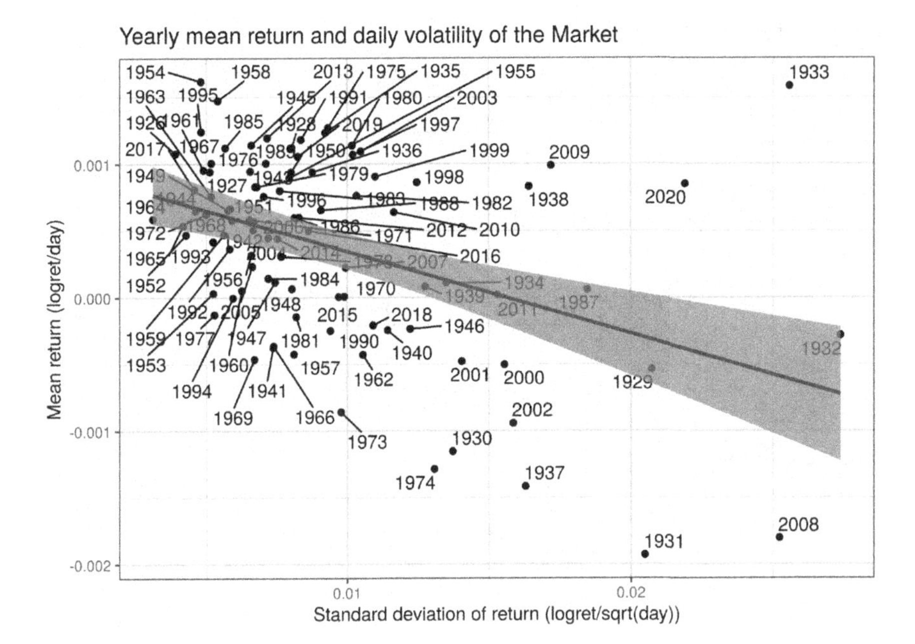

FIGURE 3.2
The mean of daily log returns of the Market are plotted versus the standard deviation of daily log returns, for each year from 1926 through 2020.

much evidence for a *positive* relationship between return and volatility in the Market, and one strongly suspects there is some relation between the two. ⊣

3.1 The Sharpe ratio for (non-*i.i.d.*) elliptical returns

We now consider the distribution of the Sharpe ratio for the case of general elliptical returns, relaxing assumptions of independence and identically distributed. That is, consider the returns $x_1, x_2, \ldots, x_n$ as an n-vector, $\boldsymbol{x}$ drawn from an elliptical distribution with mean $\boldsymbol{\mu}$, covariance Σ, and kurtosis factor κ (*cf.* Section A.2.2.). We are subsuming the Gaussian case, which is simply $\kappa = 1$.

"Independence" corresponds to a diagonal Σ, while "identically distributed" corresponds to $\boldsymbol{\mu}$ equal to some constant times $\mathbf{1}$, and Σ equal to some constant times I. The sample Sharpe ratio has the form

$$\hat{\zeta} = \sqrt{\frac{n-1}{n}} \frac{\frac{1}{n}\mathbf{1}^{\top}\boldsymbol{x} - r_0}{\sqrt{\frac{\boldsymbol{x}^{\top}\boldsymbol{x}}{n} - \left(\frac{1}{n}\mathbf{1}^{\top}\boldsymbol{x}\right)^2}} = \sqrt{\frac{n-1}{n}} f\left(\boldsymbol{v}\right),$$

where $\boldsymbol{v}$ is the 2-vector defined as

$$\boldsymbol{v} := \begin{bmatrix} \frac{1}{n}\mathbf{1}^\top \boldsymbol{x} \\ \frac{\boldsymbol{x}^\top \boldsymbol{x}}{n} - \frac{\left(\mathbf{1}^\top \boldsymbol{x}\right)^2}{n^2} \end{bmatrix},$$

and $f(\boldsymbol{v}) := (v_1 - r_0) / \sqrt{v_2}$.

We can easily find the expected value of $\boldsymbol{v}$:

$$\begin{aligned} \mathrm{E}[\boldsymbol{v}] &= \begin{bmatrix} \frac{1}{n}\mathbf{1}^\top \boldsymbol{\mu} \\ \frac{\mathrm{tr}\left(\Sigma + \boldsymbol{\mu}\boldsymbol{\mu}^\top\right)}{n} - \frac{\mathbf{1}^\top\left(\Sigma + \boldsymbol{\mu}\boldsymbol{\mu}^\top\right)\mathbf{1}}{n^2} \end{bmatrix}, \\ &= \frac{1}{n}\begin{bmatrix} \mathbf{1}^\top \boldsymbol{\mu} \\ \mathrm{tr}\left(\Sigma_c\right) + {\boldsymbol{\mu}_c}^\top \boldsymbol{\mu}_c \end{bmatrix}, \end{aligned} \tag{3.1}$$

where we define

$$\begin{aligned} \boldsymbol{\mu}_c &:= \boldsymbol{\mu} - \frac{\mathbf{1}\mathbf{1}^\top}{n}\boldsymbol{\mu}, \\ \Sigma_c &:= \Sigma - \frac{\mathbf{1}\mathbf{1}^\top}{n}\Sigma. \end{aligned} \tag{3.2}$$

We can view $\boldsymbol{\mu}_c$ and Σ_c as the (column) centered mean and covariance, respectively, where the mean values of each column have been subtracted out. (See Exercise 3.3.) Via Isserlis' theorem (and a lot of computation, *cf.* Exercise 3.4) the covariance matrix of $\boldsymbol{v}$ is [Isserlis, 1918]

$$\mathrm{Var}(\boldsymbol{v}) = \frac{1}{n^2}\begin{bmatrix} \mathbf{1}^\top\Sigma\mathbf{1} & 2{\boldsymbol{\mu}_c}^\top\Sigma\mathbf{1} \\ 2{\boldsymbol{\mu}_c}^\top\Sigma\mathbf{1} & (\kappa - 1)\,\mathrm{tr}\left(\Sigma_c\right)\mathrm{tr}\left(\Sigma_c\right) + 4{\boldsymbol{\mu}_c}^\top\Sigma\boldsymbol{\mu}_c + 2\kappa\,\mathrm{tr}\left({\Sigma_c}^2\right) \end{bmatrix}. \tag{3.3}$$

By Taylor's theorem,

$$\hat{\zeta} \approx \sqrt{\frac{n-1}{n}}\left[f\left(\mathrm{E}[\boldsymbol{v}]\right) + \left(\nabla_{\boldsymbol{v}} f(\boldsymbol{v})\big|_{\mathrm{E}[\boldsymbol{v}]}\right)^\top \left(\boldsymbol{v} - \mathrm{E}[\boldsymbol{v}]\right) + \ldots\right]$$

Taking expectations, we have

$$\begin{aligned} \mathrm{E}\left[\hat{\zeta}\right] &\approx \sqrt{\frac{n-1}{n}} f\left(\mathrm{E}[\boldsymbol{v}]\right), \\ \mathrm{Var}\left(\hat{\zeta}\right) &\approx \frac{n-1}{n}\left(\nabla_{\boldsymbol{v}} f(\boldsymbol{v})\big|_{\mathrm{E}[\boldsymbol{v}]}\right)^\top \mathrm{Var}(\boldsymbol{v})\left(\nabla_{\boldsymbol{v}} f(\boldsymbol{v})\big|_{\mathrm{E}[\boldsymbol{v}]}\right). \end{aligned} \tag{3.4}$$

For now, we leave underdefined exactly in what sense the approximations hold. After all, we have not defined how $\boldsymbol{\mu}$ and Σ might change as $n \to \infty$, rendering a discussion of the asymptotics premature. The approximate expected Sharpe ratio can be expressed compactly as

$$\mathrm{E}\left[\hat{\zeta}\right] \approx \sqrt{\frac{n-1}{n}}\frac{\frac{1}{n}\mathbf{1}^\top\boldsymbol{\mu} - r_0}{\sqrt{\left(\mathrm{tr}\left(\Sigma_c\right) + {\boldsymbol{\mu}_c}^\top\boldsymbol{\mu}_c\right)/n}}. \tag{3.5}$$

For the variance, compute the gradient of $f(\cdot)$ as

$$\nabla_{\boldsymbol{v}} f(\boldsymbol{v}) = \begin{bmatrix} \frac{1}{\sqrt{v_2}} \\ -\frac{f(\boldsymbol{v})}{2v_2} \end{bmatrix}. \tag{3.6}$$

The approximate variance of the Sharpe ratio is thus

$$\begin{aligned}\operatorname{Var}\left(\hat{\zeta}\right) \approx \frac{n-1}{n^2} \frac{\mathbf{1}^\top \Sigma \mathbf{1}}{\left(\operatorname{tr}\left(\Sigma_c\right) + {\boldsymbol{\mu}_c}^\top \boldsymbol{\mu}_c\right)} &\left[1 - 2\left(\frac{\mathbf{1}^\top \boldsymbol{\mu} - n r_0}{\operatorname{tr}\left(\Sigma_c\right) + {\boldsymbol{\mu}_c}^\top \boldsymbol{\mu}_c}\right) \frac{\mathbf{1}^\top \Sigma \boldsymbol{\mu}_c}{\mathbf{1}^\top \Sigma \mathbf{1}}\right. \\ &\left. + \left(\frac{\mathbf{1}^\top \boldsymbol{\mu} - n r_0}{\operatorname{tr}\left(\Sigma_c\right) + {\boldsymbol{\mu}_c}^\top \boldsymbol{\mu}_c}\right)^2 \frac{\frac{\kappa - 1}{4}\left(\operatorname{tr}\left(\Sigma_c\right)\right)^2 + {\boldsymbol{\mu}_c}^\top \Sigma \boldsymbol{\mu}_c + \frac{\kappa}{2} \operatorname{tr}\left({\Sigma_c}^2\right)}{\mathbf{1}^\top \Sigma \mathbf{1}}\right]. \end{aligned} \tag{3.7}$$

To get any meaningful simplification of this expression, we must assume some form for $\boldsymbol{\mu}$ and Σ and approach the problem on a case by case basis. Moreover, the model given here is completely unfalsifiable by data since it posits $n(n+3)/2$ unknown parameters (plus κ) for n observations. We must assume some form for $\boldsymbol{\mu}$ and Σ to keep ourselves honest. Also if $\boldsymbol{\mu}$ and Σ can be completely arbitrary, it is not clear why one would want to perform inference on them, since they would potentially have no relation to future returns.

Keep it normal

Despite our enthusiasm for elliptical distributions, one observes odd behavior from them when used to model returns over time of a single asset, instead of correlated contemporaneous returns of multiple assets. For example, consider a multivariate t-distribution with zero mean and covariance $\Sigma = \sigma^2 \mathsf{I}$. The marginals have zero correlation, but are not independent. If you compute the sample standard deviation over the elements, the standard error does not go to 0 as the number of elements increases. See Exercise 3.6. Perhaps real asset returns display this behavior, but it implies a dependence between returns at arbitrary time scales. So for the most part we will consider Equation 3.7 for the case of multivariate Gaussian only, with $\kappa = 1$.

3.1.1 Independent Gaussian returns

Let us simplify the model somewhat to assume independence of returns, which corresponds to a diagonal Σ matrix, say $\Sigma = \operatorname{Diag}\left(\boldsymbol{\sigma}^2\right)$, Where $\boldsymbol{\sigma}^2$ is a vector of length n whose elements are variances, not volatilities. Under this assumption, some further simplification is possible. Note that $\Sigma \mathbf{1} = \boldsymbol{\sigma}^2$, and thus $\mathbf{1}^\top \Sigma \mathbf{1} = \mathbf{1}^\top \boldsymbol{\sigma}^2$. Then

$$\begin{aligned}\operatorname{tr}\left(\Sigma_c\right) &= \frac{n-1}{n} \mathbf{1}^\top \boldsymbol{\sigma}^2, \\ \operatorname{tr}\left(\Sigma_c^2\right) &= \frac{n-1}{n} {\boldsymbol{\sigma}^2}^\top \boldsymbol{\sigma}^2 + \left(\frac{\mathbf{1}^\top \boldsymbol{\sigma}^2}{n}\right)^2.\end{aligned}$$

Then, asymptotically, Equation 3.5 simplifies to

$$\mathrm{E}\left[\hat{\zeta}\right] \approx \frac{\frac{1}{n}\mathbf{1}^{\top}\boldsymbol{\mu} - r_0}{\sqrt{\left(\mathbf{1}^{\top}\boldsymbol{\sigma}^2 + {\boldsymbol{\mu}_c}^{\top}\boldsymbol{\mu}_c\right)/n}}. \tag{3.8}$$

We can simply define

$$\zeta_i := \frac{\frac{1}{n}\mathbf{1}^{\top}\boldsymbol{\mu} - r_0}{\sqrt{\mathbf{1}^{\top}\boldsymbol{\sigma}^2/n}}, \tag{3.9}$$

as the population quantity we wish to estimate[3], as it is the mean expected return divided by the square root of the mean variance, and thus represents a kind of "long term average" signal-noise ratio. (In fact, however, the variance is inflated by variation in $\boldsymbol{\mu}$, via the ${\boldsymbol{\mu}_c}^{\top}\boldsymbol{\mu}_c$ term.) Define the geometric bias as

$$b := \left(1 + {\boldsymbol{\mu}_c}^{\top}\boldsymbol{\mu}_c/\mathbf{1}^{\top}\boldsymbol{\sigma}^2\right)^{-1/2}. \tag{3.10}$$

We have $0 < b \leq 1$, with equality only when $\boldsymbol{\mu} = \mu\mathbf{1}$ for some μ. We can then express Equation 3.8 compactly as $\mathrm{E}\left[\hat{\zeta}\right] \approx b\zeta_i$.

Equation 3.7 becomes, asymptotically,

$$\mathrm{Var}\left(\hat{\zeta}\right) \approx$$
$$\frac{b^2}{n}\left[1 - 2\zeta_i b^2\sqrt{n}\frac{{\boldsymbol{\sigma}^2}^{\top}\boldsymbol{\mu}_c}{\left(\mathbf{1}^{\top}\boldsymbol{\sigma}^2\right)^{3/2}} + \frac{n}{2}b^4\zeta_i^2\frac{2{\boldsymbol{\mu}_c}^{\top}\left(\boldsymbol{\sigma}^2 \odot \boldsymbol{\mu}_c\right) + {\boldsymbol{\sigma}^2}^{\top}\boldsymbol{\sigma}^2 + \left(\frac{\mathbf{1}^{\top}\boldsymbol{\sigma}^2}{n}\right)^2}{\left(\mathbf{1}^{\top}\boldsymbol{\sigma}^2\right)^2}\right]. \tag{3.11}$$

See Exercise 3.5.

3.1.2 Homoskedastic *i.i.d.* Gaussian returns

First, we consider the *i.i.d.* case where $\boldsymbol{\mu} = \mu\mathbf{1}$ and $\Sigma = \sigma^2\mathsf{I}$, mostly to check our work, since the answer is given to us already by Equation 2.24. In this case, because $\boldsymbol{\mu}$ is constant, we have $\boldsymbol{\mu}_c = \mathbf{0}$, simplifying the asymptotic expansion, and removing the geometric bias, *i.e.*, $b = 1$. Without geometric bias, the approximate expected value of the signal-noise ratio, from Equation 3.8, becomes

$$\mathrm{E}\left[\hat{\zeta}\right] \approx \frac{\mu - r_0}{\sigma} = \zeta. \tag{3.12}$$

Eliminating the bias term and terms involving $\boldsymbol{\mu}_c$ from Equation 3.11, and using $\zeta_i = \zeta$, we have

$$\begin{aligned}\mathrm{Var}\left(\hat{\zeta}\right) &\approx \frac{1}{n}\left[1 + \frac{n}{2}\zeta^2\frac{{\boldsymbol{\sigma}^2}^{\top}\boldsymbol{\sigma}^2 + \left(\frac{\mathbf{1}^{\top}\boldsymbol{\sigma}^2}{n}\right)^2}{\left(\mathbf{1}^{\top}\boldsymbol{\sigma}^2\right)^2}\right],\\ &= \frac{1}{n}\left[1 + \frac{n}{2}\zeta^2\frac{n\sigma^4 + \sigma^4}{n^2\sigma^4}\right].\end{aligned} \tag{3.13}$$

[3]Though it is arguable that this is really the quantity of interest. In some cases, it is clearly of interest, in others the case is not as clear.

Asymptotically in n, this is the approximate standard error of Johnson and Welch, as given in Equation 2.24. [Lo, 2002; Johnson and Welch, 1940].

3.1.3 Heteroskedastic independent Gaussian returns

Now relax the homoskedastic independent case considered previously by assuming that the observed x are homoskedastic *i.i.d.* Gaussian returns "polluted" by some variable l. Such a returns stream could arise in a number of ways:

1. As introduced in Section 1.4.2, one model of market heteroskedasticity is that price discovery unfolds along a "market clock" instead of wall clock time. In this model, if the underlying returns stream has constant signal-noise ratio when measured in market time, the observed returns will have mean and volatility equally scaled by l, the "speed" of price discovery. (Though this is not often supported by data, *cf.* Example 1.4.5 and Example 3.0.1.)
   Under this model the expected value and volatility vary from their long term mean values in lockstep. It must be noted that this is a somewhat optimistic model of market heteroskedasticity; the pessimistic alternative is that expected returns are unaffected (or even adversely impacted) by the variable l, which only scales variance.
2. Similarly, if returns of the asset were homoskedastic *i.i.d.*, but were measured over irregular time points, say weekly returns jumbled together with daily returns, similar population dynamics would arise. In this case, the expected value and *variance* rescale in unison.

The optimistic and pessimistic model mentioned above can be couched in a more general model where the variance depends linearly on l and the expected return is scaled by l^{λ} for some constant λ. The optimistic model is captured by $\lambda = \frac{1}{2}$, while $\lambda = 0$ corresponds to the pessimistic model. If underlying returns were homoskedastic *i.i.d.*, but one observed *randomly levered* returns, with leverage chosen *e.g.*, by a fund manager, then expected return and *volatility* would scale together, resulting in $\lambda = \frac{1}{2}$. The case of jumbled periodicities, where *e.g.*, weekly and daily returns are mixed together, corresponds to $\lambda = 1$.

So assume the existence of a vector $\boldsymbol{l}$ with strictly positive elements. Define M_k as the empirical raw kth moment of $\boldsymbol{l}$:

$$M_k := \frac{1}{n} \sum_{1 \le t \le n} {l_t}^k. \tag{3.14}$$

Without loss of generality, we can assume that $\boldsymbol{l}$ has been rescaled such that $M_1 = 1$.

Now suppose λ is given, and $\boldsymbol{l}$ scales variance directly, and expectation via λ, *i.e.*, $\mathsf{\Sigma} = \sigma^2 \operatorname{Diag}(\boldsymbol{l})$, and $\boldsymbol{\mu} = \mu \boldsymbol{l}^{\lambda}$, where here a vector to a power is interpreted as the Hadamard (elementwise) power. Since we are assuming independence of returns, we can rely on the simplified form of expectation and standard error. Note that

$$\boldsymbol{\mu}_c = \mu \boldsymbol{l}^{\lambda} - \mu M_{\lambda} \mathbf{1}, \quad \text{and } \boldsymbol{\sigma}^2 = \sigma^2 \boldsymbol{l}.$$

Then Equation 3.9 becomes

$$\zeta_i = \frac{\mu M_{\lambda} - r_0}{\sigma}, \tag{3.15}$$

and the geometric bias term of the sample Sharpe ratio is

$$b = \left(1 + \frac{\mu^2 \left(M_{2\lambda} - M_{\lambda}^2\right)}{\sigma^2}\right)^{-1/2}. \tag{3.16}$$

Note that when $\lambda = 0$, then $M_{2\lambda} - M_{\lambda}^2 = 0$, and there is no bias: $b = 1$. When $\lambda = 1$, the term $\left(M_{2\lambda} - M_{\lambda}^2\right)$ is effectively the squared coefficient of variation of $\boldsymbol{l}$. It balances that of the underlying returns in the geometric bias term. Again, it is not unambiguously the case that ζ_i is the population quantity of interest, since variation within $\boldsymbol{\mu}$ should probably be counted as volatility of returns, especially in the case where $\boldsymbol{l}$ is stochastic, rather than the case of jumbled periodicities. Ignoring this point, we can proceed as previously.

To compute the standard error of the Sharpe ratio, we have to compute a few more terms. We have

$$\begin{aligned}
\mathbf{1}^{\top} \boldsymbol{\sigma}^2 &= n \sigma^2 M_{1+\lambda}, \\
{\boldsymbol{\sigma}^2}^{\top} \boldsymbol{\mu}_c &= n \mu \sigma^2 \left(M_{1+\lambda} - M_1 M_{\lambda}\right) = n \mu \sigma^2 \left(M_{1+\lambda} - M_{\lambda}\right), \\
{\boldsymbol{\sigma}^2}^{\top} \boldsymbol{\sigma}^2 &= n \sigma^4 M_2, \\
{\boldsymbol{\mu}_c}^{\top} \left(\boldsymbol{\sigma}^2 \odot \boldsymbol{\mu}_c\right) &= n \mu^2 \sigma^2 M_{1+2\lambda}.
\end{aligned} \tag{3.17}$$

Plugging these into Equation 3.11 gives the approximate variance of the Sharpe ratio as

$$\operatorname{Var}\left(\hat{\zeta}\right) \approx \frac{b^2}{n}\left[1 - 2\zeta_i b^2 \frac{\mu \left(M_{1+\lambda} - M_{\lambda}\right)}{\sigma M_{1+\lambda}^{3/2}} + \frac{b^4}{2} \zeta_i^2 \frac{2\frac{\mu^2}{\sigma^2} M_{1+2\lambda} + M_2 + \mathcal{O}\left(n^{-1}\right)}{M_{1+\lambda}^2}\right]. \tag{3.18}$$

Further simplification is possible for various values of λ. Note, however, that in some situations it appears to be the case that ζ (and even the debiased version, *viz.* $\sqrt{b}\zeta$) has a *lower* standard error than in the homoskedastic *i.i.d.* case (*cf.* Equation 3.13). This can occur if the errors in estimating the mean return and volatility are positively correlated. (Equivalently, ${\boldsymbol{\sigma}^2}^{\top} \boldsymbol{\mu}_c > 0$ and sufficiently large.) Thus estimation error in the two components of the Sharpe ratio counteract each other, resulting in a (ever so slightly) decreased standard error. In practice, this will have very little effect.

Sweeping out small terms (for most cases μ^2/σ^2 will be tiny), for a few special cases of λ we have

$$
\begin{aligned}
(\lambda = 0) \;\; \operatorname{Var}\left(\hat{\zeta}\right) &\approx \frac{b^2}{n}\left[1 + \frac{b^4}{2}\zeta_i^2\frac{M_2}{M_1^2}\right] = \frac{1}{n}\left[1 + \frac{1}{2}\zeta_i^2\frac{M_2}{M_1^2}\right],\\
(\lambda = 1/2) \;\; \operatorname{Var}\left(\hat{\zeta}\right) &\approx \frac{b^2}{n}\left[1 - 2\zeta_i b^2\frac{\mu\left(M_{3/2} - M_{1/2}\right)}{\sigma M_{3/2}^{3/2}} + \frac{b^4}{2}\zeta_i^2\frac{M_2^{-1/2}}{M_{3/2}^2}\right],\\
(\lambda = 1) \;\; \operatorname{Var}\left(\hat{\zeta}\right) &\approx \frac{b^2}{n}\left[1 - 2\zeta_i b^2\frac{\mu\left(M_2 - 1\right)}{\sigma M_2^{3/2}} + \frac{b^4}{2}\zeta_i^2 M_2^{-1}\right].
\end{aligned}
\tag{3.19}
$$

It turns out that when variation in l is modest and the signal-noise ratio is realistic, the geometric bias introduced into the Sharpe ratio by heteroskedasticity has little effect. Moreover, the standard error of the Sharpe ratio is not much changed either, and can even be slightly decreased. These findings are illustrated in the following examples which use the (rescaled) VIX index as $\boldsymbol{l}$. In summary,

---

Heteroskedasticity and Sharpe ratio

Modest heteroskedasticity causes only mild bias in the Sharpe ratio and very little effect on standard error.

Example 3.1.1 (VIX and heteroskedastic bias). Consider the VIX index introduced in Example 1.4.4, with data from 1990-01-02 through 2020-12-31. Because the VIX index measures *volatility*, not variance, to use it as $\boldsymbol{l}$, we *square* it, then rescale to unit mean. For values of k between 0 and 2, empirical estimates of M_k range (non-monotonically) from around 0.92 to 2.2.

Consider the case where $\mu = 0.001\,\text{day}^{-1}$, and $\sigma = 0.013\,\text{day}^{-1/2}$. For values of λ tested between 0 and 2, the geometric bias term, b, ranged from around 0.889 to 1, with b monotonically decreasing in λ. ⊣

Example 3.1.2 (VIX and Sharpe ratio standard error). Continuing Example 3.1.1, consider the (rescaled) VIX index of Example 1.4.4, with data from 1990-01-02 through 2020-12-31. Suppose $\mu = 0.001\,\text{day}^{-1}$, and $\sigma = 0.013\,\text{day}^{-1/2}$. Take $r_0 = 0\,\text{day}^{-1}$, and suppose n is large.

For values of λ tested between 0 and 2, n/b^2 times the approximate variance from Equation 3.18 was computed, and found to range from around 0.998 to 1.003, monotonically decreasing in λ.

The equivalent value for the homoskedastic *i.i.d.* case, via Equation 3.13, is around 1.001. For values of λ greater than around 0.28, the heteroskedastic case has a (slightly) lower standard error. ⊣

Example 3.1.3 (Weekly and daily returns). Suppose you have observed 260 weeks of hypothetical weekly returns and 260 days of daily returns for a strategy, with the returns independent. Suppose the returns are Gaussian with $\mu = 0.001\,\mathrm{day}^{-1}$, and $\sigma = 0.018\,\mathrm{day}^{-1/2}$. Because of how expectation and variance scale, we have $\lambda = 1$. To get $M_1 = 1$, we define $\boldsymbol{l}$ to be 260 values of 5/3 and 260 values of 1/3. This means that if we compute the Sharpe ratio naïvely, jumbling together daily and weekly returns, we are effectively estimating μ and σ (and thus ζ_i) at the scale of 3 days.

For these values of μ and σ, the geometric bias is quite modest: $b \approx 0.998$. The standard error of the (biased) Sharpe ratio is around 0.044. This is the standard error around estimation of signal-noise ratio at the 3 days scale. If we convert the 3 days Sharpe ratio to daily scale, the standard error of the (still slightly biased) daily Sharpe ratio is $0.025\,\mathrm{day}^{-1/2}$. If we had, instead, observed 1560 days of *daily* returns, the standard error, via Equation 3.13, is approximately $0.025\,\mathrm{day}^{-1/2}$, effectively the same. We have lost little, if anything, in terms of bias or variance by pooling weekly and daily returns together.

To check these results, we perform 10,000 Monte Carlo simulations of this situation. The mean Sharpe ratio over these realizations is $0.055\,\mathrm{day}^{-1/2}$, while the population value is $0.056\,\mathrm{day}^{-1/2}$. The standard deviation of the Sharpe ratio over these 10,000 realizations is around $0.025\,\mathrm{day}^{-1/2}$, which is consistent with the approximate theoretical value of $0.025\,\mathrm{day}^{-1/2}$. See also Exercise 3.11. ⊣

Example 3.1.4 (Daily and Quarterly returns, the Market). We consider the Market returns, introduced in Example 1.1.1. We observe 172 quarterly returns from the period Mar 1927 to Dec 1969, and 12,866 daily returns from 1970-01-02 to 2020-12-31. Observing only these, we wish to compute the Sharpe ratio. As this is a "jumbled periodicity" problem, we have $\lambda = 1$. We assume 252 days per year. To get $M_1 = 1$, we define $\boldsymbol{l}$ to be 172 values of 63/1.818 and then 12,866 values of 1/1.818. If we compute the Sharpe ratio naïvely on the quarterly and daily returns, we are effectively estimating μ and σ (and thus ζ_i) at the scale of 1.818 days. Recast in annual units we compute the Sharpe ratio as $0.535\,\mathrm{yr}^{-1/2}$. In Example 2.3.4, we estimated the Sharpe ratio to be $0.541\,\mathrm{yr}^{-1/2}$ based on quarterly values.

Plugging in the long term sample values for μ and σ, the geometric bias is $b \approx 0.979$. Dividing the Sharpe ratio by this bias gives us a geometrically nearly unbiased Sharpe ratio of $0.546\,\mathrm{yr}^{-1/2}$. The standard error of the (biased) Sharpe ratio is around 0.009, at the 1.818 days scale. Annualized the standard error is $0.136\,\mathrm{yr}^{-1/2}$. ⊣

See also Exercise 3.9 through Exercise 3.12.

Remark (Weighted estimation?). For the case where the $\boldsymbol{l}$ are known (*e.g.*, jumbled weekly and daily returns) or can be roughly estimated, one is tempted to perform some kind of weighted estimation to correct for the heteroskedasticity. It is not clear, however, that by so doing, one reduces the standard error of the Sharpe ratio. Indeed, in some cases, the approximate standard error in the heteroskedastic case seems lower than that in the homoskedastic case. Perhaps a better approach would be to correct for the approximate geometric bias, though this should be very close to 1 for most applications. On the other hand, arguably the apparent reduction in standard error is due to our insistence that the population parameter of interest does not count variation in $\boldsymbol{\mu}$ towards volatility of returns.

3.1.4 Homoskedastic autocorrelated Gaussian returns

Now consider the case of constant mean and errors drawn from an AR(1) process with autocorrelation ρ. [Brockwell and Davis, 2002] Thus we have again $\boldsymbol{\mu} = \mu\mathbf{1}$, and $\Sigma_{i,j} = \sigma^2\rho^{|i-j|}$. We will assume that $|\rho|$ is bounded away from one, say smaller than one half. Because $\boldsymbol{\mu}$ is constant, we have $\boldsymbol{\mu}_c = \mathbf{0}$, and $\mathbf{1}^\top\boldsymbol{\mu} = n\mu$. We will need to compute the quantity $\mathbf{1}^\top\boldsymbol{\Sigma}\mathbf{1}$. For large n this is approximately

$$\mathbf{1}^\top\boldsymbol{\Sigma}\mathbf{1} \approx n\sigma^2\left(\frac{1+\rho}{1-\rho} + \frac{1}{n}\frac{2\rho}{(1-\rho)^2}\right) = n\sigma^2\left(\frac{1+\rho}{1-\rho} + \mathcal{O}\left(n^{-1}\right)\right), \tag{3.20}$$

cf. Exercise 3.7. Similarly,

$$\begin{aligned}
\operatorname{tr}\left(\boldsymbol{\Sigma}_c\right) &= n\sigma^2 - \frac{\mathbf{1}^\top\boldsymbol{\Sigma}\mathbf{1}}{n} \approx n\sigma^2\left(1 - \mathcal{O}\left(n^{-1}\right)\right),\\
\operatorname{tr}\left(\boldsymbol{\Sigma}_c^2\right) &\approx n\sigma^4\left(\frac{1+\rho^2}{1-\rho^2} - \mathcal{O}\left(n^{-1}\right)\right).
\end{aligned}$$

The approximation of Equation 3.5 becomes

$$\operatorname{E}\left[\hat{\zeta}\right] \approx \sqrt{\frac{n-1}{n}}\frac{\mu - r_0}{\sqrt{\sigma^2\left(1 - \mathcal{O}\left(n^{-1}\right)\right)}} \rightarrow \frac{\mu - r_0}{\sigma}. \tag{3.21}$$

Thus the Sharpe ratio is asymptotically unbiased in n. The standard error, however, is somewhat affected by the autocorrelation. Picking up from Equation 3.7,

$$\begin{aligned}
\operatorname{Var}\left(\hat{\zeta}\right) &\approx \frac{n-1}{n^2}\frac{\mathbf{1}^\top\boldsymbol{\Sigma}\mathbf{1}}{\operatorname{tr}\left(\boldsymbol{\Sigma}_c\right)}\left[1 + \left(\frac{\mathbf{1}^\top\boldsymbol{\mu} - nr_0}{\operatorname{tr}\left(\boldsymbol{\Sigma}_c\right)}\right)^2\frac{\frac{1}{2}\operatorname{tr}\left({\boldsymbol{\Sigma}_c}^2\right)}{\mathbf{1}^\top\boldsymbol{\Sigma}\mathbf{1}}\right],\\
&\approx \frac{1}{n}\frac{1}{n\sigma^2}\left(n\sigma^2\frac{1+\rho}{1-\rho} + \frac{1}{2}\left(\frac{\mu - r_0}{\sigma^2}\right)^2 n\sigma^4\left(\frac{1+\rho^2}{1-\rho^2} + \mathcal{O}\left(n^{-1}\right)\right)\right),\\
&\approx \frac{1}{n}\left(\frac{1+\rho}{1-\rho} + \frac{1}{2}\left(\frac{\mu - r_0}{\sigma}\right)^2\left(\frac{1+\rho^2}{1-\rho^2} + \mathcal{O}\left(n^{-1}\right)\right)\right),\\
&\approx \frac{1}{n}\frac{1+\rho}{1-\rho}\left(1 + \frac{1+\rho^2}{(1+\rho)^2}\frac{1}{2}\left(\frac{\mu - r_0}{\sigma}\right)^2\right).
\end{aligned} \tag{3.22}$$

Compare this with the standard error for the homoskedastic case

(Equation 3.13, equivalently Equation 2.24). For small signal-noise ratio and small ρ, the standard error of the Sharpe ratio is approximately $n^{-1/2}\sqrt{(1+\rho)/(1-\rho)}$. This corresponds to the standard error of the t statistic for AR(1) variates under the null $\mu = 0$ as described by van Belle. [van Belle, 2002, sec. 8.7] The "small angle" approximation for this correction is $\sqrt{(1+\rho)/(1-\rho)} \approx 1+\rho$, which is reasonably accurate for $|\rho| < 0.1$. In summary:

Autocorrelation and Sharpe ratio

A small autocorrelation of ρ inflates the standard error of the Sharpe ratio by about $100\rho\%$.

Since we expect ρ to be very small for real-world returns[4], autocorrelation should have little to no effect on the bias or standard error of the Sharpe ratio.

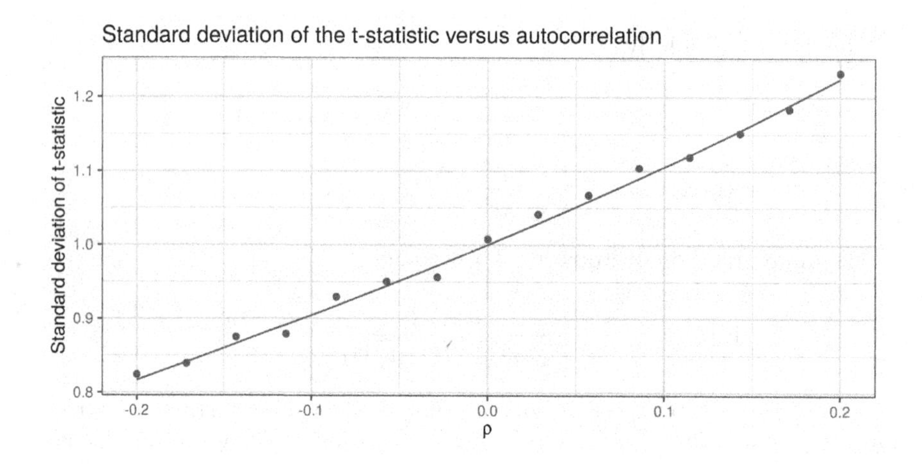

FIGURE 3.3
The empirical standard deviation of the t-statistic (*i.e.*, the rescaled Sharpe ratio), rescaled by $\sqrt{n}$, is plotted versus the autocorrelation, ρ. Each point represents 8,000 replications of approximately 4 years of daily data, with each series generated by an AR(1) process with normal innovations and $\mu = 0$. The line is $y = \sqrt{(1+\rho)/(1-\rho)}$, as given by Equation 3.22, *not* a fit of the data.

[4]While persistent significant autocorrelation should be "arb'ed out" by market participants, significant autocorrelation at long time scales may be observed in real markets, *cf.* Example 3.0.1.

Example 3.1.5 (AR(1) returns and Sharpe ratio standard error). A Monte Carlo study confirms the standard error of the Sharpe ratio given in Equation 3.22. For values of ρ ranging from -0.2 to 0.2, $8,000$ realizations of approximately 4 years of daily data generated by an AR(1) process with $\mu = 0$ were generated. The Sharpe ratio of each was computed, and the empirical standard deviation over realizations was computed. This is multiplied by $\sqrt{n}$ and then plotted versus ρ in Figure 3.3. The fit value of $\sqrt{(1+\rho)/(1-\rho)}$ is also shown. See also Exercise 1.30. ⊣

Example 3.1.6 (Autocorrelated Market returns and Sharpe ratio standard error). Consider the returns of the Market, introduced in Example 1.1.1. Based on 1,128 monthly returns from Jan 1927 to Dec 2020, we estimate the autocorrelation of returns to be 0.1. Feeding the sample estimates of ρ, μ and σ into Equation 3.22, we estimate the standard error of the Sharpe ratio to be $0.017\,\mathrm{yr}^{-1/2}$. If the autocorrelation is assumed to be zero, we would estimate the standard error as $0.014\,\mathrm{yr}^{-1/2}$. The inflation caused by autocorrelation is approximately 22%, which is close to twice the estimated autocorrelation. ⊣

Caution (Returns are autocorrelated). By the way that they are constructed, returns of an asset are typically slightly *negatively* autocorrelated. That is, since returns are defined as $x_{t+1} = \log p_{t+1}/p_t$, where p_t are the mark prices, if there is any amount of "error" in the marks, it will create negatively autocorrelated returns. This is discussed in the context of the "bounce effect," *cf.* Exercise 1.26. In general, if the "true" returns are *i.i.d.* with variance σ_m^2, and the log mark prices have a noise term with variance σ_b^2, then the autocorrelation of observed returns is

$$\rho = \frac{-\sigma_b^2}{2\sigma_b^2 + \sigma_m^2}$$

Given that σ_b^2/σ_m^2 is likely to be on the order of 0.01 or smaller, the bias introduced by this bounce effect should be very small.

3.2 Asymptotic distribution of Sharpe ratio

Here we derive the asymptotic distribution of Sharpe ratio, following Jobson and Korkie *inter alia.* [Jobson and Korkie, 1981; Lo, 2002; Mertens, 2002; Ledoit and Wolf, 2008; Leung and Wong, 2008; Wright *et al.*, 2014] Consider the case of p possibly correlated returns streams, with each observation denoted by the p-vector $\boldsymbol{x}$. Let $\boldsymbol{\mu}$ be the p-vector of population means, and let $\boldsymbol{\alpha}_2$ be the p-vector of the uncentered second moments. Let $\boldsymbol{\zeta}$ be the vector of signal-noise ratio of the assets. Let r_0 be the "risk free rate." We have

$$\boldsymbol{\zeta}_i = \frac{\boldsymbol{\mu}_i - r_0}{\sqrt{\boldsymbol{\alpha}_{2,i} - \boldsymbol{\mu}_i^{\,2}}}.$$

Consider the $2p$ vector of $\boldsymbol{x}$, "stacked" with $\boldsymbol{x}$ squared elementwise, $\left[\boldsymbol{x}^\top, {\boldsymbol{x}^2}^\top\right]^\top$. The expected value of this vector is $\left[\boldsymbol{\mu}^\top, {\boldsymbol{\alpha}_2}^\top\right]^\top$; let Ω be the variance of this vector, assuming it exists.

Given n observations of $\boldsymbol{x}$, consider the simple sample estimate

$$\left[\hat{\boldsymbol{\mu}}^\top, \hat{\boldsymbol{\alpha}}_2{}^\top\right]^\top := \frac{1}{n}\sum_t^n \left[\boldsymbol{x}_t{}^\top, {\boldsymbol{x}_t^2}^\top\right]^\top.$$

Under the multivariate central limit theorem [Wasserman, 2004]

$$\sqrt{n}\left(\left[\hat{\boldsymbol{\mu}}^\top, \hat{\boldsymbol{\alpha}}_2{}^\top\right]^\top - \left[\boldsymbol{\mu}^\top, \boldsymbol{\alpha}_2{}^\top\right]^\top\right) \rightsquigarrow \mathcal{N}\left(0, \Omega\right). \tag{3.23}$$

Let $\hat{\boldsymbol{\zeta}}$ be the sample Sharpe ratio computed from the estimates $\hat{\boldsymbol{\mu}}$ and $\hat{\boldsymbol{\alpha}}_2$: $\hat{\zeta}_i = \left(\hat{\boldsymbol{\mu}}_i - r_0\right) / \sqrt{\hat{\boldsymbol{\alpha}}_{2,i} - \hat{\boldsymbol{\mu}}_i{}^2}$. By the multivariate delta method,

$$\sqrt{n}\left(\hat{\boldsymbol{\zeta}} - \boldsymbol{\zeta}\right) \rightsquigarrow \mathcal{N}\left(0, \left(\frac{\mathrm{d}\boldsymbol{\zeta}}{\mathrm{d}[\boldsymbol{\mu}^\top, \boldsymbol{\alpha}_2{}^\top]^\top}\right)\Omega\left(\frac{\mathrm{d}\boldsymbol{\zeta}}{\mathrm{d}[\boldsymbol{\mu}^\top, \boldsymbol{\alpha}_2{}^\top]^\top}\right)^\top\right). \tag{3.24}$$

Here the derivative takes the form of two $p \times p$ diagonal matrices pasted together side by side:

$$\begin{aligned}\frac{\mathrm{d}\boldsymbol{\zeta}}{\mathrm{d}[\boldsymbol{\mu}^\top, \boldsymbol{\alpha}_2{}^\top]^\top} &= \left[\begin{array}{cc}\mathrm{Diag}\left(\frac{\boldsymbol{\alpha}_2 - \boldsymbol{\mu} r_0}{(\boldsymbol{\alpha}_2 - \boldsymbol{\mu}^2)^{3/2}}\right) & \mathrm{Diag}\left(\frac{r_0 - \boldsymbol{\mu}}{2(\boldsymbol{\alpha}_2 - \boldsymbol{\mu}^2)^{3/2}}\right)\end{array}\right],\\ &= \left[\begin{array}{cc}\mathrm{Diag}\left(\frac{\boldsymbol{\sigma} + \boldsymbol{\mu}\boldsymbol{\zeta}}{\boldsymbol{\sigma}^2}\right) & \mathrm{Diag}\left(\frac{-\boldsymbol{\zeta}}{2\boldsymbol{\sigma}^2}\right)\end{array}\right].\end{aligned} \tag{3.25}$$

where $\mathrm{Diag}\left(\boldsymbol{z}\right)$ is the matrix with vector $\boldsymbol{z}$ on its diagonal, and where the vector operations above are all performed elementwise, where we define the vector $\boldsymbol{\sigma} := \left(\boldsymbol{\alpha}_2 - \boldsymbol{\mu}^2\right)^{1/2}$, with powers taken elementwise.

In practice, the population values, $\boldsymbol{\mu}$, $\boldsymbol{\alpha}_2$, Ω are all unknown, and so the asymptotic variance has to be estimated, using the sample. Letting $\hat{\Omega}$ be some sample estimate of Ω, taken from estimating the covariance of the samples of $\boldsymbol{x}$ and $\boldsymbol{x}^2$ stacked, using Equation 3.25 we have the approximation

$$\hat{\boldsymbol{\zeta}} \approx \mathcal{N}\left(\boldsymbol{\zeta}, \frac{1}{n}\left[\begin{array}{cc}\mathrm{Diag}\left(\frac{\hat{\boldsymbol{\sigma}} + \hat{\boldsymbol{\mu}}\hat{\boldsymbol{\zeta}}}{\hat{\boldsymbol{\sigma}}^2}\right) & \mathrm{Diag}\left(\frac{-\hat{\boldsymbol{\zeta}}}{2\hat{\boldsymbol{\sigma}}^2}\right)\end{array}\right]\hat{\Omega}\left[\begin{array}{c}\mathrm{Diag}\left(\frac{\hat{\boldsymbol{\sigma}} + \hat{\boldsymbol{\mu}}\hat{\boldsymbol{\zeta}}}{\hat{\boldsymbol{\sigma}}^2}\right)\\ \mathrm{Diag}\left(\frac{-\hat{\boldsymbol{\zeta}}}{2\hat{\boldsymbol{\sigma}}^2}\right)\end{array}\right]\right), \tag{3.26}$$

where we have plugged in sample estimates. [Lo, 2002; Mertens, 2002]

Example 3.2.1 (Elliptically distributed returns). Consider the case where $\boldsymbol{x}$ is drawn from an elliptical distribution with mean $\boldsymbol{\mu}$, covariance Σ, and kurtosis factor κ. When returns are Gaussian, $\kappa = 1$. Then we have

$$\Omega = \left[\begin{array}{cc}\Sigma & 2\Sigma\,\mathrm{Diag}\left(\boldsymbol{\mu}\right)\\ 2\,\mathrm{Diag}\left(\boldsymbol{\mu}\right)\Sigma & \left(\kappa - 1\right)\mathrm{diag}\left(\Sigma\right)\left(\mathrm{diag}\left(\Sigma\right)\right)^\top + 2\kappa\Sigma\odot\Sigma + 4\,\mathrm{Diag}\left(\boldsymbol{\mu}\right)\Sigma\,\mathrm{Diag}\left(\boldsymbol{\mu}\right)\end{array}\right]. \tag{3.27}$$

cf. Exercise 3.2.

Let R be the correlation matrix of the returns, defined as

$$\mathsf{R} := \operatorname{Diag}\left(\boldsymbol{\sigma}^{-1}\right) \Sigma \operatorname{Diag}\left(\boldsymbol{\sigma}^{-1}\right), \tag{3.28}$$

where $\boldsymbol{\sigma}$ is the (positive) square root of the diagonal of Σ. Then Equation 3.26 becomes

$$\hat{\boldsymbol{\zeta}} \approx \mathcal{N}\left(\boldsymbol{\zeta}, \frac{1}{n}\left(\mathsf{R} + \frac{\kappa - 1}{4}\boldsymbol{\zeta}\boldsymbol{\zeta}^{\top} + \frac{\kappa}{2}\operatorname{Diag}\left(\boldsymbol{\zeta}\right)\left(\mathsf{R} \odot \mathsf{R}\right)\operatorname{Diag}\left(\boldsymbol{\zeta}\right)\right)\right). \tag{3.29}$$

cf. Exercise 3.39 and 3.40. Note how in the case of scalar Gaussian returns, this reduces to Equation 2.25. ⊣

3.2.1 Frequentist analysis

Equation 3.26 can be used to perform hypothesis tests and construct confidence intervals with approximately nominal coverage. To be explicit, we outline those procedures here.

test of single contrast on multiple signal-noise ratios

Suppose $\boldsymbol{v}$ is a fixed p-vector, and c a fixed scalar. Then to test the null hypothesis:

$$H_0 : \boldsymbol{v}^{\top}\boldsymbol{\zeta} = c \quad \text{versus} \quad H_1 : \boldsymbol{v}^{\top}\boldsymbol{\zeta} \leq c,$$

we first compute $\hat{\boldsymbol{\zeta}}$, then estimate $\hat{\Omega}$, either by the empirical covariance of the stacked vector of samples of $\boldsymbol{x}$ and $\boldsymbol{x}^2$, or some other method. Then reject the null if the statistic

$$z = \frac{\boldsymbol{v}^{\top}\hat{\boldsymbol{\zeta}} - c}{\sqrt{n\boldsymbol{v}^{\top}\left(\left[\begin{array}{cc}\operatorname{Diag}\left(\frac{\hat{\boldsymbol{\sigma}}+\hat{\boldsymbol{\mu}}\hat{\boldsymbol{\zeta}}}{\hat{\boldsymbol{\sigma}}^2}\right) & \operatorname{Diag}\left(\frac{-\hat{\boldsymbol{\zeta}}}{2\hat{\boldsymbol{\sigma}}^2}\right)\end{array}\right]\hat{\Omega}\left[\begin{array}{c}\operatorname{Diag}\left(\frac{\hat{\boldsymbol{\sigma}}+\hat{\boldsymbol{\mu}}\hat{\boldsymbol{\zeta}}}{\hat{\boldsymbol{\sigma}}^2}\right)\\ \operatorname{Diag}\left(\frac{-\hat{\boldsymbol{\zeta}}}{2\hat{\boldsymbol{\sigma}}^2}\right)\end{array}\right]\right)\boldsymbol{v}}}$$

is less than z_{α}, the α quantile of the standard normal distribution. Alternatively, if one assumes returns are drawn from an elliptical distribution, instead use the statistic

$$z = \frac{\boldsymbol{v}^{\top}\hat{\boldsymbol{\zeta}} - c}{\sqrt{n\boldsymbol{v}^{\top}\left(\hat{\mathsf{R}} + \frac{\hat{\kappa}-1}{4}\hat{\boldsymbol{\zeta}}\hat{\boldsymbol{\zeta}}^{\top} + \frac{\hat{\kappa}}{2}\operatorname{Diag}\left(\hat{\boldsymbol{\zeta}}\right)\left(\hat{\mathsf{R}} \odot \hat{\mathsf{R}}\right)\operatorname{Diag}\left(\hat{\boldsymbol{\zeta}}\right)\right)\boldsymbol{v}}},$$

where $\hat{\kappa}$ and $\hat{\mathsf{R}}$ are sample estimates of the kurtosis factor and correlation matrix respectively.

test of multiple signal-noise ratios

Suppose $\boldsymbol{\zeta}_0$ is some fixed p-vector. Then to test the null hypothesis:

$$H_0 : \boldsymbol{\zeta} = \boldsymbol{\zeta}_0 \quad \text{versus} \quad H_1 : \boldsymbol{\zeta} \neq \boldsymbol{\zeta}_0,$$

we first compute $\hat{\boldsymbol{\zeta}}$, then estimate $\hat{\Omega}$ as described above. Then reject the null if the statistic

$$c^2 = n\left(\hat{\boldsymbol{\zeta}} - \boldsymbol{\zeta}_0\right)^\top \left(\left[\operatorname{Diag}\left(\frac{\hat{\boldsymbol{\sigma}}+\hat{\boldsymbol{\mu}}\hat{\boldsymbol{\zeta}}}{\hat{\boldsymbol{\sigma}}^2}\right) \quad \operatorname{Diag}\left(\frac{-\hat{\boldsymbol{\zeta}}}{2\hat{\boldsymbol{\sigma}}^2}\right) \right] \hat{\Omega} \left[\begin{array}{c} \operatorname{Diag}\left(\frac{\hat{\boldsymbol{\sigma}}+\hat{\boldsymbol{\mu}}\hat{\boldsymbol{\zeta}}}{\hat{\boldsymbol{\sigma}}^2}\right) \\ \operatorname{Diag}\left(\frac{-\hat{\boldsymbol{\zeta}}}{2\hat{\boldsymbol{\sigma}}^2}\right) \end{array} \right] \right)^{-1} \left(\hat{\boldsymbol{\zeta}} - \boldsymbol{\zeta}_0\right) \tag{3.30}$$

is greater than ${\chi^2}_{1-\alpha}(p)$, the $1-\alpha$ quantile of the chi-square distribution with p degrees of freedom. Alternatively, if one assumes returns are drawn from an elliptical distribution, instead use the statistic

$$c^2 = n\left(\hat{\boldsymbol{\zeta}} - \boldsymbol{\zeta}_0\right)^\top \left(\hat{\mathsf{R}} + \frac{\hat{\kappa} - 1}{4}\hat{\boldsymbol{\zeta}}\hat{\boldsymbol{\zeta}}^\top + \frac{\hat{\kappa}}{2}\operatorname{Diag}\left(\hat{\boldsymbol{\zeta}}\right)\left(\hat{\mathsf{R}} \odot \hat{\mathsf{R}}\right)\operatorname{Diag}\left(\hat{\boldsymbol{\zeta}}\right)\right)^{-1} \left(\hat{\boldsymbol{\zeta}} - \boldsymbol{\zeta}_0\right), \tag{3.31}$$

where $\hat{\kappa}$ and $\hat{\mathsf{R}}$ are sample estimates of the kurtosis factor and correlation matrix respectively.

Similarly to construct $1 - \alpha$ confidence ellipsoids on $\boldsymbol{\zeta}$, one takes

$$\left\{ \boldsymbol{\zeta}_0 \,\middle|\, \left(\hat{\boldsymbol{\zeta}} - \boldsymbol{\zeta}_0\right)^\top \left(\boldsymbol{w}^\top \hat{\Omega} \boldsymbol{w}\right)^{-1} \left(\hat{\boldsymbol{\zeta}} - \boldsymbol{\zeta}_0\right) \leq \frac{{\chi^2}_{1-\alpha}(p)}{n} \right\},$$

where we define

$$\boldsymbol{w} = \left[\begin{array}{c} \operatorname{Diag}\left(\frac{\hat{\boldsymbol{\sigma}}+\hat{\boldsymbol{\mu}}\hat{\boldsymbol{\zeta}}}{\hat{\boldsymbol{\sigma}}^2}\right) \\ \operatorname{Diag}\left(\frac{-\hat{\boldsymbol{\zeta}}}{2\hat{\boldsymbol{\sigma}}^2}\right) \end{array} \right].$$

Example 3.2.2 (Asymptotic Sharpe ratio of SMB and HML). Consider the returns of two of the Fama French factors, SMB and HML, introduced in Example 1.1.1. Based on monthly returns from Jan 1927 to Dec 2020, we compute

$$\left[\boldsymbol{\mu}^\top, \boldsymbol{\alpha}_2{}^\top\right]^\top = \begin{bmatrix} 0.21 \\ 0.32 \\ 10.19 \\ 12.40 \end{bmatrix}, \qquad \frac{1}{n}\hat{\Omega} = \begin{bmatrix} 0.01 & 0.00 & 0.06 & 0.03 \\ 0.00 & 0.01 & 0.03 & 0.09 \\ 0.06 & 0.03 & 1.95 & 0.82 \\ 0.03 & 0.09 & 0.82 & 2.87 \end{bmatrix}.$$

Here the units of $\boldsymbol{\mu}$ are in $\%\,\text{mo.}^{-1}$, and those of $\boldsymbol{\alpha}_2$ are $\%^2\,\text{mo.}^{-1}$. We plug in the sample estimates to get the estimate of the derivative:

$$\frac{\mathrm{d}\hat{\boldsymbol{\zeta}}}{\mathrm{d}\left[\hat{\boldsymbol{\mu}}^\top, \hat{\boldsymbol{\alpha}}_2{}^\top\right]^\top} \approx \begin{bmatrix} 0.315 & 0.000 & -0.003 & 0.000 \\ 0.000 & 0.288 & 0.000 & -0.004 \end{bmatrix}.$$

The Sharpe ratios are computed as

SMB	HML
0.065	0.092

$\text{mo.}^{-1/2}$.

The estimated standard-error variance-covariance of the Sharpe ratio of the two returns is

	SMB	HML
SMB	0.00080	0.00006
HML	0.00006	0.00076

Thus, for example, taking the SMB marginal, we suppose the observed Sharpe ratio is nearly normally distributed around the true value with standard deviation approximately 0.028.

To test the null hypothesis that $+1SMB - 1HML = 0.01$ versus the alternative $+1SMB - 1HML \leq 0.01$, we estimate the test statistic as $z = -0.965$, which is a bit bigger than the critical value of $z_{0.05} = -1.645$, and we fail to reject the null hypothesis.

To test the null hypothesis that both Sharpe ratios are equal to zero, we compute the test statistic $c^2 = 15.43$ which is bigger than ${\chi^2}_{0.95}(2) = 5.99$, and we reject the null hypothesis. To instead test the null hypothesis that both Sharpe ratios are equal to $0.1\,\text{mo.}^{-1/2}$, we compute $c^2 = 1.55$, and fail to reject at the 0.05 level.

One is tempted to plot confidence ellipsoids around for the signal-noise ratios. However, in this case, the estimated standard errors of the two factors are nearly the same, and their estimated correlation is low. Thus, the confidence ellipsoids strongly resemble a circle centered at $(0.065, 0.092)$. We leave it as an exercise for the reader to draw such a circle. ⊣

Example 3.2.3 (Correlation of errors, Sharpe ratio of Fama-French factors). The Sharpe ratio of all four Fama French factors monthly returns from Example 1.1.1 were computed, using monthly returns from Jan 1927 to Dec 2020. The estimated standard error variance-covariance matrix of the computed Sharpe ratios, converted to a correlation matrix is computed as:

Mkt	SMB	HML	UMD
1.0000	0.3054	0.1510	−0.2429
0.3054	1.0000	0.0756	−0.1639
0.1510	0.0756	1.0000	−0.3506
−0.2429	−0.1639	−0.3506	1.0000

cf. Exercise 3.13. ⊣

We defer a Bayesian analysis to the more general case of functions of signal-noise ratios, given in Section 3.3.2.

3.2.2 Scalar case

For the $p = 1$ case, Ω takes the form

$$\begin{aligned}
\Omega &= \left[\begin{array}{cc} \alpha_2 - \mu^2 & \alpha_3 - \mu\alpha_2 \\ \alpha_3 - \mu\alpha_2 & \alpha_4 - {\alpha_2}^2 \end{array}\right], \\
&= \left[\begin{array}{cc} \sigma^2 & \mu_3 + 2\mu\sigma^2 \\ \mu_3 + 2\mu\sigma^2 & \mu_4 + 4\mu_3\mu + 4\sigma^2\mu^2 - \sigma^4 \end{array}\right], \\
&= \left[\begin{array}{cc} \sigma^2 & \sigma^2\left(\sigma\gamma_1 + 2\mu\right) \\ \sigma^2\left(\sigma\gamma_1 + 2\mu\right) & \sigma^4\left(\frac{\mu_4}{\sigma^4} - 3 + 2\right) + 4\sigma^3\mu\frac{\mu_3}{\sigma^3} + 4\sigma^2\mu^2 \end{array}\right], \\
&= \sigma^2\left[\begin{array}{cc} 1 & \sigma\gamma_1 + 2\mu \\ \sigma\gamma_1 + 2\mu & \sigma^2\left(\gamma_2 + 2\right) + 4\sigma\mu\gamma_1 + 4\mu^2 \end{array}\right].
\end{aligned} \tag{3.32}$$

where α_i and μ_i are the centered and uncentered moments of x, and γ_i are the skew and excess kurtosis:

$$\alpha_i := \operatorname{E}\left[x^i\right], \qquad \mu_i := \operatorname{E}\left[(x - \mu)^i\right],$$
$$\gamma_1 := \frac{\mu_3}{\sigma^3}, \qquad \gamma_2 := \frac{\mu_4}{\sigma^4} - 3.$$

In this case the derivative of the scalar signal-noise ratio with respect to the first and second moments is

$$\frac{\mathrm{d}\zeta}{\mathrm{d}{\left[\mu, \alpha_2\right]}^{\top}} = \left[\begin{array}{cc} \frac{\sigma + \mu\zeta}{\sigma^2} & -\frac{\zeta}{2\sigma^2} \end{array}\right]. \tag{3.33}$$

After much algebraic simplification (Exercise 3.26), the asymptotic variance of Sharpe ratio is given by Mertens' formula, Equation 2.28:

$$\hat{\zeta} \approx \mathcal{N}\left(\zeta, \frac{1}{n}\left(1 - \zeta\gamma_1 + \frac{\gamma_2 + 2}{4}\zeta^2\right)\right). \tag{3.34}$$

Note that Mertens' equation applies even though our definition of Sharpe ratio includes a risk-free rate, r_0. It should be stressed that the signal-noise ratio, skew and excess kurtosis appearing in Equation 3.34 are *population* values, which are unknown in practice. Typically the standard error is estimated using the plug-in method where sample estimates of these quantities are used in their place. Estimation error in those sample estimates has an unknown affect on the standard error computation, but it is likely to be very small. (See Section 3.6.) Note also that the skew and excess kurtosis are zero for the normal distribution, in which case Mertens' formula reduces to the "usual" standard error estimate given by Johnson and Welch, Equation 2.24.

Example 3.2.4 (Mertens versus vanilla standard error). Consider the returns of the Market, introduced in Example 1.1.1. Based on monthly returns from Jan 1927 to Dec 2020, the Sharpe ratio is computed as $0.61\,\mathrm{yr}^{-1/2}$. The standard error under the classical vanilla approximation, Equation 2.24, is estimated to be $0.104\,\mathrm{yr}^{-1/2}$, while under Equation 2.28, using the sample skew and excess kurtosis, it is estimated as $0.106\,\mathrm{yr}^{-1/2}$. There is no appreciable difference in these standard error estimates. ⊣

Example 3.2.5 (Hypothesis testing, the Market, 2 sample test). Consider the monthly relative returns of the Market, introduced in Example 1.1.1. The data were divided into two periods, with the dividing date 1970-01-01. The first period consists of 517 mo., starting at Jan 1927; the second 612 mo. ending on Dec 2020. The Sharpe ratio was measured to be $0.514\,\mathrm{yr}^{-1/2}$ in the first period, and $0.73\,\mathrm{yr}^{-1/2}$ in the second. To test the null hypothesis that the signal-noise ratio is equal in the two time periods, we use the normal approximation. We estimate the standard error in the two periods, via Mertens' approximation, as $0.152\,\mathrm{yr}^{-1/2}$ and $0.15\,\mathrm{yr}^{-1/2}$. Assuming independent samples, the standard error of the difference under the null should then be $0.213\,\mathrm{yr}^{-1/2}$. The difference in Sharpe ratios is computed as $-0.216\,\mathrm{yr}^{-1/2}$, comparable in magnitude to just one standard error, and we fail to reject the null hypothesis at the 0.05 level.

Contrast this with the Chow test for structural breaks, which tests for a change in mean, assuming volatility is equal in the two samples. [Chow, 1960; Kennedy, 2008] The test performed here does not require equal volatility, and tests for differences in signal-noise ratio. ⊣

Example 3.2.6 (Hypothesis testing, the Market, January Effect). Consider the monthly relative returns of the Market, introduced in Example 1.1.1. The data were divided into two samples: the 94 mo. of returns for the months of January, and the 1,034 mo. remaining months. The Sharpe ratio was measured to be $1.094\,\mathrm{yr}^{-1/2}$ for the Januaries, and $0.576\,\mathrm{yr}^{-1/2}$ for the rest of the year. The difference between these is $0.518\,\mathrm{yr}^{-1/2}$. To test the null hypothesis that the signal-noise ratio is equal in January and the rest of the year, against the alternative that it is greater in January, we estimate standard errors via Mertens' approximation to be $0.352\,\mathrm{yr}^{-1/2}$ and $0.11\,\mathrm{yr}^{-1/2}$, respectively, for Januaries and non-Januaries. Assuming independent samples, the standard error of the difference under the null should be around $0.368\,\mathrm{yr}^{-1/2}$. Given that the difference is smaller than this standard error times $z_{0.95}$, we fail to reject the null hypothesis at the 0.05 level. ⊣

3.2.3 Asymptotic bias and variance of the Sharpe ratio

Equation 3.34 gives the *asymptotic* distribution of the scalar Sharpe ratio. In particular, it claims that the Sharpe ratio is asymptotically unbiased. For small n, however, this approximation may be too coarse. One can derive approximations of the bias of the Sharpe ratio involving the (typically unknown) higher order moments of returns.

While the sample mean is unbiased, the presence of the inverse square root of the sample variance in the Sharpe ratio introduces some bias. Consider the Taylor expansion of $x^{-1/2}$:

$$\frac{1}{\sqrt{x+\epsilon}} \approx \frac{1}{\sqrt{x}} - \frac{1}{2}\frac{1}{x^{3/2}}\epsilon + \frac{3}{8}\frac{1}{x^{5/2}}\epsilon^2 + \ldots$$

Now let $\hat{\sigma}^2 = \sigma^2 (1 + z)$ for some random variable z. Then

$$\frac{1}{\sqrt{\hat{\sigma}^2}} = \frac{1}{\sigma\sqrt{1+z}} \approx \frac{1}{\sigma}\left(1 - \frac{1}{2}z + \frac{3}{8}z^2 + \ldots\right).$$

Letting $\hat{\alpha}_i$ be the sample ith moment, *e.g.*, $\hat{\alpha}_2 = \frac{1}{n}\sum_{1\le t\le n} {x_t}^2$, we can express z as

$$z = \frac{n\left(\hat{\alpha}_2 - \hat{\mu}^2\right)}{\sigma^2\left(n-1\right)} - 1.$$

Then

$$\hat{\zeta} = \frac{\hat{\mu}}{\hat{\sigma}} \approx \frac{\hat{\mu}}{\sigma}\left(1 - \frac{1}{2}\frac{n\left(\hat{\alpha}_2 - \hat{\mu}^2\right)}{\sigma^2\left(n-1\right)} + \frac{1}{2} + \frac{3}{8}\left(\frac{n\left(\hat{\alpha}_2 - \hat{\mu}^2\right)}{\sigma^2\left(n-1\right)} - 1\right)^2 + \ldots\right)$$

Taking expectations and omitting higher order terms for simplicity, we have

$$\operatorname{E}\left[\hat{\zeta}\right] = \zeta + \operatorname{E}\left[-\frac{1}{2}\frac{\hat{\mu}n\left(\hat{\alpha}_2 - \hat{\mu}^2\right)}{\sigma^3\left(n-1\right)} + \frac{\hat{\mu}}{2\sigma} + \ldots\right]$$

Using facts about expectations of products of raw sample moments (*cf.* Exercise 3.35), the bias of the Sharpe ratio can be expressed as

$$\begin{aligned}
\operatorname{E}\left[\hat{\zeta}\right] - \zeta &= -\frac{1}{2\sigma^3}\frac{n}{n-1}\left(\frac{n-1}{n^2}\mu_3 + \frac{n-1}{n}\mu\sigma^2\right) + \frac{1}{2}\zeta + \ldots, \\
&= -\frac{1}{2}\left(\frac{1}{n}\frac{\mu_3}{\sigma^3} + \zeta\right) + \frac{1}{2}\zeta + \ldots, \qquad (3.35) \\
&= -\frac{1}{2}\frac{\gamma_1}{n} + \ldots,
\end{aligned}$$

where $\gamma_1 = \mu_3/\sigma^3$ is the skew.

Higher order formulæ, due to Bao [2009], can be found by taking more terms in the Taylor expansion. These are

$$\operatorname{E}\left[\hat{\zeta}\right] - \zeta \approx -\frac{\gamma_1}{2n} + \frac{3\zeta}{8n}\left(2 + \gamma_2\right), \qquad (3.36)$$

$$\begin{aligned}
\operatorname{E}\left[\hat{\zeta}\right] - \zeta \approx &-\frac{\gamma_1}{2n} + \frac{3\zeta}{8n}\left(2 + \gamma_2\right) + \frac{3}{8n^2}\left(\gamma_3 - \gamma_1 - \frac{5}{2}\gamma_1\gamma_2\right) \qquad (3.37) \\
&+ \frac{\zeta}{32n^2}\left(49 - 10\gamma_4 - 15\gamma_2 - 40{\gamma_1}^2 + \frac{105{\gamma_2}^2}{4}\right),
\end{aligned}$$

where γ_i are the standardized higher order cumulants. In particular, γ_1 is the skew and γ_2 is the excess kurtosis of returns. The formulæ given here involve the population cumulants, which typically are not be known. Rather when the bias is estimated, sample estimates of the mean, variance, and higher order cumulants are used instead.

Example 3.2.7 (Higher order bias, Market). Consider the returns of the Market, introduced in Example 1.1.1. Based on monthly returns from Jan 1927 to Dec 2020, the Sharpe ratio is computed as $0.61\,\mathrm{yr}^{-1/2}$. Using sample estimates for the cumulants, we estimate the bias of the Sharpe ratio to be $0.002\,\mathrm{yr}^{-1/2}$ via Equation 3.36; and $0.002\,\mathrm{yr}^{-1/2}$ via Equation 3.37, which are nearly the same and both very small. For example, the latter bias is around 0.307% of the computed Sharpe ratio, and around 1.81% of the vanilla standard error estimate.

If we restrict our attention to the period of 96 months from Jan 2013 to Dec 2020, then we compute a Sharpe ratio of $1.1\,\mathrm{yr}^{-1/2}$, and compute a higher order estimate of the bias of $0.02\,\mathrm{yr}^{-1/2}$, which is around 1.8% of the computed Sharpe ratio, and around 5.45% of the standard error. ⊣

The same approach was used to find a higher order formula for the standard error,

$$
\begin{aligned}
\mathrm{Var}\left(\hat{\zeta}\right) = {} & \frac{1}{n}\left(1 + \frac{2+\gamma_2}{4}\zeta^2 - \gamma_1\zeta\right) + \frac{\zeta^2}{32n^2}\left(76 + 12\gamma_2 - 12\gamma_4 - 48{\gamma_1}^2 + 39{\gamma_2}^2\right) \\
& + \frac{5\zeta}{4n^2}\left(\gamma_3 - 3\gamma_2\gamma_1\right) + \frac{1}{4n^2}\left(8 + 7{\gamma_1}^2\right) + \ldots
\end{aligned}
\tag{3.38}
$$

The n^{-1} term is Mertens' approximation of the variance. Again, the formula involves typically unknown population cumulants which have to be estimated in practice. It is not clear under which conditions using Equation 3.38 should be preferred to using Merten's standard error, as the higher order cumulants (γ_4 is the *sixth* standardized cumulant) must be estimated from the data. Moreover, since the additional terms are multiplied by n^{-2}, they should only make a difference when n is relatively small, which is exactly when estimating higher order cumulants would be difficult.

Example 3.2.8 (Bao, Mertens, vanilla standard errors). Continuing Example 3.2.4, we look at monthly returns of the Market factor from Jan 1927 to Dec 2020. As above, the standard error under Equation 2.24, is around $0.104\,\mathrm{yr}^{-1/2}$, while by Merten's approximation, it is estimated as $0.106\,\mathrm{yr}^{-1/2}$. Via the higher order formula, Equation 3.38, it is estimated as $0.106\,\mathrm{yr}^{-1/2}$. Again, there is no appreciable difference in the standard error estimates.

If we restrict our attention to the period of 96 months from Jan 2013 to Dec 2020, then the vanilla standard error is estimated as $0.362\,\mathrm{yr}^{-1/2}$; Merten's formula gives $0.391\,\mathrm{yr}^{-1/2}$; Bao's formula gives $0.398\,\mathrm{yr}^{-1/2}$. ⊣

Confidence and Prediction Intervals

Via any of the variance formulations, Bao, Mertens or vanilla (respectively, Equation 3.38, Equation 2.28, and Equation 2.24), one can estimate the standard error of the Sharpe ratio. These require one to plug in estimates of the signal-noise ratio, and perhaps higher order moments. As outlined in Section 2.5.9, a confidence interval can be inflated to a prediction interval via a simple formula, namely the factor given in Equation 2.40.

3.2.4 † Concentration inequalities

Throughout this chapter we have considered the standard error of the Sharpe ratio under a number of deviations from the assumptions of *i.i.d.* normal returns. By showing that the standard error does not differ too much from the nominal value, we have established that hypothesis testing with moderate type I error rates is largely achievable. However, these results do not necessarily support testing with very small type I rates, as the tail distribution of the Sharpe ratio may be far from Gaussian. There are known bounds on large deviations of the t statistic which we can directly translate into equivalent facts regarding the Sharpe ratio. Under the null hypothesis, $\zeta = 0$, Hall and Wang [2004]; Wang and Hall [2009] showed that:

$$\Pr\left\{\hat{\zeta} \geq q\right\} \approx \left(1 - \Phi\left(\frac{n}{n-1}\sqrt{n}q\right)\right) \exp\left(-\frac{1}{3}\left(\frac{n}{n-1}\right)^3 q^3 n \frac{\mathrm{E}\left[x^3\right]}{\sigma^3}\right). \tag{3.39}$$

Here $\Phi\left(x\right)$ is the Gaussian distribution, and thus the approximation (which holds up to a factor in n^{-1}) compares the exceedance probability of the Sharpe ratio to the equivalent Gaussian law. Under the null, the term $\mathrm{E}\left[x^3\right]/\sigma^3$ is the skewness of returns, which appears in Bao's formula as γ_1. For moderately skewed returns and modestly sized n, we expect the correction factor to be around 1 ± 0.1 or so. This means that the type I rate assuming a normal distribution for the Sharpe ratio is "usually" within 10% of nominal, *cf.* Exercise 3.42.

It is worth noting that the deviance from the normal approximation is affected not by kurtosis *per se*, but by the skewness, which is to be expected from the Berry-Esseen theorem. It is not clear whether Equation 3.39 can be derived directly from the bias of the Sharpe ratio, which is also on the order of γ_1/n, *cf.* Equation 3.35.

In the case where $\zeta \neq 0$, a more complicated approximation holds. For a given q, let

$$\gamma = \frac{1}{2}\left(1 - \frac{n}{n-1}\frac{\zeta}{q}\right).$$

Then

$$\Pr\left\{\hat{\zeta} \geq q\right\} \approx \left(1 - \Phi\left(\frac{2\gamma n}{n-1}\sqrt{n}q\right)\right) \exp\left(\gamma^2\left(\frac{4\gamma}{3} - 2\right)\left(\frac{qn}{n-1}\right)^3 n \frac{\mathrm{E}\left[x^3\right]}{\sigma^3}\right). \tag{3.40}$$

Example 3.2.9 (Large deviations of the Sharpe ratio). We draw returns from a "Lambert W × Gaussian" distribution, with the skew parameter, δ varying from -0.4 to 0.4. [Goerg, 2010, 2009, 2016] We consider simulations with $n = 1,008$ and $n = 2,016$. For each of these we perform $1,000,000$ simulations under the null hypothesis, $\zeta = 0$, then compute the empirical probability that the Sharpe ratio exceeds some value q. We plot the empirical exceedance probabilities versus $1-\Phi\left(\frac{n}{n-1}\sqrt{n}q\right)$ in Figure 3.4, with lines for the right hand side of Equation 3.39. We see that the approximation matches the experiments fairly well. ⊣

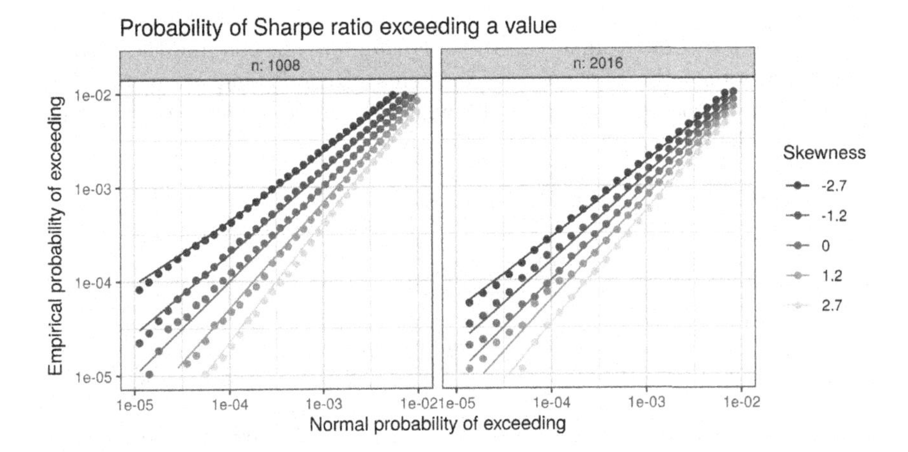

FIGURE 3.4
We draw n days of returns from a Lambert W $\times$ Gaussian distribution for various forms of the skew parameter, δ, under the null hypothesis, $\zeta = 0$. We compute the Sharpe ratio of returns, and repeat this experiment 1,000,000 times, computing the empirical probability that the Sharpe ratio exceeds given values. The empirical probability of exceeding q is plotted against $1 - \Phi\left(\frac{n}{n-1}\sqrt{n}q\right)$, the Gaussian approximation. Lines for the Wang-Hall approximation of Equation 3.39 are also plotted, and agree with the experimental results fairly well.

3.2.5 † Survivorship bias and finite moments

Among other reasons, the assumption of normal returns is questionable since it implies an infinite range for returns. Percent returns, however, are typically bounded to be no less than -100%, with some non-zero probability of hitting that lower bound. A -100% loss translates to a log return of $-\infty$; if such a loss can occur with positive probability, then the mean and all higher order moments of log returns will not exist. Needless to say, if moments of returns do not exist then the inferential techniques considered in this book are useless. They can be salvaged, though perhaps not the forward-looking use of the signal-noise ratio, by considering conditional moments.

For some fixed m, let the event that returns are less than or equal to m be called a *catastrophic event*. A catastrophic event could be caused by any number of reasons: sufficiently large loss for a strategy coupled with leverage, massive fraud by a company's leadership, global thermonuclear war, Godzilla, *etc.* Given that one is performing inference on returns to inform an investment decision, it would typically be the case that that such a catastrophic event has *not* been observed in the historical sample. Because investment returns have a finite upper bound[5], then, conditional on $x > m$, all moments of returns exist.

By analyzing returns from a *truncated distribution*, one can find the asymptotic distribution of the Sharpe ratio in terms of conditional moments. Let $v_t = 1_{\{x_t > m\}}$ be the 0/1 indicator for whether returns exceed m for some fixed m. Now define

$$\begin{aligned}
\hat{n} &= \sum_{1 \le t \le n} v_t, \\
\hat{\mu} &= \frac{\sum_{1 \le t \le n} v_t x_t}{\hat{n}}, \\
\hat{\sigma}^2 &= \frac{\sum_{1 \le t \le n} v_t \left(x_t - \hat{\mu}\right)^2}{\hat{n} - 1}.
\end{aligned}$$

When m is sufficiently negative to define a catastrophic event, one will not have observed such a loss, and it will be the case that $\hat{n} = n$ and $\hat{\mu}$ and $\hat{\sigma}^2$ match the usual definition.

Letting $\hat{\zeta} = \hat{\mu}/\hat{\sigma}$, it can be shown via the central limit theorem and delta method that

$$\sqrt{p\left(x > m\right) n}\left(\hat{\zeta} - \zeta\right) \rightsquigarrow \mathcal{N}\left(0, 1 + \frac{2 + \gamma_2}{4}\zeta^2 - \gamma_1 \zeta + \zeta^6\right), \tag{3.41}$$

where the population moments are all conditional on $x > m$. That is here we are defining

$$\begin{aligned}
\mu &= \mathrm{E}\left[x \,|\, x > m\right], \\
\sigma^2 &= \mathrm{Var}\left(x \,|\, x > m\right), \\
\gamma_1 &= \frac{\mathrm{E}\left[\left(x - \mu\right)^3 \,|\, x > m\right]}{\sigma^3}, \\
\gamma_2 &= \frac{\mathrm{E}\left[\left(x - \mu\right)^4 \,|\, x > m\right]}{\sigma^4} - 3,
\end{aligned}$$

and $\zeta = \mu/\sigma$. Again, since we assume an upper bound on returns, all these conditional moments exist.

The standard error of Equation 3.41 is essentially identical to that of Equation 2.28, up to the inclusion of $p\left(x > m\right)$ to adjust the effective number of observations, and the additional ζ^6 term, which should be so small as to be immaterial.

While we can salvage the inferential machinery, perhaps the investment

[5] A finite money supply implies an upper bound on wealth.

goal of maximizing the signal-noise ratio does not make sense when unconditional moments do not exist. Rather, perhaps one should seek to minimize the probability of a catastrophic event, or, following Roy [1952], seek to maximize the probability that returns exceed some given amount. However, estimating the probability of a catastrophic event will be difficult for some assets, and not something that can accurately done just by looking at the historical data. If the probability of catastrophe is sufficiently small, perhaps one can just seek to maximize signal-noise ratio, recognizing that, in the words of J. M. Keynes, "in the long run we are all dead."

Survivorship bias

Conditioning on not observing a catastrophic event, we have guaranteed that the conditional moments of returns exist. Moreover, the asymptotics of $\hat{\zeta}$ are essentially unchanged, swapping in the conditional moments. Assume now that the unconditional moments of returns exist, either because one is considering percent returns or because the returns are naturally bounded somehow. It remains the case that the conditional signal-noise ratio can be affected by *survivorship bias*: the returns are higher than expected because one is conditioning on survival of the asset to the present day. For fixed m define

$$\zeta_m := \frac{\mathrm{E}\left[x \,|\, x > m\right]}{\sqrt{\mathrm{Var}\left(x \,|\, x > m\right)}}.$$

It is natural to wonder whether ζ_m is larger than ζ due to the selection mechanism. We will consider the "overoptimism by selection" problem in greater detail in Chapter 4, wherein a strategy is selected based on its observed Sharpe ratio. Here we consider the overoptimism from selecting a strategy based on a floor on minimum observed return.

For general returns not much can be said. For example, the signal-noise ratio of lottery tickets conditional on only having selected winning tickets can be much larger than the unconditional signal-noise ratio. For Gaussian returns, however, the bias can be analyzed by considering conditional moments. Given $x \sim \mathcal{N}\left(\mu, \sigma^2\right)$, then, defining $z_m = \left(m - \mu\right)/\sigma$, the conditional moments are

$$\mathrm{E}\left[x \,|\, x > m\right] = \mu + \sigma \frac{\phi\left(z_m\right)}{1 - \Phi\left(z_m\right)},$$

$$\mathrm{Var}\left(x \,|\, x > m\right) = \sigma^2 \left[1 + \frac{z_m \phi\left(z_m\right)}{1 - \Phi\left(z_m\right)} - \left(\frac{\phi\left(z_m\right)}{1 - \Phi\left(z_m\right)}\right)^2\right]$$

The fraction $\phi\left(z_m\right) / \left(1 - \Phi\left(z_m\right)\right)$ is known as the *inverse Mills Ratio.*

When z_m is significantly negative, say less than -3, then the inverse Mills ratio will be very small, smaller than around 0.004. Writing ϵ for $\phi\left(z_m\right) / \left(1 - \Phi\left(z_m\right)\right)$, we have

$$\zeta_m = \zeta + \frac{\left(1 - \sqrt{1 + z_m \epsilon - \epsilon^2}\right)\zeta + \epsilon}{\sqrt{1 + z_m \epsilon - \epsilon^2}}. \tag{3.42}$$

When $\epsilon \approx 0$ this reduces to

$$\zeta_m \approx \left(1 - \frac{z_m \epsilon}{2}\right) \zeta + \epsilon \approx \zeta.$$

That is, if $x \le m$ is a very rare event, then the conditional signal-noise ratio is nearly equal to the unconditional one and the survivorship bias is negligible.

Example 3.2.10 (Effect of survivorship bias). Consider the case of some hypothetical returns stream with normally distributed returns where $\mu = 1\%\,\text{mo.}^{-1}$ and $\sigma = 5\%\,\text{mo.}^{-1/2}$. Furthermore, suppose that if this asset had ever experienced a single month's loss of -40% or worse, you would not be considering investing in the asset. While the unconditional signal-noise ratio is $0.2\,\text{mo.}^{-1/2}$, the conditional signal-noise ratio, ζ_m, is larger only by $1.821 \times 10^{-15}\,\text{mo.}^{-1/2}$. Thus while the Sharpe ratio is asymptotically unbiased for ζ_m, this is effectively equal to ζ. ⊣

3.3 Asymptotic distribution of functions of multiple Sharpe ratios

Now let $\boldsymbol{g}$ be some vector valued function of the vector $\boldsymbol{\zeta}$. Applying the delta method,

$$\sqrt{n}\left(\boldsymbol{g}\left(\hat{\boldsymbol{\zeta}}\right) - \boldsymbol{g}\left(\boldsymbol{\zeta}\right)\right) \rightsquigarrow \mathcal{N}\left(0, \left(\frac{\mathrm{d}\boldsymbol{g}}{\mathrm{d}\boldsymbol{\zeta}} \frac{\mathrm{d}\boldsymbol{\zeta}}{\mathrm{d}[\boldsymbol{\mu}^\top, \boldsymbol{\alpha_2}^\top]^\top}\right) \Omega \left(\frac{\mathrm{d}\boldsymbol{g}}{\mathrm{d}\boldsymbol{\zeta}} \frac{\mathrm{d}\boldsymbol{\zeta}}{\mathrm{d}[\boldsymbol{\mu}^\top, \boldsymbol{\alpha_2}^\top]^\top}\right)^\top\right) \tag{3.43}$$

For example, if one wanted to test the hypothesis that the signal-noise ratios of the p assets are equal, one would let $\boldsymbol{g}\left(\cdot\right)$ be the function which constructs the $p-1$ differences:

$$\boldsymbol{g}\left(\boldsymbol{\zeta}\right) = \left[\zeta_1 - \zeta_2, \ldots, \zeta_{p-1} - \zeta_p\right]^\top. \tag{3.44}$$

We will, however, consider the case of general $\boldsymbol{g}\left(\cdot\right)$.

3.3.1 Frequentist analysis

We assume $\boldsymbol{g}\left(\cdot\right)$ is defined such that it equals zero under the null. Asymptotically, under the null hypothesis that $\boldsymbol{g}\left(\boldsymbol{\zeta}\right) = \boldsymbol{0}$,

$$n\boldsymbol{g}\left(\hat{\boldsymbol{\zeta}}\right)^\top \left(\left(\frac{\mathrm{d}\boldsymbol{g}}{\mathrm{d}\boldsymbol{\zeta}} \frac{\mathrm{d}\boldsymbol{\zeta}}{\mathrm{d}[\boldsymbol{\mu}^\top, \boldsymbol{\alpha_2}^\top]^\top}\right) \Omega \left(\frac{\mathrm{d}\boldsymbol{g}}{\mathrm{d}\boldsymbol{\zeta}} \frac{\mathrm{d}\boldsymbol{\zeta}}{\mathrm{d}[\boldsymbol{\mu}^\top, \boldsymbol{\alpha_2}^\top]^\top}\right)^\top\right)^{-1} \boldsymbol{g}\left(\hat{\boldsymbol{\zeta}}\right) \sim \chi^2\left(\nu\right), \tag{3.45}$$

where ν is the rank of the matrix $\frac{\mathrm{d}\boldsymbol{g}}{\mathrm{d}\boldsymbol{\zeta}}$ at $\boldsymbol{\zeta}$.

To test the null hypothesis:

$$H_0 : \boldsymbol{g}(\boldsymbol{\zeta}) = \boldsymbol{0} \quad \text{versus} \quad H_1 : \boldsymbol{g}(\boldsymbol{\zeta}) \neq \boldsymbol{0},$$

first compute the test statistic

$$x^2 = n\boldsymbol{g}\left(\hat{\boldsymbol{\zeta}}\right)^\top \left(\left(\frac{\mathrm{d}\boldsymbol{g}}{\mathrm{d}\boldsymbol{\zeta}} \frac{\mathrm{d}\boldsymbol{\zeta}}{\mathrm{d}[\boldsymbol{\mu}^\top, \boldsymbol{\alpha_2}^\top]^\top}\right) \hat{\Omega} \left(\frac{\mathrm{d}\boldsymbol{g}}{\mathrm{d}\boldsymbol{\zeta}} \frac{\mathrm{d}\boldsymbol{\zeta}}{\mathrm{d}[\boldsymbol{\mu}^\top, \boldsymbol{\alpha_2}^\top]^\top}\right)^\top\right)^{-1} \boldsymbol{g}\left(\hat{\boldsymbol{\zeta}}\right). \tag{3.46}$$

Then reject if $x^2 \geq {\chi^2}_{1-\alpha}(\nu)$ the $1-\alpha$ quantile of the chi-square distribution with degrees of freedom ν equal to the rank of the matrix $\frac{\mathrm{d}\boldsymbol{g}}{\mathrm{d}\boldsymbol{\zeta}}$ at $\boldsymbol{\zeta}$.

This test was proposed by Wright *et al.* [2014]. The same test statistic was proposed by Leung and Wong [2008] for the $\boldsymbol{g}(\cdot)$ given in Equation 3.44. However, Leung and Wong propose that one reject if

$$\frac{(n-p+1)}{(n-1)(p-1)} x^2 \geq f_{1-\alpha}(p-1, n-1), \tag{3.47}$$

where $f_{1-\alpha}(p-1, n-1)$ is the $1-\alpha$ quantile of the F distribution with $p-1$ and $n-1$ degrees of freedom. Wright *et al.* find that the chi-square test gives closer to nominal coverage under the null than the F-test when returns are fat-tailed, even for large n. See also Exercise 3.15.

Again we compute the estimated covariance $\hat{\Omega}$ as described in Section 3.2, either by assuming elliptical returns and using Equation 3.27, or by computing the sample covariance of the vector of returns "stacked" with element-wise squared returns, $\left[\boldsymbol{x}^\top, {\boldsymbol{x}^2}^\top\right]^\top$. Ledoit and Wolf [2008] propose computing $\hat{\Omega}$ using HAC estimators or bootstrapping on the sample of stacked vectors $\left[\boldsymbol{x}^\top, {\boldsymbol{x}^2}^\top\right]^\top$.

For the case of scalar-valued g (*e.g.*, for comparing $p = 2$ assets), one can construct a two-sided test via an asymptotic t-approximation:

$$\sqrt{n} g\left(\hat{\zeta}\right) \left(\left(\frac{\mathrm{d}g}{\mathrm{d}\boldsymbol{\zeta}} \frac{\mathrm{d}\boldsymbol{\zeta}}{\mathrm{d}[\boldsymbol{\mu}^\top, \boldsymbol{\alpha_2}^\top]^\top}\right) \Omega \left(\frac{\mathrm{d}g}{\mathrm{d}\boldsymbol{\zeta}} \frac{\mathrm{d}\boldsymbol{\zeta}}{\mathrm{d}[\boldsymbol{\mu}^\top, \boldsymbol{\alpha_2}^\top]^\top}\right)^\top\right)^{-\frac{1}{2}} \sim t(n-1). \tag{3.48}$$

In all the above, one can construct asymptotic approximations of the test statistics under the alternative, allowing power analysis or computation of confidence regions on $\boldsymbol{g}(\boldsymbol{\zeta})$.

Following the rejection of the null of equal signal-noise ratios, one may be interested in testing which of the assets have different signal-noise ratios. This is via a *post hoc* test, *cf.* Section 4.2.

Example 3.3.1 (Equality of signal-noise ratios Fama-French factors). Consider the four Fama French factors from Example 1.1.1. Based on monthly returns from Jan 1927 to Dec 2020, the Sharpe ratios were computed as

$$\hat{\zeta} = \frac{\begin{bmatrix} Mkt & SMB & HML & UMD \end{bmatrix}}{\begin{bmatrix} 0.1773 & 0.0653 & 0.0919 & 0.1377 \end{bmatrix}} \text{mo.}^{-1/2},$$

and thus the *differences* in Sharpe ratios were computed as

$$\boldsymbol{g}\left(\hat{\zeta}\right) = \begin{bmatrix} 0.1120 & -0.0266 & -0.0458 \end{bmatrix} \text{mo.}^{-1/2}.$$

The covariance matrix of the error was estimated as

$$\hat{\Sigma}_{1/2} := \left(\frac{\mathrm{d}\boldsymbol{g}}{\mathrm{d}\boldsymbol{\zeta}} \frac{\mathrm{d}\boldsymbol{\zeta}}{\mathrm{d}[\boldsymbol{\mu}^\top, \boldsymbol{\alpha}_2{}^\top]^\top} \right) \hat{\Omega} \left(\frac{\mathrm{d}\boldsymbol{g}}{\mathrm{d}\boldsymbol{\zeta}} \frac{\mathrm{d}\boldsymbol{\zeta}}{\mathrm{d}[\boldsymbol{\mu}^\top, \boldsymbol{\alpha}_2{}^\top]^\top} \right)^\top \Bigg|_{\boldsymbol{\mu}=\hat{\boldsymbol{\mu}}, \boldsymbol{\alpha}_2=\hat{\boldsymbol{\alpha}}_2},$$

$$\approx \begin{bmatrix} 1.3570 & -0.6802 & 0.1928 \\ -0.6802 & 1.6200 & -0.9948 \\ 0.1928 & -0.9948 & 3.2058 \end{bmatrix} \text{mo.}^{-1}.$$

The test statistic is then computed as 11.9. Under the null hypothesis of equal signal-noise ratios, this is asymptotically distributed as a $\chi^2(3)$. This corresponds to a p-value of 0.008, and a Frequentist would reject the null hypothesis of equal signal-noise ratios.

Using the F-test, the test statistic is essentially divided by 3, resulting in a putative $F(3, 1125)$ statistic under the null. The test statistic is computed as 3.97, with a p-value of 0.008, little substantive difference.

Using the HAC estimator of covariance, the chi-squared test statistic is computed as 9.02, corresponding to a p-value under the null of 0.029. A Frequentist would then narrowly reject the null of equal signal-noise ratios. See also Example 4.2.2. ⊣

Example 3.3.2 (Equality of signal-noise ratios, HML and SMB). Continuing Example 3.2.2, consider the returns of the Fama French factors, SMB and HML. Based on monthly returns from Jan 1927 to Dec 2020, we compute the difference in Sharpe ratios of SMB and HML as $-0.027\,\text{mo.}^{-1/2}$. Under the null hypothesis of equal signal-noise ratios, the asymptotically t statistic of Equation 3.48 is computed as -0.7, which is distributed as a $t(1127)$ under the null, corresponding to a p-value of 0.76. There is little evidence to suggest either of SMB or HML has a higher signal-noise ratio than the other. ⊣

Example 3.3.3 (Equality of two signal-noise ratios, Normally distributed returns). Consider the case of two assets where $\boldsymbol{x}$ takes a bivariate normal distribution. Let ρ be the correlation between the two returns. Suppose the signal-noise ratios of the two assets are, respectively, $\zeta(1+\epsilon)$ and ζ. Suppose we are testing for equality of signal-noise ratio, so let $\boldsymbol{g}$ be the difference function.

As noted in Example 3.2.1, $\hat{\boldsymbol{\zeta}} \approx \mathcal{N}\left(\boldsymbol{\zeta}, \frac{1}{n}\left(\mathsf{R} + \frac{1}{2}\operatorname{Diag}\left(\boldsymbol{\zeta}\right)\left(\mathsf{R} \odot \mathsf{R}\right)\operatorname{Diag}\left(\boldsymbol{\zeta}\right)\right)\right)$. Then Equation 3.43 becomes

$$\left[\hat{\zeta}_1 - \hat{\zeta}_2\right] \rightsquigarrow \mathcal{N}\left(\epsilon\zeta, \frac{2}{n}\left(1-\rho\right) + \frac{\zeta^2}{2n}\left(1 + \left(1+\epsilon\right)^2 - 2\rho^2\left(1+\epsilon\right)\right)\right). \quad (3.49)$$

So for high correlation, the differences in Sharpe ratios will have small variance. See also Exercise 3.28.

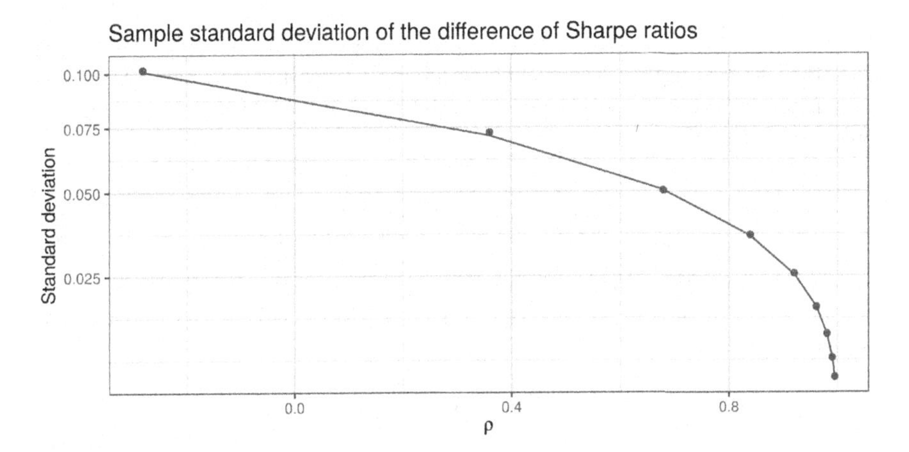

FIGURE 3.5
The empirical standard deviation for the difference of Sharpe ratios of two correlated normally distributed returns is shown for different values of the correlation, ρ. Each point represents 10,000 simulations of 252 days of normally distributed returns. The signal-noise ratios of the two strategies are $\zeta = 1\,\text{yr}^{-1/2}$ and $\zeta\left(1+\epsilon\right)$ with $\epsilon = 0.02$, assuming 252 days per year. The plotted line is the theoretical value from Equation 3.49, *not* a fit of the data. The y-axis is in square root scale.

To confirm the findings of Equation 3.49, for different values of ρ, $10,000$ simulations of 252 days of correlated returns were generated, with $\zeta = 1\,\text{yr}^{-1/2}$ and $\epsilon = 0.02$, assuming 252 days per year. The standard deviation of the differences in Sharpe ratios was computed for each value of ρ, and plotted versus ρ in Figure 3.5. The relationship of Equation 3.49 is given as a line, and fits the experimental data rather well. ⊣

Caution. The test for equality of signal-noise ratios, as described by Leung and Wong, and Wright *et al.* is designed to deal with correlation of returns. However, as demonstrated in Example 3.3.3 it can exhibit very high power when the returns series are highly positively correlated, deeming very small differences in Sharpe ratios as "significant." While a statistical test with high power sounds like a practitioner's dream, it is likely to be completely confounded by omitted variable bias. Unless you are sure that your returns are not affected by an omitted variable (and are you really?), take great care in interpreting the results of this test when returns are highly correlated.

3.3.2 Bayesian analysis

The approximation of Equation 3.43 can be used in a Bayesian framework as well. [Gelman *et al.*, 2013; Murphy, 2007] To stay within the usual Gaussian framework, consider the asymptotic distribution of $\boldsymbol{g}\left(\hat{\boldsymbol{\zeta}}\right)$ as being approximately multivariate normal. That is, if one observes the sample statistic $\hat{\boldsymbol{\zeta}}$ (either for the in-sample observations, or some future observations, say), based on a sample of size n, then simply take the following as true, even though it is only an approximation:

$$\boldsymbol{g}\left(\hat{\boldsymbol{\zeta}}\right) \sim \mathcal{N}\left(\boldsymbol{g}\left(\boldsymbol{\zeta}\right), \frac{1}{n}\mathsf{B}\right). \tag{3.50}$$

The parameter $\boldsymbol{g}\left(\boldsymbol{\zeta}\right)$ is of primary interest, and B is a nuisance. Let us suppose there are some prior beliefs on $\boldsymbol{g}\left(\boldsymbol{\zeta}\right)$ and B. We start with a Normal-Inverse Wishart model,

$$\begin{aligned} \mathsf{B} &\propto \mathcal{IW}\left(\mathsf{B}_0, m_0\right), \\ \boldsymbol{g}\left(\boldsymbol{\zeta}\right)|\mathsf{B} &\propto \mathcal{N}\left(\boldsymbol{a}_0, \frac{1}{l_0}\mathsf{B}\right). \end{aligned} \tag{3.51}$$

By "$\mathcal{IW}\left(\mathsf{B}_0, m_0\right)$," we mean an inverse Wishart distribution with matrix parameter B_0 and degrees of freedom m_0. Thus $\mathsf{B} \propto \mathcal{IW}\left(\mathsf{B}_0, m_0\right)$ means that the prior probability on B is proportional to

$$p\left(\mathsf{B}\right) \propto \left|\mathsf{B}\right|^{-(m_0+p+1)/2} \exp\left(-\frac{1}{2}\operatorname{tr}\left(\mathsf{B}_0\mathsf{B}^{-1}\right)\right), \tag{3.52}$$

where p is the number of elements in the vector $\boldsymbol{g}\left(\boldsymbol{\zeta}\right)$. Another way of stating this is that $\mathsf{X} \propto \mathcal{IW}\left(\Psi, m\right)$ means that X^{-1} takes a Wishart distribution with scale matrix Ψ^{-1} and m degrees of freedom. (See Section A.2.2.)

Via Equation A.17, the prior marginal distribution of $\boldsymbol{g}\left(\boldsymbol{\zeta}\right)$ is that of a *multivariate t*, written as

$$\boldsymbol{g}\left(\boldsymbol{\zeta}\right) \sim \mathcal{T}\left(m_0 - d + 1, \frac{\mathsf{B}_0}{l_0\left(m_0 - d + 1\right)}, \boldsymbol{a}_0\right).$$

(See Section A.2.2.)

The normal prior on $\boldsymbol{g}\left(\boldsymbol{\zeta}\right)$ means that

$$p\left(\boldsymbol{g}\left(\boldsymbol{\zeta}\right)\right)|\mathsf{B} \propto \left|\frac{\mathsf{B}}{l_0}\right|^{-1/2} \exp\left(-\frac{1}{2}\left(\boldsymbol{g}\left(\boldsymbol{\zeta}\right) - \boldsymbol{a}_0\right)^{\top}\left(\mathsf{B}/l_0\right)^{-1}\left(\boldsymbol{g}\left(\boldsymbol{\zeta}\right) - \boldsymbol{a}_0\right)\right). \tag{3.53}$$

Having observed n observations, and computing $\hat{\boldsymbol{\mu}}$, $\hat{\boldsymbol{\alpha}}_2$, and thus $\hat{\boldsymbol{\zeta}}$, as well as estimating the overall covariance $\hat{\Omega}$, consider the approximation of Equation 3.43 as being exact, and furthermore suppose that the estimated covariance from that equation takes a Wishart distribution with n degrees of freedom. Letting

$$\hat{\Sigma}_{1/2} := \left(\frac{\mathrm{d}\boldsymbol{g}}{\mathrm{d}\boldsymbol{\zeta}} \frac{\mathrm{d}\boldsymbol{\zeta}}{\mathrm{d}[\boldsymbol{\mu}^\top, \boldsymbol{\alpha}_2{}^\top]^\top} \right) \hat{\Omega} \left(\frac{\mathrm{d}\boldsymbol{g}}{\mathrm{d}\boldsymbol{\zeta}} \frac{\mathrm{d}\boldsymbol{\zeta}}{\mathrm{d}[\boldsymbol{\mu}^\top, \boldsymbol{\alpha}_2{}^\top]^\top} \right)^\top \Bigg|_{\boldsymbol{\mu}=\hat{\boldsymbol{\mu}}, \boldsymbol{\alpha}_2=\hat{\boldsymbol{\alpha}}_2}, \tag{3.54}$$

we are assuming that $n\hat{\Sigma}_{1/2}$ is Wishart with parameter B and n degrees of freedom. Thus the joint likelihood of $\boldsymbol{g}\left(\hat{\boldsymbol{\zeta}}\right)$ and $\hat{\Sigma}_{1/2}$ is

$$\begin{aligned} p\left(\boldsymbol{g}\left(\hat{\boldsymbol{\zeta}}\right), \hat{\Sigma}_{1/2}\right) \propto &\frac{\left|\hat{\Sigma}_{1/2}\right|^{\frac{n-p-2}{2}}}{|\mathsf{B}|^{n/2}} \exp\left(-\frac{1}{2}\operatorname{tr}\left(n\hat{\Sigma}_{1/2}\mathsf{B}^{-1}\right)\right) \times \\ &\exp\left(-\frac{1}{2}\left(\boldsymbol{g}(\boldsymbol{\zeta}) - \boldsymbol{g}\left(\hat{\boldsymbol{\zeta}}\right)\right)^\top \left(\frac{\mathsf{B}}{n}\right)^{-1} \left(\boldsymbol{g}(\boldsymbol{\zeta}) - \boldsymbol{g}\left(\hat{\boldsymbol{\zeta}}\right)\right)\right). \end{aligned} \tag{3.55}$$

The posterior distribution is then (*cf.* Exercise 3.30)

$$\begin{aligned} \mathsf{B} &\propto \mathcal{IW}\left(\mathsf{B}_1, m_1\right), \\ \boldsymbol{g}(\boldsymbol{\zeta})\,|\mathsf{B} &\propto \mathcal{N}\left(\boldsymbol{a}_1, \frac{1}{l_1}\mathsf{B}\right), \end{aligned} \tag{3.56}$$

where

$$\begin{aligned} l_1 &= l_0 + n, \\ \boldsymbol{a}_1 &= \frac{l_0\boldsymbol{a}_0 + n\boldsymbol{g}\left(\hat{\boldsymbol{\zeta}}\right)}{l_1}, \\ m_1 &= m_0 + n, \\ \mathsf{B}_1 &= \mathsf{B}_0 + n\hat{\Sigma}_{1/2} + \frac{l_0 n}{l_1}\left(\boldsymbol{g}\left(\hat{\boldsymbol{\zeta}}\right) - \boldsymbol{a}_0\right)\left(\boldsymbol{g}\left(\hat{\boldsymbol{\zeta}}\right) - \boldsymbol{a}_0\right)^\top. \end{aligned} \tag{3.57}$$

The posterior marginal distribution of $\boldsymbol{g}(\boldsymbol{\zeta})$ is that of a multivariate t, written as

$$\boldsymbol{g}(\boldsymbol{\zeta}) \sim \mathcal{T}\left(m_1 - d + 1, \frac{\mathsf{B}_1}{l_1\left(m_1 - d + 1\right)}, \boldsymbol{a}_1\right).$$

Example 3.3.4 (Difference of signal-noise ratios Fama-French factors, Bayesian analysis). We continue the analysis of Example 3.3.1, but as a Bayesian. We analyze the difference of signal-noise ratios, Market minus SMB, SMB minus HML, and HML minus UMD, respectively. We take the Bayesian prior to be

$$\begin{aligned} l_0 &= 60\,\text{mo.}, \\ \boldsymbol{a}_0 &= [0, 0, 0]^\top \text{mo.}^{-1/2}, \\ m_0 &= 20, \\ \mathsf{B}_0 &= \begin{bmatrix} 20.00 & 0.00 & 0.00 \\ 0.00 & 20.00 & 0.00 \\ 0.00 & 0.00 & 20.00 \end{bmatrix}. \end{aligned} \quad (3.58)$$

Having observed the Fama French monthly returns data from from Jan 1927 to Dec 2020, we estimate, as in Example 3.3.1,

$$\boldsymbol{g}\left(\hat{\zeta}\right) = [0.112, -0.027, -0.046]^\top \text{mo.}^{-1/2},$$

$$\hat{\Sigma}_{1/2} = \begin{bmatrix} 1.3570 & -0.6802 & 0.1928 \\ -0.6802 & 1.6200 & -0.9948 \\ 0.1928 & -0.9948 & 3.2058 \end{bmatrix} \text{mo.}^{-1}.$$

Combining this with the prior above, we have the posterior parameters

$$\begin{aligned} l_1 &= 1,188\,\text{mo.}, \\ \boldsymbol{a}_1 &= [0.106, -0.025, -0.044]^\top \text{mo.}^{-1/2}, \\ m_1 &= 1148, \\ \mathsf{B}_1 &= \begin{bmatrix} 1551.37 & -767.43 & 217.17 \\ -767.43 & 1847.43 & -1122.11 \\ 217.17 & -1122.11 & 3636.31 \end{bmatrix}. \end{aligned} \quad (3.59)$$

For sufficiently large m_1, the belief in B is highly concentrated around the mean value of the inverse Wishart, in this case $\mathsf{B}_1/\left(m_1 - 3 - 1\right)$. We compute this to be

$$\frac{1}{m_1 - 3 - 1}\mathsf{B}_1 = \begin{bmatrix} 1.3561 & -0.6708 & 0.1898 \\ -0.6708 & 1.6149 & -0.9809 \\ 0.1898 & -0.9809 & 3.1786 \end{bmatrix} \text{mo.}^{-1}.$$

As expected, this is not substantively different from $\hat{\Sigma}_{1/2}$, as the number of pseudo-observations in the prior is small compared to the number of actual observations.

To explore the implications of this posterior, consider the Market minus SMB element. Our posterior belief is that the difference in signal-noise ratios of Market minus SMB is approximately normally distributed with mean $0.106\,\text{mo.}^{-1/2}$ and standard deviation around $0.034\,\text{mo.}^{-1/2}$. In Example 3.3.1, we estimated the difference to be $0.112\,\text{mo.}^{-1/2}$, with estimated standard error around $0.035\,\text{mo.}^{-1/2}$. We draw 1,000 draws from the posterior, finding the minimal value of the difference in signal-noise ratios drawn is $0.018\,\text{mo.}^{-1/2}$. ⊣

3.4 † The ex-factor Sharpe ratio for (non-*i.i.d.*) Gaussian and elliptical returns

We will now consider the sensitivity of the ex-factor Sharpe ratio of Equation 1.10 to *i.i.d.* assumptions, but still assuming Gaussian returns. This is rather more complicated than the Sharpe ratio. Here we abuse the "augmented form" of vectors, which we will use later to analyze the distribution of portfolios. [Pav, 2013]

Recall the setup leading to Equation 1.10: we have a scalar return x_t, which is aligned with a l-vector, $\boldsymbol{f}_t$, which typically contains one element that is a constant 1. We will consider $\boldsymbol{f}$ to be random, which matches typical usage. Define

$$\tilde{\boldsymbol{x}}_t := \left[x_t, \boldsymbol{f}_t^{\top}\right]^{\top}. \tag{3.60}$$

Define the uncentered second moment of $\tilde{\boldsymbol{x}}$ as

$$\Theta := \mathrm{E}\left[\tilde{\boldsymbol{x}}\tilde{\boldsymbol{x}}^{\top}\right]. \tag{3.61}$$

(We will discuss Θ, in a slightly different form, in greater detail in Chapter 7.)

Given a sample of size n of returns and corresponding factors, we can define $\tilde{\mathsf{X}}$ as the $n \times (1 + l)$ matrix whose rows are $\tilde{\boldsymbol{x}}_1, \tilde{\boldsymbol{x}}_2, \ldots, \tilde{\boldsymbol{x}}_n$. The vanilla sample estimate[6] of Θ is

$$\hat{\Theta} := \frac{1}{n}\sum_t \tilde{\boldsymbol{x}}_t\tilde{\boldsymbol{x}}_t^{\top} = \frac{1}{n}\tilde{\mathsf{X}}^{\top}\tilde{\mathsf{X}}. \tag{3.62}$$

With some work we can extract the ex-factor Sharpe ratio from some transformations of $\hat{\Theta}$. Note that since $\hat{\Theta}$ is a simple average, it is unbiased.

Now note that

$$\Theta = \begin{bmatrix} \sigma^2 + \beta^{\top}\Gamma_f\beta & \beta^{\top}\Gamma_f \\ \Gamma_f\beta & \Gamma_f \end{bmatrix}, \tag{3.63}$$

where $\Gamma_f := \mathrm{E}\left[\boldsymbol{f}\boldsymbol{f}^{\top}\right]$. Simple matrix multiplication (Exercise 3.21) confirms that the inverse of Θ is

$$\Theta^{-1} = \begin{bmatrix} \sigma^{-2} & -\beta^{\top}\sigma^{-2} \\ -\beta\sigma^{-2} & {\Gamma_f}^{-1} + \sigma^{-2}\beta\beta^{\top} \end{bmatrix}, \tag{3.64}$$

and the Cholesky factor of that inverse is

$$\Theta^{-1/2} = \begin{bmatrix} \sigma^{-1} & 0 \\ -\beta\sigma^{-1} & {\Gamma_f}^{-1/2} \end{bmatrix}. \tag{3.65}$$

[6] *i.e.*, the one with n in the denominator, instead of, say $n - l$. Typical estimators of the regression coefficients may use a different denominator, an immaterial difference for sufficiently large n.

The ex-factor signal-noise ratio (*cf.* Equation 1.10) can thus be expressed as

$$\zeta_g = \frac{\boldsymbol{\beta}^\top \boldsymbol{v} - r_0}{\sigma} = -\operatorname{tr}\left(\boldsymbol{e}_1 \left[r_0, \boldsymbol{v}^\top\right] \Theta^{-1/2}\right). \tag{3.66}$$

The sample ex-factor Sharpe ratio takes the same form in the sample analogue:

$$-\operatorname{tr}\left(\boldsymbol{e}_1 \left[r_0, \boldsymbol{v}^\top\right] \hat{\Theta}^{-1/2}\right) \to \hat{\zeta}_g.$$

For fixed r_0, $\boldsymbol{v}$, define the function $g(\cdot)$ by

$$g\left(\hat{\Theta}\right) := -\operatorname{tr}\left(\boldsymbol{e}_1 \left[r_0, \boldsymbol{v}^\top\right] \hat{\Theta}^{-1/2}\right). \tag{3.67}$$

Example 3.4.1 (ex-factor Sharpe ratio on Fama-French factors). Consider the four Fama French factors from Example 1.1.1. We will model UMD as a linear combination of Market, SMB, HML, and an intercept term. Given the 1,128 months of data from Jan 1927 to Dec 2020, we compute

$$\hat{\Theta} = \left[\begin{array}{rrrrr} \mathit{UMD} & \mathit{Mkt} & \mathit{SMB} & \mathit{HML} & \mathit{intercept} \\ \hline 22.54 & -8.00 & -2.24 & -6.61 & 0.65 \\ -8.00 & 29.54 & 5.71 & 4.80 & 0.95 \\ -2.24 & 5.71 & 10.19 & 1.50 & 0.21 \\ -6.61 & 4.80 & 1.50 & 12.40 & 0.32 \\ 0.65 & 0.95 & 0.21 & 0.32 & 1.00 \end{array}\right] \text{bps mo.}^{1}$$

The Cholesky factor of the inverse is

$$\hat{\Theta}^{-1/2} = \left[\begin{array}{rrrrr} \mathit{UMD} & \mathit{Mkt} & \mathit{SMB} & \mathit{HML} & \mathit{intercept} \\ \hline 0.24 & 0.00 & 0.00 & 0.00 & 0.00 \\ 0.05 & 0.20 & 0.00 & 0.00 & 0.00 \\ 0.01 & -0.10 & 0.32 & 0.00 & 0.00 \\ 0.11 & -0.06 & -0.04 & 0.29 & 0.00 \\ -0.25 & -0.15 & -0.05 & -0.09 & 1.00 \end{array}\right] \%^{-1}\,\text{mo.}^{-1/2}$$

Now consider the idiosyncratic return of UMD. That is, we isolate the intercept portion of the attribution by taking $\boldsymbol{v} = \left[0, 0, 0, 1\right]^\top \%$, and we assume $r_0 = 0\%$. Via Equation 3.66 we compute,

$$\zeta_g = 0.247\,\text{mo.}^{-1/2}$$

Via the sample estimate of Ω, and using the delta method, we estimate the standard error of this to be $0.033\,\text{mo.}^{-1/2}$. In Example 2.5.9 under the assumption that the factor returns were deterministic, we computed a slightly smaller estimate of the standard error, $0.03\,\text{mo.}^{-1/2}$. ⊣

Now we will consider the effect of the distribution of $\hat{\zeta}_g$ on assumptions of independence, homoskedasticity, *etc.*, but keeping normality of returns. To tame the computations, some simplifying assumptions are made. We assume that

$$\tilde{\mathsf{X}} \sim \mathcal{MN}_{n,1+l}\left(\mathsf{M}, \mathsf{H}, \Sigma_f\right), \tag{3.68}$$

for some $n \times (1+l)$ matrix M, symmetric positive definite $n \times n$ matrix H and symmetric positive semidefinite[7] $(1+l) \times (1+l)$ matrix Σ_f. Recall (Section A.2.3) that this means

$$\operatorname{vec}\left(\tilde{\mathsf{X}}\right) \sim \mathcal{N}\left(\operatorname{vec}\left(\mathsf{M}\right), \Sigma_f \otimes \mathsf{H}\right).$$

For this characterization to be comparable to the vanilla case, we require $\mathsf{M}^\top \mathbf{1}_n = n\boldsymbol{\mu}$ and $\operatorname{tr}\left(\mathsf{H}\right) = n$. Note that the homoskedastic *i.i.d.* case corresponds to $\mathsf{M} = \mathbf{1}_n \boldsymbol{\mu}^\top$ and $\mathsf{H} = \mathsf{I}$.

From these we can compute the expected value and covariance of $\hat{\Theta}$. This will require heavy use of Isserlis' theorem. [Isserlis, 1918] Tedious computation (Exercise 3.20) yields

$$\begin{aligned} \operatorname{E}\left[\hat{\Theta}\right] &= \frac{1}{n}\left(\mathsf{M}^\top \mathsf{M} + \operatorname{tr}\left(\mathsf{H}\right)\Sigma_f\right), \\ \operatorname{Var}\left(\operatorname{vec}\left(\hat{\Theta}\right)\right) &= \frac{1}{n^2}\left(\mathsf{I} + \mathsf{K}\right)\left\{\mathsf{M}^\top \mathsf{H} \mathsf{M} \otimes \Sigma_f + \Sigma_f \otimes \mathsf{M}^\top \mathsf{H} \mathsf{M}\right. \\ &\qquad \left. + \operatorname{tr}\left(\mathsf{H}^2\right)\Sigma_f \otimes \Sigma_f\right\}. \end{aligned} \tag{3.69}$$

cf. Lemma 9 of Magnus and Neudecker [1986]. We will again insist, perhaps pigheadedly, that the sample estimate only be unbiased in the absence of variation of the mean value. That is, we take Θ to be the expected value of the expected second moment under uniform sampling of the $\tilde{\boldsymbol{x}}_t$, or

$$\Theta = \operatorname{E}_t\left[\operatorname{E}\left[\tilde{\boldsymbol{x}}_t \tilde{\boldsymbol{x}}_t^\top\right]\right],$$

where we sample the $\tilde{\boldsymbol{x}}_t$ uniformly. This implies that

$$\Theta = \Sigma_f + \boldsymbol{\mu}\boldsymbol{\mu}^\top, \tag{3.70}$$

and thus $\hat{\Theta}$ is only unbiased when $\mathsf{M} = \mathbf{1}_n \boldsymbol{\mu}^\top$, *cf.* Exercise 3.22.

Now we proceed as previously, using Taylor's theorem to claim the following approximations:

$$\begin{aligned} \operatorname{E}\left[\hat{\zeta}_g\right] &\approx g\left(\operatorname{E}\left[\hat{\Theta}\right]\right), \\ \operatorname{Var}\left(\hat{\zeta}_g\right) &\approx \left(\left.\frac{\mathrm{d}g\left(\Theta\right)}{\mathrm{d}\operatorname{vech}\left(\Theta\right)}\right|_{\Theta=\operatorname{E}\left[\hat{\Theta}\right]}\right) \operatorname{Var}\left(\operatorname{vech}\left(\hat{\Theta}\right)\right) \left(\left.\frac{\mathrm{d}g\left(\Theta\right)}{\mathrm{d}\operatorname{vech}\left(\Theta\right)}\right|_{\Theta=\operatorname{E}\left[\hat{\Theta}\right]}\right)^\top, \end{aligned} \tag{3.71}$$

where, via Equation A.33, the derivative can be computed as

$$\frac{\mathrm{d}g\left(\Theta\right)}{\mathrm{d}\operatorname{vech}\left(\Theta\right)} = \left(\mathsf{K}\operatorname{vec}\left(\boldsymbol{e}_1\left[r_0, \boldsymbol{v}^\top\right]\right)\right)^\top \left(\mathsf{L}\left(\mathsf{I}+\mathsf{K}\right)\left(\left(\left(\Theta^{-1/2}\right)^\top\right)^{-1} \otimes \Theta\right)\mathsf{L}^\top\right)^{-1}, \tag{3.72}$$

[7] We can allow elements of $\boldsymbol{f}$ to be deterministic by taking corresponding rows and columns of Σ_f to be all zeros.

where K is the commutation matrix of Definition A.1.3.

Before proceeding, we check analytically whether this formulation reduces to Equation 3.13 for the trivial case.

Example 3.4.2 (Homoskedastic, independent returns and Sharpe ratio standard error). Consider the case where returns are homoskedastic and independent (and thus $\mathsf{M} = \mathbf{1}\boldsymbol{\mu}^{\top}$ and $\mathsf{H} = \mathsf{I}$), and where there is a single "factor" identically equal to one. In this case

$$\Theta = \begin{bmatrix} \sigma^2 + \mu^2 & \mu \\ \mu & 1 \end{bmatrix}, \quad \text{and} \quad \Theta^{-1/2} = \begin{bmatrix} \sigma^{-1} & 0 \\ -\zeta & 1 \end{bmatrix}.$$

The variance of $\hat{\Theta}$ from Equation 3.69 becomes

$$\operatorname{Var}\left(\operatorname{vech}\left(\hat{\Theta}\right)\right) = \frac{1}{n} \begin{bmatrix} \sigma^4\left(4\zeta^2 + 2\right) & 2\sigma^3\zeta & 0 \\ 2\sigma^3\zeta & \sigma^2 & 0 \\ 0 & 0 & 0 \end{bmatrix}.$$

The derivative of Equation 3.72 becomes

$$\frac{\mathrm{d}g\left(\Theta\right)}{\mathrm{d}\operatorname{vech}\left(\Theta\right)} = \begin{bmatrix} -\frac{\zeta}{2\sigma^2} & \frac{1}{\sigma}\left(1 + \zeta^2\right) & -\frac{\zeta^3}{2} - \zeta \end{bmatrix}.$$

The approximate variance of Equation 3.71 becomes

$$\operatorname{Var}\left(\hat{\zeta}_g\right) \approx \frac{1}{n}\left(1 + \frac{\zeta^2}{2}\right),$$

i.e., the classical approximate standard error of Equation 2.24. Note that this computation is unchanged in the case where $r_0 \neq 0$ as long as one defines $\zeta = \left(\mu - r_0\right)/\sigma$. ⊣

Independent elliptical returns

We now derive similar results but instead of assuming normal returns, as in Equation 3.68, we assume returns follow an elliptical distribution. As noted in Section 3.1, we should avoid assuming returns follow an elliptical distribution across time, as this can cause even uncorrelated returns to be dependent at arbitrary time separation, and can lead to statistics which do not converge to the population values.

However, it is attractive to be able to describe the vector $\tilde{\boldsymbol{x}}_t$ as following an elliptical distribution (or distributions), with parameters varying over time. So assume that $\tilde{\boldsymbol{x}}_t$ follows an elliptical distribution with kurtosis factor κ_t, mean vector $\boldsymbol{\mu}_t$ and covariance matrix Σ_t. Note that in the cases where $\tilde{\boldsymbol{x}}_t$ contains a deterministic 1, the corresponding row and column of Σ_t will be all zero. Define $\Theta_t = \boldsymbol{\mu}_t{\boldsymbol{\mu}_t}^\top + \boldsymbol{\Sigma}_t$. Then Equation 3.69 is replaced by [Pav, 2013]

$$\begin{aligned}
\mathrm{E}\left[\hat{\Theta}\right] &= \frac{1}{n}\sum_t^n \Theta_t,\\
\mathrm{Var}\left(\mathrm{vec}\left(\hat{\Theta}\right)\right) &= \frac{1}{n^2}\sum_t^n \Big\{(\kappa_t - 1)\left[\mathrm{vec}\left(\boldsymbol{\Sigma}_t\right)\mathrm{vec}\left(\boldsymbol{\Sigma}_t\right)^\top + (\mathsf{I}+\mathsf{K})\,\boldsymbol{\Sigma}_t\otimes\boldsymbol{\Sigma}_t\right]\\
&\qquad + (\mathsf{I}+\mathsf{K})\left[\Theta_t\otimes\Theta_t - \boldsymbol{\mu}_t{\boldsymbol{\mu}_t}^\top\otimes\boldsymbol{\mu}_t{\boldsymbol{\mu}_t}^\top\right]\Big\}.
\end{aligned} \tag{3.73}$$

3.4.1 Market term, constant expectation

The most general form of the expectation and variance-covariance for the ex-factor Sharpe ratio presented above yields no simple closed form solution. As a simplifying assumption, in this section we assume that the factor returns consist of a constant one and the returns of a "Market," which has mean and volatility of μ_m and σ_m, respectively. Define $\zeta_m = \mu_m/\sigma_m$. Suppose that the asset has non-zero beta against the market, and define ζ_g as the idiosyncratic mean return divided by the residual volatility. We further require that M be constant over time, as otherwise the variance-covariance is too complicated.

$$\begin{aligned}
\mathsf{M} &= \mathbf{1}_n\boldsymbol{\mu}^\top,\\
\boldsymbol{\mu}^\top &= \left[\mu + \beta\mu_m, 1, \mu_m\right],\\
\boldsymbol{\Sigma}_f &= \begin{bmatrix} \beta^2\sigma_m^2 + \sigma^2 & 0 & \beta\sigma_m^2\\ 0 & 0 & 0\\ \beta\sigma_m^2 & 0 & \sigma_m^2 \end{bmatrix},\\
\zeta_m &:= \frac{\mu_m}{\sigma_m}, \qquad \text{and} \qquad \zeta_g := \frac{\mu - r_0}{\sigma}.
\end{aligned}$$

With $\Theta = \boldsymbol{\Sigma}_f + \boldsymbol{\mu}\boldsymbol{\mu}^\top$, we have

$$\Theta^{-1/2} = \begin{bmatrix} \frac{1}{\sigma} & 0 & 0\\ -\zeta_g & \sqrt{\zeta_m^2+1} & 0\\ -\frac{\beta}{\sigma} & -\frac{\zeta_m}{\sigma_m\sqrt{\zeta_m^2+1}} & \frac{1}{\sigma_m\sqrt{\zeta_m^2+1}} \end{bmatrix}.$$

Without variation in M over time, $\hat{\Theta}$ is unbiased for Θ. Via Equation 3.71, $\hat{\zeta}_g$ is approximately unbiased for ζ_g.

Via Equation 3.71, and a lot of algebra, the variance of $\hat{\zeta}_g$ is

$$\mathrm{Var}\left(\hat{\zeta}_g\right) \approx \frac{1}{n}\left[\frac{\mathrm{tr}\left(\mathsf{H}^2\right)}{n}\left(\frac{\zeta_g^2}{2} + \zeta_m^2\right) + \frac{\mathbf{1}^\top\mathsf{H}\mathbf{1}}{n}\right]. \tag{3.74}$$

3.4.2 Homoskedastic *i.i.d.* elliptical returns

For the completely "vanilla" case of homoskedastic, *i.i.d.* elliptically distributed returns, plugging the covariance of Equation 3.73 into Equation 3.71 one arrives at the following equation:

$$\operatorname{Var}\left(\hat{\zeta}_g\right) \approx \frac{1}{n}\left[\frac{3(\kappa-1)}{4}\zeta_g^2 + \frac{\zeta_g^2}{2} + \kappa\zeta_m^2 + 1\right]. \tag{3.75}$$

Note that $3(\kappa - 1)$ is the excess kurtosis of the marginal returns. One can derive this equation in the Gaussian case ($\kappa = 1$) by taking H in Equation 3.74 to be the identity matrix.

This nicely generalizes the approximate standard error of Equation 2.24 in the Gaussian case, and Equation 2.28 for unskewed returns. Note that one expects that ζ_m^2 is very small or nearly zero[8], so the addition of the Market term and the regression against it does little to increase the variance in estimated Sharpe ratio.

Example 3.4.3 (Standard error of ex-factor Sharpe ratio with Market). To verify Equation 3.75 in the Gaussian case, for selected values of ζ_m ranging from $0\,\mathrm{yr}^{-1/2}$ to $5\,\mathrm{yr}^{-1/2}$, 50,000 simulations were performed. In each simulation, 1,260 days of daily returns of an asset with fixed ex-factor signal-noise ratio of $0.79\,\mathrm{yr}^{-1/2}$ and "beta" of 1 against the Market term were drawn; the ex-factor Sharpe ratio was computed, and the standard deviation over the 50,000 simulations was computed. The experiment was repeated for returns drawn from an elliptical distribution, a multivariate t distribution with 6 degrees of freedom, corresponding to $\kappa = 2$. The empirical standard errors are plotted versus ζ_m in Figure 3.6, along with the theoretical value from Equation 3.75.

See also Exercise 3.25. ⊣

3.4.3 Homoskedastic autocorrelated Gaussian returns

The homoskedastic, simple autoregressive case of Section 3.1.4 corresponds to $\mathsf{H}_{i,j} = \rho^{|i-j|}$. Again, the ex-factor Sharpe ratio is approximately unbiased. The variance of $\hat{\zeta}_g$ becomes

$$\begin{aligned}\operatorname{Var}\left(\hat{\zeta}_g\right) &\approx \frac{1}{n}\left[\left(1 + 2\frac{n-1}{n}\rho^2 + \ldots\right)\left(\frac{\zeta_g^2}{2} + \zeta_m^2\right) + \left(1 + 2\frac{n-1}{n}\rho + \ldots\right)\right],\\ &\approx \frac{1}{n}\left[\left(\frac{\zeta_g^2}{2} + \zeta_m^2\right) + \frac{1+\rho}{1-\rho}\right],\end{aligned} \tag{3.76}$$

where we assume that ρ is small enough that ρ^n is negligible. The variance here is equivalent to Equation 3.22 for the case of the unattributed model

[8] If ζ_m^2 were large, presumably investors would merely invest in the Market instrument instead.

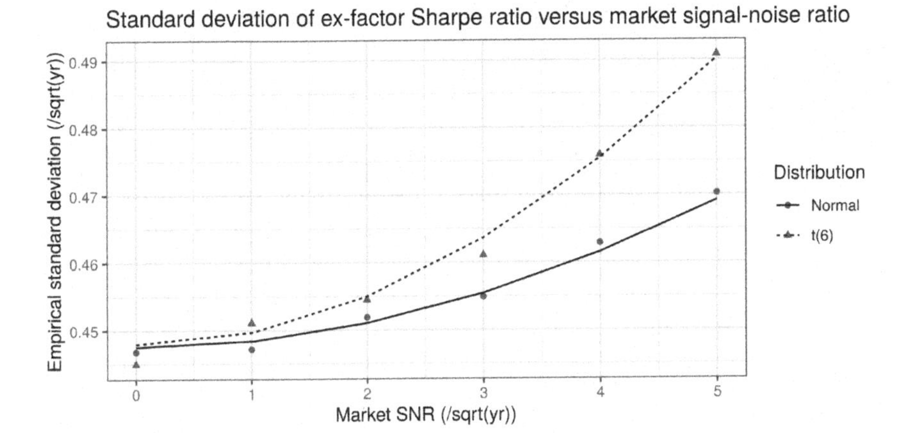

FIGURE 3.6
The empirical standard error of the ex-factor Sharpe ratio is plotted versus the Market signal-noise ratio for the case of an asset with a beta of 1 against the Market, and a ex-factor signal-noise ratio of $0.79\,\mathrm{yr}^{-1/2}$. The standard error is estimated from 50,000 simulations. Returns are drawn from a normal distribution, and from a multivariate t distribution with 6 degrees of freedom. The theoretical values from Equation 3.75 are plotted as a line for each value of κ.

under the small ρ case, and we see again that the Market term introduces a ζ_m^2/n term to the variance. Again, we see, as in the unattributed case, that a small autocorrelation of ρ inflates the standard error of the Sharpe ratio by about $100\,\rho\%$.

Example 3.4.4 (Standard error of ex-factor Sharpe ratio with Market, with autocorrelation). To verify Equation 3.76, for selected values of ρ ranging from -0.15 to 0.15, 50,000 simulations were performed. In each simulation, 1,260 days of daily returns of an asset with fixed ex-factor signal-noise ratio of $0.79\,\mathrm{yr}^{-1/2}$ and "beta" of 1 against the Market term were drawn, where the Market had signal-noise ratio of $1\,\mathrm{yr}^{-1/2}$. The ex-factor Sharpe ratio was computed, and the standard deviation was computed and plotted in Figure 3.7. We see good agreement of theoretical and empirical standard errors. ⊣

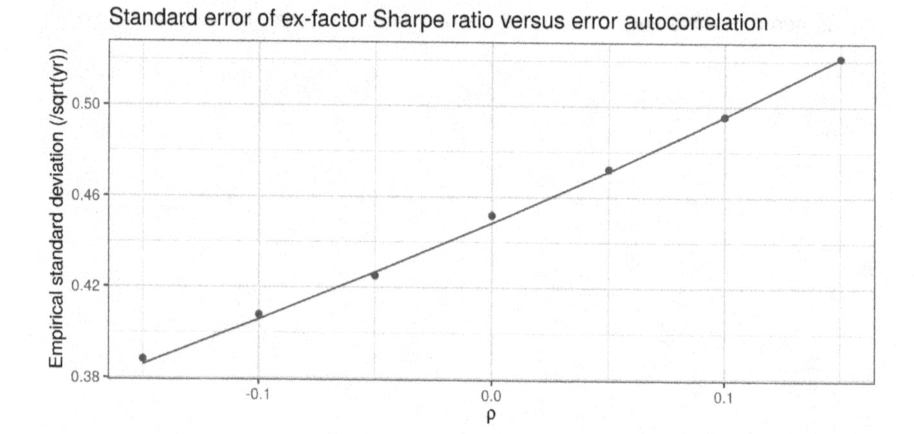

FIGURE 3.7
The empirical standard error of the ex-factor Sharpe ratio is plotted versus ρ, the autocorrelation of errors. The asset has a beta of 1 against the Market, and a ex-factor signal-noise ratio of $0.79\,\mathrm{yr}^{-1/2}$. The Market has a signal-noise ratio of $1\,\mathrm{yr}^{-1/2}$. Empirical standard errors are estimated over 50,000 simulations. The theoretical value from Equation 3.76 is plotted as a line.

3.5 † Asymptotic distribution of ex-factor Sharpe ratio

Continuing on the work from the previous section, here we consider the asymptotic distribution of the ex-factor Sharpe ratio, independent of assumptions on the distribution the returns take.

Note that since $\hat{\Theta}$ is a simple average, under mild conditions we can rely on the central limit theorem to claim that

$$\sqrt{n}\left(\operatorname{vech}\left(\hat{\Theta}\right) - \operatorname{vech}\left(\Theta\right)\right) \rightsquigarrow \mathcal{N}\left(0, \Omega\right), \tag{3.77}$$

where

$$\Omega = \operatorname{Var}\left(\operatorname{vech}\left(\hat{\Theta}\right)\right).$$

In general this variance-covariance matrix, Ω, is unknown, but can be consistently estimated from the data. As described in Section 3.2, Ω can be estimated via the sample covariance of $\operatorname{vech}\left(\tilde{\boldsymbol{x}}_i \tilde{\boldsymbol{x}}_i^{\top}\right)$; or $\hat{\Omega}$ can be constructed using Equation 3.73 and sample estimates of $\boldsymbol{\mu}$, Σ and κ.

By the multivariate delta method, we now claim

$$\sqrt{n}\left(g\left(\hat{\Theta}\right) - g\left(\Theta\right)\right) \rightsquigarrow \mathcal{N}\left(0, \left(\left.\frac{\mathrm{d}g\left(\Theta\right)}{\mathrm{d}\,\mathrm{vech}\left(\Theta\right)}\right|_{\Theta=\mathrm{E}\left[\hat{\Theta}\right]}\right) \Omega \left(\left.\frac{\mathrm{d}g\left(\Theta\right)}{\mathrm{d}\,\mathrm{vech}\left(\Theta\right)}\right|_{\Theta=\mathrm{E}\left[\hat{\Theta}\right]}\right)^{\top}\right), \quad (3.78)$$

where the derivative is as given in Equation 3.72.

3.5.1 Testing the ex-factor Sharpe ratio

From the variance computation, we can compute a Wald statistic to perform hypothesis testing. To be concrete, for scalar-valued function $g\left(\cdot\right)$, to test the hypothesis

$$H_0 : g\left(\Theta\right) = g_0 \quad \text{versus} \quad H_1 : g\left(\Theta\right) \neq g_0,$$

one computes the Wald statistic

$$W = \frac{\sqrt{n}\left(g\left(\hat{\Theta}\right) - g_0\right)}{\sqrt{\left(\left.\frac{\mathrm{d}g(\Theta)}{\mathrm{d}\,\mathrm{vech}(\Theta)}\right|_{\Theta=\mathrm{E}\left[\hat{\Theta}\right]}\right) \hat{\Omega} \left(\left.\frac{\mathrm{d}g(\Theta)}{\mathrm{d}\,\mathrm{vech}(\Theta)}\right|_{\Theta=\mathrm{E}\left[\hat{\Theta}\right]}\right)^{\top}}},$$

where $\hat{\Omega}$ is some estimate of $\mathrm{Var}\left(\mathrm{vech}\left(\hat{\Theta}\right)\right)$, and rejects at the α level when W is greater than $z_{1-\alpha}$, the 1-α quantile of the standard normal distribution.

Example 3.5.1 (ex-factor Sharpe ratio on Fama-French factors (continued)). Continuing Example 3.4.1, Consider the four Fama French factors from Example 1.1.1. We model UMD as a linear combination of Market, SMB, HML and an intercept term. Now consider the idiosyncratic return of UMD. That is, we isolate the intercept portion of the attribution by taking $\boldsymbol{v} = \left[0, 0, 0, 1\right]^{\top}\%$, and we assume $r_0 = 0\%$. Via Equation 3.66 we compute,

$$\zeta_g = 0.247\,\mathrm{mo.}^{-1/2}$$

Via the sample estimate of Ω, and using the delta method, we estimate the standard error of this to be $0.033\,\mathrm{mo.}^{-1/2}$. The resulting Wald statistic is then $W = 7.447$, and one would reject the null hypothesis that the idiosyncratic part is zero (or negative). ⊣

Example 3.5.2 (Technology ex-factor signal-noise ratio (continued)). Consider the ex-factor signal-noise ratio of monthly Technology industry returns against the Market, as described in Example 2.5.7. Taking $\boldsymbol{v} = \left[0, 1\right]^{\top}\%$ and $r_0 = 0\%$, we compute, via Equation 3.66

$$\zeta_g = 0.041\,\mathrm{mo.}^{-1/2}$$

First we estimate Ω by the sample estimate of the covariance of stacked vectors $\left[\tilde{\boldsymbol{x}}^{\top}, \tilde{\boldsymbol{x}}^{2\top}\right]^{\top}$. With this estimate, using the delta method, we estimated the standard error of ζ_g to be $0.03\,\mathrm{mo.}^{-1/2}$. The resulting Wald statistic is then $W = 1.361$, and we fail to reject the null hypothesis that the idiosyncratic part is zero (or negative).

If instead we use Equation 3.73 and sample estimates of $\boldsymbol{\mu}$, Σ, with $\kappa = 1$ (assuming Gaussian returns), the results change very little. We estimate the standard error of ζ_g to be $0.03\,\mathrm{mo.}^{-1/2}$, and the Wald statistic to be $W = 1.358$. If we feed in $\kappa = 2.828$, which is the average kurtosis of Technology and Market, divided by three, then the standard error estimate becomes $0.031\,\mathrm{mo.}^{-1/2}$, and the Wald statistic 1.321. Using the delta method with Equation 3.73 should give the same results as using Equation 3.75 directly. Plugging $n = 1,128\,\mathrm{mo.}$, $\zeta_g = 0.041\,\mathrm{mo.}^{-1/2}$, $\zeta_m = 0.177\,\mathrm{mo.}^{-1/2}$, and $\kappa = 2.828$ into Equation 3.75, we get a standard error estimate of $0.031\,\mathrm{mo.}^{-1/2}$, which is indeed the same.

Recall that in Example 2.5.10, we estimated the standard error of ζ_g to be $0.03\,\mathrm{mo.}^{-1/2}$. In that exercise we assumed that errors were Gaussian, that the returns of the Market were deterministic, and then adapted the approximate standard error of the t-distribution. This is nearly the same as the values we compute here via the delta method using three different estimators of Ω. In this case the extra computation has not changed our conclusions. For smaller n, however, one might see greater difference between the estimated standard errors. ⊣

3.5.2 Ex-factor Sharpe ratio prediction intervals

As in Section 2.5.9, we can construct an approximate prediction interval on functions of $\hat{\Theta}$. So suppose that for $i = 1, 2$, $\hat{\Theta}_i$ is computed on n_i independent observations drawn from the same population. Starting from Equation 3.77, we claim that

$$\left(\operatorname{vech}\left(\hat{\Theta}_1\right) - \operatorname{vech}\left(\hat{\Theta}_2\right)\right) \rightsquigarrow \mathcal{N}\left(0, \left(\frac{1}{n_1} + \frac{1}{n_2}\right)\Omega\right), \tag{3.79}$$

where convergence is jointly in n_1 and n_2.

Then by the multivariate delta method, we now have

$$\left(\frac{1}{n_1} + \frac{1}{n_2}\right)^{-\frac{1}{2}} \left(g\left(\hat{\Theta}_1\right) - g\left(\hat{\Theta}_2\right)\right) \rightsquigarrow \mathcal{N}\left(0, \left(\left.\frac{\mathrm{d}g\left(\Theta\right)}{\mathrm{d}\operatorname{vech}\left(\Theta\right)}\right|_{\Theta=\mathrm{E}[\hat{\Theta}]}\right)\Omega\left(\left.\frac{\mathrm{d}g\left(\Theta\right)}{\mathrm{d}\operatorname{vech}\left(\Theta\right)}\right|_{\Theta=\mathrm{E}[\hat{\Theta}]}\right)^{\top}\right), \tag{3.80}$$

where the derivative is as given in Equation 3.72. This is the same inflation factor given by Equation 2.40.

3.6 Sharpe ratio and non-normality, an empirical study

The results of this chapter largely exonerate the sample Sharpe ratio under non-normality: moderate autocorrelation and heteroskedasticity do not seem to cause great bias in the Sharpe ratio, nor do they greatly inflate its standard error. The formula for the standard error of ζ which applies under normality, Equation 2.24, is commonly used for real world returns, even though Mertens' formula, Equation 2.28, is more accurate. However, Mertens' formula relies on higher order moments which must be estimated, and thus might be *less* accurate because of this additional estimation burden. Leaving behind the theoretical findings of the rest of this chapter, we present an empirical study. Roughly, the goals of this study are to compare these two estimates of the standard error and present practical guidelines and limitations for their use.

In this study, we spawn returns according to one of a number of distributions, of a given sample size and signal-noise ratio, then we use each of three different methods to construct one-sided lower confidence intervals of a given nominal type I rate, α. we then record whether the signal-noise ratio falls within the computed confidence intervals. If the test maintains nominal coverage we should see the signal-noise ratio fall outside each confidence interval at the type I rate, that is, the empirical type I rate should match the nominal. For each distribution, sample size, and signal-noise ratio, we repeat this process $1e + 06$ times, recording the empirical average type I rate. Each simulated returns series is tested with each combination of method and α.

The probability distributions used are given in Table 3.1, along with their population skews and excess kurtoses. The distributions are:

1. The normal distribution, with zero skew and excess kurtosis.
2. Student's t distribution, considered twice, with degrees of freedom $\nu = 10$ and $\nu = 4$. These are both unskewed. The $\nu = 10$ distribution has moderate kurtosis, while in the $\nu = 4$ case the kurtosis is infinite.
3. The daily returns of the S & P 500 from 1970-01-05 through 2015-12-31, shifted and rescaled to be zero mean and unit standard deviation. As shown in Table 3.1, this distribution has mild negative skew and modest kurtosis. We also sample from a "symmetrized" S & P 500 distribution, where the absolute value of the daily S & P returns are given a random sign, which removes the skew, but leaves most of the kurtosis. When sampling from these distributions, all autocorrelation is removed.
4. Tukey's h-distribution, a special case of the "g and h distribution" when $g = 0$, for varying values of h. If z is a standard unit normal, then $y = \mu + ze^{hz^2/2}$ takes a Tukey h distribution. We consider the cases $h \in \{0.1, 0.2, 0.3\}$, which yield unskewed, but progressively more kurtotic distributions.
5. "Trio," a discrete distribution with three values, a low probability, high-value payout of v which occurs with probability p, and equally probable

payouts of l and h, each with probability $(1-p)/2$. Given p and v (with $v^2 \leq (1-p)/p$), the values of l and h are chosen so the process has zero mean and unit variance. The distribution is then shifted and scaled to achieve the target signal-noise ratio and variance.

6. Draws from the "Lambert W × Gaussian" distribution, with different values of the skew parameter, δ. [Goerg, 2010, 2009, 2016] If z takes a standard zero mean, unit variance normal distribution, then $y = \mu + ze^{\delta z}$ takes a Lambert W × Gaussian distribution with parameter δ. We consider the cases $\delta \in \{0.4, 0.2, -0.2, -0.4, -0.6\}$ with progressively more negative skews, and with kurtosis monotonic in $|\delta|$.

nickname	family	param	skew	kurtosis
normal	Gaussian		0	0
t10	Student's t	$\nu = 10$	0	1
t4	Student's t	$\nu = 4$	0	∞
sp500	SP500		−1.01	25.4
symsp500	Symmetric SP500		0	25.3
trio2	Trio	$p = 0.02, v = 2.5$	0.179	−1.43
trio1	Trio	$p = 0.01, v = 5$	1.14	3.83
trio05	Trio	$p = 0.005, v = 10$	4.92	47.3
tukey1	Tukey's h	$h = 0.1$	0	5.51
tukey2	Tukey's h	$h = 0.2$	0	36.2
tukey3	Tukey's h	$h = 0.3$	0	∞
lambertp4	Lambert × Gaussian	$\delta = 0.4$	2.74	17.8
lambertp2	Lambert × Gaussian	$\delta = 0.2$	1.24	5.69
lambertn2	Lambert × Gaussian	$\delta = -0.2$	−1.24	5.69
lambertn4	Lambert × Gaussian	$\delta = -0.4$	−2.74	17.8
lambertn6	Lambert × Gaussian	$\delta = -0.6$	−4.9	58.1

TABLE 3.1
The distributions used in the empirical are listed, with theoretical skew and *excess* kurtosis of each. The S&P daily returns are sampled from 1970-01-05 through 2015-12-31.

Again, during the simulations the returns distributions are shifted and rescaled to have a fixed signal-noise ratio. The signal-noise ratio in the simulations varies from $0\,\mathrm{yr}^{-1/2}$ through $2\,\mathrm{yr}^{-1/2}$. We simulate the case of observing between 1 through 48 months of *daily* returns, where a month always consists of exactly 21 days of returns.

We consider three methods of computing confidence intervals:

1. Approximating the standard error by "plugging in" to the form of Equation 2.24, and assuming approximate normality of the Sharpe ratio. We refer to this as "vanilla" for the vanilla standard error, Equation 2.24

2. Doing the same, but using Mertens' correction, Equation 2.28, called "mertens."
3. Assuming returns are drawn from a normal and using the quantile function of the non-central t-distribution to compute confidence intervals. This is referred to as the "t" method.

In all, We considered 5 values of signal-noise ratio, 7 values of the number of months, and 16 different returns distributions. This translates into 560 sets of simulations, each of which consist of 10^6 replications of generating the returns stream and then computing confidence intervals for 4 different values of the type I rate for each of the 3 methods considered. The result is a collection of $6,720$ empirical type I rejection rates.

It is not immediately clear how to evaluate and compare different methods of constructing confidence intervals. We compute the "conservatism" of each set of simulations, defined as the nominal type I rate divided by the empirical rate. This should be exactly equal to one: a larger value indicates the confidence interval is conservative; values smaller than 1 indicate the test is anti-conservative. Having computed the conservatism, we boxplot them in Figure 3.8 versus distribution skew. In Figure 3.9 the conservatism is boxplotted against the population excess kurtosis. In Figure 3.10, it is boxplotted versus n, in months.

One clear trend visible in Figure 3.8 is that all three methods seem to be anti-conservative for negative skew, and conservative for large positive skew. This is consistent with what we know of the true standard error of the Sharpe ratio via Equation 2.28, *i.e.*, that it is smaller when the population skew and the signal-noise ratio have the same size, and thus the we see a lower type I rate in practice. This is "solved" to some degree by using the Mertens' method for estimating standard errors, which sees near-nominal coverage for large absolute skew. Conversely, when the true skew is near or exactly zero, and thus Equation 2.24 gives the standard error of the Sharpe ratio to a good approximation, Mertens' method suffers from its ability to mis-estimate the skew.

A similar pattern is seen in Figure 3.9, with Mertens' method giving better coverage for large excess kurtosis, though not universally, and perhaps worse coverage when the true population excess kurtosis is zero. Again, this is based on the observation that if what you assume away when simplifying Mertens' form to the vanilla form happens to be true, you come out ahead, otherwise you lose control of your type I rate. Mertens' method tends to be anti-conservative for large excess kurtosis, though it seems more reliably so, as the other two methods tend to be quite poor in that region.

Finally, in Figure 3.10, we plot the conservatism versus the number of months of daily data observed. Here we see that Mertens' method is somewhat anti-conservative for sample sizes up to around two years. Beyond that sample size, it has negligible bias and much less spread than the other two methods. We generally prescribe Merten's method for inference when the sample size is this large.

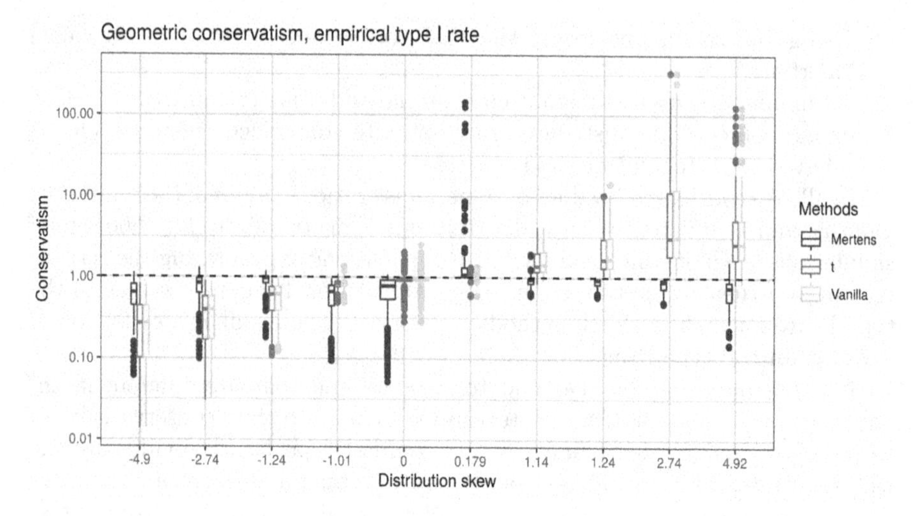

FIGURE 3.8
The geometric conservatism for each confidence interval is plotted versus the skew of the returns distribution. Conservatism is defined as the nominal type I rate divided by the empirical rate. Each box is over multiple values of signal-noise ratio, type I rate, sample size, and possibly returns distributions. The x-axis is not to scale. For non-zero skews Mertens' method does a better job of maintaining near nominal coverage than the other two methods, while it seems to perform worse for zero skew.

Mertens Correction

Mertens correction is appropriate when population skew or kurtosis are known to be large, or the sample size is more than around 500, otherwise the usual standard error is acceptable.

Note that this rule is mostly a realization of the fact that if you can assume away something true (*e.g.*, that the population skew is truly zero), you gain inferential power, and, conversely, if what you assume is false you lose power. We will witness this rule again.

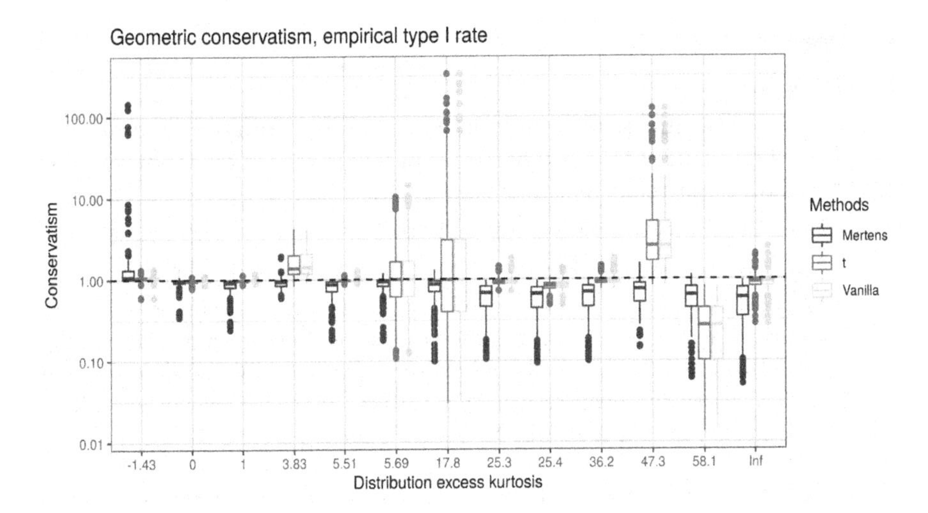

FIGURE 3.9
The geometric conservatism for each confidence interval is plotted versus the excess kurtosis of the returns distribution. Conservatism is defined as the nominal type I rate divided by the empirical rate. Each box is over multiple values of signal-noise ratio, type I rate, sample size, and possibly returns distributions. The x-axis is not to scale.

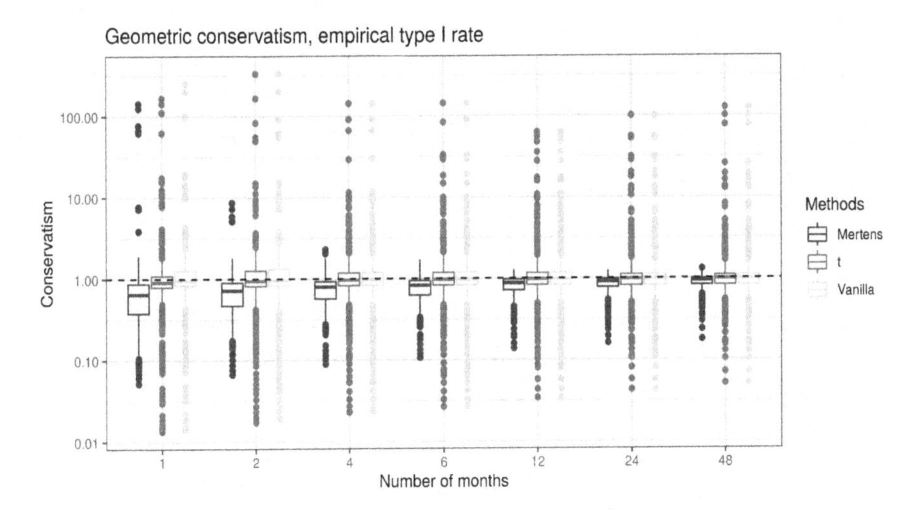

FIGURE 3.10
The geometric conservatism for each confidence interval is plotted versus the number of months observed. Conservatism is defined as the nominal type I rate divided by the empirical rate. Each box is over multiple values of signal-noise ratio, type I rate, and possibly returns distributions. The x-axis is not to scale.

3.7 Exercises

Ex. 3.1 The market, heteroskedasticity by quarter Repeat the analysis of Figure 3.2, but aggregating mean and standard deviation of daily log returns over quarters.

1. Does the negative relationship still hold? Is it statistically significant?
2. Compute the autocorrelation of quarterly aggregated volatilities.
3. Since volatility seems autocorrelated, and current quarter volatility seems to be correlated with negative returns, look for a market timing model based on vol. Lag the quarterly volatilities by one quarter, and scatter quarterly mean returns against last quarter's volatility. Is there still a negative relationship? Is it significant?

** **Ex. 3.2 Expectation of products of elliptical variables** Let $\boldsymbol{x}$ follow an elliptical distribution with parameters $\boldsymbol{\mu}$, Σ, and kurtosis factor κ (*cf.* Section A.2.2.). In the Gaussian case, we have $\kappa = 1$.
An extension of Isserlis' theorem to elliptical distributions gives us

$$\begin{aligned} \mathrm{E}\left[\left(x_i - \mu_i\right)\left(x_j - \mu_j\right)\left(x_k - \mu_k\right)\right] &= 0, \\ \mathrm{E}\left[\left(x_i - \mu_i\right)\left(x_j - \mu_j\right)\left(x_k - \mu_k\right)\left(x_l - \mu_l\right)\right] &= \kappa\left(\Sigma_{i,j}\Sigma_{k,l} + \Sigma_{i,k}\Sigma_{j,l} + \Sigma_{i,l}\Sigma_{j,k}\right), \end{aligned}$$

where there may be replication in the indices i, j, k, l. [Vignat and Bhatnagar, 2007]

1. Show that

$$\mathrm{E}\left[x_i x_j x_k\right] = \mu_i\mu_j\mu_k + \mu_i\Sigma_{j,k} + \mu_j\Sigma_{i,k} + \mu_k\Sigma_{i,j}.$$

2. Show that

$$\begin{aligned} \mathrm{E}\left[x_i x_j x_k x_l\right] = &\ \kappa\left(\Sigma_{i,j}\Sigma_{k,l} + \Sigma_{i,k}\Sigma_{j,l} + \Sigma_{i,l}\Sigma_{j,k}\right) \\ &+ \mu_i\mu_j\Sigma_{k,l} + \mu_i\mu_k\Sigma_{j,l} + \mu_i\mu_l\Sigma_{j,k} \\ &+ \mu_j\mu_k\Sigma_{i,l} + \mu_j\mu_l\Sigma_{i,k} + \mu_k\mu_l\Sigma_{i,j} \\ &+ \mu_i\mu_j\mu_k\mu_l \end{aligned}$$

3. Show that

$$\mathrm{E}\left[\left(\boldsymbol{x}^{\top}\boldsymbol{x}\right)\right] = \boldsymbol{\mu}^{\top}\boldsymbol{\mu} + \mathrm{tr}\left(\Sigma\right) = \mathrm{tr}\left(\boldsymbol{\mu}\boldsymbol{\mu}^{\top} + \Sigma\right).$$

4. Show that

$$\mathrm{E}\left[\boldsymbol{x}^{\top}\boldsymbol{x}\boldsymbol{x}^{\top}\mathbf{1}\right] = \boldsymbol{\mu}^{\top}\boldsymbol{\mu}\boldsymbol{\mu}^{\top}\mathbf{1} + 2\boldsymbol{\mu}^{\top}\Sigma\mathbf{1} + \mathrm{tr}\left(\Sigma\right)\boldsymbol{\mu}^{\top}\mathbf{1}.$$

5. Show that
$$\mathrm{E}\left[\left(\boldsymbol{x}^{\top}\mathbf{1}\right)^{3}\right] = \left(\boldsymbol{\mu}^{\top}\mathbf{1}\right)^{3} + 3\boldsymbol{\mu}^{\top}\mathbf{1}\mathbf{1}^{\top}\Sigma\mathbf{1}.$$

6. Show that
$$\mathrm{Cov}\left(x_i, x_j^2\right) = 2\mu_j\Sigma_{i,j}.$$

7. Show that
$$\begin{aligned}\mathrm{Cov}\left(x_i x_j, x_j x_k\right) = {} & (\kappa - 1)\,\Sigma_{i,j}\Sigma_{k,l} + \kappa\left(\Sigma_{i,k}\Sigma_{j,l} + \Sigma_{i,l}\Sigma_{j,k}\right) \\ & + \mu_i\mu_k\Sigma_{j,l} + \mu_i\mu_l\Sigma_{j,k} + \mu_j\mu_k\Sigma_{i,l} + \mu_j\mu_l\Sigma_{i,k}.\end{aligned}$$
   Use this to prove that
$$\mathrm{Cov}\left(x_i^2, x_j^2\right) = (\kappa - 1)\,\Sigma_{i,i}\Sigma_{j,j} + 2\kappa\Sigma_{i,j}\Sigma_{i,j} + 4\mu_i\mu_j\Sigma_{i,j}.$$

8. Let $\boldsymbol{y} = \mathrm{vec}\left(\boldsymbol{x}\boldsymbol{x}^{\top}\right) = \boldsymbol{x}\otimes\boldsymbol{x}$. Show that
$$\mathrm{E}\left[\boldsymbol{y}\right] = \boldsymbol{\mu}\otimes\boldsymbol{\mu} + \mathrm{vec}\left(\Sigma\right).$$
   Show that
$$\begin{aligned}\mathrm{Var}\left(\boldsymbol{y}\right) = {} & (\kappa - 1)\left[\mathrm{vec}\left(\Sigma\right)\mathrm{vec}\left(\Sigma\right)^{\top} + (\mathsf{I} + \mathsf{K})\,\Sigma\otimes\Sigma\right] \\ & + (\mathsf{I} + \mathsf{K})\left(\left(\boldsymbol{\mu}\boldsymbol{\mu}^{\top} + \Sigma\right)\otimes\left(\boldsymbol{\mu}\boldsymbol{\mu}^{\top} + \Sigma\right) - \left(\boldsymbol{\mu}\boldsymbol{\mu}^{\top}\right)\otimes\left(\boldsymbol{\mu}\boldsymbol{\mu}^{\top}\right)\right),\end{aligned}$$
   where K is the commutation matrix of Definition A.1.3.

9. Prove Equation 3.27: let $\boldsymbol{y} = \left[\boldsymbol{x}^{\top}, {\boldsymbol{x}^2}^{\top}\right]^{\top}$, and let Ω be the covariance matrix of $\boldsymbol{y}$. Show that Ω equals
$$\begin{bmatrix}\Sigma & 2\Sigma\mathsf{U} \\ 2\mathsf{U}\Sigma & (\kappa - 1)\,\boldsymbol{s}\boldsymbol{s}^{\top} + 2\kappa\left(\Sigma\odot\Sigma\right) + 4\mathsf{U}\Sigma\mathsf{U}\end{bmatrix},$$
   where U is the diagonal matrix with vector μ on its diagonal, and $\boldsymbol{s}$ is the column vector of the diagonal of Σ.

Ex. 3.3 Sweeping column means Let A be a matrix with n rows (or a column vector of length n). Show that $\frac{\mathbf{1}\mathbf{1}^{\top}}{n}\mathsf{A}$ is the row vector of means of each column of A. Show that each column of $\mathsf{A} - \frac{\mathbf{1}\mathbf{1}^{\top}}{n}\mathsf{A}$ is zero-mean.

** **Ex. 3.4 Moments of products of elliptical variables II** Let $\boldsymbol{x}$ follow an elliptical distribution with parameters $\boldsymbol{\mu}$, Σ, and kurtosis factor κ. Use results from Exercise 3.2 to prove the relationship in Equation 3.3.

1. Show that
$$\mathrm{E}\left[\left(\boldsymbol{x}^{\top}\mathbf{1}\right)^{4}\right] = \left(\boldsymbol{\mu}^{\top}\mathbf{1}\right)^{4} + 3\kappa\left(\mathbf{1}^{\top}\Sigma\mathbf{1}\right)^{2} + 6\left(\boldsymbol{\mu}^{\top}\mathbf{1}\right)^{2}\mathbf{1}^{\top}\Sigma\mathbf{1}.$$
   Use this to prove that
$$\mathrm{Var}\left(\left(\boldsymbol{x}^{\top}\mathbf{1}\right)^{2}\right) = (3\kappa - 1)\left(\mathbf{1}^{\top}\Sigma\mathbf{1}\right)^{2} + 4\left(\boldsymbol{\mu}^{\top}\mathbf{1}\right)^{2}\mathbf{1}^{\top}\Sigma\mathbf{1}.$$

2. Show that

$$\mathrm{E}\left[\left(\boldsymbol{x}^{\top}\boldsymbol{x}\right)^{2}\right] = \left(\boldsymbol{\mu}^{\top}\boldsymbol{\mu}\right)^{2} + \kappa\left(\mathrm{tr}\left(\Sigma\right)\right)^{2} + 2\kappa\,\mathrm{tr}\left(\Sigma\Sigma\right) + 2\boldsymbol{\mu}^{\top}\boldsymbol{\mu}\,\mathrm{tr}\left(\Sigma\right) + 4\boldsymbol{\mu}^{\top}\Sigma\boldsymbol{\mu}.$$

Use this to prove that

$$\mathrm{Var}\left(\left(\boldsymbol{x}^{\top}\boldsymbol{x}\right)\right) = \left(\kappa - 1\right)\left(\mathrm{tr}\left(\Sigma\right)\right)^{2} + 2\kappa\,\mathrm{tr}\left(\Sigma\Sigma\right) + 4\boldsymbol{\mu}^{\top}\Sigma\boldsymbol{\mu}.$$

3. Show that

$$\mathrm{E}\left[\left(\boldsymbol{x}^{\top}\mathbf{1}\right)^{2}\boldsymbol{x}^{\top}\boldsymbol{x}\right] = \left(\kappa\mathbf{1}^{\top}\Sigma\mathbf{1} + \boldsymbol{\mu}^{\top}\boldsymbol{\mu}\right)\mathrm{tr}\left(\Sigma\right) + 2\kappa\mathbf{1}^{\top}\Sigma^{2}\mathbf{1} + \left(\boldsymbol{\mu}^{\top}\boldsymbol{\mu} + \mathrm{tr}\left(\Sigma\right)\right)\left(\boldsymbol{\mu}^{\top}\mathbf{1}\right)^{2} + 4\left(\boldsymbol{\mu}^{\top}\mathbf{1}\right)\boldsymbol{\mu}^{\top}\Sigma\mathbf{1}.$$

Use this to prove that

$$\mathrm{Cov}\left(\boldsymbol{x}^{\top}\boldsymbol{x}, \left(\boldsymbol{x}^{\top}\mathbf{1}\right)^{2}\right) = \left(\kappa - 1\right)\mathbf{1}^{\top}\Sigma\mathbf{1}\,\mathrm{tr}\left(\Sigma\right) + 2\kappa\mathbf{1}^{\top}\Sigma^{2}\mathbf{1} + 4\left(\boldsymbol{\mu}^{\top}\mathbf{1}\right)\boldsymbol{\mu}^{\top}\Sigma\mathbf{1}.$$

** 4. Use the results of the previous three questions to show that

$$\mathrm{Var}\left(\boldsymbol{x}^{\top}\boldsymbol{x} - \frac{\left(\boldsymbol{x}^{\top}\mathbf{1}\right)^{2}}{n}\right) = \left(\kappa - 1\right)\mathrm{tr}\left(\Sigma_{c}\right)\mathrm{tr}\left(\Sigma_{c}\right) + 2\kappa\,\mathrm{tr}\left(\Sigma_{c}\Sigma_{c}\right) + 4\,\mathrm{tr}\left(\boldsymbol{\mu}_{c}\boldsymbol{\mu}_{c}^{\top}\Sigma\right),$$

where $\boldsymbol{\mu}_c := \boldsymbol{\mu} - \frac{\mathbf{1}\mathbf{1}^{\top}}{n}\boldsymbol{\mu}$ and $\Sigma_c := \Sigma - \frac{\mathbf{1}\mathbf{1}^{\top}}{n}\Sigma$ are the column centered mean and covariance. This is the bottom corner of Equation 3.3.

Ex. 3.5 Proofs

1. Derive Equation 3.11 from Equation 3.7 assuming $\kappa = 1$.

* 2. Derive Equation 3.11 from Equation 3.7 for the case of general κ.

Ex. 3.6 Elliptical marginals over time Let $\boldsymbol{x}$ be an n-vector which follows an elliptical distribution with mean zero and covariance parameter $\Sigma = \sigma^2\mathsf{I}$, and kurtosis factor κ. Let $s^2 = \boldsymbol{x}^{\top}\boldsymbol{x}/n$

1. Show that $\mathrm{E}\left[s^2\right] = \sigma^2$.
2. Show that $\mathrm{Var}\left(s^2\right) = \sigma^4\left(\kappa - 1 + \frac{2\kappa}{n}\right)$.
3. Confirm that the standard error of s^2 does not go to zero as $n \to \infty$ by way of Monte Carlo simulations using draws from a multivariate t distribution with 5 degrees of freedom.

Ex. 3.7 Autocorrelation matrix computations Let Σ be the $n \times n$ correlation matrix associated with simple autocorrelation, from Section 3.1.4. That is, $\Sigma_{i,j} = \sigma^2\rho^{|i-j|}$.

1. Show that $\mathbf{1}^\top \mathsf{\Sigma} \mathbf{1} \approx \sigma^2 \left(n\frac{1+\rho}{1-\rho} + \frac{2\rho}{(1-\rho)^2}\right)$ for large n. (*Hint:* write $\mathbf{1}^\top \mathsf{\Sigma} \mathbf{1}/\sigma^2 = n + 2(n-1)\rho^1 + 2(n-2)\rho^2 + \ldots + 2\rho^{n-1}$. Then derive approximations for $n\sum_i \rho^i$ and $\sum_i i\rho^{i-1}$. For the latter, use calculus.)

2. Show that $\operatorname{tr}\left(\mathsf{\Sigma}^2\right) \approx \sigma^4 \left(n\frac{1+\rho^2}{1-\rho^2} + \frac{2\rho^2}{(1-\rho^2)^2}\right)$ for large n. (*Hint:* Use the previous approximation.)

Ex. 3.8 Autocorrelation and standard error A positive (negative) autocorrelation inflates (deflates) the standard error of the Sharpe ratio. Can you give an intuitive explanation for why this is so? How should an investor buy a strongly mean reverting asset? A strongly trend-following asset?

*** Ex. 3.9 Bounds on heteroskedasticity bias** Suppose you observe otherwise homoskedastic returns levered by a fund manager (*i.e.*, the $\lambda = 1$ case). The fund manager can choose leverage l between $\frac{1}{2}$ and 2. What is the largest possible value of the squared coefficient of variation of l? (Note: this corresponds to normalizing to unit mean, and then computing M_2.) Suppose the underlying homoskedastic returns have signal-noise ratio between 0 and $0.15\,\mathrm{day}^{-1/2}$.

1. What is the smallest possible value of the geometric bias, b?
2. What is the largest possible value of the (approximate) variance of the Sharpe ratio, given in Equation 3.19?

Ex. 3.10 Homoskedastic varying mean returns In Section 3.1.3, we considered the case where $\boldsymbol{\mu}$ and $\boldsymbol{\sigma}^2$ vary together. Here consider the homoskedastic case, *i.e.*, where $\mathsf{\Sigma} = \sigma^2\mathsf{I}$, with varying mean: $\boldsymbol{\mu} = \mu\boldsymbol{l}$, where $\frac{1}{n}\mathbf{1}^\top\boldsymbol{l} = 1$. Again, M_2 is the squared coefficient of variation of $\boldsymbol{l}$.

1. Assuming $0 \le \mu/\sigma \le 0.2$, what is the smallest possible value of b, expressed in terms of M_2?
2. Simplify Equation 3.11 for this case. Express the variance in terms of μ, σ, b and M_2. Is it much different than in the *i.i.d.* homoskedastic case?

Ex. 3.11 Returns at different frequencies Repeat the exercise of Example 3.1.3, but under different situations:

1. Assume you have 10 years of monthly data and 2 years of daily data (at a rate of 252 days per year). Assume $r_0 = 0$, $\mu = 0.0012\,\mathrm{day}^{-1}$, $\sigma = 0.02\,\mathrm{day}^{-1/2}$. What is the bias in the Sharpe ratio? Converting the Sharpe ratio back to daily units, what is the approximate standard error? If, instead, one had observed 12 years of daily data, what would the standard error be? Perform 1,000 Monte Carlo simulations to test your work.
2. Assume you have 100 years of yearly data and 20 years of monthly data

on normal returns. Assume $r_0 = 0$, $\mu = 0.014\,\text{mo.}^{-1}$, $\sigma = 0.05\,\text{mo.}^{-1/2}$. What is the bias in the Sharpe ratio? Converting the Sharpe ratio back to monthly units, what is the approximate standard error? If, instead, one had observed 120 years of monthly data, what would the standard error be? Perform 1,000 Monte Carlo simulations to test your work.

Ex. 3.12 Any unskewed distribution Suppose returns are drawn *i.i.d.* from the product of a normal and an independent positive random variable. That is, let

$$x_t = z_t l_t + \mu,$$

where z_t are *i.i.d.* standard normal random variables independent of l_t which are *i.i.d.* and positive. Let M_k be the raw kth population moment of l_t:

$$M_k := \mathrm{E}\left[l^k\right].$$

Under these assumptions, the analysis of Section 3.1.3 is relevant with $\lambda = 0$, except here we are using M_k to be the *population* moments whereas we had previously defined M_k to be the *empirical* moment for a given sample. For large n, the empirical moments will approach the population moments.

1. Find the mean and variance of x. What is the signal-noise ratio of x?
2. Find the skew and kurtosis of x, in terms of the M_k.
3. Using Equation 3.16, what is the expected geometric bias of $\hat{\zeta}$?
4. Using Equation 3.19, write the approximate standard error of $\hat{\zeta}$.
5. Draw $n = 10^4$ realizations of a log normal l_i with $M_1 = 1$ and $M_2 = e^1$. Multiply these by independent normals z_t and add $\mu = 0.001$ to get x_t, and compute the signal-noise ratio of the x series. Repeat this 100 times. Confirm the theoretical bias and standard errors of $\hat{\zeta}$.

* 6. Compute the centered moments of x with respect to M_k. Is it the case that *any* unskewed probability distribution admits a representation as the product of a normal and an independent positive random variable? (The point here is that the results of Section 3.1.3 are not universally applicable to all unskewed returns distributions.)

Ex. 3.13 Correlation Fama-French Sharpe ratio, elliptical Repeat the analysis of Example 3.2.3, but assume the Fama French returns follow an elliptical distribution, and use Equation 3.29 to compute the variance-covariance of $\hat{\boldsymbol{\zeta}}$. Take the four Fama French factors monthly returns from Jan 1927 to Dec 2020.

1. Using those returns, estimate the correlation matrix R via the usual sample estimator.
2. Using those returns, estimate the kurtosis of each series separately. Take the median value and divide by three to get an estimate of κ.
3. Use Equation 3.29 to compute the variance-covariance of $\hat{\boldsymbol{\zeta}}$.

4. Turn that variance-covariance estimate into a correlation estimate and compare to the results of Example 3.2.3.
5. Is elliptical distribution a good assumption for the Fama French returns? Check by computing the skewness of the "HML" returns, and bootstrapping for significance. elliptical distributions have zero skewness.

Ex. 3.14 Equality of industry signal-noise ratios Use the Leung-Wong test (Equation 3.45) on the Fama French 5 Industry returns data to test the hypothesis that the 5 industries have equal signal-noise ratios.

Ex. 3.15 Comparing tests on signal-noise ratios Wright *et al.* find that the chi-square test (Inequality 3.46) has closer to nominal coverage than the F-test (Inequality 3.47) for fat-tailed distributions. Replicate their work. Set $n = 2,000$, and let $p = 5, 10, 20$. Draw returns from multivariate normal, and from the elliptical multivariate t distribution with $4, 6,$ and 8 degrees of freedom. Let the returns have zero mean. Test the hypothesis that all assets have the same signal-noise ratio at the $\alpha = 0.05$ level. Use Equation 3.27 to compute $\hat{\Omega}$, assuming normally distributed returns. Compare these with table I of Wright *et al.* [2014], which finds, for example, that for $t\,(4)$ returns and $p = 20$, the F test has an empirical rejection rate of around 74.7%, while the chi-square test has a rejection rate of around 17.3%.

1. Repeat the experiment using the chi-square statistic, but use Equation 3.27 assuming the true value of κ is known to compute $\hat{\Omega}$.
2. Repeat the experiment using the chi-square statistic, but use HAC estimators on the sample of stacked vectors $\left[\boldsymbol{x}^{\top}, {\boldsymbol{x}^{2}}^{\top}\right]^{\top}$ to compute $\hat{\Omega}$, as in Ledoit and Wolf [2008].

Ex. 3.16 Equality of unpaired sample signal-noise ratios The method for testing multiple signal-noise ratios outlined in Section 3.3 and Equation 3.45 apply to the case of a paired sample, *i.e.*, contemporaneous returns with possible correlation among them.

1. Formulate the test for equality of 2 signal-noise ratios with an unpaired sample by abusing Equation 3.45. (Simply compute Ω as a diagonal matrix.) Your test should be capable of testing against a one-sided alternative.
2. Use your test on the returns of the Market, dividing the sample into two periods using the cutoff date of 1970-01-01. *cf.* Exercise 2.9, where the same test is performed assuming Gaussian returns.
3. Formulate the test for equality of multiple signal-noise ratios with an unpaired sample.

Ex. 3.17 Testing signal-noise ratios, mixed samples Suppose you observe two returns streams with some period of overlap. So for example, suppose you observe returns x_t for $t = 1, \ldots, n_1$ and returns y_t for $t = m, m+1, \ldots, m+n_2-1$ with $t < m \le n_1 < m + n_2 - 1$. Formulate the test for equality of 2 signal-noise ratios in this case.

Ex. 3.18 Overlapping returns Suppose you observe overlapping returns. That is, x_t correspond to returns over a time period of length q, measured in years, observed at a rate of $1/r$ per year, evenly spaced through the year, with $q \ge r$. For example, suppose you observe rolling quarterly returns every month, in which case $q = \frac{1}{4}$ and $r = \frac{1}{12}$. *cf.* Valkanov [2003]; Britten-Jones and Neuberger [2007]; Boudoukh *et al.* [2018]

1. Show that the autocorrelation of returns is $\rho = 1 - \frac{r}{q}$.
2. Using Equation 3.22, show that the standard error of the Sharpe ratio is inflated by a factor of approximately $\sqrt{2\frac{q}{r} - 1}$.
3. Confirm the inflated standard error empirically by computing 5 years of rolling quarterly returns observed monthly of a zero mean random variable, and computing the Sharpe ratio, repeating the experiment thousands of times.

*** Ex. 3.19 Augmented form moments, Gaussian returns** Let $\tilde{\mathsf{X}}$ be an normally distributed $n \times l$ matrix, *i.e.*, $\operatorname{vec}\left(\tilde{\mathsf{X}}\right)$ is normal with mean $\operatorname{vec}(\mathsf{M})$, covariance $\Sigma_f \otimes \mathsf{H}$. As in Equation 3.62, define

$$\hat{\Theta} := \frac{1}{n}\tilde{\mathsf{X}}^\top\tilde{\mathsf{X}}.$$

Using results from Exercise 3.2, confirm Equation 3.69:

1. Show that

$$\mathrm{E}\left[\hat{\Theta}\right] = \frac{1}{n}\left(\mathsf{M}^\top\mathsf{M} + \operatorname{tr}(\mathsf{H})\,\Sigma_f\right).$$

2. Show that

$$\operatorname{Var}\left(\operatorname{vec}\left(\hat{\Theta}\right)\right) = \frac{1}{n^2}(\mathsf{I} + \mathsf{K})\left\{\mathsf{M}^\top\mathsf{H}\mathsf{M} \otimes \Sigma_f + \Sigma_f \otimes \mathsf{M}^\top\mathsf{H}\mathsf{M} + \operatorname{tr}\left(\mathsf{H}^2\right)\Sigma_f \otimes \Sigma_f\right\}.$$

*** Ex. 3.20 Augmented form moments, elliptical returns** Establish the variance equation of Equation 3.73. [Pav, 2013] Let $\tilde{\boldsymbol{x}}$ take an elliptical distribution with kurtosis factor κ, mean vector $\boldsymbol{\mu}$ and covariance matrix Σ.

1. Using results from Exercise 3.2, prove

$$\operatorname{Var}\left(\operatorname{vec}\left(\tilde{\boldsymbol{x}}\tilde{\boldsymbol{x}}^\top\right)\right) = \left\{(\kappa_i - 1)\left[\operatorname{vec}(\Sigma_i)\operatorname{vec}(\Sigma_i)^\top + (\mathsf{I} + \mathsf{K})\Sigma_i \otimes \Sigma_i\right] + (\mathsf{I} + \mathsf{K})\left[\Theta_i \otimes \Theta_i - \boldsymbol{\mu}_i{\boldsymbol{\mu}_i}^\top \otimes \boldsymbol{\mu}_i{\boldsymbol{\mu}_i}^\top\right]\right\}.$$

Ex. 3.21 The ex-factor Sharpe ratio distribution

1. Confirm that Equation 3.64 gives the inverse of the matrix Θ from Equation 3.63.
2. Confirm that the matrix in Equation 3.65 is the Cholesky factor of the matrix from Equation 3.64.
3. Confirm the identity of Equation 3.66.

Ex. 3.22 Moments under general returns
Suppose

$$\mathrm{E}\left[\tilde{\mathsf{X}}\right] = \mathsf{M},$$
$$\mathrm{Var}\left(\mathrm{vec}\left(\tilde{\mathsf{X}}\right)\right) = \Sigma_f \otimes \mathsf{H},$$

1. Let the ith row of $\tilde{\mathsf{X}}$ be $\tilde{\boldsymbol{x}}_i{}^{\top}$. Show that

$$\mathrm{E}\left[\tilde{\boldsymbol{x}}_i \tilde{\boldsymbol{x}}_i{}^{\top}\right] = \mathsf{M}_{i,}{}^{\top}\mathsf{M}_{i,} + \mathsf{H}_{i,i}\Sigma_f,$$

   where $\mathsf{M}_{i,}$ is the ith row of M.
2. From $\mathbf{1}_n{}^{\top}\mathsf{M} = n\boldsymbol{\mu}$, $\mathrm{tr}\left(\mathsf{H}\right) = n$, confirm Equation 3.70, *i.e.*,

$$\Theta = \Sigma_f + \boldsymbol{\mu}\boldsymbol{\mu}^{\top}.$$

3. Show that $\hat{\Theta}$ is an unbiased estimator of Θ only when $\mathsf{M} = \mathbf{1}\boldsymbol{\mu}^{\top}$.

Ex. 3.23 Market ex-factor Sharpe ratio variance

1. What happens to the variance of Equation 3.75 as the returns of the Market factor approach a non-zero constant (over time) value?
2. Justify *why* that should happen to the variance.

Ex. 3.24 Market ex-factor Sharpe ratio standard error
Prove Equation 3.74.

Ex. 3.25 Multiple attribution ex-factor Sharpe ratio
Consider what happens to the standard error of the ex-factor Sharpe ratio expressed in Equation 3.75 as *multiple* market instruments are added to the attribution model. Assume $\kappa = 1$. The math is probably too complicated to find an analytical solution, so explore it empirically. Repeat the empirical experiments of Example 3.4.3, but expand the number of "Market" terms against which one performs attribution. Look at $1, 2, 4, 8, 16$, and 32 terms in a factor model. For each choice of number of factors keep the sum of squared signal-noise ratios constant at a value of, say, $1.0\,\mathrm{yr}^{-1}$. Fix the beta of the

asset at 1 against each Market term, and keep the ex-factor signal-noise ratio and volatility as in Example 3.4.3. The Market terms should have *independent* returns. Plot the empirical standard error versus the number of factors.

1. Repeat the experiment, but make the Market returns highly positively correlated. How does the standard error change?

Ex. 3.26 Mertens correction Show that Merten's correction of Equation 3.34 follows from the Delta method form of Equation 3.26.

Ex. 3.27 Mertens contradiction It appears that if the skew of the returns distribution is sufficiently large and positive, the standard error of the Sharpe ratio under Mertens' formula, Equation 3.34, can become negative.

1. Show that this is not the case.
2. Let $\zeta = 0.1\,\mathrm{day}^{-1/2}$. What is the smallest value that $\left(1 - \zeta\gamma_1 + \frac{\gamma_2+2}{4}\zeta^2\right)$ can attain for skew γ_1 and excess kurtosis γ_2? Do you know of a distribution that achieves this value?

Ex. 3.28 Test of Sharpe ratios, correlated returns Replicate the Monte Carlo simulations of Example 3.3.3.

1. Confirm that the empirical mean of the difference in Sharpe ratios is consistent with the value suggested by Equation 3.49.
2. Generalize Equation 3.49 to the case of bivariate elliptical returns. (*cf.* Example 3.2.1.)
3. The form of the standard deviation of the difference in Sharpe ratios given in Equation 3.49 is predicated on bivariate Gaussian returns. Modify your simulation so that returns marginally follow a scaled, shifted $t(4)$ distribution. Does the empirical standard deviation still follow the form given by Equation 3.49? Is it more accurately described by the formula you found for elliptical returns?

Ex. 3.29 Asymptotic prediction intervals on Sharpe ratio In Section 2.5.9, Frequentist prediction intervals on the Sharpe ratio for the case of *i.i.d.* Gaussian returns were studied. Here consider *i.i.d.* returns without an assumption of normality. That is, suppose you observe $\hat{\zeta}_1$ on n_1 observations of *i.i.d.* returns, then observe $\hat{\zeta}_2$ on n_2 observations from the same returns stream.

1. Find some interval that is a function of $\hat{\zeta}_1, n_1, n_2$ such that with probability $1-\alpha$, $\hat{\zeta}_2$ falls within the interval, where the probability is under replication of the entire experiment. Start from Equation 3.43.

* 2. Now assume that the returns are Gaussian, but follow an AR(1) process (see Exercise 1.30), with autocorrelation ρ. Assume, however, that the two samples, of size n_1 and n_2, are independent. Construct a prediction interval on $\hat{\zeta}_2$.

Ex. 3.30 Bayesian update Multiply the prior probabilities of Equation 3.53 and Equation 3.52 by the joint likelihood of Equation 3.55 to arrive at the posterior updating rules of Equation 3.57.

Ex. 3.31 Kalman filter, functional signal-noise ratio Consider the update step of functions of signal-noise ratios as

$$\boldsymbol{g}\left(\boldsymbol{\zeta}\right) \mapsto \mathsf{A}\boldsymbol{g}\left(\boldsymbol{\zeta}\right) + \boldsymbol{b} + \boldsymbol{z},$$

where $\boldsymbol{z} \sim \mathcal{N}\left(\boldsymbol{0}, \Omega\right)$.

1. From Equation 3.51, derive the following prior as part of the predict step:

$$\begin{aligned} \mathsf{B} &\propto \mathcal{IW}\left(\mathsf{B}_0, m_0\right), \\ \boldsymbol{g}\left(\boldsymbol{\zeta}\right)^{-} | \mathsf{B} &\propto \mathcal{N}\left(\mathsf{A}\boldsymbol{a}_0 + \boldsymbol{b}, \Omega + \frac{1}{l_0}\mathsf{A}\mathsf{B}\mathsf{A}^{\top}\right). \end{aligned}$$

2. Derive the changes this causes to the posterior belief on $\boldsymbol{g}\left(\boldsymbol{\zeta}\right)$.

Ex. 3.32 Bayesian difference of signal-noise ratios Perform the analysis of Example 3.3.4, but use an uninformative prior.

Ex. 3.33 Functional Sharpe ratio Suppose one has backtested the returns of a number of different trading strategies, say k of them. For each strategy one observes some vector of "features" about the strategy, call it $\boldsymbol{f}_i$, for $i = 1, 2, \ldots, k$. For example, one feature might be the length of a sliding window over which to perform some machine learning mumbo jumbo to construct a portfolio, or whether a certain signal is used in constructing the portfolio, *etc.*
Let ζ_i be the signal-noise ratio of the ith strategy. A simple linear model posits that $\zeta_i = {\boldsymbol{f}_i}^{\top}\boldsymbol{\beta}$. Suppose you compute the Sharpe ratio of the backtested returns, along with the estimated variance-covariance of the same, call it $\hat{\Omega}$.

1. How would you estimate $\boldsymbol{\beta}$?
2. How would you estimate the asymptotic variance-covariance of your estimate?

Ex. 3.34 Double Sharpe ratio Morey and Vinod [2001] describe a "double Sharpe ratio," defined as $\hat{\zeta}$ divided by some estimate of the standard error of $\hat{\zeta}$. [Bao, 2009] Critique this idea. Is the double Sharpe ratio useful for estimation of ζ? Is it useful for hypothesis testing? (And what is the implicit hypothesis?)

* **Ex. 3.35 Expectations of sums** Suppose that x are *i.i.d.* draws from some distribution whose ith (uncentered) moment is α_i. Let $\hat{\alpha}_i$ be the ith sample moment, $\hat{\alpha}_i := \frac{1}{n}\sum_{1 \le t \le n} {x_t}^i$. (Note that $\mu = \alpha_1$ and $\hat{\mu} = \hat{\alpha}_1$.)

By definition $\mathrm{E}\left[\hat{\alpha}_i\right] = \alpha_i$, but consider the expectation of products of these sample sums:

1. Show that
$$\mathrm{E}\left[\hat{\alpha}_1^2\right] = \frac{1}{n}\alpha_2 + \frac{n-1}{n}{\alpha_1}^2.$$
(*Hint:* express $\sum_s x_s \sum_t x_t$ as $\sum_s {x_s}^2 + \sum_{s\neq t} x_s x_t$, then use independence of the x_t.)
Using this result, show that
$$\mathrm{E}\left[\hat{\alpha}_2 - \hat{\alpha}_1^2\right] = \frac{n-1}{n}\alpha_2 - \frac{n-1}{n}{\alpha_1}^2,$$
and thus $\mathrm{E}\left[\hat{\sigma}^2\right] = \sigma^2$.
2. Show that
$$\mathrm{E}\left[\hat{\alpha}_2\hat{\alpha}_1\right] = \frac{1}{n}\alpha_3 + \frac{n-1}{n}\alpha_1\alpha_2.$$
3. Assume that $n \geq 2$. Show that
$$\mathrm{E}\left[\hat{\alpha}_1^3\right] = \frac{1}{n^2}\alpha_3 + 3\frac{n-1}{n^2}\alpha_2\alpha_1 + \frac{(n-1)(n-2)}{n^2}{\alpha_1}^3.$$
4. Assume that $n \geq 2$. Using the previous two results show that
$$\begin{aligned}\mathrm{E}\left[\hat{\alpha}_2\hat{\alpha}_1 - \hat{\alpha}_1^3\right] &= \frac{n-1}{n^2}\left(\alpha_3 - 3\alpha_2\alpha_1 + {\alpha_1}^3\right) + \frac{n-1}{n}\left(\alpha_2\alpha_1 - {\alpha_3}^3\right),\\ &= \frac{n-1}{n^2}\mu_3 + \frac{n-1}{n}\mu{\sigma}^2,\end{aligned} \tag{3.81}$$
where $\mu_3 = \mathrm{E}\left[(x-\mu)^3\right]$. (*Hint:* expand the centered μ_3 in terms of the raw moments α_i.)

Ex. 3.36 Sharpe ratio bias & variance, Gaussian returns Consider the case of Gaussian returns:

1. Use Equation 3.37 to show that the geometric bias of the Sharpe ratio for Gaussian returns is approximately $1 + \frac{3}{4n} + \frac{49}{32n^2}$. (*cf.* Approximation 2.14.)
2. Use Equation 3.38 to derive an expression for the standard error of the Sharpe ratio for Gaussian returns.

Ex. 3.37 Correlated sample moments Suppose that, for a given set of returns, $\hat{\mu}$ and $\hat{\sigma}^2$ are unbiased for μ and σ^2, but positively correlated with correlation ρ. Will this cause a bias in $\hat{\zeta}$? Will it be a positive or negative bias?

* **Ex. 3.38 Less biased estimation of signal-noise ratio** Try to

build a "better" estimator of the signal-noise ratio. Consider an estimator of the form

$$a_0 + \frac{a_1 + (1 + a_2)\hat{\mu} + a_3\hat{\mu}^2}{\hat{\sigma}},$$

for constants a_i to be determined. Note that the usual Sharpe ratio corresponds to $a_0 = a_1 = a_2 = a_3 = 0$.

1. Use the Taylor's expansion $(1 + x)^{-1/2} \approx 1 - \frac{x}{2}$ to approximate this estimator in terms of the uncentered sample moments $\hat{\alpha}_i$.
2. From this approximation, derive the approximate expected value and variance of this estimator in terms of the moments and cumulants of the returns distribution, and the coefficients a_i.
3. Compute the expected squared error of the estimator, again as a function of the moments and cumulants, and a_i.
4. Take the partial derivative of the mean squared error with respect to each a_i and show that the squared error is minimized for $a_0 = a_1 = a_2 = a_3 = 0$, *i.e.*, the vanilla estimator.

Ex. 3.39 Asymptotic Sharpe ratio, elliptical returns
From Equation 3.27 and Equation 3.26, derive Equation 3.29.

Ex. 3.40 Sharpe ratio correlation, uncorrelated returns
We claim in Equation 3.29 that the Sharpe ratios of *uncorrelated* elliptical returns streams can be correlated for $\kappa \neq 1$.

1. Confirm this numerically, for two assets whose returns follow a multivariate t distribution.
2. How can uncorrelated returns have correlated Sharpe ratios?

Ex. 3.41 Markets following U.S. elections It has been noted that, since World War II, the U.S. stock market has experienced higher returns following midterm elections than following presidential elections. Consider this hypothesis: that the signal-noise ratio of the Market is higher post-midterms than post-presidential elections. For the purposes of this exercise, define "post-midterm" as the 12 month period starting in November of even years which are not divisible by four (*e.g.*, 2018), and define "post-presidential elections" as the 12 month period starting in November of even years which *are* divisible by four (*e.g.*, 2016).
Use the monthly returns of the Fama-French factor data from **tsrsa**, using code as given in Example 1.1.1.

1. Compute the Sharpe ratio of the Market for the two periods, along with confidence interval using Mertens' correction.
2. Perform the frequentist hypothesis test for equality of signal-noise ratio of the two periods.
3. Perform a *paired* test on the difference in signal-noise ratio of the two

periods, pairing midterm periods with the following presidential periods. Do you results change if you pair midterm periods with the *preceding* presidential periods?

Ex. 3.42 Scale of large deviance approximation Suppose that the skew of returns is less than 5 in absolute value. Find bounds on q and n such that the term

$$\exp\left(-\frac{1}{3}\left(\frac{n}{n-1}\right)^3 q^3 n \frac{\mathrm{E}\left[x^3\right]}{\sigma^3}\right)$$

from Equation 3.39 is within 10% of 1.

Ex. 3.43 Large deviance and kurtosis Determine experimentally whether excess kurtosis of returns affects the large-deviation probability of the Sharpe ratio. Repeat the experiments of Example 3.2.9 but drawing returns from an unskewed, though highly kurtotic distribution. (For example, draw returns from a t distribution with degrees of freedom $4+\epsilon$.) Do you see empirical evidence for deviation from the normal approximation?

Ex. 3.44 Large deviance, adjusting for moments Examine whether the asymptotic bias of the Sharpe ratio explains the large-scale deviation probability. Repeat the experiments of Example 3.2.9 but plot the empirical exceedance probability against $1 - \Phi(z)$ where z is the z-scored value, using actual mean and standard deviation from Equation 3.36 and Equation 3.38. That is, compare the empirical $\Pr\left\{\hat{\zeta} \geq q\right\}$ against

$$1 - \Phi\left(\frac{q - \mathrm{E}\left[\hat{\zeta}\right]}{\sqrt{\mathrm{Var}\left(\hat{\zeta}\right)}}\right).$$

Ex. 3.45 Non-catastrophic asymptotics Establish the standard error of Equation 3.41. Fix finite m, let $v_t = 1_{\{x_t > m\}}$, and define $\pi = p\left(x > m\right)$.

1. Show that

$$\mathrm{E}\left[\sum_{1 \leq t \leq n} v_t {x_t}^k\right] = n\pi\,\mathrm{E}\left[x^k \,|\, x > m\right].$$

Show that

$$\mathrm{E}\left[\sum_{1 \leq s \leq n} v_s {x_s}^k \sum_{1 \leq t \leq n} v_t {x_t}^j\right] = n\pi\,\mathrm{E}\left[x^{j+k} \,|\, x > m\right]$$
$$+ \left(n^2 - n\right)\pi^2\,\mathrm{E}\left[x^j \,|\, x > m\right]\mathrm{E}\left[x^k \,|\, x > m\right].$$

2. Let $\boldsymbol{y}$ be the vector
$$\sum_{1 \le t \le n} v_t \left[1, x_t, {x_t}^2\right]^\top.$$
Compute the mean and covariance of $\boldsymbol{y}$.
3. Given that
$$\hat{\zeta} = \frac{y_1/y_0}{\sqrt{\left(y_2 - y_1^2/y_0\right)/\left(y_0 - 1\right)}},$$
compute the gradient of $\hat{\zeta}$ with respect to $\boldsymbol{y}$. (Here it is convenient to zero-index the vector.) Plug in the expected value of $\boldsymbol{y}$.
4. Using the delta method, compute the asymptotic variance of $\hat{\zeta}$ with respect to the conditional moments. Discard all $\mathcal{O}\left(n^{-2}\right)$ terms.

*** Ex. 3.46 Research problem: omitted variable bias** ⚡

Omitted variable bias is the bias in the Sharpe ratio or ex-factor Sharpe ratio caused by omitting a relevant variable from the linear regression. Omitted variable bias is like the weather: everyone talks about it, but nobody does anything about it. For this open research problem, prove that omitted variable bias is not too large, subject to some reasonable conditions on the omitted variable. For example, one might reasonably assume that the omitted variable is only slowly varying over time (has a low autocorrelation); based on this and the autocorrelation of observed returns, one should be able to bound the size of the omitted variable bias. *cf.* the "reasonable timing" argument of Equation 9.19.

4

Overoptimism

Everybody dies
Frustrated and sad
And that is beautiful

THEY MIGHT BE GIANTS
Don't Let's Start

Suppose you will enter a coin-flipping contest: whoever flips the most Heads wins. You have at your disposal a chest full of seemingly identical coins of unknown provenance. To prepare you select one coin, flip it some number of times, noting the proportion of Heads. You repeat the process with another and another, finally selecting, through some procedure, one "lucky coin" to take to the contest. You naïvely estimate the probability of landing Heads by the in-sample proportion.

The difference between the estimated quality of the lucky coin based on historical data and the actual probability of landing Heads is the "overfit bias" of this experiment. If all the coins in the box are identical and fair, unbeknownst to you, the overfit bias is likely to be very large, *i.e.*, you have seriously overestimated the probability of landing Heads. If, on the other hand, there is a large variation in the coins, perhaps some of them *almost always* land Heads, say, then the overfit bias is probably smaller.

Constructing and testing trading strategies is, perhaps uncharitably, analogous to selecting lucky coins, but with the following differences:

1. While you could flip a given coin more times to collect more data, often a quantitative strategist is stuck with a fixed amount of historical data, and can only collect more data at a rate of one day per day.
2. Presumably tests of different coins have errors independent of each other, while the quant is usually observing historical returns of multiple strategies which are dependent on each other at some (backtested) point in time.
3. Coins are just coins. They lack parameters. Often a quant is refining a model with free parameters, either by an optimization procedure, brute force, or sequential knob twiddling. Thus the simulated historical returns of the strategies are not only dependent, they are dependent in a way driven by these parameters. Moreover, the model can be too complex for the data, which can deteriorate actual performance.
4. In our thought experiment, the coins appear identical and are selected at random. Often trading strategies are passed along by word of mouth, or

DOI: 10.1201/9781003181057-4

are discovered via the media (an article, book, white-paper, blog, *etc.*), and the quant does not know how much overfitting was involved in the original "discovery" of the strategy, and there is likely to be very little historical 'out of sample' data.

5. A coin can be physically inspected for bias, while quantitative strategies often leak very little information that can be tested independently of returns.
6. Coin flips have an unambiguous outcome. Quantitative strategies are often backtested, and so are subject to the biases and imprecision of simulated fill prices, market impact, and market reaction. The "ground truth" of backtested strategies is often uncertain.

Some of these issues can be approached statistically, and we will attempt to address them here, while others are beyond the scope of this text.

Remark (Overoptimism and Overfit). Though the definitions are a little nebulous, we will use *overoptimism* to refer to the case where one has a (positively) biased estimate of the "performance" of some model caused by selection among many competing models or fitting of parameters. For example, if one selects among many trading strategies based on the observed Sharpe ratios, then the Sharpe ratio of the best strategy will be biased upwards by this selection, and may not be a good estimate of its signal-noise ratio.

We will use the term *overfitting* to describe the case where by selecting a too complex model one causes a decrease in the performance. In this case there is still a bias in the estimated performance, as otherwise one presumably would pick a better model. An example might be using more and more factors to forecast the returns of some asset. The Sharpe ratio will increase with more factors, though likely the signal-noise ratio is decreasing.

We illustrate these with Figure 4.1, where we plot overoptimism and overfit as the signal-noise ratio, the Sharpe ratio, and the bias between them versus some undefined "effort," which could be number of strategies tested, or amount of complexity added to the trading strategy, *etc.*

In this chapter we will consider overoptimism, and will consider overfitting in the sequel.

4.1 Overoptimism by selection

Here we consider perhaps the simplest model of strategy development, which we call *overoptimism by selection.* Suppose you observe returns from k different assets (or "strategies") each over n independent periods. The returns among assets may be correlated, but in the simplest realization we assume they are independent. We compute the Sharpe ratios of each asset then select the one with the highest Sharpe ratio, which presumably we will trade out-of-sample.

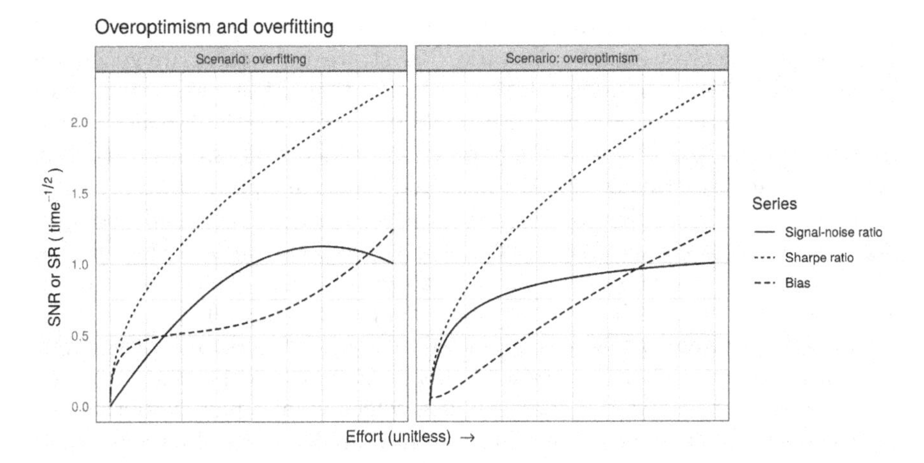

FIGURE 4.1
Overoptimism and overfitting are illustrated as the signal-noise ratio, Sharpe ratio and their difference versus some "effort." In the case of overfitting, the signal-noise ratio peaks and increased effort contributes to declining out-of-sample performance. In the case of overoptimism, the signal-noise ratio continues to grow with increased effort, albeit slowly.

The gap between the Sharpe ratio and signal-noise ratio of the chosen asset is the overoptimism.

Aronson [2006, chapter 6] presents a very accessible description of essentially this problem, calling our overoptimism, "data-mining bias." He cites five factors affecting overoptimism, namely

1. the number of strategies tested, k,
2. the length of backtests, n,
3. the correlation of returns among strategies,
4. the presence of "outliers,"
5. the spread in the population expected returns.

All five factors are relevant in the context considered by Aronson, where strategies are apparently selected by maximum sample mean return. However, by using the Sharpe ratio to measure quality of a strategy we are somewhat unaffected by outliers, though they might be a sign of backtest biases. Also we have put a fair amount of work in the preceeding chapters to understand the error of the Sharpe ratio with respect to n, and can adjust for sample size as needed. The correlation of returns and the spread in population signal-noise ratios are the two relevant, unobservable factors that we must consider in our analyses.

Example 4.1.1 (Overoptimism by selection). We create a population of k strategies, with normally distributed signal-noise ratios:

$$\zeta_i \sim \mathcal{N}\left(0, 0.403\,\mathrm{yr}^{-1}\right).$$

Then for each strategy we sample 504 days (2 years) of normally distributed returns with the given signal-noise ratio. The strategies' returns are generated independently. We compute the k Sharpe ratios, then select the strategy with the highest. We record its Sharpe ratio and signal-noise ratio. We also record the highest signal-noise ratio in the population. We let k vary from 1 to $10,000$. We repeat this experiment 20,000 times, and compute empirical averages.

In Figure 4.2, we plot the Sharpe ratio, signal-noise ratio and overoptimism as a function of k. We see that the signal-noise ratio of the selected strategy appears to be growing in k, albeit very weakly. Thus the extra effort associated with testing more strategies appears to pay off in this case. In Figure 4.3, we again plot the signal-noise ratio of the selected strategy, but also the maximum signal-noise ratio of all k strategies. The difference between them is a kind of "regret," or overoptimism, which we also plot. Note that the regret grows with k, which seems to be an inescapable curse of decision making with limited information. For $k = 10,000$ the expected signal-noise ratio of the selected strategy is around 2.6 standard deviations above the mean, while the population maximum is around 3.9 standard deviations. ⊣

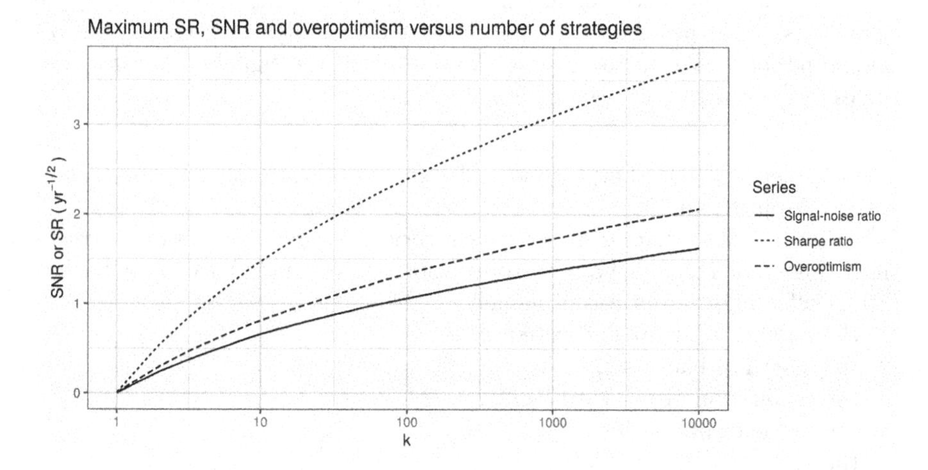

FIGURE 4.2
The maximum Sharpe ratio and the associated signal-noise ratio of the selected strategy is plotted versus k for the case of overoptimism by selection with normally distributed signal-noise ratio. The x-axis is in log scale.

Does fitting work?

Aronson claims, citing White, that the selection procedure outlined here "works" in the sense that as $n \to \infty$, it selects the optimal model almost surely. [Aronson, 2006; White, 2000] This intuitively makes sense since the error of the Sharpe ratio around the signal-noise ratio goes to zero in large n. Under the (admittedly unrealistic) assumption of independent returns and

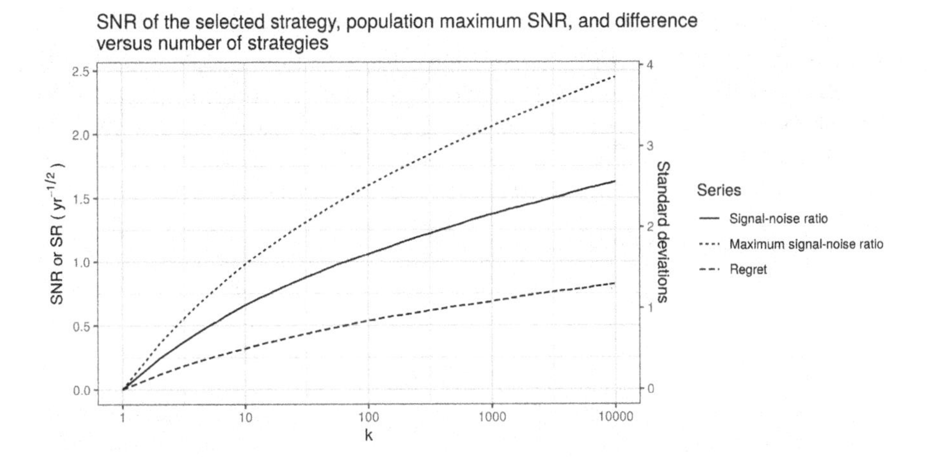

FIGURE 4.3
The signal-noise ratio of the selected strategy and the population maximum signal-noise ratio, and their difference are plotted versus k for the case of overoptimism by selection. The x-axis is in log scale. The right side expresses the signal-noise ratio in standard deviations above the mean, based on a standard devation of $0.04\,\mathrm{day}^{-1/2}$.

non-zero spread in signal-noise ratios, the procedure also works in the other asymptotic direction. That is, as k increases the signal-noise ratio of the strategy selected for having maximal Sharpe ratio is increasing, albeit potentially very slowly in k.

Example 4.1.2 (Overoptimism by selection (again)). Continuing Example 4.1.1, since $\hat{\zeta}_i$ is approximately normally distributed around ζ_i, we can make the approximation

$$\left[\begin{array}{c}\zeta_i\\ \hat{\zeta}_i\end{array}\right] \sim \mathcal{N}\left(\left[\begin{array}{c}0\\ 0\end{array}\right], \left[\begin{array}{cc}\sigma_1^2 & \sigma_1^2\\ \sigma_1^2 & \sigma_1^2+\sigma_2^2\end{array}\right]\right),$$

where σ_1^2 is the variance in the generation of ζ_i, and σ_2^2 is approximated as n^{-1} via Equation 2.24. Then conditional on observing $\hat{\zeta}_i$ we have

$$\zeta_i \left| \hat{\zeta}_i \sim \mathcal{N}\left(\rho^2 \hat{\zeta}_i, \left(1-\rho^2\right)\sigma_1^2\right),\right.$$

where $\rho = \sigma_1/\sqrt{\sigma_1^2+\sigma_2^2}$. That is, we expect the signal-noise ratio associated with the asset with maximal Sharpe ratio to be ρ^2 times that maximum Sharpe ratio. [Huber, 2019] If

$$\frac{\rho^2 \hat{\zeta}_i}{\sigma_1\sqrt{1-\rho^2}} = \frac{\rho\hat{\zeta}_i}{\sigma_2} \gg 2,$$

then we expect that ζ_i is likely to be bigger than zero. The ratio ρ/σ_2 is something like a "signal-noise ratio of signal-noise ratios" in this example.

In our example we have $\sigma_1^2 = 0.403\,\mathrm{yr}^{-1}$, and $\sigma_2^2 \approx \frac{1}{n} = 0.5\,\mathrm{yr}^{-1}$, and therefore $\rho = 0.668$. In this case we have $\rho/\sigma_2 \approx 0.945\,\mathrm{yr}^{1/2}$, and we are unlikely to select a strategy with negative signal-noise ratio if $\hat{\zeta}_i \gg 2$. We plot the empirical expected value of the ζ associated with the strategy with maximal Sharpe ratio in Figure 4.4, along with ρ^2 times the empirical expected maximum Sharpe ratio. We find these lines to be in good agreement.

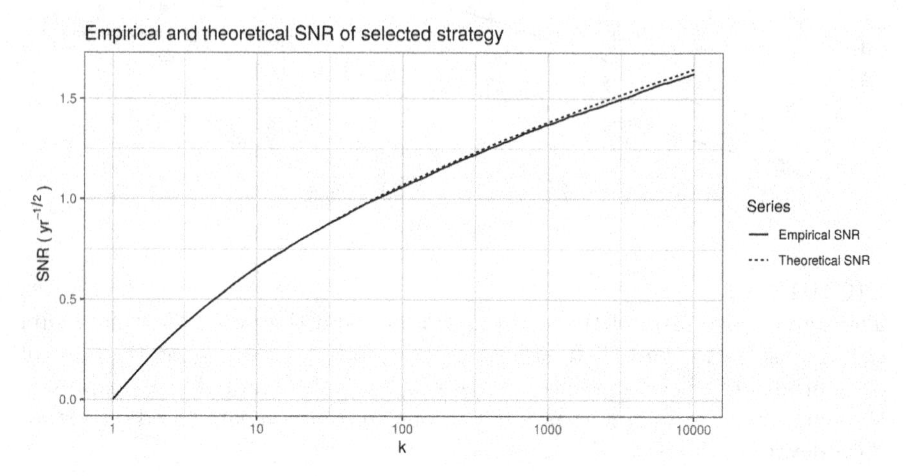

FIGURE 4.4
The signal-noise ratio of the selected strategy, and its approximate expected value are plotted against k for the case of overoptimism by selection on normally distributed signal-noise ratio.

We note that in real use the spread of the ζ_i, which we have called σ_1, is unknown. However, one could estimate it from the sample. That is, observing many $\hat{\zeta}_i$, one could compute their sample variance, then subtract σ_2^2 to get an estimate for σ_1^2. It is not clear what one should do in the case this estimate is negative, however. ⊣

4.1.1 Multiple hypothesis testing

The most basic approach to overoptimism is to appeal to traditional corrections for *multiple hypothesis testing*, or MHT[1]. [Bretz *et al.*, 2016; White, 2000; Harvey and Liu, 2015] So suppose we have observed Sharpe ratios $\hat{\zeta}_1, \hat{\zeta}_2, \ldots, \hat{\zeta}_k$. Consider the reindexing of the assets in order of their observed Sharpe ratio. That is, consider the permutation of $1, 2, \ldots, k$, denoted $(1), (2), \ldots, (k)$ such that

$$\hat{\zeta}_{(1)} \le \hat{\zeta}_{(2)} \le \cdots \hat{\zeta}_{(k-1)} \le \hat{\zeta}_{(k)}.$$

[1]This section is adapted from the author's paper on inference on the asset with maximal Sharpe ratio. [Pav, 2019b]

We will use the same indexing on the signal-noise ratios, thus $\zeta_{(k)}$ will refer to the signal-noise ratio associated with the strategy that has maximum Sharpe ratio, and not the maximum of the ζ_i.

To test the null hypothesis

$$H_0 : \zeta_{(k)} = \zeta_0 \quad \text{versus} \quad H_1 : \zeta_{(k)} > \zeta_0,$$

one can instead perform a *union intersection test.* In contrast to a intersection union test (*cf.* Section 2.5.4), here the null hypothesis is the intersection of hypotheses, and the alternative hypothesis (and the critical region, where one rejects) is a union. That is, to test the hypothesis above, one instead tests

$$H_0 : \forall i\, \zeta_i = \zeta_0 \quad \text{versus} \quad H_1 : \exists i : \zeta_i > \zeta_0. \tag{4.1}$$

To test this hypothesis via the *Bonferroni correction* at the α level, perform the k separate hypothesis tests for $\zeta_i = \zeta_0$, but at the α/k level, and reject H_0 if any of these reject. [Wasserman, 2004; Harvey and Liu, 2015; Bretz *et al.*, 2016]

The Bonferroni correction is equivalent to the observation that one can attribute at most α/k probability of a type I error to each of the separate tests and arrive at a total type I rate of no more than α. A slightly more powerful correction is possible under the (generally suspect) assumption that returns of the strategies are independent. In this case, the $\hat{\zeta}_i$ are then also independent, as are the outcomes of the k separate hypothesis tests. In this case, one can use the *Šidák correction,* where one tests $\zeta_i = \zeta_0$ at the $1 - (1 - \alpha)^{1/k}$ level, rejecting the intersection H_0 if you reject for any ζ_i. The Šidák correction is equivalent to finding the probability that we commit a type I error for *none* of the separate sub-tests, and setting it to $1 - \alpha$, then using independence.

Note that in our case the null hypothesis posits the same value for each ζ_i. Moreover, because we observe n returns for each strategy, the distribution of each $\hat{\zeta}_i$ is identical under the null. Thus to use the Bonferroni or Šidák corrections we need not test every ζ_i, rather we need only test $\zeta_{(k)}$. Thus to test the null hypothesis above using the Bonferroni correction, reject if $\hat{\zeta}_{(k)} > SR_{1-\alpha/k}\left(\zeta_0, n\right)$, where $SR_p\left(\zeta, n\right)$ is the pth quantile of the Sharpe ratio distribution with signal-noise ratio ζ on n samples; under the Šidák correction reject if $\hat{\zeta}_{(k)} > SR_{(1-\alpha)^{1/k}}\left(\zeta_0, n\right)$.

Another way to arrive at the Šidák correction is via the beta distribution: if $p_1, p_2, \ldots, p_k$ are independent random variables each uniform on $[0, 1)$, then the jth largest of them, call it $p_{(j)}$ follows a beta distribution:

$$p_{(j)} \sim \mathcal{B}\left(j, k + 1 - j\right).$$

In particular, the smallest is distributed as $\mathcal{B}\left(1, k\right)$. Thus to test the intersection null hypothesis $\forall i\, \zeta_i = \zeta_0$ at the α level under the assumption of independence, one should compute the p-values of each of the hypotheses $\zeta_i = \zeta_0$, then compare the smallest to the α quantile of the beta distribution $\mathcal{B}(1, k)$. The smallest p-value will correspond to the largest Sharpe ratio, and the α quantile is $\beta_{\alpha}(1, k) = 1 - (1 - \alpha)^{1/k}$, which is the Šidák correction.

In practice the nominal gain in power from using the Šidák correction is negligible, as illustrated in Example 4.1.3. For large k, the Šidák correction offers little additional power for the assumption of independence and one typically sees the Bonferroni correction used instead.

Example 4.1.3 (Sharpe ratio, Bonferroni and Šidák). In Table 4.1, we display cutoffs computed by the Bonferroni and Šidák corrections for the case of testing the null hypothesis at the 0.05 level that all signal-noise ratios are equal to zero, for the case of $n = 504$ days. The cutoffs are in annualized terms and are quantiles of the Sharpe ratio distribution. Note how little difference there is between the two corrections. ⊣

k	Bonferroni cutoff	Šidák cutoff
1	1.165	1.165
10	1.828	1.823
100	2.341	2.335
1,000	2.773	2.769
10,000	3.156	3.152
100,000	3.502	3.499

TABLE 4.1
The Bonferroni and Šidák corrected cutoff values are presented for increasing k for the case of testing the hypothesis that all strategies have zero signal-noise ratio. The cutoffs are for $\alpha = 0.05$ and $n = 504$, and have units $\text{yr}^{-1/2}$.

4.1.2 Multiple hypothesis testing, correlated returns

A glaring problem with the simple corrections for multiple hypothesis testing are that they do not take into account possible correlation of strategies' returns. The Bonferroni correction maintains an *upper bound* on the type I error, but is too conservative in the case where returns are mostly positively correlated. (And it is impossible to have many mutually negatively correlated returns, *cf.* Exercise 4.3.)

We illustrate this problem in Example 4.1.4, where we empirically estimate the mean Sharpe ratio of mutually correlated returns. In the presence of large positive mutual correlation of returns, the type I rate will be much lower than the nominal rate. Equivalently, the Bonferroni correction will have low power, since a large signal-noise ratio will have to overcome the "penalty" paid by the correlation. Effectively, the correlation structure has reduced the number of independent hypothesis tests being performed.

Example 4.1.4 (Overoptimism by selection, correlated returns). We draw k-variate daily returns from a normal distribution with zero mean and covariance

$$(1-\rho)\,\mathsf{I} + \rho\left(\mathbf{1}\mathbf{1}^{\top}\right)$$

Returns are drawn independently for each of 504 days. We compute the Sharpe ratio of each of the k columns and then compute the maximum. We repeat this 10,000 times, and compute the empirical 0.95 quantile of the Sharpe ratios. We let k vary from 1 to 100. We repeat for several values of ρ.

In Figure 4.5, we plot the empirical quantiles of the maximum Sharpe ratio versus k. The line for $\rho = 0$ should correspond approximately to the Šidák column from Table 4.1, which is the cutoff for the null hypothesis that all strategies have signal-noise ratio equal to zero. When ρ is large, the maximum Sharpe ratio is very far from the 0.95 cutoff, meaning the rate of type I errors is much smaller than 0.05. *cf.* Exercise 4.4. ⊣

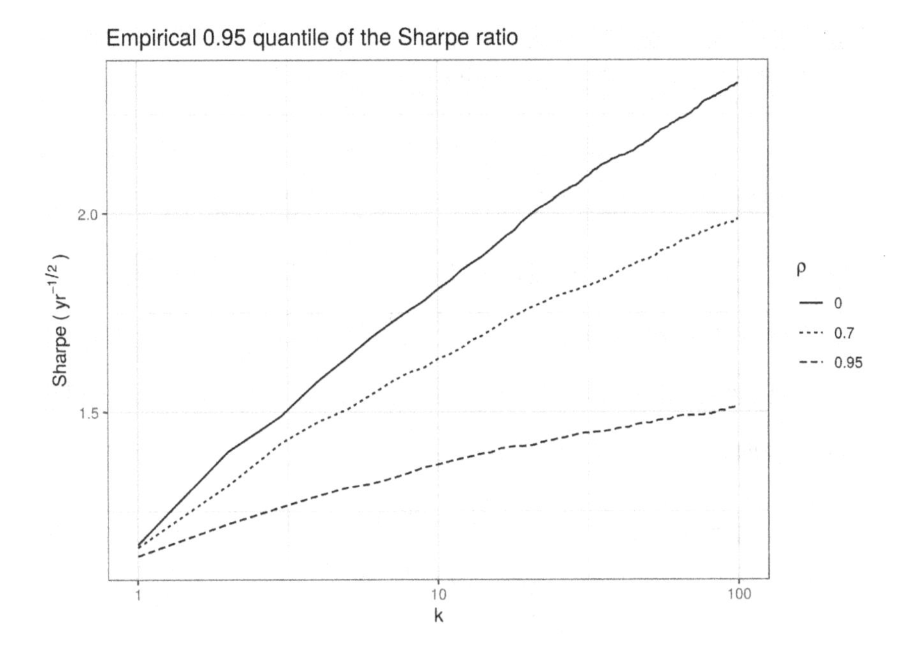

FIGURE 4.5
The empirical 0.95 quantile of the maximum Sharpe ratio of k strategies is plotted versus k for returns drawn from a normal distribution with zero mean and covariance $(1-\rho)\,\mathsf{I} + \rho\left(\mathbf{1}\mathbf{1}^{\top}\right)$. Quantiles are computed over 10,000 simulations of 504 days of returns.

To deal with correlation among returns, consider the normal approximation to the vector of Sharpe ratios,

$$\hat{\zeta} \approx \mathcal{N}\left(\zeta, \frac{1}{n}\Omega\right), \tag{4.2}$$

where Ω will not, in general, be diagonal. To test the null hypothesis that $\boldsymbol{\zeta} = \boldsymbol{\zeta}_0$, convert the above approximation to a standard normal by an inverse square root of Ω:

$$\sqrt{n}\Omega^{-1/2}\left(\hat{\boldsymbol{\zeta}} - \boldsymbol{\zeta}_0\right) \approx \mathcal{N}\left(\mathbf{0}, \mathsf{I}\right). \tag{4.3}$$

There are many ways we could test this hypothesis, but we are interested in a one-sided alternative, and wish to focus on the asset which demonstrated maximal Sharpe ratio. Thus a general chi-square test, as suggested in Section 3.2.1, seems inappropriate. Moreover, to apply the technique to the case of very large k, we wish to avoid having to computationally invert $\Omega^{1/2}$.

If each period's returns are elliptical with kurtosis factor κ, and R is the correlation of returns then, recalling Equation 3.29,

$$\Omega = \mathsf{R} + \frac{\kappa - 1}{4}\boldsymbol{\zeta}\boldsymbol{\zeta}^{\top} + \frac{\kappa}{2}\operatorname{Diag}\left(\boldsymbol{\zeta}\right)\left(\mathsf{R} \odot \mathsf{R}\right)\operatorname{Diag}\left(\boldsymbol{\zeta}\right).$$

To tame the computations, consider the simple *rank one updated correlation model*, where we assume that returns have correlation

$$\mathsf{R} = \left(1 - \rho\right)\mathsf{I} + \rho\left(\mathbf{1}\mathbf{1}^{\top}\right), \tag{4.4}$$

for $|\rho| < 1$. For this R, under the null hypothesis that $\boldsymbol{\zeta} = \zeta_0\mathbf{1}$,

$$\begin{aligned}\Omega &= \left(1 - \rho\right)\mathsf{I} + \rho\left(\mathbf{1}\mathbf{1}^{\top}\right) + \frac{\kappa - 1}{4}\zeta_0^2\mathbf{1}\mathbf{1}^{\top} + \frac{\kappa}{2}\zeta_0^2\left(\left(1 - \rho^2\right)\mathsf{I} + \rho^2\left(\mathbf{1}\mathbf{1}^{\top}\right)\right),\\ &= \left[1 - \rho + \frac{\kappa}{2}\zeta_0^2\left(1 - \rho^2\right)\right]\mathsf{I} + \left[\rho + \frac{\kappa - 1}{4}\zeta_0^2 + \frac{\kappa}{2}\zeta_0^2\rho^2\right]\left(\mathbf{1}\mathbf{1}^{\top}\right),\\ &= a_0\,\mathsf{I} + a_2\left(\mathbf{1}\mathbf{1}^{\top}\right).\end{aligned} \tag{4.5}$$

It then follows (*cf.* Exercise 4.6) that

$$\begin{aligned}\Omega^{-1/2} &= a_0^{-1/2}\,\mathsf{I} - \left[\frac{\sqrt{a_0 + a_2\mathbf{1}^{\top}\mathbf{1}} - \sqrt{a_0}}{\sqrt{a_0^2 + a_0a_2\mathbf{1}^{\top}\mathbf{1}}}\right]\frac{\left(\mathbf{1}\mathbf{1}^{\top}\right)}{\mathbf{1}^{\top}\mathbf{1}},\\ &= b_0\,\mathsf{I} + b_2\left(\mathbf{1}\mathbf{1}^{\top}\right).\end{aligned} \tag{4.6}$$

Note that as a transform, this $\Omega^{-1/2}$ is order-preserving: if $y_i \le y_j$ and $\boldsymbol{z} = \Omega^{-1/2}\boldsymbol{y}$, then $z_i \le z_j$. In particular, it preserves the identity of the maximum element. Thus if the ith element of $\hat{\boldsymbol{\zeta}} - \boldsymbol{\zeta}_0$ is the largest, then it is also the largest element of $\sqrt{n}\Omega^{-1/2}\left(\hat{\boldsymbol{\zeta}} - \boldsymbol{\zeta}_0\right)$. Note also that $\mathbf{1}/\mathbf{1}^{\top}\mathbf{1}$ computes the mean, thus we can view the transform of $\Omega^{-1/2}$ as a shrinkage to (or away) from the mean.

Thus to test the null in Hypothesis 4.1, let

$$
\begin{aligned}
a_0 &= 1 - \rho + \frac{\kappa}{2}\zeta_0^2\left(1 - \rho^2\right),\\
a_2 &= \rho + \frac{\kappa - 1}{4}\zeta_0^2 + \frac{\kappa}{2}\zeta_0^2\rho^2,\\
b_0 &= a_0^{-1/2},\\
b_2 &= \frac{1}{k}\frac{\sqrt{a_0} - \sqrt{a_0 + k a_2}}{\sqrt{a_0^2 + k a_0 a_2}},
\end{aligned}
$$

then compute

$$
z = \sqrt{n}\left[b_0\left(\hat{\zeta}_{(k)} - \zeta_0\right) + b_2 \sum_{1 \le i \le k}\left(\hat{\zeta}_i - \zeta_0\right)\right], \tag{4.7}
$$

and reject at the α level if $z \ge z_{1-\alpha/k}$, the $1 - \alpha/k$ quantile of the normal distribution. Of course, κ and ρ are unknown and have to be estimated from the sample. Note we can rewrite Equation 4.7 as

$$
\begin{aligned}
z &= \sqrt{n}\left[b_0\left(\hat{\zeta}_{(k)} - \zeta_0\right) + b_2 k\left(\bar{\zeta} - \zeta_0\right)\right],\\
&= \sqrt{n}\left[b_0\left(\hat{\zeta}_{(k)} - \bar{\zeta}\right) + \left(b_0 + b_2 k\right)\left(\bar{\zeta} - \zeta_0\right)\right],
\end{aligned} \tag{4.8}
$$

where $\bar{\zeta}$ is the average of the sample Sharpe ratios.

See Exercise 4.11 for a "direct" approach to testing Hypothesis 4.1 that does not require one to compute the sample mean, but which only has approximate type I rate.

In Figure 4.6, we plot b_0, b_2 versus ρ where $\Omega^{-1/2} = b_0\,\mathsf{I} + b_2\left(\mathbf{1}\mathbf{1}^\top\right)$, for the case where $k = 100, \kappa = 1$, and $\zeta_0 = 0$. We note that when $\rho \approx 1$ the values of b_0 and b_2 are highly sensitive to ρ. Moreover, as $\rho \to 1$, the procedure would appear to compute z as a large multiple times the difference between $\hat{\zeta}_{(k)}$ and the mean of the $\hat{\zeta}_i$, meaning it would *always* reject for ρ sufficiently close to 1. Noting that ρ has to be estimated from the sample, to keep the test conservative one should bound one's estimate of ρ away from 1. Moreover, as we shall see below, the maximum type I rate is maintained by assuming a smaller ρ, at the expense of power.

Example 4.1.5 (MHT Correction, Correlated Returns). We perform simulations under the null hypothesis, $\zeta = \mathbf{0}$, with $\mathsf{R} = (1 - \rho)\,\mathsf{I} + \rho\left(\mathbf{1}\mathbf{1}^\top\right)$. We set $k = 100$, and simulate $n = 1{,}008$ days of returns, compute the maximum Sharpe ratio, use a Bonferroni correction to test the null hypothesis, and tabulate the empirical rejection rate at the nominal 0.05 level over 5,000 simulations. We allow the correlation correction method to use the population ρ, rather than estimate it from the data.

In Figure 4.7, we plot the empirical rejection rate versus ρ at the nominal

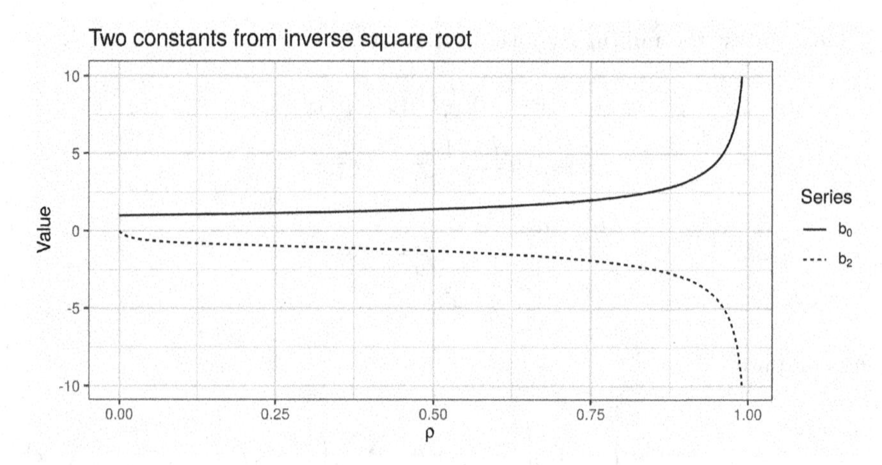

FIGURE 4.6
Letting $\Omega^{-1/2} = b_0\,\mathsf{I} + b_2\left(\mathbf{1}\mathbf{1}^\top\right)$ for $\Omega = a_0\,\mathsf{I} + a_2\left(\mathbf{1}\mathbf{1}^\top\right)$, we plot b_0, b_2 versus ρ for $k = 100$.

0.05 type I level. The vanilla MHT test is conservative, with near zero rejection rates for large ρ, while the correlation correction yields nominal rejection rates. $\dashv$

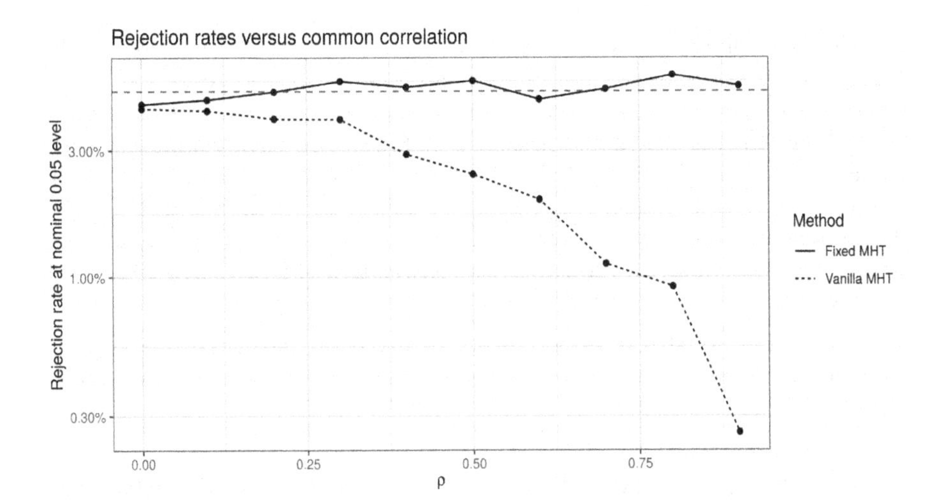

FIGURE 4.7
The empirical type I rate under the null hypothesis is plotted versus ρ for the case where $\mathsf{R} = (1-\rho)\,\mathsf{I} + \rho\left(\mathbf{1}\mathbf{1}^\top\right)$, for the vanilla Bonferroni correction, and Bonferroni correction with fix for common correlation. Tests are performed with Gaussian returns, for 100 assets over 1,008 days. Tests were performed at the 0.05 level, which appears to be maintained by the fixed Bonferroni procedure but not by the regular Bonferroni procedure. Empirical rates are over 5,000 simulations. The y-axis is in log scale.

Alternative Correlation Models

Inasmuch as we would like to apply the same analysis to a broader class of correlation matrices, they do not lead to an order-preserving $\Omega^{-1/2}$. Thus it is hard to test $\boldsymbol{z} = \sqrt{n}\Omega^{-1/2}\left(\hat{\boldsymbol{\zeta}} - \boldsymbol{\zeta}_0\right)$ via Bonferroni correction, as the largest element cannot easily be identified.

However, we can use the analysis above in the case where R consists of non-negative elements to arrive at a test with bounded type I rate. Consider multivariate normal vectors $\boldsymbol{x}, \boldsymbol{y}$ with $\boldsymbol{x} \sim \mathcal{N}(\mathbf{0}, \mathsf{R})$ and $\boldsymbol{y} \sim \mathcal{N}\left(\mathbf{0}, (1-\rho_0)\mathsf{I} + \rho_0\left(\mathbf{1}\mathbf{1}^\top\right)\right)$, $\mathsf{R}_{i,j} \geq \rho_0 \geq 0$ for all $i \neq j$. By Slepian's lemma, $\Pr\left\{\max_i x_i \geq t\right\} \leq \Pr\left\{\max_i y_i \geq t\right\}$ for all t. [Slepian, 1962; Zeitouni, 2017]

Thus we can appeal to analysis above by assuming the correlation matrix has the form $(1-\rho_0)\mathsf{I} + \rho_0\left(\mathbf{1}\mathbf{1}^\top\right)$ and accepting a biased test, that is one with type I rate (approximately) no greater than α. Thus to test the null in Hypothesis 4.1, supose that $\mathsf{R}_{i,j} \geq \rho_0 \geq 0$ for $i \neq j$. Then compute z as in Equation 4.7, plugging in ρ_0 in the definitions of a_0 and a_2, and reject at the nominal α level if $z \geq z_{1-\alpha/k}$, the $1-\alpha/k$ quantile of the normal distribution. This procedure has (approximate) type I rate *no greater* than α. By "approximate" here, we allude to the normal approximation of Equation 3.29 and the approximate variance-covariance therein.

Example 4.1.6 (MHT Correction, AR(1) Correlated Returns). We perform simulations under the null hypothesis, $\boldsymbol{\zeta} = \mathbf{0}$, where R takes the form of an $AR(1)$ matrix:

$$\mathsf{R}_{i,j} = \rho^{|i-j|},$$

for some ρ assumed known. We set $k = 100$, and simulate $n = 1,008$ days of returns, compute the maximum Sharpe ratio, use a Bonferroni correction to test the null hypothesis, and tabulate the empirical rejection rate at the nominal 0.05 level over 10,000 simulations. We allow the correlation correction method to use the population ρ, rather than estimate it from the data.

In Figure 4.8, we plot the empirical rejection rate versus ρ at the nominal 0.05 type I level. We apply the procedure for Bonferroni correction with correlation matrix $(1-\rho_0)\mathsf{I} + \rho_0\left(\mathbf{1}\mathbf{1}^\top\right)$ where $\rho_0 = \rho^{k-1}$. The procedure has access to ρ, and need not estimate it. We get near-nominal coverage up until around $\rho \approx 0.75$, after which the procedure is conservative.

See also Exercise 4.12. ⊣

The results of Example 4.1.6 indicate that when only a few non-diagonal elements of R are bigger than zero (*e.g.*, the AR(1) correlation matrix when ρ is not near 1), one can still get near-nominal coverage assuming that $\mathsf{R} = \mathsf{I}$. Slepian's lemma only goes one way, giving us a test with type I error at most a fixed value; however, there are known bounds on the other side. For example, Li and Shao [2002] prove the following: Suppose that $\boldsymbol{x} \sim \mathcal{N}\left(\mathbf{0}, \mathsf{R}^0\right)$ and

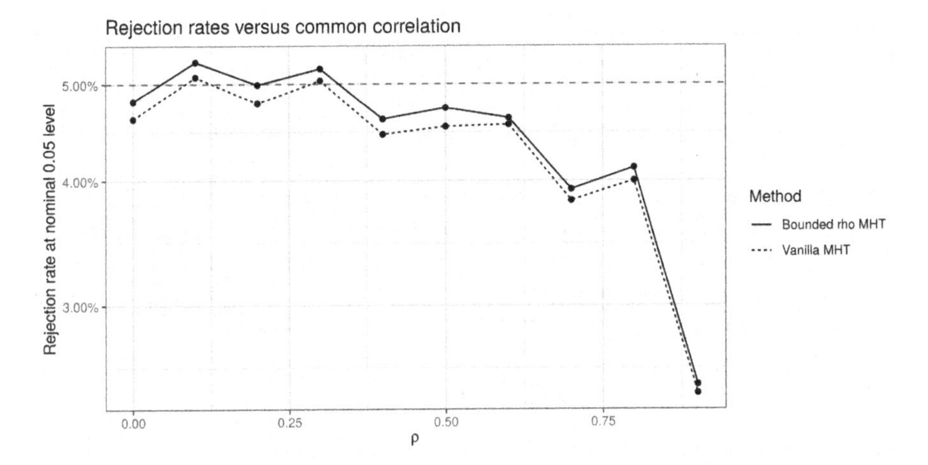

FIGURE 4.8
The empirical type I rate under the null hypothesis is plotted versus ρ for the case where R is an $AR(1)$ matrix. That is, $\mathsf{R}_{i,j} = \rho^{|i-j|}$. The test applies a Bonferroni correction with fix for common correlation of ρ^{k-1}. We also plot rejection rates for the Bonferroni test with no correction for correlation. Tests are performed with Gaussian returns, for 100 assets over 1,008 days. Tests were performed at the nominal 0.05 level, which appears to be maintained by the bounded correlation Bonferroni procedure for modest, but not large values of ρ. Empirical rates are over 10,000 simulations. The y-axis is in log scale.

$\boldsymbol{y} \sim \mathcal{N}\left(\mathbf{0}, \mathsf{R}^1\right)$, where the diagonal elements of R^0 and R^1 are all one. Then

$$\begin{aligned}\Pr\left\{\max_i x_i > t\right\} \le \Pr\left\{\max_i y_i > t\right\} \\ + \sum_{1\le i<j\le k} \frac{\left(\operatorname{asin} R^1_{i,j} - \operatorname{asin} R^0_{i,j}\right)^+}{2\pi} \exp\left(\frac{-t^2}{1+max\left(\left|R^0_{i,j}\right|, \left|R^1_{i,j}\right|\right)}\right).\end{aligned} \tag{4.9}$$

Thus if there are only $2m$ elements such that $R^1_{i,j} > R^0_{i,j}$ we can roughly transform this bound into

$$\Pr\left\{\max_i x_i > t\right\} \le \Pr\left\{\max_i y_i > t\right\} + \frac{m}{2} e^{-t^2/2}. \tag{4.10}$$

See also Exercise 4.13.

4.1.3 One-sided alternatives

Another approach to the problem of overoptimism by selection is via testing against (multivariate) one-sided alternatives. As with the Bonferroni correction, in this approach we do not test the signal-noise ratio of the asset with maximal Sharpe ratio, *i.e.*, $\zeta_{(k)}$, but instead test the population as a whole. That is, we wish to test the following hypothesis:

$$H_0 : \forall_i \zeta_i = \zeta_0 \quad \text{versus} \quad H_1 : \forall_i \zeta_i \geq \zeta_0 \text{ and } \exists_j \zeta_j > \zeta_0. \tag{4.11}$$

There has been considerable research on testing hypotheses of this form but when the parameter is the expected value of a multivariate normal. [Silvapulle and Sen, 2005; Vock, 2007; Perlman, 1969; Perlman and Wu, 1999; Tang, 1994] By appealing to the normal approximation of Equation 3.29, we can use these tests for tackling overoptimism by selection. That is, assume

$$\hat{\boldsymbol{\zeta}} \approx \mathcal{N}\left(\boldsymbol{\zeta}, \frac{1}{n}\Omega\right), \tag{4.12}$$

where Ω is some function of R, κ and $\boldsymbol{\zeta}$, as in Equation 3.29.

The classical procedure for testing equality of $\boldsymbol{\zeta} = \boldsymbol{\zeta}_0$ in this case, without regard to the form of the alternatives, is via Hotelling's T^2 test. [Anderson, 2003; Press, 2012] In our case since we have assumed a normal approximation, we are essentially assuming Ω is known, and so a χ^2 test would be used instead. The most basic test against a one-sided alternative is similar in nature, but involves projection to the positive orthant.

We first present the general solution, which is probably only of use when k is small and when Ω can be inverted; then we will consider the large k case for the simple rank-one model of correlation of Equation 4.4. The general case proceeds by computing

$$\bar{x}^2 = n\left[\left(\hat{\boldsymbol{\zeta}} - \boldsymbol{\zeta}_0\right)^\top \Omega^{-1}\left(\hat{\boldsymbol{\zeta}} - \boldsymbol{\zeta}_0\right) - \min_{\boldsymbol{x} \geq \boldsymbol{\zeta}_0}\left(\hat{\boldsymbol{\zeta}} - \boldsymbol{x}\right)^\top \Omega^{-1}\left(\hat{\boldsymbol{\zeta}} - \boldsymbol{x}\right)\right]. \tag{4.13}$$

Here $\boldsymbol{\zeta}_0 = \zeta_0 \mathbf{1}$, and $\boldsymbol{x} \geq \boldsymbol{\zeta}_0$ is taken to be element-wise. Compare Equation 4.13 to Equation 3.31 for testing essentially the same null against an unrestricted alternative.

Under the null hypothesis, $\bar{x}^2$ will follow not a chi-square distribution, but rather a *chi-bar-square distribution.* To test Hypothesis 4.11, reject at the α level if

$$\sum_{0 \leq i \leq k} w_i\left(\Omega\right) F_{\chi^2}\left(\bar{x}^2; i\right) \geq 1 - \alpha, \tag{4.14}$$

where $F_{\chi^2}\left(x; i\right)$ is the cumulative distribution of the χ^2 distribution with i degrees of freedom, and $w_i\left(\Omega\right)$ are the *chi-bar-square weights.*

The $w_i\left(\Omega\right)$, which are a function of Ω, are tricky to describe analytically except for some simple cases of Ω: for example, when Ω is diagonal, we have

$$w_i\left(\Omega\right) = \binom{k}{i} 2^{-k}.$$

For non-diagonal Ω, the recommended course of action is to estimate the w_i via simulation. [Silvapulle and Sen, 2005] The simulations proceed by repeating the following steps some large number of times (say $N = 10^4$):

1. Generate $\boldsymbol{z} \sim \mathcal{N}\left(\boldsymbol{\zeta}_0, \Omega\right)$.
2. Find the $\boldsymbol{x} \geq \boldsymbol{\zeta}_0$ that minimizes $\left(\boldsymbol{z} - \boldsymbol{x}\right)^\top \Omega^{-1} \left(\boldsymbol{z} - \boldsymbol{x}\right)$.
3. Count the number of elements of $\boldsymbol{x}$ which are strictly greater than ζ_0, call it s.

Estimate $w_i\left(\Omega\right)$ as the proportion of the N simulations where the s equals i.

Example 4.1.7 (Testing signal-noise ratios Fama-French factors). Consider the four Fama French factors from Example 1.1.1. Based on monthly returns from Jan 1927 to Dec 2020, the Sharpe ratios were computed as

$$\hat{\zeta} = \begin{bmatrix} Mkt & SMB & HML & UMD \\ \hline 0.1773 & 0.0653 & 0.0919 & 0.1377 \end{bmatrix} \text{mo.}^{-1/2}.$$

First we test the null hypothesis $H_0 : \forall_i \zeta_i = \zeta_0$ for $\zeta_0 = 0.12\,\text{mo.}^{-1/2}$, against an unrestricted alternative. This we do via the χ^2 test of Equation 3.30.

We estimate the covariance of the Sharpe ratios (up to n) as

$$\hat{\Omega} \approx \begin{bmatrix} 1.0504 & 0.2970 & 0.1429 & -0.3097 \\ 0.2970 & 0.9005 & 0.0662 & -0.1936 \\ 0.1429 & 0.0662 & 0.8520 & -0.4027 \\ -0.3097 & -0.1936 & -0.4027 & 1.5485 \end{bmatrix} \text{mo.}^{-1}.$$

This is estimated by the procedure described in Section 3.2: namely we stack $\boldsymbol{x}$ with $\boldsymbol{x}^2$, compute the sample mean and covariance of this vector, then apply the delta method.

The test statistic is then computed as

$$x^2 = n\left(\hat{\zeta} - \zeta_0 \mathbf{1}\right)^\top \hat{\Omega}^{-1} \left(\hat{\zeta} - \zeta_0 \mathbf{1}\right) = 12.1.$$

Under the null hypothesis, this is asymptotically distributed as a $\chi^2\left(4\right)$, which corresponds to a p-value of of 0.017, and a Frequentist would reject the null hypothesis.

Now we test the null $H_0 : \forall_i \zeta_i = \zeta_0 = 0.12\,\text{mo.}^{-1/2}$ against the one-sided alternative $H_1 : \forall_i \zeta_i \geq \zeta_0$ and $\exists_j \zeta_j > \zeta_0$. First we estimate the chi-bar-square weights using 100,000 simulations using the estimated $\hat{\Omega}$. The weights are estimated as

$$\begin{bmatrix} w_0 & w_1 & w_2 & w_3 & w_4 \\ \hline 0.025 & 0.128 & 0.342 & 0.428 & 0.077 \end{bmatrix}.$$

The projection of $\hat{\zeta}$ onto the alternative orthant is

$$\hat{\zeta} = \begin{bmatrix} Mkt & SMB & HML & UMD \\ \hline 0.1980 & 0.1200 & 0.1200 & 0.1200 \end{bmatrix} \text{mo.}^{-1/2}.$$

We compute the statistic as

$$\bar{x}^2 = n\left[\left(\hat{\boldsymbol{\zeta}} - \zeta_0\mathbf{1}\right)^\top \Omega^{-1}\left(\hat{\boldsymbol{\zeta}} - \zeta_0\mathbf{1}\right) - \min_{\boldsymbol{x}\geq\zeta_0\mathbf{1}}\left(\hat{\boldsymbol{\zeta}} - \boldsymbol{x}\right)^\top \Omega^{-1}\left(\hat{\boldsymbol{\zeta}} - \boldsymbol{x}\right)\right]$$
$$= 12.108 - 4.545 = 7.563.$$

We then compute the p-value using the simulated weights as 0.041, and we narrowly reject the null at the 0.05 level. ⊣

The chi-bar-square test for general R will not scale well to large k, since it requires solving a quadratic program to perform a projection. Worse, the computation of the chi-bar-square weights requires that we solve that quadratic program thousands of times. Here we describe a simpler form of the test under the assumption that R follows the rank one updated correlation model of Equation 4.4, namely

$$\mathsf{R} = (1-\rho)\,\mathsf{I} + \rho\left(\mathbf{1}\mathbf{1}^\top\right), \tag{4.15}$$

for $\rho \geq 0$.

Again, we assume a normal approximation for $\hat{\boldsymbol{\zeta}}$. Under the null hypothesis that $\boldsymbol{\zeta} = \zeta_0\mathbf{1}$, we have

$$\hat{\boldsymbol{\zeta}} \approx \mathcal{N}\left(\boldsymbol{\zeta}, \frac{1}{n}\Omega\right), \text{ where } \Omega = a_0\,\mathsf{I} + a_2\left(\mathbf{1}\mathbf{1}^\top\right). \tag{4.16}$$

As noted above (and via Exercise 4.6) the inverse square root of this Ω takes the form

$$\Omega^{-1/2} = b_0\,\mathsf{I} + b_2\left(\mathbf{1}\mathbf{1}^\top\right).$$

Letting $\hat{\boldsymbol{\xi}} = \Omega^{-1/2}\hat{\boldsymbol{\zeta}}$ and $\boldsymbol{\xi} = \Omega^{-1/2}\boldsymbol{\zeta}$, the normal approximation can be rewritten as

$$\sqrt{n}\hat{\boldsymbol{\xi}} \approx \mathcal{N}\left(\sqrt{n}\boldsymbol{\xi}, \mathsf{I}\right).$$

Note that $\mathbf{1}$ is an eigenvector of $\Omega^{-1/2}$ with

$$\Omega^{-1/2}\mathbf{1} = c\mathbf{1},$$

for $c = b_0 + kb_2$. Note that the b_2 can be chosen such that $c > 0$. From the order-preserving nature of $\Omega^{-1/2}$, the null and alternative of Hypothesis 4.11 can be expressed as

$$H_0 : \forall_i \xi_i = c\zeta_0 \quad \text{versus} \quad H_1 : \forall_i \xi_i \geq c\zeta_0 \text{ and } \exists_j \xi_j > c\zeta_0.$$

We can now appeal to the simple chi-bar square test. [Silvapulle and Sen, 2005; Wolak, 1987] First transform the vector of Sharpe ratios to $\hat{\boldsymbol{\xi}}$ via

$$\hat{\boldsymbol{\xi}} = \mathsf{R}^{-1/2}\hat{\boldsymbol{\zeta}} = c\bar{\zeta}\mathbf{1} + (1-\rho)^{-1/2}\left(\hat{\boldsymbol{\zeta}} - \bar{\zeta}\mathbf{1}\right), \tag{4.17}$$

where $\bar{\zeta}$ is the average of the sample Sharpe ratios. Then compute

$$\bar{x}^2 = n\sum_i \left(\hat{\xi}_i - c\zeta_0\right)_+^2, \tag{4.18}$$

where y_+ is the positive part of y, *i.e.*, $y_+ = y$ if $y > 0$ and zero otherwise. Then compute the CDF of the corresponding chi-bar square distribution as

$$Q = \sum_{i=0}^{k} w_i F_{\chi^2}\left(\bar{x}^2; i\right), \tag{4.19}$$

where $F_{\chi^2}(i; x)$ is the cumulative distribution of the χ^2 distribution with i degrees of freedom, and w_i are the chi-bar square weights, which for diagonal matrix take the value

$$w_i = \binom{k}{i} 2^{-k}.$$

Reject the null hypothesis at the α level if $1 - Q \leq \alpha$.

Note that, as with the Bonferroni correction, the test statistic $\bar{x}^2$ is computed on all elements of $\hat{\boldsymbol{\zeta}}$, and thus the decision to reject the null may not be "about" $\hat{\zeta}_1$. In testing we will see that the one-sided test is highly susceptible to distribution of the $\boldsymbol{\zeta}$, moreso than the Bonferroni correction.

We note that under this setup it is also easy to use Follman's test, which is a very simple procedure with increased power against one-sided alternatives. [Follmann, 1996] Here we would compute

$$g^2 = nkc^2\left(\bar{\zeta} - \zeta_0\right)^2 + nb_0^2 \sum_i \left(\hat{\zeta}_i - \bar{\zeta}\right)^2, \tag{4.20}$$

and reject at the α level if both $1 - F_{\chi^2}\left(g^2; k\right) \leq 2\alpha$ and $\bar{\zeta} > \zeta_0$.

Example 4.1.8 (One-Sided Tests, Five Industry Portfolios). We consider the monthly returns of five industry portfolios, as introduced in Example 1.2.5. These include 1,128 months of data on five industries from Jan 1927 to Dec 2020. We compute the Sharpe ratio of the returns of each as follows:

Other	*Manufacturing*	*Technology*	*Consumer*	*Healthcare*
0.142	0.172	0.178	0.193	0.195

$\text{mo.}^{-1/2}$.

We have reordered the industries in increasing Sharpe ratio. The industry portfolio with the highest Sharpe ratio was Healthcare with a Sharpe ratio of around $0.195\,\text{mo.}^{-1/2}$ which is approximately $0.676\,\text{yr}^{-1/2}$.

The correlation of industry returns is high: the pairwise sample correlations range from 0.709 to 0.892 with a median value of 0.803. We assume $\mathsf{R} = (1-\rho)\,\mathsf{I} + \rho\left(\mathbf{1}\mathbf{1}^{\top}\right)$ and assume this median value as ρ. We compute the median kurtosis factor of industry returns to be 3.469.

We test the null $H_0 : \forall_i \zeta_i = \zeta_0 = 0.15\,\text{mo.}^{-1/2}$ against the one-sided alternative $H_1 : \forall_i \zeta_i \geq \zeta_0$ and $\exists_j \zeta_j > \zeta_0$. Under the null we compute

$$a_0 = 0.21, \qquad a_2 = 0.842,$$
$$b_0 = 2.18, \qquad b_2 = -0.341.$$

From these we compute $c = 0.475$. We compute the statistic from Equation 4.18 as $\bar{x}^2 = 6.343$, corresponding to a p-value of 0.081. We fail to reject at the 0.05 level.

We compute Follman's statistic (Equation 4.20) as $g^2 = 10.627$. This corresponds to the upper 0.059 quantile of the χ^2 distribution with 5 degrees of freedom, which is smaller than 0.10. Moreover, since $\bar{\zeta} = 0.176\,\text{mo.}^{-1/2} \geq 0.15\,\text{mo.}^{-1/2} = \zeta_0$, we reject the null hypothesis at the 0.05 level. ⊣

4.1.4 Hansen's asymptotic correction

One failing of many of the approaches considered above is the problem of irrelevant alternatives. That is, instead of testing under the null of equality (Hypothesis 4.11 above) we should test the following

$$H_0 : \forall_i \zeta_i \leq c \quad \text{versus} \quad H_1 : \exists_i \zeta_i > c.$$

Testing such a composite null hypothesis is typically via a *non-similar* test, *i.e.*, one which has a type I rate no greater than the nominal rate for all $\boldsymbol{\zeta}$ in the null, and which achieves that nominal rate for some $\boldsymbol{\zeta}$ under the null. Such tests achieve the nominal rate under the *least favorable configuration,* which in our case is the null of equality given in Hypothesis 4.11. [Silvapulle and Sen, 2005]

Hansen [2003, 2005] describes a procedure which avoids this problem. The idea is elegant, and ultimately simple to implement. For the problem of overoptimism by selection, it amounts to assuming that the null takes the form

$$H_0 : \forall_i \, \zeta_i \leq c \text{ and } \left|\zeta_i - \hat{\zeta}_i\right| \leq g_n \quad \text{versus} \quad H_1 : \exists_i \zeta_i > c,$$

for some g_n. Note this seems odd since the sample Sharpe ratio appears in the null hypothesis to be tested. Nonetheless, Hansen describes how such a test can be performed while maintaining a maximum type I rate asymptotically, and achieving higher power.

Hansen [2003, 2005] applied this correction to the chi-bar-square statistic of Equation 4.18, and later to a Studentized version of White's Reality Check statistic, which is rather like the corrected Bonferroni statistic computed in Equation 4.7. Applying Hansen's correction to our problem is simple: compute $\hat{\boldsymbol{\xi}}$ as in Equation 4.17. Let $\tilde{k}$ be the number of elements of $\hat{\boldsymbol{\xi}}$ greater than $c\zeta_0 - \sqrt{\left(2 \log \log n\right)/n}$, where $c = \left(1 + \left(k - 1\right)\rho\right)^{-1/2}$. If $\tilde{k} = 0$ fail to reject. Otherwise compute the chi-bar-square statistic $\bar{x}^2$ as in Equation 4.18 and reject if

$$\sum_{i=0}^{\tilde{k}} \binom{\tilde{k}}{i} 2^{-\tilde{k}} F_{\chi^2}\left(\bar{x}^2; i\right) \geq 1 - \alpha.$$

This is the same as the chi-bar-square test considered above, but with reduced degrees of freedom *which depend on the observed.*

The same correction is easily applied to the Bonferroni maximum test: again, compute $\hat{\boldsymbol{\xi}}$ and $\tilde{k}$. If $\tilde{k} = 0$ fail to reject. Otherwise reject at the α level if

$$\max_i \hat{\boldsymbol{\xi}}_i - c\zeta_0 \geq z_{1-\alpha/\tilde{k}}.$$

This is similar in spirit to Hansen's SPA test, except it does not use the bootstrap procedure to estimate the standard error as described by Hansen [2005]; it is similar in every other regard.

Example 4.1.9 (One-Sided Tests with Hansen's Correction, Five Industry Portfolios). We return to the testing of the five industry portfolios considered in Example 4.1.8. There we used the chi-bar-square test to test the null $H_0 : \forall_i \zeta_i = 0.15\,\text{mo.}^{-1/2}$, and failed to reject at the 0.05 level.

We now apply Hansen's correction to the chi-bar-square test. As above, we compute $c = 0.475$, and $\bar{x}^2 = 6.343$. The cutoff for the $\hat{\boldsymbol{\xi}}$ is $c\zeta_0 - \sqrt{2 \log\log n/n} = 0.013$. We find that 4 of the elements of $\hat{\boldsymbol{\xi}}$ are above the cutoff. Based on this many degrees of freedom, we compute the p-value of the chi-bar-square statistic to be 0.054 and we again fail to reject the null at the 0.05 level. ⊣

4.1.5 Conditional inference

Taming overoptimism via testing Hypothesis 4.1 seems like overkill. One is typically only interested in performing inference on $\zeta_{(k)}$, the signal-noise ratio associated with the strategy that has maximum Sharpe ratio, rather than on *all* the ζ_i. We can do this directly via *conditional inference*. The idea is to perform inference on $\zeta_{(k)}$ conditional on having observed that $\hat{\zeta}_{(k)}$ is the largest Sharpe ratio. One suspects that by directly testing the quantity of interest, one could gain statistical power over the MHT corrections considered above.

Briefly the conditional probability of event A conditional on event B is the probability that both A and B occur, divided by the probability that B occurs:

$$\Pr\{A \mid B\} = \frac{\Pr\{A \cap B\}}{\Pr\{B\}}.$$

First we will use conditional probability to analyze some simpler problems, before returning to overoptimism by selection.

Recall the opportunistic strategy introduced in Section 2.5.2: you observe the Sharpe ratio of an asset, then decide to hold it long if positive, otherwise you will hold the asset short. You then wish to perform inference on your strategy, taking into account that the sign depends on the sample. Previously we used symmetric confidence intervals to approach this problem, but it is easily described as a conditional inference problem.

Adjusting the sign of returns to match whether we hold the asset long or short, we are performing inference on ζ conditional on $\hat{\zeta} > 0$. The Sharpe ratio in this case has the (conditional) distribution function

$$F_{SR}\left(\hat{\zeta} \,\middle|\, \hat{\zeta} > 0 \,; \zeta, n\right) = \frac{F_{SR}\left(\hat{\zeta}; \zeta, n\right) - F_{SR}\left(0; \zeta, n\right)}{1 - F_{SR}\left(0; \zeta, n\right)}.$$

More generally, to test the null hypothesis

$$H_0 : \zeta_{(k)} = \zeta_0 \,\middle|\, \hat{\zeta} \geq \zeta_1 \quad \text{versus} \quad H_1 : \zeta_{(k)} > \zeta_0, \tag{4.21}$$

reject at the α level if $F_{SR}\left(\hat{\zeta} \,\middle|\, \hat{\zeta} > \zeta_1 \,; \zeta_0, n\right) \geq 1 - \alpha$. Equivalently, to construct an α lower confidence bound, find ζ_0 such that $F_{SR}\left(\hat{\zeta} \,\middle|\, \hat{\zeta} > \zeta_1 \,; \zeta_0, n\right) = 1 - \alpha$.

Example 4.1.10 (Confidence intervals, Conditional Inference). As in Example 2.5.3, consider the case of 504 daily observations of some hypothetical asset's returns, which result in a measured Sharpe ratio of exactly $0.6\,\text{yr}^{-1/2}$, assuming 252 days per year. The "exact" 95% confidence interval is computed as $[-0.787, 1.986)\ \text{yr}^{-1/2}$.

Assuming that we are considering holding this asset long because we observed $\hat{\zeta} \geq 0$, we compute 95% conditional confidence interval on ζ as $[-2.627, 1.954)\ \text{yr}^{-1/2}$. Note this confidence interval is much wider than the nearly symmetric unconditional confidence interval, which, as outlined in Section 2.5.2, we are justified in applying to the opportunistic strategy.

As a consolation prize, by using conditional inference, we can compute a one-sided conditional confidence interval to this problem. In this case, we compute the 95% one sided conditional confidence interval on ζ as $[-2.014, \infty)\,\text{yr}^{-1/2}$. This is still lower than the lower bound suggested by the symmetric confidence interval. ⊣

At risk of overgeneralizing from this example, we often find that conditional inference has lower power than other inferential approaches. In this case, the conditional inference procedure could not exploit the fact that had we observed $\hat{\zeta} < 0$, we would have opportunistically flipped the sign of the asset and performed another test. That is, the conditioning event does not fully capture our testing procedure and instead assumes that had we observed $\hat{\zeta} < 0$ we would not have performed any test.

Conditional inference for overoptimism by selection

To simplify the exposition, we will suppose that, conditional on observing the vector $\hat{\boldsymbol{\zeta}}$, one rearranges the indices in increasing order, so that the first index refers to the asset with the smallest observed Sharpe ratio, and the last index, k, to the asset with the largest. This is to avoid the cumbersome notation of $\hat{\zeta}_{(k)}$, and we instead can just write $\hat{\zeta}_k$. We note this maximum condition can be written in the form $\mathsf{A}\hat{\boldsymbol{\zeta}} \leq \boldsymbol{b}$ for $(k-1) \times k$ matrix A defined by

$$\mathsf{A} = \begin{bmatrix} 1 & 0 & \dots & 0 & -1 \\ 0 & 1 & \dots & 0 & -1 \\ \vdots & \vdots & \ddots & \vdots & \vdots \\ 0 & 0 & \dots & 1 & -1 \end{bmatrix},$$

and where $\boldsymbol{b}$ is the $(k-1)$-dimensional zero vector. Also note that we are interested in performing inference on ζ_k, which we can express as $\boldsymbol{\eta}^\top \boldsymbol{\zeta}$ for $\boldsymbol{\eta} = \boldsymbol{e}_k$.

Since, by Equation 3.29, the vector $\hat{\boldsymbol{\zeta}}$ is approximately normally distributed, we can use the following "polyhedral" theorem of Lee *et al.* [2013] [Pav, 2019b] :

Theorem 4.1.11 (Lee *et al.*, Theorem 5.2 [Lee *et al.*, 2013]). *Suppose* $\boldsymbol{y} \sim \mathcal{N}(\boldsymbol{\mu}, \Sigma)$. *Define* $\boldsymbol{c} = \Sigma\boldsymbol{\eta}/\boldsymbol{\eta}^\top\Sigma\boldsymbol{\eta}$, *and* $\boldsymbol{z} = \boldsymbol{y} - \boldsymbol{c}\boldsymbol{\eta}^\top\boldsymbol{y}$. *Let* $\Phi(x)$ *be the CDF of a standard normal, and let Let* $F(x; a, b, 0, 1)$ *be the CDF of a standard normal truncated to* $[a, b)$:

$$F(x; a, b, 0, 1) := \frac{\Phi(x) - \Phi(a)}{\Phi(b) - \Phi(a)}.$$

Let $F(x; a, b, \mu, \sigma^2)$ *be the CDF of a general truncated normal, defined by*

$$F(x; a, b, \mu, \sigma^2) = F\left(\frac{x-\mu}{\sigma}; \frac{a-\mu}{\sigma}, \frac{b-\mu}{\sigma}, 0, 1\right).$$

Then, conditional on $\mathsf{A}\boldsymbol{y} \leq \boldsymbol{b}$, *the random variable*

$$F\left(\boldsymbol{\eta}^\top\boldsymbol{y}; \mathcal{V}^-, \mathcal{V}^+, \boldsymbol{\eta}^\top\boldsymbol{\mu}, \boldsymbol{\eta}^\top\Sigma\boldsymbol{\eta}\right)$$

is Uniform on $[0, 1)$, *where* $\mathcal{V}^-$ *and* $\mathcal{V}^+$ *are given by*

$$\mathcal{V}^- = \max_{j:(\mathsf{A}\boldsymbol{c})_j < 0} \frac{\boldsymbol{b}_j - (\mathsf{A}\boldsymbol{z})_j}{(\mathsf{A}\boldsymbol{c})_j}, \quad \mathcal{V}^+ = \min_{j:(\mathsf{A}\boldsymbol{c})_j > 0} \frac{\boldsymbol{b}_j - (\mathsf{A}\boldsymbol{z})_j}{(\mathsf{A}\boldsymbol{c})_j}.$$

This theorem gives us a way to test the null hypothesis

$$H_0 : \zeta_{(k)} = \zeta_0 \quad \text{versus} \quad H_1 : \zeta_{(k)} > \zeta_0. \tag{4.22}$$

To test this null, set $\boldsymbol{y} = \hat{\boldsymbol{\zeta}}$ and compute $\boldsymbol{c}, \boldsymbol{z}, \mathcal{V}^-$, and $\mathcal{V}^+$ as in the theorem; estimate the covariance, Ω via Equation 3.29; finally, reject the null hypothesis at the α level if $F\left(\hat{\zeta}_{(k)}; \mathcal{V}^-, \mathcal{V}^+, \zeta_0, \frac{1}{n}\Omega_{(k),(k)}\right) > 1 - \alpha$.

Note that we do not have to compute the entire matrix Ω, we only need one column of it to compute the vector $\mathbf{c}$ (and one element of that is $\Omega_{(k),(k)}$). That column is

$$\Omega_{:,(k)} = \mathsf{R}_{:,(k)} + \frac{\kappa - 1}{4}\boldsymbol{\zeta}\zeta_0 + \frac{\kappa}{2}\,\mathrm{Diag}\left(\boldsymbol{\zeta}\right)\left(\mathsf{R}_{:,(k)} \odot \mathsf{R}_{:,(k)}\right)\zeta_0. \tag{4.23}$$

The vector $\boldsymbol{\zeta}$ is unknown; under the null one can estimate it as either $\zeta_0\mathbf{1}$ or $\zeta_0\boldsymbol{e}_{(k)}$.

Example 4.1.12 (Five Industry Portfolios). We consider the monthly returns of five industry portfolios, as introduced in Example 1.2.5. These include 1,128 months of data on five industries from Jan 1927 to Dec 2020. We compute the Sharpe ratio of the returns of each as follows:

$$\begin{bmatrix} \mathit{Other} & \mathit{Manufacturing} & \mathit{Technology} & \mathit{Consumer} & \mathit{Healthcare} \\ \hline 0.142 & 0.172 & 0.178 & 0.193 & 0.195 \end{bmatrix} \mathrm{mo.}^{-1/2}.$$

We have reordered the industries in increasing Sharpe ratio. The industry portfolio with the highest Sharpe ratio was Healthcare with a Sharpe ratio of around 0.195 $\mathrm{mo.}^{-1/2}$ which is approximately 0.676 $\mathrm{yr}^{-1/2}$.

We are interested in computing 95% upper confidence intervals on the signal-noise ratio of the Healthcare portfolio. We are only considering this portfolio as it is the one with maximum Sharpe ratio in our sample. If we had been interested in testing Healthcare without our conditional selection, we would compute the confidence interval $\left[0.146\ \mathrm{mo.}^{-1/2}, \infty\right)$ based on inverting the non-central t-distribution. If instead we approximate the standard error by plugging in 0.195 $\mathrm{mo.}^{-1/2}$ as the signal-noise ratio of Healthcare into Equation 3.29, we estimate the standard error of the Sharpe ratio to be 0.03 $\mathrm{mo.}^{-1/2}$. Based on this we can compute the naïve confidence interval of the measured Sharpe ratio plus $z_{0.05} = -1.645$ times the standard error. This also gives the confidence interval $\left[0.146\ \mathrm{mo.}^{-1/2}, \infty\right)$.

Using the simple Bonferroni correction, however, since we selected Healthcare only for having the maximum Sharpe ratio, we should compute the confidence interval by adding $z_{0.01} = -2.326$ times the standard error. This yields the confidence interval $\left[0.125\ \mathrm{mo.}^{-1/2}, \infty\right)$.

The correlation of industry returns is high, however. The pairwise sample correlations range from 0.709 to 0.892 with a median value of 0.803. Plugging this value in as ρ, we find the value ζ_0 such that the z_1 from Equation 4.7 is equal to $z_{0.01} = -2.326$. This leads to the confidence interval $\left[0.129\ \mathrm{mo.}^{-1/2}, \infty\right)$.

We use this estimate of ρ to compute the chi-bar square test. We invert the test to find the 95% upper confidence interval $\left[0.145\ \mathrm{mo.}^{-1/2}, \infty\right)$.

Finally we use the conditional estimation procedure, inverting the hypothesis test to find the corresponding population value. We compute Ω by assuming returns are Gaussian, so $\kappa = 1$, and plugging in the sample $\hat{\boldsymbol{\zeta}}$ for $\boldsymbol{\zeta}$. This yields the confidence interval $\left[-0.051\ \mathrm{mo.}^{-1/2}, \infty\right)$. ⊣

This example is consistent with our previous findings suggesting conditional estimation is less powerful than an MHT correction. In the following example we will examine this question directly via simulations under the alternative hypothesis. However, "the" alternative can take many forms. One interpretation is that we condition on $\zeta_k > 0$, where again the indexing is such that $\hat{\zeta}_k$ was the maximum over k assets; then we estimate the probability of (correctly) rejecting $\zeta_k = 0$ versus the value that ζ_k takes. However, we suspect that the power, as described in this way, would depend on the distribution of values of $\boldsymbol{\zeta}$.

We consider three forms for the alternative: one where all k elements of ζ are equal ("all-equal"), and two where m of k elements of ζ are equal to some positive value, and the remaining $k - m$ are negative that value, for the case $m = 1$ ("one-good"), and $k = 2m$, which we call "half-good."

Note that in the all equal case, since every asset has the same signal-noise ratio, whichever we select will have the same signal-noise ratio, and the test should have the same power as the t-test for a single asset. The conditional estimation procedure, however, may suffer in this case as we may condition on a $\hat{\zeta}_k$ that is very close to being non-optimal, resulting in a small test statistic for which we do not reject. On the other hand, for the one-good case, as the $k-1$ assets may have considerably negative signal-noise ratio, they are unlikely to exhibit the largest Sharpe ratio, and so the MHT is merely testing a single asset, but at the α/k level instead of the α level, resulting in lower power. The conditional estimation procedure, however, should not suffer in this test.

Example 4.1.13 (Power of conditional inference and MHT corrections). We perform simulations under all equal, one-good, and half-good configurations, letting the "good" signal-noise ratio vary from 0 to $0.15\,\text{day}^{-1/2}$, which corresponds to an "annualized" signal-noise ratio of around $2.4\,\text{yr}^{-1/2}$. We draw Gaussian returns with diagonal covariance for 100 assets, with $n = 1{,}008$. For each setting we perform $10{,}000$ simulations then compute the empirical rejection rate of the test at the 0.05 level, conditional on the signal-noise ratio of the *selected* asset, which is to say the one with the largest Sharpe ratio. Note that in some simulations the largest Sharpe ratio is observed in an asset with a negative signal-noise ratio.

In Figure 4.9, we plot the empirical power of the MHT Bonferroni test and the chi-bar square test, both with Hansen's log-log correction, Follman's test, and the conditional estimation procedure versus the signal-noise ratio of the selected asset. We present facet columns for the three configurations, *viz.* all-equal, one-good, half-good. A horizontal line at 0.05 gives the nominal rate under the null, which occurs at $\zeta = 0$ in these plots. As expected from the above explanation, the chi-bar square test has the highest power for the all-equal alternative, followed by Follman's test, then the MHT, then the conditional estimation test. These relationships are reversed for the one-good case. The chi-bar square test and Follman's test have very low power against the one-good alternative. All tests perform similarly in the half-good and all-equal alternatives, with the exception of Follman's test, which achieves a maximum power of 1/2 in the half-good case, as is to be expected since this is the probability that $\bar{\zeta} > 0$ in the half-good case.

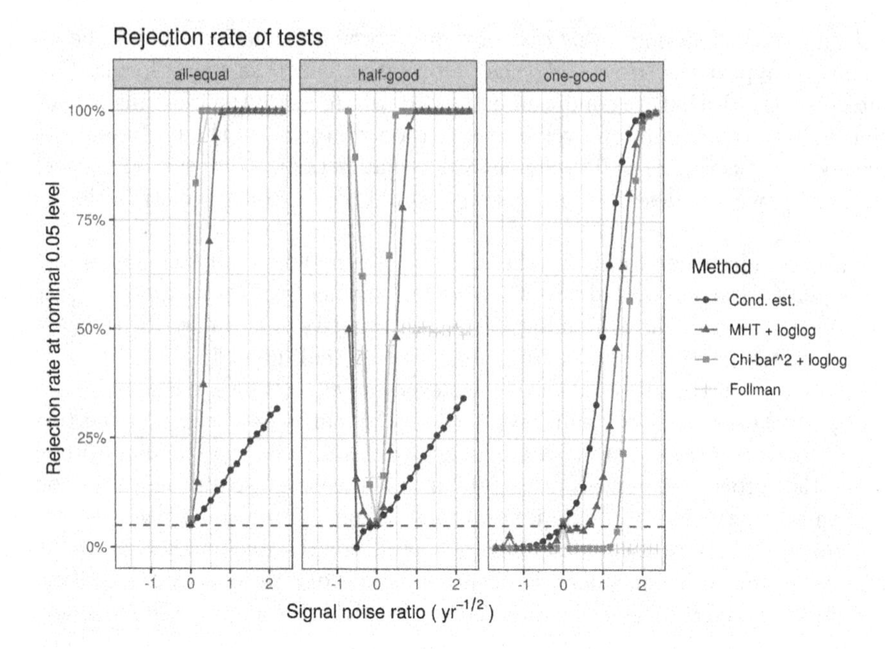

FIGURE 4.9
The empirical power of the conditional estimation test, Follman's test, and the chi-bar-square and MHT tests with Hansen's log log correction are shown versus the signal-noise ratio of the asset with maximum Sharpe ratio under different arrangements of the vector $\boldsymbol{\zeta}$.

The power of the conditional estimation procedure for the all equal case is rather disappointing. For the case where all assets have a signal-noise ratio of $2.4\,\mathrm{yr}^{-1/2}$, the test has a power of only around a half. The test suffers from low power because we are conditioning on "$\hat{\zeta}_k$ is the largest Sharpe ratio," where we should actually be conditioning on "the asset with the largest Sharpe ratio."

Note the odd plot in the half-good facet: the MHT correction and one-sided tests have greater than 0.05 rejection rate for *negative* signal-noise ratio. The plot is somewhat misleading in this case, however. We have performed 10,000 simulations for each setting of the "good" signal-noise ratio; in some very small number of them for the half-good case, an asset with negative signal-noise ratio exhibits the maximum Sharpe ratio. We are plotting the rejection rate for the test in this case. But note that the null hypothesis that MHT and the one-sided test are testing *is* violated in this case, because half the assets have positive signal-noise ratio, and the alternative procedures test the null that all assets have zero or lower signal-noise ratio. We have not shown the probability that a "bad" asset has the highest Sharpe ratio, but note that when the "good" signal-noise ratio is greater than $0.05\,\mathrm{day}^{-1/2}$ we do not observe this occuring even once over the 10,000 simulations performed for each setting.

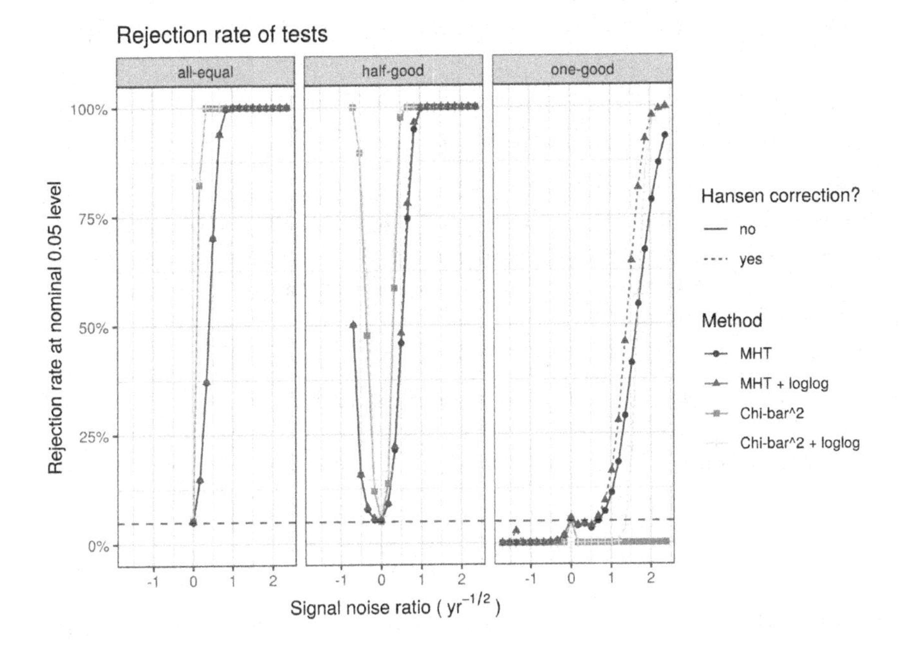

FIGURE 4.10
The empirical power of the MHT corrected test, the chi-bar-square test, and the chi-bar-square and MHT tests with Hansen's log log correction are shown versus the signal-noise ratio of the asset with maximum Sharpe ratio under different arrangements of the vector $\boldsymbol{\zeta}$.

We note that the chi-bar-square test appears to have higher power than the other tests for the all-equal and half-good populations, but fails to reject at all in the one-good case, except under the null. This is to be expected, since the chi-bar-square test depends strongly on all the observed Sharpe ratios, and in this case we expect many of them to be negative.

In Figure 4.10, we plot the power versus the signal-noise ratio of the selected asset for the MHT Bonferroni and the chi-bar square tests, both with and without the log-log corrections. The log log adjustment has little effect in the all-equal and half-good case, but greatly improves the power of the MHT and chi-bar square tests in the one-good case. For this sample size, Hansen's adjustment nearly maintains the nominal type I rate under the null, though this is unlikely to hold for smaller n.

It is interesting to note that while the conditional estimation procedure generally has lower power than the other tests (except in the one-good case), it appears to have monotonic rejection probability with respect to the signal-noise ratio of the *selected* asset. That is, in the half-good case, it has low rejection probability in the odd simulations where a "bad" asset is selected.

We doubt that the simple experiments performed here have revealed all the relevant differences between the various tests or when one dominates the others. ⊣

4.1.6 † Subspace approximation

Another potential approach to the problem which may be useful in the case where returns are highly correlated, as one expects when returns are from backtested quantitative trading strategies, is via a subspace approximation. First we assume that the $n \times k$ matrix of returns, X can be approximated by a p-dimensional subspace

$$\mathsf{X} \approx \mathsf{Y}\mathsf{W},$$

where Y is a $n \times p$ matrix of "latent" returns, and W is a $k \times k$ "loading" matrix.

Now the column of X with maximal $\hat{\zeta}$ has Sharpe ratio that is smaller than

$$\hat{\zeta}_* := \max_{\boldsymbol{\nu}} \frac{\hat{\boldsymbol{\mu}}^\top \boldsymbol{\nu}}{\sqrt{\boldsymbol{\nu}^\top \hat{\Sigma} \boldsymbol{\nu}}},$$

where $\hat{\boldsymbol{\mu}}$ is the p-vector of the (sample) means of columns of Y and $\hat{\Sigma}$ is the sample covariance matrix. This maximum takes value

$$\hat{\zeta}_* = \sqrt{\hat{\boldsymbol{\mu}}^\top \hat{\Sigma}^{-1} \hat{\boldsymbol{\mu}}},$$

which is, up to scaling, Hotelling's T^2 statistic.

We shall see in the sequel that under the null hypothesis that the rows of Y are independent draws from a Gaussian random variable with zero mean, then

$$\frac{(n-p)\,\hat{\zeta}_*^2}{p\,(n-1)}$$

follows an F distribution with p and $n-p$ degrees of freedom. Under the alternative it follows a non-central F distribution. [Anderson, 2003; Press, 2012] Via this upper bound $\hat{\zeta}_1 \le \hat{\zeta}_*$, one can then perform tests on the null hypothesis $\forall_i \zeta_i = 0$.

However, this approach requires that one estimate p, the dimensionality of the latent subspace. Moreover, the subspace approximation may not be very good. It would seem that to get near equality of $\hat{\zeta}_1$ and $\hat{\zeta}_*$, the columns of X would have to contain both positive and negative exposure to the columns of Y. This in turn should result in mixed correlation of asset returns, which we may not observe in practice. Finally, empirical testing indicates this approach requires further development, as in the following example.

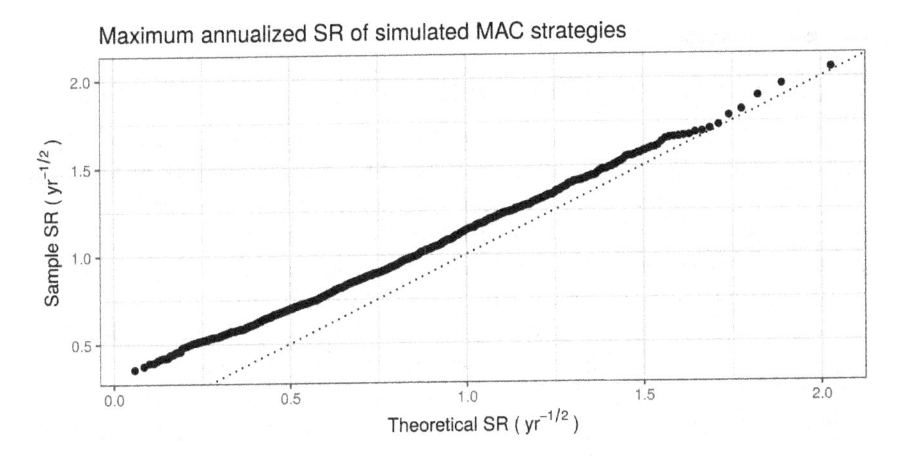

FIGURE 4.11
Q-Q plot of the Sharpe ratio of the optimal 2-window MAC in a simulated backtest under the null. Simulations are over 1,100 days. Theoretical quantiles are computed via the F distribution assuming $k = 3$.

Example 4.1.14 (Overfit of MAC Strategy). Here we analyze, via simulation, the Sharpe ratios of Moving Average Crossover (MAC) strategies. A MAC strategy is simple to describe: for a single asset, compute two moving averages of the price series with different windows. When one moving average is greater than the other, hold the asset long, otherwise hold it short. One selects the two windows by (over)fitting to the available data.

We perform simulations of that process under the null hypothesis, where generated returns are zero mean and independent. Any realization of a MAC strategy in these simulations, with any choice of the windows, will have zero mean return and thus zero Sharpe. We draw 1,100 days of returns independently with zero mean. We turn the returns into a price series, and then test MAC strategies with a "short" window size varying from 4 to 36 days, and the "long" window size varying from 40 to 280 days. We select the window combination with highest Sharpe ratio and record that Sharpe ratio. We do not force the strategy to be long or short either windowed average, rather we test both. We repeat this process 1,024 times.

In Figure 4.11, we Q-Q plot the annualized Sharpe ratio of the selected MAC strategy against the theoretical distribution based on an F distribution with 3 and $1100 - 3$ degrees of freedom. Note that the choice of p is intuitive, but perhaps a different value is more consistent with the F distribution approximation for this data. The Q-Q plot indicates that our approximation is off by an affine shift: we observe higher Sharpe ratios than our subspace approximation supports, or we have underestimated the p. ⊣

4.2 A *post hoc* test on the Sharpe ratio

In Section 3.3.1, we described the test for the null hypothesis that k different assets have the same signal-noise ratio[2]. This is analogous to the classical ANOVA procedure, which is for testing whether a variable has the same mean across k different groups. In contrast to the ANOVA, we are usually performing a paired test, where returns across the k assets are observed contemporaneously, and we need not assume common volatility of returns.

In the classical setting, if one rejects the null in an ANOVA, a *post hoc* test is then prescribed. [Bretz *et al.*, 2016] This test, also known as *Tukey's Range Test* or the *Honest Significant Difference* test, is based on the range of a number of independent Gaussians, divided by a rescaled chi-square. Tukey's test can easily be adapted to the case of testing the signal-noise ratios of multiple assets. [Pav, 2019c]

As in Section 4.1.2, we first consider the case of rank one updated correlation model,

$$\mathsf{R} = (1-\rho)\,\mathsf{I} + \rho\left(\mathbf{1}\mathbf{1}^{\top}\right), \tag{4.24}$$

where $|\rho| \leq 1$. Under this correlation structure, recalling Equation 4.3,

$$\sqrt{n}\Omega^{-1/2}\left(\hat{\boldsymbol{\zeta}} - \boldsymbol{\zeta}_0\right) \approx \mathcal{N}\left(\mathbf{0}, \mathsf{I}\right).$$

We then showed (*cf.* Equation 4.6 and Exercise 4.6) that

$$\mathsf{R}^{-1/2} = \left((1-\rho)\left(1 + \frac{\kappa}{2}\zeta_0^2\left(1+\rho\right)\right)\right)^{-1/2}\mathsf{I} + c\left(\mathbf{1}\mathbf{1}^{\top}\right). \tag{4.25}$$

for some constant c.

Now we consider the difference in Sharpe ratios of two assets, indexed by i and j. Let $\boldsymbol{v} = \boldsymbol{e}_i - \boldsymbol{e}_j$, where $\boldsymbol{e}_i$ is the ith column of the identity matrix. From Equation 4.2 we have

$$\begin{aligned}
\boldsymbol{v}^{\top}\boldsymbol{z} &= \boldsymbol{v}^{\top}\sqrt{n}\mathsf{R}^{-1/2}\left(\hat{\boldsymbol{\zeta}} - \boldsymbol{\zeta}_0\right),\\
&= \sqrt{n}\boldsymbol{v}^{\top}\left[\left((1-\rho)\left(1 + \frac{\kappa}{2}\zeta_0^2\left(1+\rho\right)\right)^{-1/2}\right)\mathsf{I} + c\left(\mathbf{1}\mathbf{1}^{\top}\right)\right]\left(\hat{\boldsymbol{\zeta}} - \boldsymbol{\zeta}_0\right),\\
&= \sqrt{\frac{n}{(1-\rho)\left(1 + \frac{\kappa}{2}\zeta_0^2\left(1+\rho\right)\right)}}\boldsymbol{v}^{\top}\hat{\boldsymbol{\zeta}},\\
&\approx \sqrt{\frac{n}{1-\rho}}\boldsymbol{v}^{\top}\hat{\boldsymbol{\zeta}}.
\end{aligned}$$

Here we have used that $\boldsymbol{v}^{\top}\mathbf{1} = 0$ and under the null hypothesis, $\boldsymbol{\zeta}_0$ is some

[2]This section is adapted from the author's paper on a *post hoc* test on the Sharpe ratio. [Pav, 2019c]

constant times **1**. The last approximation follows because ζ_0^2/n is likely to be very small for most practical work. Then

$$\hat{\zeta}_i - \hat{\zeta}_j = \sqrt{\frac{1-\rho}{n}}\left(z_i - z_j\right). \tag{4.26}$$

Now note that the $\boldsymbol{z}$ is distributed as a standard multivariate normal. So the *range* of $\hat{\boldsymbol{\zeta}}$, which is to say $\max_{i,j}\left(\hat{\zeta}_i - \hat{\zeta}_j\right)$, is distributed as $\sqrt{\left(1-\rho\right)/n}$ times the range of a standard k-variate normal.

To quote this as a hypothesis test,

$$\max_{i,j}\left|\hat{\zeta}_i - \hat{\zeta}_j\right| \geq HSD = q_{1-\alpha,k,\infty}\sqrt{\frac{\left(1-\rho\right)}{n}}, \tag{4.27}$$

with probability α, where the $q_{1-\alpha,k,l}$ is the upper α-quantile of the Tukey distribution with k and l degrees of freedom. In the R language, this quantile may be computed via the `qtukey` function. [R Core Team, 2015; Odeh and Evans, 1974] With $l = \infty$, the cutoff HSD is the rescaled upper α quantile of the range of k independent Gaussians. That is, $q_{1-\alpha,k,\infty}$ is the number such that

$$1 - \alpha = k\int_{-\infty}^{\infty}\phi\left(x\right)\left(\Phi\left(x + q_{1-\alpha,k,\infty}\right) - \Phi\left(x\right)\right)^{k-1}\,\mathrm{d}x.$$

The normal approximation of Equation 3.29 may be too rough of an approximation for computation of HSD, even if the covariance is approximately correct. The distributional shape of $\hat{\boldsymbol{\zeta}}$ may be far enough from multivariate normal that we cannot use Tukey's distribution for a cutoff, especially when n is small and k is large. In that case, one is tempted to compare the observed range to

$$HSD = q_{1-\alpha,k,n-1}\sqrt{\frac{\left(1-\rho\right)}{n-1}}. \tag{4.28}$$

The reasoning behind this heuristic is that we are computing the range of (non-independent) t statistics, up to scaling, which is almost the same as the Tukey distribution, which is the ratio of the range of normals divided by a pooled χ variable. We will refer to the cutoff of Equation 4.27 as "$df = \infty$" and the cutoff of Equation 4.28 as the "$df = n-1$" cutoff. While we note the $df = n-1$ cutoff stands on perhaps shakier theoretical grounds, experimental evidence suggests that it maintains the nominal type I rate much better than the $df = \infty$ cutoff, especially for the small n case. [Pav, 2019c]

Bonferroni Cutoff:

We note that an alternative calculation provides a very similar cutoff value. Consider two assets with correlation ρ, and with signal-noise ratios of $\zeta\left(1+\epsilon\right)$ and ζ. Recalling Equation 3.49, the difference in Sharpe ratios can be shown to be approximately normal with

$$\left[\hat{\zeta}_1 - \hat{\zeta}_2\right] \rightsquigarrow \mathcal{N}\left(\epsilon\zeta, \frac{2}{n}\left(1-\rho\right) + \frac{\zeta^2}{2n}\left(1 + \left(1+\epsilon\right)^2 - 2\rho^2\left(1+\epsilon\right)\right)\right). \tag{4.29}$$

Again, assuming that ζ^2/n will be very small for most practical work, one can compute a "Bonferroni cutoff" as

$$BC = \sqrt{\frac{2\,(1-\rho)}{n}}\, z_{1-\alpha/\binom{k}{2}}, \tag{4.30}$$

where z_{α} is the α quantile of the standard normal distribution. This cutoff is based on a Bonferroni correction that recognizes we are performing $\binom{k}{2}$ pairwise comparison tests. The cutoff BC is typically very similar to HSD (for $df = \infty$) or slightly smaller (and is easier to compute), *cf.* Exercise 4.18. We note that since BC is based on a normal approximation, it may suffer from the same issues that the HSD cutoff does for small samples. However, there is hope one can compute an exact small n Bonferroni cutoff.

Compact Letter Displays:

One way to summarize the results of the *post hoc* test is via a compact letter display. [Bretz *et al.*, 2016; Piepho, 2004] Here one assigns each asset to one or more groups, with the groups identified by a letter. The group assignment is chosen such that two assets with Sharpe ratio greater than HSD are not assigned to the same group. While this could be done trivially, usually an assignment with few groups is achievable. We illustrate in Example 4.2.1.

Arbitrary correlation structure:

The test outlined above is strictly only applicable to the rank-one correlation matrix, $\mathsf{R} = (1-\rho)\,\mathsf{I} + \rho\left(\mathbf{1}\mathbf{1}^{\top}\right)$. To apply the test to assets with arbitrary correlation matrices, one would like to appeal to a stochastic dominance result. For example, if one could adapt Slepian's lemma to the distribution of the *range*, then the above analysis could be applied where ρ is the smallest off-diagonal correlation, to give a test with maximum type I rate of α. However, it is not immediately clear that Slepian's lemma can be so modified. [Slepian, 1962; Zeitouni, 2017; Yin, 2019] The Bonferroni Cutoff, however, is easily adapted to this kind of worst-case analysis, however.

Example 4.2.1 (Five Industry Portfolios, *post hoc* test). We consider the returns of the 5 industry portfolios introduced in Example 1.2.5. The data consist of 1,128 months of returns, from Jan 1927 to Dec 2020. Over this period we estimate the correlation of returns to be

R =

	Consum	*Manufa*	*Techno*	*Health*	*Other*
Consum	1.00	0.87	0.81	0.78	0.88
Manufa	0.87	1.00	0.81	0.75	0.89
Techno	0.81	0.81	1.00	0.71	0.80
Health	0.78	0.75	0.71	1.00	0.74
Other	0.88	0.89	0.80	0.74	1.00

This is fairly well approximated by $(1-\rho)\,\mathsf{I} + \rho\left(\mathbf{1}\mathbf{1}^{\top}\right)$, with $\rho = 0.8$. As listed in Example 1.2.6, the Sharpe ratios range from $0.492\,\text{yr}^{-1/2}$ for Other to $0.676\,\text{yr}^{-1/2}$ for Healthcare.

First we perform the hypothesis test of equality of signal-noise ratios, via the χ^2 test of Equation 3.46. We compute a statistic of 13.6 which should be distributed as a $\chi^2(4)$ under the null. This corresponds to a p-value of 0.009, and we reject the null of equality of all signal-noise ratios.

Using the $df = n-1$ formulation and the estimated ρ, we compute $HSD = 0.177\,\mathrm{yr}^{-1/2}$ for $\alpha = 0.05$, and narrowly reject the equality of signal-noise ratios for Other and Healthcare. Based on the observed Sharpe ratios and this HSD cutoff, a compact letter display of the five industries might look as follows:

Healthcare	*Consumer*	*Technology*	*Manufacturing*	*Other*
a	*ab*	*ab*	*ab*	*b*

The assignment to groups recognizes that we found a significant difference between Other and Healthcare, but not between any of the other pairs of assets. ⊣

Example 4.2.2 (Fama-French factors, *post hoc* test). Consider the four Fama French factors from Example 1.1.1. In Example 3.3.1, based on monthly returns from Jan 1927 to Dec 2020, we narrowly rejected the null hypothesis of equal signal-noise ratios among the four factors. The correlation of returns was estimated as

$\mathsf{R} =$

	Mkt	*SMB*	*HML*	*UMD*
Mkt	1.00	0.32	0.24	−0.34
SMB	0.32	1.00	0.13	−0.16
HML	0.24	0.13	1.00	−0.41
UMD	−0.34	−0.16	−0.41	1.00

Given the negative correlation of UMD to the other three factors, $(1-\rho)\,\mathsf{I} + \rho\left(\mathbf{1}\mathbf{1}^\top\right)$ is not a good model for R. Moreover, we cannot easily use such a correlation matrix as a lower bound. Instead, we perform the $\binom{4}{2}$ tests for equality of signal-noise ratio via the χ^2 test of Wright *et al.*, as given in Equation 3.46. We perform a Bonferroni correction on these, rejecting only if the p-value is less than $\alpha/6$. Under this test, we only reject equality of Mkt and SMB. The Sharpe ratios and compact letter display of the four factors then might look as follows:

factor	*Sharpe*	*letters*
Mkt	0.61	*a*
UMD	0.48	*ab*
HML	0.32	*ab*
SMB	0.23	*b*

⊣

4.3 Exercises

Ex. 4.1 The Šidák correction Establish the correctness of the Šidák correction.

1. Let $p_1, p_2, \ldots, p_k$ be independent random variables uniformly on $[0, 1)$. Prove that
$$\Pr\left\{\min_i p_i \le x\right\} = 1 - (1 - x)^k .$$
2. Find x such that $1 - (1 - x)^k = \alpha$.

Ex. 4.2 Bonferroni and Šidák The Šidák correction yields a slightly larger critical region than the Bonferroni correction. What happens to the ratio of their sub-test type I errors,

$$\frac{1 - (1 - \alpha)^{1/k}}{\alpha / k},$$

for $\alpha \approx 0.05$ as $k \to \infty$?

Ex. 4.3 Limits to anti-correlation Let R be a $k \times k$ correlation matrix.

1. Prove that R is positive semi-definite.
2. Prove that the average off-diagonal element of R is no less than $-1/(k-1)$.

Ex. 4.4 Bonferroni power loss Extend Example 4.1.4 by finding the signal-noise ratio such that the Bonferroni correction achieves the nominal 0.05 type I rate for the k, n and ρ considered in that example.

Ex. 4.5 Rank one updated correlation

1. Show that the matrix $\mathsf{R} = (1 - \rho)\mathsf{I} + \rho\left(\mathbf{1}\mathbf{1}^{\top}\right)$ is a valid correlation matrix if and only if $|\rho| \le 1$. (A valid correlation matrix is symmetric and positive semidefinite, has elements in $[-1, 1)$ and unit diagonal.)
2. Let $\boldsymbol{q}$ be a k-vector whose elements are ± 1. Show that the matrix $\mathsf{R} = (1 - \rho)\mathsf{I} + \rho\left(\boldsymbol{q}\boldsymbol{q}^{\top}\right)$ is a valid correlation matrix if and only if $|\rho| \le 1$.
3. Let $\boldsymbol{q}$ be a k-vector whose elements are no greater than 1 in absolute value. Show that
$$\mathrm{Diag}\left(\mathbf{1} - \boldsymbol{q} \odot \boldsymbol{q}\right) + \boldsymbol{q}\boldsymbol{q}^{\top}$$
is a valid correlation matrix.

Ex. 4.6 Rank one update matrix powers Let $\boldsymbol{q}$ be a k-vector whose elements are -1, 0 or 1.

1. Show that the vectors $\mathbf{1} - |\boldsymbol{q}|$ and $\boldsymbol{q}$ are othogonal to each other, and thus
$$\text{Diag}\left(\mathbf{1} - |\boldsymbol{q}|\right) \boldsymbol{q}\boldsymbol{q}^\top = \mathsf{0}.$$

2. Show that $\text{Diag}\left(\mathbf{1} - |\boldsymbol{q}|\right)$ is idempotent:
$$\text{Diag}\left(\mathbf{1} - |\boldsymbol{q}|\right) \text{Diag}\left(\mathbf{1} - |\boldsymbol{q}|\right) = \text{Diag}\left(\mathbf{1} - |\boldsymbol{q}|\right).$$

3. Let
$$\begin{aligned} \mathsf{A} &= a_0 \mathsf{I} + a_1 \text{Diag}\left(\mathbf{1} - |\boldsymbol{q}|\right) + a_2 \boldsymbol{q}\boldsymbol{q}^\top, \\ \mathsf{B} &= b_0 \mathsf{I} + b_1 \text{Diag}\left(\mathbf{1} - |\boldsymbol{q}|\right) + b_2 \boldsymbol{q}\boldsymbol{q}^\top, \end{aligned}$$
with $a_0 > 0$. Show that $\mathsf{BA} = \mathsf{I}$ if
$$\begin{aligned} b_0 &= a_0^{-1}, \\ b_1 &= \frac{-a_1}{a_0^2 + a_0 a_1}, \\ b_2 &= \frac{-a_2}{a_0^2 + a_0 a_2 \boldsymbol{q}^\top \boldsymbol{q}}. \end{aligned}$$

4. Let
$$\begin{aligned} \mathsf{A} &= a_0 \mathsf{I} + a_1 \text{Diag}\left(\mathbf{1} - |\boldsymbol{q}|\right) + a_2 \boldsymbol{q}\boldsymbol{q}^\top, \\ \mathsf{B} &= b_0 \mathsf{I} + b_1 \text{Diag}\left(\mathbf{1} - |\boldsymbol{q}|\right) + b_2 \boldsymbol{q}\boldsymbol{q}^\top, \end{aligned}$$
with $a_0 > 0$. Show that $\mathsf{BB} = \mathsf{A}$ if
$$\begin{aligned} b_0 &= \pm\sqrt{a_0}, \\ b_1 &= -b_0 \pm \sqrt{a_0 + a_1}, \\ b_2 &= \frac{-b_0 + \sqrt{a_0 + a_2 \boldsymbol{q}^\top \boldsymbol{q}}}{\boldsymbol{q}^\top \boldsymbol{q}}. \end{aligned}$$

5. Let
$$\begin{aligned} \mathsf{A} &= a_0 \mathsf{I} + a_1 \text{Diag}\left(\mathbf{1} - |\boldsymbol{q}|\right) + a_2 \boldsymbol{q}\boldsymbol{q}^\top, \\ \mathsf{B} &= b_0 \mathsf{I} + b_1 \text{Diag}\left(\mathbf{1} - |\boldsymbol{q}|\right) + b_2 \boldsymbol{q}\boldsymbol{q}^\top, \end{aligned}$$

with $a_0 > 0$. Show that $\mathsf{BBA} = \mathsf{I}$ if

$$
\begin{aligned}
b_0 &= a_0^{-1/2},\\
b_1 &= -\frac{1}{\sqrt{a_0}} \pm \sqrt{\frac{1}{a_0 + a_1}} = -\frac{\sqrt{a_0 + a_1} \mp \sqrt{a_0}}{\sqrt{a_0^2 + a_0 a_1}},\\
b_2 &= -\frac{1}{\boldsymbol{q}^\top \boldsymbol{q} \sqrt{a_0}} \pm \frac{1}{\boldsymbol{q}^\top \boldsymbol{q}} \sqrt{\frac{1}{a_0 + a_2 \boldsymbol{q}^\top \boldsymbol{q}}}\\
&= -\frac{1}{\boldsymbol{q}^\top \boldsymbol{q}} \frac{\sqrt{a_0 + a_2 \boldsymbol{q}^\top \boldsymbol{q}} \mp \sqrt{a_0}}{\sqrt{a_0^2 + a_0 a_2 \boldsymbol{q}^\top \boldsymbol{q}}}.
\end{aligned}
$$

Ex. 4.7 **Order preserving matrices** Say that a square matrix A is *order preserving* if for every $\boldsymbol{y} = \mathsf{A}\boldsymbol{x}$ if $x_i \le x_j$ then $y_i \le y_j$

1. Prove that A is order preserving if and only if all eigenvalues are non-negative and equal, with the possible exception of the eigenvalue associated with the eigenvector $\mathbf{1}$.
2. Prove that the matrix $a_0 \mathsf{I} + a_1 \left(\mathbf{1}\mathbf{1}^\top\right)$ is order preserving if and only if $a_0 \ge 0$.

Ex. 4.8 **More power testing** Repeat Example 4.1.13, but under different configurations of the alternative. For some fixed μ and σ draw the population ζ_i from a normal distribution with mean μ and variance σ^2. Plot the rejection probabilities as a function of the signal-noise ratio of the selected strategy. Qualitatively do you observe the same patterns as in the example? What if you vary μ and σ?

Ex. 4.9 **Monotonicity of inference** Show that when using the rank-one model of correlation, it is conservative to keep ρ bounded away from 1. Let

$$
\begin{aligned}
a_0 &= 1 - \rho + \frac{\kappa}{2}\zeta_0^2\left(1 - \rho^2\right), & a_2 &= \rho + \frac{\kappa - 1}{4}\zeta_0^2 + \frac{\kappa}{2}\zeta_0^2\rho^2,\\
b_0 &= a_0^{-1/2}, & b_2 &= \frac{1}{k}\frac{\sqrt{a_0} - \sqrt{a_0 + k a_2}}{\sqrt{a_0^2 + k a_0 a_2}},
\end{aligned}
$$

for $\rho \ge 0, \kappa \ge 1$.

1. Show that $\frac{\mathrm{d}a_0}{\mathrm{d}\rho} \le 0$ and $\frac{\mathrm{d}a_2}{\mathrm{d}\rho} \ge 0$.
2. Show that $b_2 \le 0$, and that $\frac{\mathrm{d}b_0}{\mathrm{d}\rho} \ge 0$ and $\frac{\mathrm{d}b_2}{\mathrm{d}\rho} \le 0$.
3. Let $c = b_0 + k b_2$. Show that $c > 0$ and $\frac{\mathrm{d}c}{\mathrm{d}\rho} \le 0$.
4. Show that
$$\lim_{\rho \to 1^-} b_0 = \infty.$$
5. Show that the z in Equation 4.8 goes to ∞ as $\rho \to 1$.

6. Show that the summands of Equation 4.18 take the form

$$b_0\left(\hat{\zeta}_i - \bar{\zeta}\right) + \left(b_0 + b_2 k\right)\left(\bar{\zeta} - \zeta_0\right).$$

Show that the $\bar{x}^2$ of Equation 4.18 goes to ∞ as $\rho \to 1$.

7. Show that Follman's statistic, the g^2 of Equation 4.20, goes to ∞ as $\rho \to 1$.

Ex. 4.10 Follman's test, half-good case In Example 4.1.13, we saw that Follman's test achieved maximum power of around 1/2 for the half-good case.

1. Why does this happen?
2. Suppose you expected that 1/4 of the assets' signal-noise ratios were greater than ζ_0. How would you adapt Follman's test to have higher power under this alternative?

Ex. 4.11 Bonferroni correction, direct estimation Suppose that $\boldsymbol{y} = \left(b_0 \mathsf{I} + b_1 \left(\mathbf{1}\mathbf{1}^{\top}\right)\right)\boldsymbol{z}$, where $\boldsymbol{z} \sim \mathcal{N}\left(\mathbf{0}, \mathsf{I}\right)$. In this exercise you will construct an approximate CDF for the maximum element $y_{(k)}$.

1. Show empirically that the correlation between $z_{(k)}$ and $\mathbf{1}^{\top}\boldsymbol{z}$ goes to zero for large k.
2. Suppose that $\boldsymbol{y} = b_0\boldsymbol{z} + b_1 k z_0$, where $z_0 \sim \mathcal{N}\left(0, 1\right)$ is independent of $\boldsymbol{z} \sim \mathcal{N}\left(\mathbf{0}, \mathsf{I}\right)$. Show that

$$\Pr\left\{y_{(k)} \le t\right\} = \int \Phi^k\left(\frac{t - b_1 k x}{b_0}\right)\phi\left(x\right)\,\mathrm{d}x$$

3. Implement code to approximately compute that integral given k, b_0, and b_1. Use a basic quadrature scheme like Trapezoid rule.
4. Test your code empirically: spawn such a $\boldsymbol{y}$ for large k, compute $y_{(k)}$, compute the probability that you would see a value so large using your code, and repeat a thousand times. Are your putative p-values approximately uniform?

Ex. 4.12 Bonferroni correction, bounded correlation Repeat the experiment of Example 4.1.6:

1. Repeat the experiment, but assume that the correlation of returns is R with

$$\mathsf{R}_{i,j} = \begin{cases} \rho^{|i-j|} & \text{if } |i-j| \le 1, \\ \frac{\rho}{2} & \text{otherwise.} \end{cases}$$

2. Repeat the experiment assuming that

$$\mathsf{R} = (1-\rho)\,\mathsf{I} + \rho\left(\boldsymbol{q}\boldsymbol{q}^{\top}\right),$$

where exactly half the elements of $\boldsymbol{q}$ are -1 and the other half are $+1$.

Ex. 4.13 Slepian type bounds Suppose $\boldsymbol{x} \sim \mathcal{N}\left(\mathbf{0}, (1-\rho)\,\mathsf{I} + \rho\left(\boldsymbol{q}\boldsymbol{q}^{\top}\right)\right)$, and $\boldsymbol{y} \sim \mathcal{N}\left(\mathbf{0}, (1-\rho)\,\mathsf{I} + \rho\left(\mathbf{1}\mathbf{1}^{\top}\right)\right)$, for some $\rho > 0$, where l of the k elements of $\boldsymbol{q}$ are -1 and the rest are $+1$.
From the Li and Shao bound of Equation 4.9, prove that

$$\Pr\left\{\max_i x_i > t\right\} \leq \Pr\left\{\max_i y_i > t\right\} + \frac{l\,(k-l)\,\operatorname{asin}\rho}{\pi} e^{-t^2/(1+\rho)}. \tag{4.31}$$

Ex. 4.14 Fama French six factors Consider the Fama French Six factor returns, which are available via the **tsrsa** package as the dataset `mff6`:

```
library(tsrsa)
data(mff6)
```

1. Test the null hypothesis that all six returns have identical signal-noise ratio via the Bonferroni, one-sided, Follman, and conditional inference procedures.
2. Perform the *post hoc* test on these returns, computing the compact letter displays.

Ex. 4.15 Conditional CI, survivorship bias Suppose that you consider investing in an asset only because its observed historical Sharpe ratio was larger than ζ_0. Having observed $\hat{\zeta}$, use the polyhedral theorem of Theorem 4.1.11 to construct a $1-\alpha$ confidence interval on ζ.

Ex. 4.16 Conditional CI, opportunistic strategy Consider how large a Sharpe ratio you have to observe such that a lower one-sided conditional confidence bound is exactly zero.

1. Show that $[0, \infty)$ is a conditional $1-\alpha$ interval for the signal-noise ratio of the opportunistic strategy exactly when
$$F_{SR}\left(\hat{\zeta}; 0, n\right) = 1 - \frac{\alpha}{2}.$$
2. Confirm that $[0, \infty)$ is an unconditional $1-\alpha$ interval for the signal-noise ratio of an asset exactly when
$$F_{SR}\left(\hat{\zeta}; 0, n\right) = 1 - \alpha.$$

Ex. 4.17 Overoptimism by opportunistic selection Consider a mashup of overoptimism by selection and the opportunistic strategy. That is, suppose you observe the Sharpe ratios of k strategies, then select the asset with the largest *absolute* Sharpe ratio, which you will hold long or short depending on the sign of its Sharpe ratio. How would you use Theorem 4.1.11 for inference on this problem?

Ex. 4.18 BC and HSD bounds Compare the HSD and BC cutoffs from Equation 4.27 and Equation 4.30. Compute both for the $\alpha = 0.05$ level, $n = 504$ days, $\rho = 0.8$ and vary the number of assets, k from 4 to 200. Plot both cutoffs.

* **Ex. 4.19 Research problem: shrinkage after selection** ☨
Hwang [1993] describes an estimator of the mean of selected populations for the case of independent multivariate normal errors that has reduced Bayesian Risk. [Fuentes and Gopal, 2017] This could easily be used to describe the problem of Bayesian inference on the signal-noise ratio of the asset with maximal Sharpe ratio, via the normal approximation of Equation 3.29. However, it would only apply for $\mathsf{R} = \sigma^2\mathsf{I}$. Generalize Hwang's procedure to the case of the rank-one correlation matrix of Equation 4.4, $\mathsf{R} = (1-\rho)\,\mathsf{I} + \rho\left(\mathbf{1}\mathbf{1}^{\top}\right).$

* **Ex. 4.20 Research problem: CIs on selected SNRs** ☨ Fuentes *et al.* [2018] describe a procedure for computing confidence intervals on the means of selected populations for the case where errors are independent and homoskedastic. This could easily be used to compute confidence intervals on the signal-noise ratios of, say, the top p assets as selected by Sharpe ratio, via the normal approximation of Equation 3.29 for the case where $\mathsf{R} = \sigma^2\mathsf{I}$. Generalize this procedure to the case of the rank-one correlation matrix of Equation 4.4, $\mathsf{R} = (1-\rho)\,\mathsf{I} + \rho\left(\mathbf{1}\mathbf{1}^{\top}\right).$

* **Ex. 4.21 Research problem: Slepian's lemma for range** ☨ Extend Slepian's lemma to the range: Let $\boldsymbol{x} \sim \mathcal{N}\left(\mathbf{0}, \Sigma^{x}\right)$ and $\boldsymbol{y} \sim \mathcal{N}\left(\mathbf{0}, \Sigma^{y}\right)$ where $\Sigma^{x}_{i,j} \leq \Sigma^{y}_{i,j}$ for every i,j. Let $r\left(\boldsymbol{x}\right) = \max_{i,j} x_i - x_j$. Prove that $\Pr\left\{r\left(\boldsymbol{x}\right) \geq t\right\} \geq \Pr\left\{r\left(\boldsymbol{y}\right) \geq t\right\}.$ [Yin, 2019]

Part II

Maximizing the Signal-Noise Ratio

5

Maximizing the Sharpe Ratio

> I wish my wish would not be granted!
>
> DOUGLAS R. HOFSTADTER
> *Gödel, Escher, Bach: An Eternal Golden Braid*

Caution. In the previous chapters, we considered the statistics of analyzing asset returns without prescribing a course of action. In what follows we will often assume an investment objective, find a course of action for an asset manager to optimize that objective, then consider inference on the relevant population parameters given sample estimates. It should be stressed that the investment decision is almost always guided by legal considerations: securities law, the terms of the manager's license, the offering material and stated objective of the fund, any contract with the investor, and so on. Inasmuch as these may be vaguely couched, there remains a moral and ethical obligation to act in the best interests of the investor upon whose behalf you may be managing. In case of any ambiguity, an investor should communicate their financial goals, and the asset manager should clarify which objective they seek to optimize.

It is taken as dogma throughout this text that an asset manager ought to maximize signal-noise ratio, and that this may be in the best interests of the investor. This religion is wide enough to admit some schisms, however. One is the *term* or *horizon* over which we define this metric.

Definition 5.0.1 (multi-period signal-noise ratio). The *single-period signal-noise ratio* is the signal-noise ratio measured over a single investment decision return cycle. That is, if an investment manager rebalances weekly, the single-period signal-noise ratio is measured over the week following their investment. It is usually considered conditional on the information available at the decision time.

The *multi-period signal-noise ratio* is the signal-noise ratio measured over a time frame longer than a single investment decision horizon, perhaps over an infinite time frame, spanning several investment decisions. It is unconditional.

Note that for us to analyze the performance of a manager optimizing multi-period signal-noise ratio, their investment decisions should be systematic and dependent on the conditioning information available. For this reason, we have

DOI: 10.1201/9781003181057-5

not considered the multi-period form in previous chapters, as there simply was no conditioning information. An example may illustrate the difference.

Example 5.0.2 (Alice and Bob time the market). Consider a single asset with returns x_t, and a single binary feature $f_{t-1} \in \{-1, +1\}$ observed prior to the time required to capture the returns. Suppose that $f_{t-1} = 1$ with probability 0.2. Moreover, suppose that the distribution of x_t is conditional on f_{t-1} via:

$$\mathrm{E}\left[x_t\right] = \begin{cases} 0.01\,\text{mo.}^{-1} & \text{if } f_{t-1} = -1, \\ 0.03\,\text{mo.}^{-1} & \text{otherwise.} \end{cases}$$

$$\mathrm{Var}\left(x_t\right) = \begin{cases} 0.001\,\text{mo.}^{-1} & \text{if } f_{t-1} = -1, \\ 0.008\,\text{mo.}^{-1} & \text{otherwise.} \end{cases}$$

In this case, the feature f_{t-1} is a kind of "market clock" indicator, triggering in an environment of heightened volatility and expected return, *cf.* Section 3.1.3.

Suppose that somehow these population parameters are known to market participants. Consider two asset managers: Bob always keeps his entire portfolio invested long in the asset[1]. Alice, however, observes f_{t-1}: when it takes value $+1$ she down-levers to hold only one third of her wealth in the asset, with the remainder in cash; otherwise she holds her entire portfolio in the asset.

Both managers have the same single-period signal-noise ratio, namely $0.33\,\text{mo.}^{-1/2}$. However, by down-levering in periods of higher volatility, Alice has "smoothed out" her returns and has a multi-period signal-noise ratio of $0.33\,\text{mo.}^{-1/2}$, whereas Bob has a multi-period signal-noise ratio of $0.29\,\text{mo.}^{-1/2}$. ⊣

In this text, we will mostly consider the case of a manager who maximizes her multi-period signal-noise ratio[2]. When the analysis proves too difficult we will at times switch to analyzing the single-period decision, or perhaps even the *expected value* of the single-period signal-noise ratio as measured over multiple periods, effectively integrating out the conditioning information. Again these choices are purely pragmatic. It seems, however, that maximizing and analyzing the multi-period signal-noise ratio is the most general and practically useful case.

Caution. Continuing Example 5.0.2, still assuming $f_{t-1} = 1$ with probability 0.2, Bob's expected return is $0.014\,\text{mo.}^{-1}$ but Alice's expected return is $0.01\,\text{mo.}^{-1}$. By engaging in market timing, Alice has improved her signal-noise ratio modestly but decreased her expected returns by around 29% compared to Bob. While we advocate in general for optimization of the signal-noise ratio, arguably Alice is not acting in the best interests of her clients. However,

[1]You will often find that some "active" managers are most active on the golf course.

[2]An interesting question is whether Alice has indeed *maximized* her multi-period signal-noise ratio, or whether another strategy would have higher signal-noise ratio; see Exercise 5.24.

if Alice had promised her investors to keep volatility below a certain level, and communicated her volatility forecasts to investors on an timely basis, it is unlikely to cause an issue. When in doubt, consult a securities lawyer, not some book written by an academic.

5.1 As an optimization problem

We now consider the general optimization problem

$$\max_{\theta} \zeta\left(\theta\right), \text{where } \zeta\left(\theta\right) := \frac{\mu\left(\theta\right) - r_0}{\sigma\left(\theta\right)}, \tag{5.1}$$

where we consider the mean return and volatility as *functions* of some parameter, θ, which is somehow under the control of the asset manager. We have not specified whether this is single-period or multi-period, or how θ informs the managers decisions.

We can easily solve this problem by taking the derivative of $\zeta\left(\theta\right)$ with respect to θ. We have

$$\begin{aligned} \nabla_{\theta}\zeta\left(\theta\right) &= \frac{1}{\sigma\left(\theta\right)}\nabla_{\theta}\mu\left(\theta\right) - \frac{\mu\left(\theta\right) - r_0}{\sigma^2\left(\theta\right)}\nabla_{\theta}\sigma\left(\theta\right), \\ &= \frac{1}{\sigma\left(\theta\right)}\left(\nabla_{\theta}\mu\left(\theta\right) - \zeta\left(\theta\right)\nabla_{\theta}\sigma\left(\theta\right)\right). \end{aligned} \tag{5.2}$$

Note that $\sigma\left(\theta\right)$ is always positive, so $\zeta\left(\theta\right)$ is always increasing in increasing μ. When μ is greater than the disastrous rate, r_0, then ζ is increasing in *decreasing* σ. However, if μ is smaller than r_0, then ζ increases when σ increases. This is an often lamented fact about the Sharpe ratio as a metric: when expected return is negative, it "aims for the fences," increasing volatility. (See also Exercise 1.21.)

The second derivative, or Hessian, of the signal-noise ratio is

$$H_{\theta}\zeta\left(\theta\right) = \frac{H_{\theta}\mu\left(\theta\right) - \zeta\left(\theta\right)H_{\theta}\sigma\left(\theta\right) - \left(\nabla_{\theta}\zeta\left(\theta\right)\nabla_{\theta}^{\top}\sigma\left(\theta\right) + \nabla_{\theta}\sigma\left(\theta\right)\nabla_{\theta}^{\top}\zeta\left(\theta\right)\right)}{\sigma\left(\theta\right)}. \tag{5.3}$$

The *first-order necessary conditions*, which must hold at an optimum (maximum or minimum) are

$$\nabla_{\theta}\mu\left(\theta\right) = \zeta\left(\theta\right)\nabla_{\theta}\sigma\left(\theta\right). \tag{5.4}$$

The *second-order necessary conditions* for the problem state that the Hessian must be negative definite at a local maximum. That is,

$$\frac{H_{\theta}\mu\left(\theta\right) - \zeta\left(\theta\right)H_{\theta}\sigma\left(\theta\right)}{\sigma\left(\theta\right)} \preceq 0 \tag{5.5}$$

at the local maximum. If this condition holds for all θ, then the function $\zeta(\theta)$ is concave and so the local maximum is the *global* maximum.

In the case, when θ is a scalar, the first- and second-order necessary conditions translate to

$$\zeta(\theta) = \frac{\mathrm{d}\mu(\theta)}{\mathrm{d}\theta} / \frac{\mathrm{d}\sigma(\theta)}{\mathrm{d}\theta}, \text{ and}$$
$$\frac{\mathrm{d}^2\mu(\theta)}{\mathrm{d}\theta^2} \le \zeta(\theta) \frac{\mathrm{d}^2\sigma(\theta)}{\mathrm{d}\theta^2}.$$

Alternative expressions

At times it may be more convenient to consider derivatives with respect to the *variance*, or even the second moment of returns. As functions of θ let these be denoted as $\sigma^2(\theta)$ and $\alpha_2(\theta)$, respectively. The gradient can then be expressed as

$$\begin{aligned}\nabla_\theta\zeta(\theta) &= \frac{1}{\sigma(\theta)}\left(\nabla_\theta\mu(\theta) - \frac{\zeta(\theta)}{2\sigma(\theta)}\nabla_\theta\sigma^2(\theta)\right),\\ &= \frac{1}{\sigma(\theta)}\left(\left(1+\zeta^2(\theta)+\frac{r_0\zeta(\theta)}{\sigma(\theta)}\right)\nabla_\theta\mu(\theta) - \frac{\zeta(\theta)}{2\sigma(\theta)}\nabla_\theta\alpha_2(\theta)\right).\end{aligned} \tag{5.6}$$

The Hessian can be re-expressed as

$$\begin{aligned}H_\theta\zeta(\theta) = {}& \frac{H_\theta\mu(\theta)}{\sigma(\theta)} - \frac{\zeta(\theta)H_\theta\sigma^2(\theta)}{2\sigma^2(\theta)} + \frac{\zeta(\theta)\nabla_\theta\sigma^2(\theta)\nabla_\theta^\top\sigma^2(\theta)}{4\left(\sigma^2(\theta)\right)^2}\\ &- \frac{\left(\nabla_\theta\zeta(\theta)\nabla_\theta^\top\sigma^2(\theta) + \nabla_\theta\sigma^2(\theta)\nabla_\theta^\top\zeta(\theta)\right)}{2\sigma^2(\theta)}\end{aligned} \tag{5.7}$$

(*cf.* Exercise 5.1.)

5.2 Portfolio optimization

Caution. In this chapter we will describe portfolio construction as if the population parameters, *e.g.*, the mean and covariance of returns, were known to the manager. As they generally are not, in the sequel we will consider how to perform inference on the portfolio weights, the total effect size, and so on.

It is commonly the case that the parameters to be optimized are dollarwise portfolio weights, giving you linear exposure to some returns. In this case, we use the symbol $\boldsymbol{\nu}$ to represent the unknown portfolio weights, and we have

$$\mu(\boldsymbol{\nu}) = \boldsymbol{\nu}^\top\boldsymbol{\mu},$$
$$\sigma(\boldsymbol{\nu}) = \sqrt{\boldsymbol{\nu}^\top\Sigma\boldsymbol{\nu}},$$

for some $\boldsymbol{\mu}$ and Σ, with Σ symmetric and positive semi-definite. For example these equations hold when $\boldsymbol{\mu}$ and Σ are the mean and covariance of the relative returns of the assets. The optimization problem Equation 5.1 becomes

$$\max_{\boldsymbol{\nu}} \frac{\boldsymbol{\nu}^{\top}\boldsymbol{\mu} - r_0}{\sqrt{\boldsymbol{\nu}^{\top}\Sigma\boldsymbol{\nu}}}. \tag{5.8}$$

Then the first-order conditions of Equation 5.4 are equivalent to

$$\boldsymbol{\mu} = \frac{\zeta(\boldsymbol{\nu})}{\sigma(\boldsymbol{\nu})}\Sigma\boldsymbol{\nu},$$

which is solved by

$$\boldsymbol{\nu}_* = c\Sigma^{-1}\boldsymbol{\mu}, \tag{5.9}$$

for some $c > 0$ which controls the leverage. We refer to $\Sigma^{-1}\boldsymbol{\mu}$ as the *Markowitz portfolio.*

The first-order conditions are *necessary*, but not sufficient for a solution. We now find conditions on c that determine a solution. The mean, standard deviation, and signal-noise ratio of the portfolio $c\Sigma^{-1}\boldsymbol{\mu}$ are:

$$\mu\left(c\Sigma^{-1}\boldsymbol{\mu}\right) = c\boldsymbol{\mu}^{\top}\Sigma^{-1}\boldsymbol{\mu}, \tag{5.10}$$

$$\sigma\left(c\Sigma^{-1}\boldsymbol{\mu}\right) = |c|\sqrt{\boldsymbol{\mu}^{\top}\Sigma^{-1}\boldsymbol{\mu}}, \tag{5.11}$$

$$\zeta\left(c\Sigma^{-1}\boldsymbol{\mu}\right) = \operatorname{sign}(c)\sqrt{\boldsymbol{\mu}^{\top}\Sigma^{-1}\boldsymbol{\mu}} - \frac{r_0}{|c|\sqrt{\boldsymbol{\mu}^{\top}\Sigma^{-1}\boldsymbol{\mu}}}. \tag{5.12}$$

When $r_0 = 0$, the signal-noise ratio is maximized for all $c > 0$. If $r_0 > 0$, then the signal-noise ratio is not maximized for any finite c, rather one should take $c \to \infty$. If for some reason $r_0 < 0$, the signal-noise ratio is maximized as $c \to 0$, *i.e.*, one can lock in a gain of $-r_0$ with arbitrarily small risk. The $r_0 \neq 0$ case only makes sense with further constraints on the portfolio to avoid these two pathological "solutions."

It is sometimes more convenient to express these results in terms of the signal-noise ratios of the assets and their correlation. Let R be the correlation matrix, defined as

$$\mathsf{R} := \operatorname{Diag}\left(\boldsymbol{\sigma}^{-1}\right)\Sigma\operatorname{Diag}\left(\boldsymbol{\sigma}^{-1}\right),$$

where $\boldsymbol{\sigma}$ is the (positive) square root of the diagonal of Σ. If $\boldsymbol{\zeta}$ is the vector of the signal-noise ratios of the assets, then the Markowitz portfolio of Equation 5.9 can be rewritten as

$$\boldsymbol{\nu}_* = c\operatorname{Diag}\left(\boldsymbol{\sigma}^{-1}\right)\mathsf{R}^{-1}\boldsymbol{\zeta}.$$

Similarly, the signal-noise ratio of this portfolio can be shown to be

$$\zeta_* = \operatorname{sign}(c)\sqrt{\boldsymbol{\zeta}^{\top}\mathsf{R}^{-1}\boldsymbol{\zeta}} - \frac{r_0}{|c|\sqrt{\boldsymbol{\zeta}^{\top}\mathsf{R}^{-1}\boldsymbol{\zeta}}}.$$

Example 5.2.1 (Markowitz portfolio on correlated assets). Suppose that returns are correlated with rank one updated correlation matrix $\mathsf{R} = (1-\rho)\,\mathsf{I} + \rho\left(\mathbf{1}\mathbf{1}^{\top}\right)$ as in Equation 4.4. Then using Exercise 4.6,

$$\boldsymbol{\nu}_{*} = c\,\mathrm{Diag}\left(\boldsymbol{\sigma}^{-1}\right)\frac{1}{1-\rho}\left(\boldsymbol{\zeta} - \frac{\rho k}{(1-\rho)+\rho k}\tilde{\zeta}\mathbf{1}\right),$$

where $\tilde{\zeta}$ is the average signal-noise ratio of the k assets. We also have

$$\begin{aligned}\boldsymbol{\zeta}^{\top}\mathsf{R}^{-1}\boldsymbol{\zeta} &= \frac{1}{1-\rho}\left(\sum_{i}^{k}\zeta_i^2 - \frac{\rho k^2}{(1-\rho)+\rho k}\tilde{\zeta}^2\right),\\ &= \frac{1}{1-\rho}\sum_{i}^{k}\left(\zeta_i - \tilde{\zeta}\right)^2 + \frac{k}{(1-\rho)+\rho k}\tilde{\zeta}^2.\end{aligned}$$

The first term is k times the population variance of the signal-noise ratios, while the second term is the squared average signal-noise ratio up to some scaling.

So suppose that one draws assets from a fixed population with finite mean and variance of signal-noise ratio, $\tilde{\zeta}$ and σ_{ζ}^2, respectively, and with common correlation ρ, and performs portfolio optimization. As k grows, the limit value is

$$\lim_{k\to\infty}\boldsymbol{\zeta}^{\top}\mathsf{R}^{-1}\boldsymbol{\zeta} = \frac{k\sigma_{\zeta}^2}{1-\rho} + \frac{\tilde{\zeta}^2}{\rho}.$$

Thus the maximal signal-noise ratio, ζ_{*}, is roughly proportional to $\sqrt{k}\sigma_{\zeta}$ as k grows. This is an example of the *fundamental law of active management.* However, in the case that $\sigma_{\zeta}^2 = 0$, we have $\lim_{k\to\infty}\zeta_{*} \approx \tilde{\zeta}/\sqrt{\rho}$, and there is *no diversification benefit*, rather the common correlation absorbs all the signal-noise ratio. Thus we can view this "fundamental law" as an assumption (or assumptions) rather than a law. ⊣

Example 5.2.2 (Markowitz portfolio on two assets). It is simple to completely describe the Markowitz portfolio and maximal signal-noise ratio in the two asset case. Letting ρ be the correlation of the two assets, and following from the correlation form, one has

$$\boldsymbol{\nu}_{*} = \frac{c}{1-\rho^2}\left[\begin{array}{c}\frac{\zeta_1-\rho\zeta_2}{\sigma_1}\\ \frac{\zeta_2-\rho\zeta_1}{\sigma_2}\end{array}\right]. \tag{5.13}$$

Assuming $r_0 = 0$, the signal-noise ratio of this portfolio is

$$\zeta_{*} = \mathrm{sign}\left(c\right)\sqrt{\frac{\zeta_1^2 - 2\rho\zeta_1\zeta_2 + \zeta_2^2}{1-\rho^2}}. \tag{5.14}$$

See also Smith [2019] for a further analysis of $\boldsymbol{\nu}_{*}$ in terms of ρ and the ratio ζ_2/ζ_1.

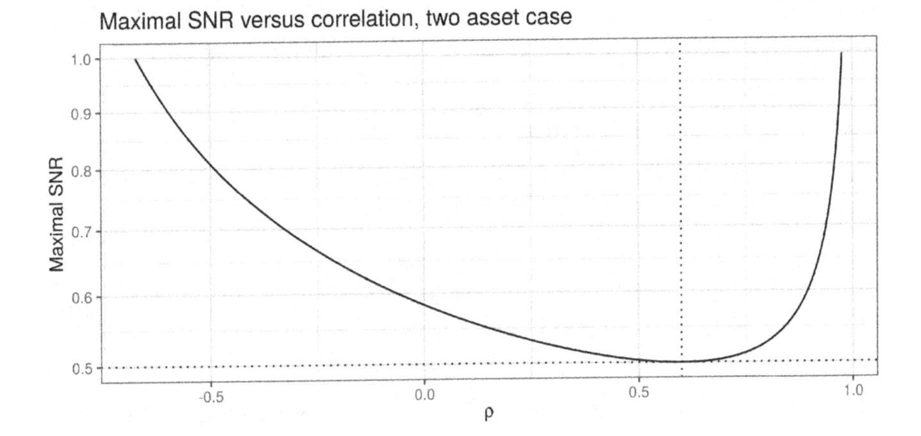

FIGURE 5.1
The maximal signal-noise ratio is plotted versus ρ for the two asset case, assuming $r_0 = 0$. The assets have signal-noise ratios of $\zeta_1 = 0.3$ and $\zeta_2 = 0.5$. We plot a vertical line at the correlation ζ_1/ζ_2, and a horizontal line at $max\left(\zeta_1, \zeta_2\right)$.

Supposing, without loss of generality, that $\zeta_1 \leq \zeta_2$, we note that if $\rho = \zeta_1/\zeta_2$, there is zero weight on the first asset. In this case, since there is zero weight on the first asset, the signal-noise ratio equals that of the second asset, ζ_2, and there is zero diversification benefit. In Figure 5.1, we plot the signal-noise ratio versus ρ for some values of ζ_1, ζ_2. We indeed see no diversification benefit at the "unfortunate value" where $\rho = \zeta_1/\zeta_2$.

Note that when ζ_1 and ζ_2 are both positive, you see higher ζ_* for negative ρ than for positive ρ. (For example, you see higher ζ_* for $\rho = -\left|r\right|$ than for $\rho = \left|r\right|$ for every r.) This is consistent with the rule of thumb that one should seek assets with positive signal-noise ratio, but which are negatively correlated, or hedge against each other.

$\dashv$

The failure of diversification seen in Example 5.2.2 is not confined to the two asset case. Indeed, it is simple to show that for any Σ, there is a $\boldsymbol{\mu}$ such that $\Sigma^{-1}\boldsymbol{\mu} = \boldsymbol{e}_i$ for any i, in which case $\boldsymbol{\mu}^\top \Sigma^{-1}\boldsymbol{\mu} = \zeta_i^2$. Moreover, it can also be shown that for any $\boldsymbol{\mu}$, there is a Σ for which this holds, see Exercise 5.17. While this breakdown in diversification is certainly possible, it is not clear whether such situations are likely to occur in practice. We note, however, that in practice, when the population values are unknown, the signal-noise ratio will need to grow at a rate faster than $k^{1/4}$, as we explain in the sequel.

Note that when the assets are uncorrelated, Σ is diagonal, thus $\mathsf{R} = \mathsf{I}$, and

$$\sqrt{\boldsymbol{\zeta}^\top \mathsf{R}^{-1}\boldsymbol{\zeta}} = \left\|\boldsymbol{\zeta}\right\|_2.$$

Thus the signal-noise ratios of *independent* assets grow *geometrically,* rather than *arithmetically.* This is a restatement of the "fundamental law."

5.2.1 Basic risk constraint

It is mathematically convenient to amend the problem of Equation 5.8 to add an upper bound on the risk, and quote the problem as

$$\max_{\nu:\nu^\top\Sigma\nu\le R^2} \frac{\nu^\top\mu - r_0}{\sqrt{\nu^\top\Sigma\nu}}, \tag{5.15}$$

where $R > 0$ is some "risk budget." For $r_0 > 0$, this is uniquely solved by

$$\nu_* = \frac{R}{\sqrt{\mu^\top\Sigma^{-1}\mu}}\Sigma^{-1}\mu. \tag{5.16}$$

This portfolio has signal-noise ratio

$$\zeta\left(\nu_*\right) = \zeta_* = \sqrt{\mu^\top\Sigma^{-1}\mu} - \frac{r_0}{R}. \tag{5.17}$$

Note that the optimal portfolio does not depend on r_0, but the signal-noise ratio does.

Example 5.2.3 (Markowitz portfolio on toy universe). Imagine a toy universe of 4 assets where

$$\mu = \begin{bmatrix} 0.000 \\ 0.050 \\ -0.050 \\ 0.100 \end{bmatrix}, \quad \Sigma = \begin{bmatrix} 0.0400 & 0.0240 & -0.0160 & 0.0100 \\ 0.0240 & 0.1600 & 0.1120 & -0.0400 \\ -0.0160 & 0.1120 & 0.1600 & 0.0400 \\ 0.0100 & -0.0400 & 0.0400 & 0.2500 \end{bmatrix}.$$

Suppose the risk-free rate is $r_0 = 0.01$ and the risk budget is $R = 0.1$. The re-scaled Markowitz portfolio on this universe is

$$\nu_*{}^\top = \begin{bmatrix} -0.713 & 0.652 & -0.631 & 0.277 \end{bmatrix}.$$

Note that even though the first element of μ is zero, the first element of ν_* is not; we hold that asset as a hedge against the other assets. This portfolio has signal-noise ratio

$$\zeta\left(\nu_*\right) = \sqrt{\mu^\top\Sigma^{-1}\mu} - \frac{r_0}{R} = 0.919 - 0.1 = 0.819.$$

⊣

5.2.2 Basic subspace constraint

Let G be an $k_g \times k$ matrix of rank $k_g \le k$. Let G^C be the matrix whose rows span the null space of the rows of G, *i.e.*, $\mathsf{G}^C \mathsf{G}^\top = 0$. Consider the constrained optimization problem

$$\max_{\substack{\boldsymbol{\nu}: \; \mathsf{G}^C \boldsymbol{\nu} = \mathbf{0}, \\ \boldsymbol{\nu}^\top \Sigma \boldsymbol{\nu} \le R^2}} \frac{\boldsymbol{\nu}^\top \boldsymbol{\mu} - r_0}{\sqrt{\boldsymbol{\nu}^\top \Sigma \boldsymbol{\nu}}}, \tag{5.18}$$

where, as previously, $\boldsymbol{\mu}$, Σ are the sample mean vector and covariance matrix, r_0 is the risk-free rate, and $R > 0$ is a risk "budget."

The gist of this constraint is that feasible portfolios must be some linear combination of the rows of G, or $\boldsymbol{\nu} = \mathsf{G}^\top \boldsymbol{\nu}_g$, for some unknown vector $\boldsymbol{\nu}_g$. When viewed in this light, the constrained problem reduces to that of optimizing the portfolio on k_g assets with sample mean $\mathsf{G}\boldsymbol{\mu}$ and sample covariance $\mathsf{G}\Sigma\mathsf{G}^\top$. This problem has unique solution

$$\boldsymbol{\nu}_{*,\mathsf{G}} := \frac{R}{\sqrt{(\mathsf{G}\boldsymbol{\mu})^\top \left(\mathsf{G}\Sigma\mathsf{G}^\top\right)^{-1} (\mathsf{G}\boldsymbol{\mu})}} \mathsf{G}^\top \left(\mathsf{G}\Sigma\mathsf{G}^\top\right)^{-1} \mathsf{G}\boldsymbol{\mu} \tag{5.19}$$

when $r_0 > 0$. The signal-noise ratio of this portfolio is

$$\zeta_{*,\mathsf{G}} := \zeta\left(\boldsymbol{\nu}_{*,\mathsf{G}}\right) = \sqrt{(\mathsf{G}\boldsymbol{\mu})^\top \left(\mathsf{G}\Sigma\mathsf{G}^\top\right)^{-1} (\mathsf{G}\boldsymbol{\mu})} - \frac{r_0}{R}. \tag{5.20}$$

Example 5.2.4 (Markowitz portfolio on toy universe, subspace constraint). Continuing Example 5.2.3, suppose

$$\mathsf{G}^C = \begin{bmatrix} 1.00 & 0.00 & 0.00 & 0.00 \end{bmatrix}.$$

In this case, the optimal portfolio is

$$\boldsymbol{\nu}_{*,\mathsf{G}}{}^\top = \begin{bmatrix} 0.000 & 0.350 & -0.350 & 0.187 \end{bmatrix}.$$

There are now no holdings on the first asset, since we have restricted the portfolio to be a linear combination of the columns of G. This portfolio has signal-noise ratio

$$\zeta\left(\boldsymbol{\nu}_{*,\mathsf{G}}\right) = 0.536 - 0.1 = 0.436.$$

This is quite a bit lower than the unrestricted value we saw in Example 5.2.3. This is really remarkable because the constraint has caused us to drop from our portfolio an asset with zero mean return. ⊣

5.2.3 Spanning and hedging

Consider the constrained portfolio optimization problem on k assets,

$$\max_{\substack{\mathsf{G}\Sigma\boldsymbol{\nu}=\boldsymbol{g},\\ \boldsymbol{\nu}^{\top}\Sigma\boldsymbol{\nu}\le R^2}} \frac{\boldsymbol{\nu}^{\top}\boldsymbol{\mu} - r_0}{\sqrt{\boldsymbol{\nu}^{\top}\Sigma\boldsymbol{\nu}}}, \tag{5.21}$$

where G is an $k_g \times k$ matrix of rank k_g, and, as previously, $\boldsymbol{\mu}$, Σ are the mean vector and covariance matrix, r_0 is the risk-free rate, and $R > 0$ is a "risk budget." We can interpret the G constraint as stating that the covariance of the returns of a feasible portfolio with the returns of a portfolio whose weights are in a given row of G shall equal the corresponding element of $\boldsymbol{g}$. In the garden variety application of this problem, G consists of k_g rows of the identity matrix, and $\boldsymbol{g}$ is the zero vector; in this case, feasible portfolios are *hedged* with respect to the k_g assets selected by G (although they may hold some position in the hedged assets). We use "hedged" to mean a portfolio with some fixed covariance (typically zero) against some other portfolio(s).

Assuming that the G constraint and risk budget can be simultaneously satisfied, the solution to this problem, via the Lagrange multiplier technique, is

$$\begin{aligned}\boldsymbol{\nu}_* &= c\left(\Sigma^{-1}\boldsymbol{\mu} - \mathsf{G}^{\top}\left(\mathsf{G}\Sigma\mathsf{G}^{\top}\right)^{-1}\mathsf{G}\boldsymbol{\mu}\right) + \mathsf{G}^{\top}\left(\mathsf{G}\Sigma\mathsf{G}^{\top}\right)^{-1}\boldsymbol{g},\\ \text{where } c^2 &= \frac{R^2 - \boldsymbol{g}^{\top}\left(\mathsf{G}\Sigma\mathsf{G}^{\top}\right)\boldsymbol{g}}{\boldsymbol{\mu}^{\top}\Sigma^{-1}\boldsymbol{\mu} - \left(\mathsf{G}\boldsymbol{\mu}\right)^{\top}\left(\mathsf{G}\Sigma\mathsf{G}^{\top}\right)^{-1}\left(\mathsf{G}\boldsymbol{\mu}\right)},\end{aligned} \tag{5.22}$$

where the numerator in the last equation must be positive for the problem to be feasible.

The case where $\boldsymbol{g} \neq 0$ is "pathological," as it requires a fixed non-zero covariance of the target portfolio with some other portfolio's returns. Setting $\boldsymbol{g} = 0$ ensures the problem is feasible, and we will make this assumption hereafter. Under this assumption, the optimal portfolio is

$$\boldsymbol{\nu}_{*,\mathsf{I}\backslash\mathsf{G}} := c\left(\Sigma^{-1}\boldsymbol{\mu} - \mathsf{G}^{\top}\left(\mathsf{G}\Sigma\mathsf{G}^{\top}\right)^{-1}\mathsf{G}\boldsymbol{\mu}\right) = c_1\boldsymbol{\nu}_{*,\mathsf{I}} - c_2\boldsymbol{\nu}_{*,\mathsf{G}},$$

using the notation from Equation 5.19. The constants c_1, c_2 adjust for the risk budget. Note that the $\boldsymbol{\nu}_{*,\mathsf{I}}$ denoted here uses the identity matrix I as the "projection" matrix, but that imposes no constraints and so we simply have $\boldsymbol{\nu}_{*,\mathsf{I}} = \boldsymbol{\nu}_*$.

Note that, up to scaling, $\Sigma^{-1}\boldsymbol{\mu}$ is the unconstrained optimal portfolio, and thus the imposition of the G constraint only changes the unconstrained portfolio in assets corresponding to columns of G containing non-zero elements. In the garden variety application where G is a single row of the identity matrix, the imposition of the constraint only changes the holdings in the asset to be hedged (modulo changes in the leading constant to satisfy the risk budget).

The squared signal-noise ratio of the optimal portfolio is

$$\Delta_{\mathsf{I}\setminus\mathsf{G}}\zeta_*^2 = \boldsymbol{\mu}^\top\Sigma^{-1}\boldsymbol{\mu} - (\mathsf{G}\boldsymbol{\mu})^\top\left(\mathsf{G}\Sigma\mathsf{G}^\top\right)^{-1}(\mathsf{G}\boldsymbol{\mu}) = \zeta_{*,\mathsf{I}}^2 - \zeta_{*,\mathsf{G}}^2, \tag{5.23}$$

using the notation from Equation 5.20, and setting $r_0 = 0$.

The quantity ζ_*^2 in Equation 5.23 is the drop in squared signal-noise ratio incurred by imposing the G hedge constraint. This population quantity is the subject of inference in tests of *portfolio spanning.* [Kan and Zhou, 2012; Huberman and Kandel, 1987] A spanning test considers whether the optimal portfolio on a pre-fixed subset of k_g assets has the same Sharpe ratio as the optimal portfolio on all k assets, *i.e.*, whether those k_g assets "span" the set of all assets. We will consider those in the sequel.

Example 5.2.5 (Markowitz portfolio on toy universe, hedging constraint). Continuing Example 5.2.3, suppose you wish to hedge out

$$\mathsf{G} = \begin{bmatrix}1.00 & 1.00 & 1.00 & 1.00\end{bmatrix}.$$

That is, you want to have no correlation to the portfolio consisting of equal dollar exposure to the 4 assets. In this case, the optimal portfolio is

$${\boldsymbol{\nu}_*}^\top = \begin{bmatrix}-0.731 & 0.644 & -0.648 & 0.267\end{bmatrix}.$$

Assuming $r_0 = 0$, this portfolio has squared signal-noise ratio

$$\zeta_*^2 = 0.844 - 0.011 = 0.833.$$

In this case, the loss of signal-noise ratio due to the hedging constraint is rather small. ⊣

Note (On Hedging). At first glance the formulation of Equation 5.21 seems arbitrary, as if the problem were designed to make the solution work out cleanly, and the description as a "hedged portfolio" seems like an afterthought. One interpretation of "portfolio A is hedged against portfolio B" is that their returns are statistically independent. Statistical independence is a hard condition to establish, however. Independence of returns implies zero covariance of returns, which is a much easier condition to describe. In fact, zero covariance is the gist of the constraint $\mathsf{G}\Sigma\boldsymbol{\nu} = \mathbf{0}$.

Spanning Decomposition

We note that the relationship in Equation 5.23 can be rewritten as the "spanning decomposition" relationship

$$\left(\max_{\boldsymbol{\nu}} \frac{\left(\boldsymbol{\nu}^\top\boldsymbol{\mu}\right)^2}{\boldsymbol{\nu}^\top\Sigma\boldsymbol{\nu}}\right) = \left(\max_{\boldsymbol{\nu}:\mathsf{G}\Sigma\boldsymbol{\nu}=\mathbf{0}} \frac{\left(\boldsymbol{\nu}^\top\boldsymbol{\mu}\right)^2}{\boldsymbol{\nu}^\top\Sigma\boldsymbol{\nu}}\right) + \left(\max_{\boldsymbol{\nu}:\boldsymbol{\nu}=\mathsf{G}\boldsymbol{\beta}} \frac{\left(\boldsymbol{\nu}^\top\boldsymbol{\mu}\right)^2}{\boldsymbol{\nu}^\top\Sigma\boldsymbol{\nu}}\right). \tag{5.24}$$

That is, the unconstrained optimal squared signal-noise ratio decomposes as the sum of the squared signal-noise ratios of the hedged problem and the subspace-constrained problem.

5.2.4 Inequality constraints

A fairly general form of the portfolio optimization problem takes the form

$$\max_{\substack{\mathsf{A}\nu=a,\\ \mathsf{B}\nu\leq b,\\ \mathsf{C}|\nu|\leq c}} \frac{\nu^{\top}\mu - r_0}{\sqrt{\nu^{\top}\Sigma\nu}}, \tag{5.25}$$

for conformable matrices $\mathsf{A}, \mathsf{B}, \mathsf{C}$ and vectors $\boldsymbol{a}, \boldsymbol{b}, \boldsymbol{c}$. There is no general form expression for the solution to this problem, rather it must be found numerically. Thus we will have fairly little to say about the mathematics of the solution.

Such an optimization problem arises from many of the kinds of constraints which are imposed on portfolios in practice, including:

- Long-only constraint, or some constraint on shorts. The long-only constraint can be implemented as an element-wise lower bound on the elements of the portfolio vector to be optimized.
- Maximum position constraints, which you can implement as "box constraints" on the elements of the portfolio vector.
- Hedging constraints, where some linear combinations of portfolio weights have to equal zero, or be small in absolute value. These include dollar-neutrality, market- (or beta-) neutrality, zero exposure to certain factors like momentum or value, or industries or geographical regions.
- Maximum sector concentration constraints, wherein the sums of absolute values of some subset of elements of the portfolio vector must be less than a certain amount.
- Element-wise trade constraints, where the delta from the current position to the target position can be no larger than a certain amount. Again this is a box constraint on the elements. Imposition of such a constraint can cause the zero vector to not be feasible, which we consider an anti-pattern. Instead we recommend that one constrain trades *away* from the current position, but always allow zero to be a feasible solution.
- Total trade budget constraints, wherein the total absolute value of the change in position is constrained to be less than a certain amount. Again, we recommend instead that one allow zero to be feasible as this guarantees the existence of some feasible solution.

As a practical matter one will likely find that while each such constraint sounds sensible, many of these constraints will have very little effect on the optimal portfolio for a particular problem. For example, the imposition of a "beta-neutrality" constraint might change the allocation to only one asset in the portfolio, and by a negligible amount, especially if a dollar-neutrality constraint is already imposed. It is more likely that constraints as such are not really improving the risk of the portfolio, but instead reduce the degrees of freedom of the problem, which reduces overfitting risk. For more on the balance of signal-noise ratio and degrees of freedom, see Chapter 8.

Reducing to quadratic programming with this one weird trick:

The objective we have considered here does not fit neatly into the usual optimization paradigms. However, there is a trick for optimizing the signal-noise ratio via a quadratic optimization routine. [Cornuejols and Tütüncü, 2006, Proposition 8.1] The trick consists of adding another unknown to be optimized, call it K. Letting $\boldsymbol{x} = K\boldsymbol{\nu}$, constrain $\boldsymbol{x}$ to have unit expected return: $\boldsymbol{x}^\top\boldsymbol{\mu} = 1$, and thus maximizing the signal-noise ratio under this constraint consists of minimizing the risk, which is quadratic in the unknowns. Upon finding the optimal $\boldsymbol{x}$ and K, transform them back to the optimal $\boldsymbol{\nu}$ via $\boldsymbol{\nu} = \boldsymbol{x}/K$.

That is, instead of solving

$$\max_{\substack{\mathsf{A}\boldsymbol{\nu}=\boldsymbol{a}\\ \mathsf{B}\boldsymbol{\nu}\le\boldsymbol{b}\\ \boldsymbol{\nu}^\top\Sigma\boldsymbol{\nu}\le R^2}} \frac{\boldsymbol{\nu}^\top\boldsymbol{\mu} - r_0}{\sqrt{\boldsymbol{\nu}^\top\Sigma\boldsymbol{\nu}}},$$

one instead solves

$$\min_{\substack{K>0\\ \boldsymbol{\mu}^\top\boldsymbol{x}=1\\ \mathsf{A}\boldsymbol{x}=K\boldsymbol{a}\\ \mathsf{B}\boldsymbol{x}\le K\boldsymbol{b}\\ \boldsymbol{x}^\top\Sigma\boldsymbol{x}\le K^2R^2}} \boldsymbol{x}^\top\Sigma\boldsymbol{x}$$

in the unknowns $\boldsymbol{x}$ and K.

5.3 Conditional portfolio optimization

Above we treated the portfolio problem in an unconditional setting where the mean and covariance of assets are fixed across all time. Here we consider conditional portfolio problems. The simplest model is the *conditional Markowitz* model, where the expected returns are linear in some observable features while the covariance is fixed.

We assume $\boldsymbol{f}_{t-1}$ is some f-vector of factors observable before the time required to capture returns $\boldsymbol{x}_t$, the corresponding k-vector of asset returns. The model is expressed as

$$\begin{aligned} \operatorname{E}\left[\boldsymbol{x}_t \,\middle|\, \boldsymbol{f}_{t-1}\right] &= \mathsf{B}\boldsymbol{f}_{t-1}, \\ \operatorname{Var}\left(\boldsymbol{x}_t \,\middle|\, \boldsymbol{f}_{t-1}\right) &= \Sigma. \end{aligned} \tag{5.26}$$

Here B is the $k \times f$ linear coefficient matrix. Again, in this section we will pretend that B and Σ are known.

A *linear strategy* under this model is to allocate dollars proportional to $\mathsf{N}\boldsymbol{f}_{t-1}$ upon observing $\boldsymbol{f}_{t-1}$, for some $k \times f$ "passthrough" matrix, N. This passthrough matrix is the analogue of the $\boldsymbol{\nu}$ to be determined in the unconditional case.

One can select N to optimize the conditional single-period signal-noise ratio, or the unconditional multi-period signal-noise ratio. While the latter sounds more appealing, the analysis is a bit more involved. In Section 5.3.2 we will show that to maximize the multi-period signal-noise ratio, one should instead hold a *non-linear* portfolio, though in practice the distinction will be very small.

The conditional case

Conditional on observing $\boldsymbol{f}_{t-1}$, one has the opportunity to invest in assets with expected return $\mathsf{B}\boldsymbol{f}_{t-1}$ and covariance Σ. This easily reduces to the analysis above in Section 5.2.1. Thus the single-period signal-noise ratio maximization problem

$$\max_{\boldsymbol{\nu}:\,\boldsymbol{\nu}^{\top}\Sigma\boldsymbol{\nu}\le R^2} \frac{\boldsymbol{\nu}^{\top}\mathsf{B}\boldsymbol{f}_{t-1} - r_0}{\sqrt{\boldsymbol{\nu}^{\top}\Sigma\boldsymbol{\nu}}}. \tag{5.27}$$

is solved by the portfolio

$$\boldsymbol{\nu}_* = c\Sigma^{-1}\mathsf{B}\boldsymbol{f}_{t-1}.$$

This suggests a passthrough matrix of $\mathsf{N} = c\Sigma^{-1}\mathsf{B}$. The matrix N is the (population) *Markowitz coefficient* or *Markowitz passthrough matrix*. The signal-noise ratio conditional on observing $\boldsymbol{f}_{t-1}$ is

$$\zeta_* \,|\, \boldsymbol{f}_{t-1} = \operatorname{sign}(c)\sqrt{\left(\mathsf{B}\boldsymbol{f}_{t-1}\right)^{\top}\Sigma^{-1}\left(\mathsf{B}\boldsymbol{f}_{t-1}\right)},$$

assuming $r_0 = 0$.

Now let $\mathsf{\Gamma}_f$ be the $f \times f$ second moment matrix of the factors: $\mathsf{\Gamma}_f := \operatorname{E}\left[\boldsymbol{f}_{t-1}{\boldsymbol{f}_{t-1}}^{\top}\right]$. Then the expected value of the squared signal-noise ratio for this passthrough is

$$\operatorname{E}_{\boldsymbol{f}}\left[\zeta_*^2\right] = \operatorname{tr}\left(\mathsf{B}^{\top}\Sigma^{-1}\mathsf{B}\mathsf{\Gamma}_f\right). \tag{5.28}$$

This quantity is the population analogue of the *Hotelling Lawley trace*. We will discuss the (sample) Hotelling Lawley trace in greater detail in Section 6.1.3, but you can think of the population value as simply the expected squared maximal signal-noise ratio, and thus just measures the population "effect size" for the single-period optimization problem.

Example 5.3.1 (Conditional expectations on toy universe, single-period signal-noise ratio). Imagine a toy universe of 4 assets and 2 factors with

$$\mathsf{B} = \begin{bmatrix} 0.000 & 0.040 \\ 0.050 & -0.010 \\ -0.030 & 0.060 \\ 0.000 & -0.020 \end{bmatrix}, \quad \Sigma = \begin{bmatrix} 0.0400 & 0.0240 & -0.0160 & 0.0100 \\ 0.0240 & 0.1600 & 0.1120 & -0.0400 \\ -0.0160 & 0.1120 & 0.1600 & 0.0400 \\ 0.0100 & -0.0400 & 0.0400 & 0.2500 \end{bmatrix}.$$

This leads to a passthrough matrix

$$\mathsf{N} = \begin{bmatrix} -3.807 & 6.702 \\ 3.565 & -5.204 \\ -3.380 & 5.191 \\ 1.263 & -2.011 \end{bmatrix}.$$

Suppose one observes

$$\boldsymbol{f}_{t-1}{}^{\top} = \begin{bmatrix} 1.000 & 0.200 \end{bmatrix}.$$

Then the suggested portfolio is

$$\left(\mathsf{N}\boldsymbol{f}_{t-1}\right)^{\top} = \begin{bmatrix} -2.466 & 2.524 & -2.341 & 0.861 \end{bmatrix}.$$

The signal-noise ratio conditional on this $\boldsymbol{f}_{t-1}$ is 0.374. Supposing the second moment matrix is

$$\mathsf{\Gamma}_f = \begin{bmatrix} 1.100 & 0.200 \\ 0.200 & 1.400 \end{bmatrix},$$

then the population Hotelling-Lawley trace is

$$\operatorname{tr}\left(\mathsf{B}^{\top}\mathsf{\Sigma}^{-1}\mathsf{B}\mathsf{\Gamma}_f\right) = 1.082.$$

⊣

Conditionality via flattening

To find the passthrough that maximizes the multi-period unconditional signal-noise ratio, we rely on the *flattening trick* for turning conditional problems into unconditional problems. [Brandt and Santa-Clara, 2006] First note that the returns collected at time t for a given passthrough are

$$x_t = \boldsymbol{x}_t^{\top}\mathsf{N}\boldsymbol{f}_{t-1} = \operatorname{tr}\left(\boldsymbol{f}_{t-1}\boldsymbol{x}_t^{\top}\mathsf{N}\right) = \operatorname{vec}\left(\boldsymbol{x}_t\boldsymbol{f}_{t-1}^{\top}\right)^{\top}\operatorname{vec}\left(\mathsf{N}\right).$$

Here the outer product of features and returns have been "flattened" into the vector $\operatorname{vec}\left(\boldsymbol{x}_t\boldsymbol{f}_{t-1}^{\top}\right)$. We can then use the results on maximization of signal-noise ratio in the unconditional case to solve for $\operatorname{vec}\left(\mathsf{N}\right)$.

Because we will consider higher order moments of $\boldsymbol{f}_{t-1}$, we need to impose some distributional assumption. We will assume that $\boldsymbol{f}_{t-1}$ takes an elliptical distribution with kurtosis factor κ, and with expected value $\boldsymbol{\mu}_f$ and covariance $\mathsf{\Sigma}_f$. Let $\mathsf{\Gamma}_f = \operatorname{E}\left[\boldsymbol{f}_{t-1}\boldsymbol{f}_{t-1}^{\top}\right] = \mathsf{\Sigma}_f + \boldsymbol{\mu}_f\boldsymbol{\mu}_f^{\top}$ be the second moment of the features. The expectation and covariance of $\operatorname{vec}\left(\boldsymbol{x}_t\boldsymbol{f}_{t-1}^{\top}\right)$, the latter following from Question 8 of Exercise 3.2, are

$$\begin{aligned}
\mathrm{E}\left[\mathrm{vec}\left(\boldsymbol{x}_t\boldsymbol{f}_{t-1}^{\top}\right)\right] &= \mathrm{vec}\left(\mathsf{B}\Gamma_f\right),\\
\mathrm{Var}\left(\mathrm{vec}\left(\boldsymbol{x}_t\boldsymbol{f}_{t-1}^{\top}\right)\right) &= \Gamma_f\otimes\Sigma + (\kappa-1)\left(\mathsf{I}_f\otimes\mathsf{B}\right)\left[\mathrm{vec}\left(\Sigma_f\right)\mathrm{vec}\left(\Sigma_f\right)^{\top}\right]\left(\mathsf{I}_f\otimes\mathsf{B}^{\top}\right)\\
&\quad + (\kappa-1)\left[\Sigma_f\otimes\mathsf{B}\Sigma_f\mathsf{B}^{\top} + \mathsf{K}_{f,k}\left(\mathsf{B}\Sigma_f\otimes\Sigma_f\mathsf{B}^{\top}\right)\right]\\
&\quad + \Gamma_f\otimes\mathsf{B}\Gamma_f\mathsf{B}^{\top} + \mathsf{K}_{f,k}\left(\mathsf{B}\Gamma_f\otimes\Gamma_f\mathsf{B}^{\top}\right)\\
&\quad - \left(\boldsymbol{\mu}_f\boldsymbol{\mu}_f^{\top}\right)\otimes\left(\mathsf{B}\boldsymbol{\mu}_f\boldsymbol{\mu}_f^{\top}\mathsf{B}^{\top}\right)\\
&\quad - \mathsf{K}_{f,k}\left(\left(\mathsf{B}\boldsymbol{\mu}_f\boldsymbol{\mu}_f^{\top}\right)\otimes\left(\boldsymbol{\mu}_f\boldsymbol{\mu}_f^{\top}\mathsf{B}^{\top}\right)\right),
\end{aligned} \tag{5.29}$$

where $\mathsf{K}_{f,k}$ is the commutation matrix that transforms $\mathrm{vec}\left(\mathsf{A}\right)$ to $\mathrm{vec}\left(\mathsf{A}^{\top}\right)$ for $f \times k$ matrix A. [Magnus and Neudecker, 2007, 1979]

Thus to find the Markowitz passthrough matrix, one should set $\boldsymbol{\mu}$ and Σ to be the mean and covariance from Equation 5.29 and then use the results of Section 5.2.1 above to solve for $\mathrm{vec}\left(\mathsf{N}\right)$. One can then also easily compute the maximal squared signal-noise ratio in the same way. It does not seem, however, that there is a simple representation of that maximal signal-noise ratio, as that covariance matrix seems too hairy to admit a tractable inverse.

Example 5.3.2 (Conditional expectations on toy universe, multi-period signal-noise ratio). Continuing Example 5.3.1, suppose you wanted to maximize the multi-period signal-noise ratio. Suppose further that

$$\boldsymbol{\mu}_f = \begin{bmatrix}1.000\\0.000\end{bmatrix},\qquad \Sigma_f = \begin{bmatrix}0.1000 & 0.2000\\0.2000 & 1.4000\end{bmatrix},$$

and $\kappa = 1.33$.

The expectation and variance of $\mathrm{vec}\left(\boldsymbol{x}_t\boldsymbol{f}_{t-1}^{\top}\right)$ are computed via Equation 5.29 as

$$\mathsf{E}^{\top} = \begin{bmatrix}0.008 & 0.053 & -0.021 & -0.004 & 0.056 & -0.004 & 0.078 & -0.028\end{bmatrix},$$

$$\mathsf{V} = \begin{bmatrix}
0.047 & 0.027 & -0.014 & 0.010 & 0.009 & 0.008 & -0.003 & 0.001\\
0.027 & 0.177 & 0.123 & -0.044 & 0.005 & 0.032 & 0.022 & -0.008\\
-0.014 & 0.123 & 0.181 & 0.042 & -0.001 & 0.026 & 0.032 & 0.007\\
0.010 & -0.044 & 0.042 & 0.276 & 0.001 & -0.009 & 0.008 & 0.050\\
0.009 & 0.005 & -0.001 & 0.001 & 0.065 & 0.033 & -0.009 & 0.009\\
0.008 & 0.032 & 0.026 & -0.009 & 0.033 & 0.228 & 0.154 & -0.056\\
-0.003 & 0.022 & 0.032 & 0.008 & -0.009 & 0.154 & 0.244 & 0.049\\
0.001 & -0.008 & 0.007 & 0.050 & 0.009 & -0.056 & 0.049 & 0.352
\end{bmatrix}.$$

This yields a passthrough matrix of

$$\mathsf{N} = \begin{bmatrix}-1.469 & 1.804\\1.755 & -1.287\\-1.572 & 1.316\\0.563 & -0.519\end{bmatrix}.$$

The multi-period squared signal-noise ratio in this case is 0.335. Note this is quite a bit smaller than the Hotelling-Lawley trace on the same universe, which we computed in Example 5.3.1 as 1.082. The additional variance in returns due to changes in the signal have caused this outsized decrease in the signal-noise ratio.

If, as in Example 5.3.1, one observes

$$\boldsymbol{f}_{t-1}^{\top} = \begin{bmatrix} 1.000 & 0.200 \end{bmatrix},$$

then the suggested portfolio is

$$\left(\mathsf{N}\boldsymbol{f}_{t-1}\right)^{\top} = \begin{bmatrix} -1.108 & 1.497 & -1.309 & 0.459 \end{bmatrix}.$$

⊣

Unfettered flattening

We note that the covariance matrix of Equation 5.29 is highly structured. In fact, it is a consequence of the conditional expectation model of Equation 5.26 and the elliptical distribution imposed on the $\boldsymbol{f}_{t-1}$. One could instead take the flattening trick to its logical extreme and simply define

$$\begin{aligned} \boldsymbol{\mu} &= \operatorname{E}\left[\operatorname{vec}\left(\boldsymbol{x}_t \boldsymbol{f}_{t-1}^{\top}\right)\right], \\ \Sigma &= \operatorname{Var}\left(\operatorname{vec}\left(\boldsymbol{x}_t \boldsymbol{f}_{t-1}^{\top}\right)\right), \end{aligned}$$

then use the results of Section 5.2.1 to solve for $\operatorname{vec}(\mathsf{N})$ and to compute its signal-noise ratio. The resultant portfolio optimization problem is less structured than the conditional expectation problem, as it discards the assumptions of constant covariance. In particular, it allows the covariance of returns to depend on the $\boldsymbol{f}_{t-1}$.

For example, if one assumes that the stacked vector $\left[\boldsymbol{f}_{t-1}^{\top}, \boldsymbol{x}_t^{\top}\right]^{\top}$ is elliptically distributed with mean and covariance

$$\left[{\boldsymbol{\mu}_f}^{\top}, \mathsf{B}{\boldsymbol{\mu}_f}^{\top}\right]^{\top}, \qquad \text{and} \qquad \begin{bmatrix} \Sigma_f & \Sigma_f \mathsf{B}^{\top} \\ \mathsf{B}\Sigma_f & \Sigma + \mathsf{B}\Sigma_f \mathsf{B}^{\top} \end{bmatrix},$$

then Question 8 of Exercise 3.2 gives the expectation and variance of the stacked vector. From this one can subselect elements to arrive at the expectation and variance of $\operatorname{vec}\left(\boldsymbol{x}_t \boldsymbol{f}_{t-1}^{\top}\right)$. The analytic expressions are fairly ugly, and not terribly informative. They are *not* the same as given by Equation 5.29 except in the case where $\kappa = 1$, which includes Gaussian returns.

Example 5.3.3 (Conditional expectations on jointly elliptical toy universe, multi-period signal-noise ratio). Continuing Example 5.3.2, suppose you wanted to maximize the multi-period signal-noise ratio assuming that the stacked vector $\left[\boldsymbol{f}_{t-1}^{\top}, \boldsymbol{x}_t^{\top}\right]^{\top}$ is elliptically distributed. From the mean and variance of

$\left[\boldsymbol{f}_{t-1}^{\top}, \boldsymbol{x}_t^{\top}\right]^{\top}$, we compute the mean and variance of $\operatorname{vec}\left(\boldsymbol{x}_t \boldsymbol{f}_{t-1}^{\top}\right)$, which we compute as

$$\mathsf{E}^{\top} = \begin{bmatrix} 0.008 & 0.053 & -0.021 & -0.004 & 0.056 & -0.004 & 0.078 & -0.028 \end{bmatrix},$$

$$\mathsf{V} = \begin{bmatrix} 0.048 & 0.027 & -0.015 & 0.010 & 0.012 & 0.009 & -0.004 & 0.002 \\ 0.027 & 0.182 & 0.127 & -0.045 & 0.006 & 0.043 & 0.030 & -0.011 \\ -0.015 & 0.127 & 0.186 & 0.044 & -0.002 & 0.034 & 0.043 & 0.010 \\ 0.010 & -0.045 & 0.044 & 0.284 & 0.002 & -0.012 & 0.011 & 0.067 \\ 0.012 & 0.006 & -0.002 & 0.002 & 0.084 & 0.044 & -0.017 & 0.014 \\ 0.009 & 0.043 & 0.034 & -0.012 & 0.044 & 0.302 & 0.205 & -0.074 \\ -0.004 & 0.030 & 0.043 & 0.011 & -0.017 & 0.205 & 0.317 & 0.068 \\ 0.002 & -0.011 & 0.010 & 0.067 & 0.014 & -0.074 & 0.068 & 0.468 \end{bmatrix}.$$

This yields a passthrough matrix of

$$\mathsf{N} = \begin{bmatrix} -1.484 & 1.653 \\ 1.752 & -1.217 \\ -1.574 & 1.233 \\ 0.564 & -0.483 \end{bmatrix}.$$

The multi-period squared signal-noise ratio in this case is 0.319.

We note that the expectation, E is the same as in Example 5.3.2, but the variance V, passthrough N, and the signal-noise ratio are different, see also Exercise 5.16. ⊣

5.3.1 † Conditional heteroskedasticity models

Above we considered conditional models where features predict expected returns, but not volatility. However, it is often the case that the volatility of returns is not constant, a condition referred to as *heteroskedasticity.* Moreover the volatility of returns is often easier to forecast than the expected value, as volatility is often clustered. To model this situation, consider a strictly positive scalar random variable, s_{t-1}, observable at the time the investment decision is required to capture $\boldsymbol{x}_t$. For reasons to be obvious later, it is more convenient to think of s_{t-1} as a "quietude" indicator, or a "weight" for a weighted regression.

Two of the simple models for conditional heteroskedasticity introduced in Section 1.4.2 are

$$\text{(constant):}\quad \operatorname{E}\left[\boldsymbol{x}_t \,|\, s_{t-1}\right] = s_{t-1}^{-1}\boldsymbol{\mu}, \qquad \operatorname{Var}\left(\boldsymbol{x}_t \,|\, s_{t-1}\right) = s_{t-1}^{-2}\Sigma, \tag{5.30}$$

$$\text{(floating):}\quad \operatorname{E}\left[\boldsymbol{x}_t \,|\, s_{t-1}\right] = \boldsymbol{\mu}, \qquad \operatorname{Var}\left(\boldsymbol{x}_t \,|\, s_{t-1}\right) = s_{t-1}^{-2}\Sigma. \tag{5.31}$$

To maximize the single-period signal-noise ratio under either of these models, one should hold

$$\boldsymbol{\nu}_* = c\Sigma^{-1}\boldsymbol{\mu},$$

where c is chosen to satisfy the risk budget. Assuming $r_0 = 0$, under the model in Equation 5.30, the optimal single-period signal-noise ratio is $\sqrt{\boldsymbol{\mu}^{\top}\Sigma^{-1}\boldsymbol{\mu}}$, independent of s_{t-1}; under the model of Equation 5.31, it is $s_{t-1}\sqrt{\boldsymbol{\mu}^{\top}\Sigma^{-1}\boldsymbol{\mu}}$. Thus the model names reflect whether or not the maximal signal-noise ratio varies conditional on s_{t-1}.

Now consider maximizing the multi-period signal-noise ratio. Under the constant model of Equation 5.30, one can use the flattening trick. The returns on the pseudo-assets $s_{t-1}\boldsymbol{x}_t$ have expected value $\boldsymbol{\mu}$ and covariance Σ, and thus the optimal portfolio is $cs_{t-1}\Sigma^{-1}\boldsymbol{\mu}$ with multi-period signal-noise ratio $\sqrt{\boldsymbol{\mu}^{\top}\Sigma^{-1}\boldsymbol{\mu}}$. Under the floating model of Equation 5.31, the portfolio that maximizes the multi-period signal-noise ratio is

$$c\frac{\Sigma^{-1}\boldsymbol{\mu}}{{s_{t-1}}^{-2} + \boldsymbol{\mu}^{\top}\Sigma^{-1}\boldsymbol{\mu}},$$

as given in general form in Equation 5.47.

Again we note that assuming optimal signal-noise ratio is either fixed or floating with respect to volatility may be unwarranted. They are both generalized by the mixed model containing both characteristics:

$$\text{(mixed):}\quad \mathrm{E}\left[\boldsymbol{x}_t \,|\, s_{t-1}\right] = s_{t-1}^{-1}\boldsymbol{\mu}_0 + \boldsymbol{\mu}_1,\quad \mathrm{Var}\left(\boldsymbol{x}_t \,|\, s_{t-1}\right) = s_{t-1}^{-2}\Sigma. \tag{5.32}$$

Conditional Expectation and Heteroskedasticity

In some situations one may have forecasts of both volatility and expected return. That is, suppose you observe random variables $s_{t-1} > 0$, and f-vector $\boldsymbol{f}_{t-1}$ at some time prior to when the investment decision is required to capture $\boldsymbol{x}_t$. The variables s and $\boldsymbol{f}$ need not be independent. The general model is

$$\text{(bi-conditional):}\ \mathrm{E}\left[\boldsymbol{x}_t \,\middle|\, s_{t-1}, \boldsymbol{f}_{t-1}\right] = \mathsf{B}\boldsymbol{f}_{t-1},\quad \mathrm{Var}\left(\boldsymbol{x}_t \,\middle|\, s_{t-1}, \boldsymbol{f}_{t-1}\right) = s_{t-1}^{-2}\Sigma, \tag{5.33}$$

where B is a $k \times f$ matrix. Without the s_{t-1} term, these are the "predictive regression" equations commonly used in Tactical Asset Allocation. [Connor, 1997; Herold and Maurer, 2004; Brandt, 2009] By letting $\boldsymbol{f}_{t-1} = \left[{s_{t-1}}^{-1}, 1\right]^{\top}$ we recover the mixed model in Equation 5.32.

The single-period signal-noise ratio maximization problem is expressed as

$$\max_{\boldsymbol{\nu}:\, {s_{t-1}}^{-2}\boldsymbol{\nu}^{\top}\Sigma\boldsymbol{\nu} \le R^2} \frac{\boldsymbol{\nu}^{\top}\mathsf{B}\boldsymbol{f}_{t-1} - r_0}{{s_{t-1}}^{-1}\sqrt{\boldsymbol{\nu}^{\top}\Sigma\boldsymbol{\nu}}}. \tag{5.34}$$

For $r_0 \ge 0, R > 0$ this problem is solved for

$$\boldsymbol{\nu}_* = \frac{s_{t-1}R}{\sqrt{{\boldsymbol{f}_{t-1}}^{\top}\mathsf{B}^{\top}\Sigma^{-1}\mathsf{B}\boldsymbol{f}_{t-1}}}\Sigma^{-1}\mathsf{B}\boldsymbol{f}_{t-1}. \tag{5.35}$$

As above, conditional on observing $\boldsymbol{f}_{t-1}$ and s_{t-1}, the squared maximal signal-noise ratio is

$$\zeta_*^2 \,|\, s_{t-1}, \boldsymbol{f}_{t-1} = \left(\frac{\mathrm{E}\left[{\boldsymbol{\nu}_*}^{\top}\boldsymbol{x}_t \,\middle|\, s_{t-1}, \boldsymbol{f}_{t-1}\right]}{\sqrt{\mathrm{Var}\left({\boldsymbol{\nu}_*}^{\top}\boldsymbol{x}_t \,\middle|\, s_{t-1}, \boldsymbol{f}_{t-1}\right)}}\right)^2 = {\boldsymbol{f}_{t-1}}^{\top}\mathsf{B}^{\top}\Sigma^{-1}\mathsf{B}\boldsymbol{f}_{t-1}.$$

One would then likely consider the expected value of this ζ_*^2 to arrive at the Hotelling-Lawley trace. The multi-period signal-noise ratio maximization problem is solved by flattening, as described above; the non-linear optimal strategy is given in Section 5.3.2.

5.3.2 Optimal conditional portfolios

Above we considered a linear passthrough for the linear conditional expectation model. Here we maximize the multi-period signal-noise ratio under a *nonparametric model.* Here "nonparametric" means we will assume the moments of returns follow some arbitrary function of the observed, and seek to find a functional form for the optimal portfolio. Thus, suppose you observe some features $\boldsymbol{f}_{t-1}$, conditional on which the first and second moments can be expressed as functions:

$$\begin{aligned} \mathrm{E}\left[\boldsymbol{x}_t \,\middle|\, \boldsymbol{f}_{t-1}\right] &= \boldsymbol{\mu}\left(\boldsymbol{f}_{t-1}\right), \\ \mathrm{E}\left[\boldsymbol{x}_t \boldsymbol{x}_t^{\top} \,\middle|\, \boldsymbol{f}_{t-1}\right] &= \mathsf{A}_2\left(\boldsymbol{f}_{t-1}\right). \end{aligned} \tag{5.36}$$

Furthermore write the density of the random variable $\boldsymbol{f}_{t-1}$ as $g\left(\boldsymbol{f}\right)$. Upon observing $\boldsymbol{f}_{t-1}$, you allocate $\boldsymbol{\nu}\left(\boldsymbol{f}_{t-1}\right)$ proportion of your wealth into the assets. The unconditional first and second moment of returns of this strategy are

$$\begin{aligned} \alpha_1 &= \int \left(\boldsymbol{\nu}\left(\boldsymbol{x}\right)\right)^{\top} \boldsymbol{\mu}\left(\boldsymbol{x}\right) g\left(\boldsymbol{x}\right) \,\mathrm{d}\boldsymbol{x}, \\ \alpha_2 &= \int \left(\boldsymbol{\nu}\left(\boldsymbol{x}\right)\right)^{\top} \mathsf{A}_2\left(\boldsymbol{x}\right) \boldsymbol{\nu}\left(\boldsymbol{x}\right) g\left(\boldsymbol{x}\right) \,\mathrm{d}\boldsymbol{x}. \end{aligned} \tag{5.37}$$

With these definitions we seek to solve the optimization problem:

$$\max_{\boldsymbol{\nu}(\boldsymbol{f}):\alpha_2-\alpha_1^2\le R^2} \frac{\alpha_1 - r_0}{\sqrt{\alpha_2 - \alpha_1^2}}, \tag{5.38}$$

for $R > 0$. Here we have imposed a risk-free rate and a risk budget constraint as in Section 5.2.1.

Because maximizing the signal-noise ratio is the same as maximizing the ratio of first to second moment of returns (*cf.* the "TAS function" of Equation 1.24), it suffices to maximize the Hansen ratio $\alpha_1/\sqrt{\alpha_2}$. [Černý, 2020] Again this problem is homogeneous of order zero: we can "lever up" any functional solution to arrive at another solution, so without loss of generality we can constrain the second moment to equal some arbitrary value, say 1. We then have a *isoperimetric problem* of the form [MacCluer, 2005]

$$\max \int L\left(\boldsymbol{x}, \vec{y}\left(\boldsymbol{x}\right)\right) \mathrm{d}\boldsymbol{x}, \quad \text{subject to} \quad \int M\left(\boldsymbol{x}, \vec{y}\left(\boldsymbol{x}\right)\right) \mathrm{d}\boldsymbol{x} = 1. \tag{5.39}$$

Here the function $\vec{y}\left(\boldsymbol{x}\right)$ is the allocation to be solved, the integral of $L\left(\cdot\right)$ gives the expected value of the allocation, and the integral of $M\left(\cdot\right)$ is the squared second moment. The canonical form of this problem has a dependance on the derivative of $\vec{y}$ with respect to $\boldsymbol{x}$ that our problem lacks, making the analysis much simpler. The necessary condition for a solution reduces to the existence of a constant λ such that

$$\frac{\mathrm{d}(L + \lambda M)}{\mathrm{d}\vec{y}} = \mathbf{0}.$$

This leads to the solution

$$\boldsymbol{\nu}_* \left(\boldsymbol{f}_{t-1}\right) = c\mathsf{A}_2^{-1} \left(\boldsymbol{f}_{t-1}\right) \boldsymbol{\mu} \left(\boldsymbol{f}_{t-1}\right), \tag{5.40}$$

for some c. When the risk-free rate is positive there is a unique c that maximizes the signal-noise ratio,

$$c = \frac{R}{\sqrt{q - q^2}}, \quad \text{for } q = \int \left(\boldsymbol{\mu}\left(\boldsymbol{x}\right)\right)^\top \mathsf{A}_2^{-1} \left(\boldsymbol{x}\right) \boldsymbol{\mu}\left(\boldsymbol{x}\right) g\left(\boldsymbol{x}\right) \,\mathrm{d}\boldsymbol{x}. \tag{5.41}$$

The signal-noise ratio of this allocation is equal to

$$\zeta_* = \operatorname{sign}\left(c\right) \sqrt{\frac{q}{1 - q}} - \frac{r_0}{R},$$

where the q is as in Equation 5.41.

It is helpful to think of the moments of the strategy $\boldsymbol{\nu}$ in terms of an *inner product.* For given portfolio-valued functions of $\boldsymbol{f}_t$, $\boldsymbol{a}\left(\boldsymbol{f}\right)$ and $\boldsymbol{b}\left(\boldsymbol{f}\right)$, define their inner product by

$$\left\langle \boldsymbol{a}, \boldsymbol{b} \right\rangle := \int \boldsymbol{a}^\top \left(\boldsymbol{x}\right) \boldsymbol{b}\left(\boldsymbol{x}\right) g\left(\boldsymbol{x}\right) \,\mathrm{d}\boldsymbol{x}. \tag{5.42}$$

To make this a proper inner product, we take functions modulo equality under the norm implied by this inner product. We can rewrite the optimization problem as

$$\max_{\boldsymbol{\nu}: \left\langle \boldsymbol{\nu}, \mathsf{A}_2 \boldsymbol{\nu} \right\rangle - \left\langle \boldsymbol{\nu}, \boldsymbol{\mu} \right\rangle^2 \le R^2} \frac{\left\langle \boldsymbol{\nu}, \boldsymbol{\mu} \right\rangle - r_0}{\sqrt{\left\langle \boldsymbol{\nu}, \mathsf{A}_2 \boldsymbol{\nu} \right\rangle - \left\langle \boldsymbol{\nu}, \boldsymbol{\mu} \right\rangle^2}} \tag{5.43}$$

We note that the q factor can be expressed as the inner product $q = \left\langle \boldsymbol{\mu}, \mathsf{A}_2^{-1} \boldsymbol{\mu} \right\rangle$.

Example 5.3.4 (Correlation profiles). Consider the case of scalar returns and feature, where the feature is zero mean, unit variance, and the correlation between returns and feature is ρ. Thus

$$\begin{aligned} \mathrm{E}\left[x_t \,|\, f_{t-1}\right] &= \mu + \rho\sigma f_{t-1}, \\ \mathrm{E}\left[x_t^2 \,|\, f_{t-1}\right] &= \left(\mu + \rho\sigma f_{t-1}\right)^2 + \left(1 - \rho^2\right) \sigma^2, \end{aligned}$$

where μ and σ^2 are the unconditional mean and variance of returns. The optimal allocation in response to observing f_{t-1} is then

$$c \frac{\mu + \rho\sigma f_{t-1}}{\left(\mu + \rho\sigma f_{t-1}\right)^2 + \left(1 - \rho^2\right) \sigma^2}.$$

To find the signal-noise ratio of the optimal allocation, compute

$$q = \int \frac{(\mu + \rho\sigma x)^2}{(\mu + \rho\sigma x)^2 + (1 - \rho^2)\sigma^2} g(x)\, \mathrm{d}x,$$

then $\zeta_* = \sqrt{q/(1-q)}$.

If we furthermore assume that f_{t-1} is normally distributed, μ is zero, and ρ is small, then

$$q \approx \frac{\rho^2}{1 - \rho^2},$$

and so $\zeta_* \approx |\rho| / \sqrt{1 - 2\rho^2} \approx |\rho|$. This justifies the rule of thumb that the signal-noise ratio is approximately equal to the correlation of feature and returns. ⊣

Example 5.3.5 (Market timing, conditional exponential heteroskedasticity). Consider the case of scalar returns and feature, where returns follow the conditional heteroskedasticity model

$$\begin{aligned} \mathrm{E}\left[x_t \,|\, f_{t-1}\right] &= \mu, \\ \mathrm{Var}\left(x_t \,|\, f_{t-1}\right) &= f_{t-1}\sigma^2. \end{aligned} \tag{5.44}$$

Suppose that f_{t-1} follows an exponential distribution with rate $\lambda = 1$. The exponential distribution has density function $\lambda e^{-\lambda x}$ for non-negative x. The unconditional signal-noise ratio of the *buy-and-hold* strategy (*i.e.*, always allocate 100% of your wealth long in the asset) is simply μ/σ as the unconditional variance is σ^2. (*cf.* Exercise 5.19.)

The optimal leverage is $w_*(f_{t-1}) = c/f_{t-1}$. The q factor is computed as

$$q = \zeta^2 e^{\zeta^2} \int_{\zeta^2}^{\infty} \frac{e^{-x}}{x}\, \mathrm{d}x,$$

where $\zeta = \mu/\sigma$. This is known as the *exponential integral.* For example, if $\zeta = 0.1$ then we compute $q \approx 0.041$ and the signal-noise ratio of the optimal strategy is around 0.206. See also Exercise 5.25 and Exercise 5.26. ⊣

Example 5.3.6 (Discrete state market timing). Consider the case of scalar returns and feature, where the feature takes one of a finite number of values or "states," $\mathcal{S}_1, \mathcal{S}_2, \ldots, \mathcal{S}_J$. Note that this subsumes the case where one observes a vector of discrete features, since their Cartesian product could be converted into a single discrete feature[3].

We let π_j be the probability the feature takes the jth state:

$$\pi_j = \Pr\left\{f_{t-1} = \mathcal{S}_j\right\}.$$

[3]There is a kind of middle ground, where imposes some kind of structure on the expected returns and volatility based on the discrete features. For example, one might suppose that returns depend only on your first discrete feature, but volatility depends only on the second. This considerably complicates the exposition, and we pursue this possibility no further.

We assume that probability is independent of t.

Let μ_j and σ_j^2 be the mean and variance of x_t conditional on observing the jth state:

$$\mu_j = \mathrm{E}\left[x_t \,|\, f_{t-1} = \mathcal{S}_j\right],$$
$$\sigma_j^2 = \mathrm{Var}\left(x_t \,|\, f_{t-1} = \mathcal{S}_j\right).$$

Then the optimal allocation conditional on observing $f_{t-1} = \mathcal{S}_j$ is

$$\boldsymbol{\nu}_*\left(\mathcal{S}_j\right) = \frac{c\left|\mu_j\right|}{\sigma_j^2 + {\mu_j}^2}. \tag{5.45}$$

The q factor is computed as

$$q = \sum_{j \le J} \pi_j \frac{{\mu_j}^2}{\sigma_j^2 + {\mu_j}^2}.$$

⊣

Conditional expectation model revisited

Now we return to the conditional expectation model of Equation 5.26, but without *a priori* specifying a linear passthrough strategy. Using the nonparametric result, the portfolio that maximizes the multi-period signal-noise ratio is the portfolio

$$\boldsymbol{\nu} \,|\, \boldsymbol{f}_{t-1} = c\left(\Sigma + \left(\mathsf{B}\boldsymbol{f}_{t-1}\right)\left(\mathsf{B}\boldsymbol{f}_{t-1}\right)^\top\right)^{-1} \mathsf{B}\boldsymbol{f}_{t-1} = \frac{c\Sigma^{-1}\mathsf{B}\boldsymbol{f}_{t-1}}{1 + \left(\mathsf{B}\boldsymbol{f}_{t-1}\right)^\top \Sigma^{-1}\left(\mathsf{B}\boldsymbol{f}_{t-1}\right)}. \tag{5.46}$$

The latter equality follows from the Sherman-Morrison-Woodbury identity, *cf.* Exercise 5.22. While this portfolio is proportional to the linear passthrough portfolio $\Sigma\mathsf{B}\boldsymbol{f}_{t-1}$ conditional on $\boldsymbol{f}_{t-1}$, it cannot be expressed as a linear passthrough of $\boldsymbol{f}_{t-1}$ because of the quadratic form in the denominator, which acts to slightly downscale the portfolio when $\left\|\boldsymbol{f}_{t-1}\right\|_2$ is large. We note however that in practice one is likely to see small values of $\left(\mathsf{B}\boldsymbol{f}_{t-1}\right)^\top \Sigma^{-1}\left(\mathsf{B}\boldsymbol{f}_{t-1}\right)$, which is the squared signal-noise ratio conditional on $\boldsymbol{f}_{t-1}$, and thus the strategy implied here is *almost* the same as the linear passthrough strategy.

To compute the signal-noise ratio of this strategy compute

$$q = \int \left(\mathsf{B}\boldsymbol{x}\right)^\top \left(\Sigma + \left(\mathsf{B}\boldsymbol{x}\right)\left(\mathsf{B}\boldsymbol{x}\right)^\top\right)^{-1} \left(\mathsf{B}\boldsymbol{x}\right) g\left(\boldsymbol{x}\right) \mathrm{d}\boldsymbol{x},$$
$$= \mathrm{E}_{\boldsymbol{f}}\left[\mathrm{tr}\left(\left(\Sigma + \left(\mathsf{B}\boldsymbol{f}\right)\left(\mathsf{B}\boldsymbol{f}\right)^\top\right)^{-1} \left(\mathsf{B}\boldsymbol{f}\right)\left(\mathsf{B}\boldsymbol{f}\right)^\top\right)\right],$$

then compute $\zeta_* = \sqrt{q/\left(1-q\right)}$. In a sense, q looks like the expected Pillai-Bartlett trace which is another measure of total effect size like the Hotelling Lawley trace. We will consider the Pillai-Bartlett trace in greater detail in Section 6.1.3.

Conditional expectation and heteroskedasticity revisited

Consider now the model of Equation 5.33, which has conditional expectation and heteroskedasticity in the variables $\boldsymbol{f}_{t-1}$ and s_{t-1}. As above the multi-period signal-noise ratio is maximized by holding the portfolio

$$\boldsymbol{\nu} \,\middle|\, s_{t-1}, \boldsymbol{f}_{t-1} = c \frac{\Sigma^{-1} \mathsf{B} \boldsymbol{f}_{t-1}}{{s_{t-1}}^{-2} + \left(\mathsf{B} \boldsymbol{f}_{t-1}\right)^{\top} \Sigma^{-1} \left(\mathsf{B} \boldsymbol{f}_{t-1}\right)}. \tag{5.47}$$

To compute the multi-period signal-noise ratio of this strategy one must compute

$$q = \mathrm{E}_{s,\boldsymbol{f}}\left[\mathrm{tr}\left(\left(s^{-2}\Sigma + (\mathsf{B}\boldsymbol{f})(\mathsf{B}\boldsymbol{f})^{\top}\right)^{-1} (\mathsf{B}\boldsymbol{f})(\mathsf{B}\boldsymbol{f})^{\top}\right)\right],$$

then let $\zeta_* = \sqrt{q/(1-q)}$.

5.3.3 Constrained optimal conditional portfolios

We now seek to impose the same kinds of constraints considered in Section 5.2 onto the nonparametric optimization problem. That is, we consider the problem

$$\max_{\substack{\boldsymbol{\nu}: \langle \boldsymbol{\nu}, \mathsf{A}_2 \boldsymbol{\nu}\rangle - \langle \boldsymbol{\nu}, \boldsymbol{\mu}\rangle^2 \le R^2,\\ \forall i \in I\ \langle \boldsymbol{B}_i, \boldsymbol{\nu}\rangle = b_i.}} \frac{\langle \boldsymbol{\nu}, \boldsymbol{\mu}\rangle - r_0}{\sqrt{\langle \boldsymbol{\nu}, \mathsf{A}_2 \boldsymbol{\nu}\rangle - \langle \boldsymbol{\nu}, \boldsymbol{\mu}\rangle^2}}, \tag{5.48}$$

Where I is some set indexing the allocations $\boldsymbol{B}_i(\boldsymbol{f})$, and the b_i are scalars. We note that this problem encompasses hedging constraints of the form

$$\langle \mathsf{A}_2 \boldsymbol{G}_j, \boldsymbol{\nu}\rangle - \langle \boldsymbol{G}_j, \boldsymbol{\mu}\rangle \langle \boldsymbol{\nu}, \boldsymbol{\mu}\rangle = g_j,$$

since they can be written as

$$g_j = \langle \mathsf{A}_2 \boldsymbol{G}_j, \boldsymbol{\nu}\rangle - \langle \boldsymbol{G}_j, \boldsymbol{\mu}\rangle \langle \boldsymbol{\nu}, \boldsymbol{\mu}\rangle = \langle \mathsf{A}_2 \boldsymbol{G}_j - \boldsymbol{\mu} \langle \boldsymbol{G}_j, \boldsymbol{\mu}\rangle, \boldsymbol{\nu}\rangle,$$

and thus can be expressed by taking $\boldsymbol{B}_j(\boldsymbol{f}) = \mathsf{A}_2(\boldsymbol{f}) \boldsymbol{G}_j(\boldsymbol{f}) - \langle \boldsymbol{G}_j, \boldsymbol{\mu}\rangle \boldsymbol{\mu}(\boldsymbol{f})$ as a constraint.

The solution to Equation 5.48, again via the calculus of variations, is

$$\boldsymbol{\nu}_*(\boldsymbol{f}) = c\left(\mathsf{A}_2^{-1}(\boldsymbol{f}) \boldsymbol{\mu}(\boldsymbol{f}) + \sum_{i \in I} \lambda_i \mathsf{A}_2^{-1}(\boldsymbol{f}) \boldsymbol{B}_i(\boldsymbol{f})\right), \tag{5.49}$$

for some constants c and λ_i. The constant c is chosen to satisfy the risk budget, while the λ_i are fixed by the hedging constraints. That is, the solution must satisfy the equations

$$\langle \boldsymbol{B}_j, \mathsf{A}_2^{-1} \boldsymbol{\mu}\rangle + \sum_{i \in I} \lambda_i \langle \boldsymbol{B}_j, \mathsf{A}_2^{-1} \boldsymbol{B}_i\rangle = \frac{b_j}{c},\ \forall j \in I. \tag{5.50}$$

This uniquely determines the λ_i once c is known. If the b_j are all zero, or in the case where the risk budget is arbitrary, one can set the $c = 1$.

Hedged against buy-and-hold

A common realization of the constraints is a single constraint of zero correlation against the buy-and-hold allocation. Again, the buy-and-hold allocation is the unconditional allocation of 100% of wealth long in the asset. We can write the constraint as

$$\left\langle \mathsf{A}_2\mathbf{1}, \boldsymbol{\nu}\right\rangle - \left\langle \mathbf{1}, \boldsymbol{\mu}\right\rangle \left\langle \boldsymbol{\nu}, \boldsymbol{\mu}\right\rangle = 0. \tag{5.51}$$

The optimal allocation is

$$\boldsymbol{\nu}_*\left(\boldsymbol{f}\right) = \lambda_1\mathbf{1} + \left(1 - \lambda_1\left\langle \mathbf{1}, \boldsymbol{\mu}\right\rangle\right)\mathsf{A}_2^{-1}\left(\boldsymbol{f}\right)\boldsymbol{\mu}\left(\boldsymbol{f}\right), \tag{5.52}$$

where λ_1 is solved by

$$\lambda_1 = -\frac{\left\langle \mathbf{1}, \boldsymbol{\mu}\right\rangle \left(1 - \left\langle \boldsymbol{\mu}, \mathsf{A}_2^{-1}\boldsymbol{\mu}\right\rangle\right)}{\left\langle \mathbf{1}, \mathsf{A}_2\mathbf{1}\right\rangle - \left\langle \mathbf{1}, \boldsymbol{\mu}\right\rangle^2 - \left\langle \mathbf{1}, \boldsymbol{\mu}\right\rangle^2\left(1 - \left\langle \boldsymbol{\mu}, \mathsf{A}_2^{-1}\boldsymbol{\mu}\right\rangle\right)}. \tag{5.53}$$

We note that $\left\langle \mathbf{1}, \boldsymbol{\mu}\right\rangle$ is the expected return of the buy-and-hold strategy and $\left\langle \mathbf{1}, \mathsf{A}_2\mathbf{1}\right\rangle - \left\langle \mathbf{1}, \boldsymbol{\mu}\right\rangle^2$ is the variance of those returns. Also $\left\langle \boldsymbol{\mu}, \mathsf{A}_2^{-1}\boldsymbol{\mu}\right\rangle$ is the q for the unconstrained problem.

Example 5.3.7 (Hedged Market timing, conditional exponential heteroskedasticity). We return to problem of Example 5.3.5. First we consider the case where we optimize the signal-noise ratio conditional on

$$\left\langle \mathbf{1}, \boldsymbol{\nu}\right\rangle = 0.$$

That is, the expected value of the allocation is zero. Under this constraint the optimal allocation is the all zeros allocation! This is not surprising given that the expected return is strictly positive. See also Exercise 5.27.

Consider then optimizing the signal-noise ratio under the hedged constraint

$$\left\langle \mathsf{A}_2\mathbf{1}, \boldsymbol{\nu}\right\rangle - \left\langle \mathbf{1}, \boldsymbol{\mu}\right\rangle \left\langle \boldsymbol{\nu}, \boldsymbol{\mu}\right\rangle = 0.$$

In this case, λ_1 takes the form

$$\lambda_1 = -\frac{\mu\left(1 - q\right)}{\sigma^2 - \mu^2\left(1 - q\right)},$$

with $q = \left\langle \boldsymbol{\mu}, \mathsf{A}_2^{-1}\boldsymbol{\mu}\right\rangle$, and so

$$\boldsymbol{\nu}_*\left(\boldsymbol{f}\right) = c\left(\frac{\sigma^2}{\boldsymbol{f}\sigma^2 + \mu^2} - \left(1 - q\right)\right).$$

⊣

Reduced to a parametric problem

One consequence of the Lagrange multiplier solution in Equation 5.49 is that we can simply reduce the nonparametric problem of Equation 5.48 to a parametric problem in the variables c and λ_i. Moreover, the resultant parametric problem looks just like the vanilla portfolio problem. Changing the notation slightly we rewrite Equation 5.49 as

$$\boldsymbol{\nu}_* (\boldsymbol{f}) = \lambda_0 \mathsf{A}_2^{-1} (\boldsymbol{f}) \boldsymbol{\mu} (\boldsymbol{f}) + \sum_{i \le |I|} \lambda_i \mathsf{A}_2^{-1} (\boldsymbol{f}) \boldsymbol{B}_i (\boldsymbol{f}). \tag{5.54}$$

To simplify the analysis, express $\boldsymbol{\mu}$ as $\boldsymbol{B}_0$. Then note that

$$\begin{aligned}
\langle \boldsymbol{\nu}_*, \boldsymbol{\mu} \rangle &= \sum_{0 \le i \le |I|} \lambda_i \langle \boldsymbol{\mu}, \mathsf{A}_2^{-1} \boldsymbol{B}_i \rangle, \\
\langle \boldsymbol{\nu}_*, \mathsf{A}_2 \boldsymbol{\nu}_* \rangle &= \sum_{0 \le i,j \le |I|} \lambda_i \lambda_j \langle \boldsymbol{B}_j, \mathsf{A}_2^{-1} \boldsymbol{B}_i \rangle.
\end{aligned}$$

So consider the vector $\boldsymbol{m}$ and matrix A defined by

$$\begin{aligned}
\boldsymbol{m}_i &= \langle \boldsymbol{\mu}, \mathsf{A}_2^{-1} \boldsymbol{B}_i \rangle, \\
\mathsf{A}_{i,j} &= \langle \boldsymbol{B}_j, \mathsf{A}_2^{-1} \boldsymbol{B}_i \rangle.
\end{aligned}$$

Furthermore define the matrix $\mathsf{S} = \mathsf{A} - \boldsymbol{m}\boldsymbol{m}^{\top}$. The λ_i then solve the following optimization problem

$$\max_{\substack{\boldsymbol{\lambda}: \ \boldsymbol{\lambda}^{\top} \mathsf{S} \boldsymbol{\lambda} \le R^2, \\ \forall i \in I \ \boldsymbol{e}_i^{\top} \mathsf{A} \boldsymbol{\lambda} = b_i.}} \frac{\boldsymbol{m}^{\top} \boldsymbol{\lambda} - r_0}{\sqrt{\boldsymbol{\lambda}^{\top} \mathsf{S} \boldsymbol{\lambda}}}. \tag{5.55}$$

This is of course simply a parametric portfolio optimization problem with an equality constraint. Note, however, it does not look like Equation 5.18 because of how we defined the subspace constraint there. Let U_{-1} be the "remove first" matrix, whose size should be inferred in context: it is a matrix of all rows but the zeroth of the identity matrix. We can express the equality constraint on $\boldsymbol{\lambda}$ as $\mathsf{U}_{-1} \mathsf{A} \boldsymbol{\lambda} = \boldsymbol{b}$. Then the solution to the problem of Equation 5.55 is

$$\begin{aligned}
\boldsymbol{\lambda}_* = c &\left(\mathsf{I} - \mathsf{S}^{-1} \mathsf{A} \mathsf{U}_{-1}^{\top} \left(\mathsf{U}_{-1} \mathsf{A} \mathsf{S}^{-1} \mathsf{A} \mathsf{U}_{-1}^{\top} \right)^{-1} \mathsf{U}_{-1} \mathsf{A} \right) \mathsf{S}^{-1} \boldsymbol{m} \\
&+ \mathsf{S}^{-1} \mathsf{A} \mathsf{U}_{-1}^{\top} \left(\mathsf{U}_{-1} \mathsf{A} \mathsf{S}^{-1} \mathsf{A} \mathsf{U}_{-1}^{\top} \right)^{-1} \boldsymbol{b}.
\end{aligned} \tag{5.56}$$

Again the constant c serves to hit the risk budget.

We note that when the b_i are not all zeros, it may be impossible to satisfy the risk budget constraint, as the equality constraint may specify a riskier position. This was avoided in Section 5.2.2 and Section 5.2.3 by specifying that all the b_i be zero. This is the most common kind of equality constraint for portfolios, as it specifies zero expected exposure to some factor or zero covariance against another allocation. Assuming $\boldsymbol{b} = \boldsymbol{0}$, the constant takes form

$$c = \frac{R}{\boldsymbol{m}^\top \mathsf{S}^{-1}\boldsymbol{m} - \boldsymbol{m}^\top \mathsf{S}^{-1}\mathsf{A}\mathsf{U}_{-1}^\top\left(\mathsf{U}_{-1}\mathsf{A}\mathsf{S}^{-1}\mathsf{A}\mathsf{U}_{-1}^\top\right)^{-1}\mathsf{U}_{-1}\mathsf{A}\mathsf{S}^{-1}\boldsymbol{m}}.$$

The signal-noise ratio of this allocation is then

$$\zeta_* = \sqrt{\boldsymbol{m}^\top \mathsf{S}^{-1}\boldsymbol{m} - \boldsymbol{m}^\top \mathsf{S}^{-1}\mathsf{A}\mathsf{U}_{-1}^\top\left(\mathsf{U}_{-1}\mathsf{A}\mathsf{S}^{-1}\mathsf{A}\mathsf{U}_{-1}^\top\right)^{-1}\mathsf{U}_{-1}\mathsf{A}\mathsf{S}^{-1}\boldsymbol{m}} - \frac{r_0}{R}. \tag{5.57}$$

By definition, the unconstrained version of this problem is solved by $\boldsymbol{\lambda} = c\boldsymbol{e}_0$ for some positive c, which has signal-noise ratio equal to $\left(\boldsymbol{m}_0/\sqrt{\mathsf{S}_{0,0}}\right) - (r_0/R)$ when $r_0 \geq 0$. That is $\boldsymbol{m}^\top\mathsf{S}^{-1}\boldsymbol{m} = \boldsymbol{m}_0^2/\mathsf{S}_{0,0}$. Thus it seems that in this nonparametric setting there may be a kind of spanning decomposition as in the parametric hedged setting, *cf.* Equation 5.24.

5.4 Exercises

Ex. 5.1 Gradient of signal-noise ratio, other ways Derive the alternative expressions for gradient and Hessian via the Chain Rule of calculus.

1. First show that $\nabla_{\theta}\sigma^2(\theta) = 2\sigma(\theta)\nabla_{\theta}\sigma(\theta)$.
2. Derive the expressions of Equation 5.6 from Equation 5.2.
3. Derive the expressions of Equation 5.7 from Equation 5.3.

Ex. 5.2 Achieved Sharpe ratio with history Suppose you take over management of a fund which has experienced $n-1$ periods of returns, $x_1, \ldots, x_{n-1}$. Let $\hat{\mu}_{n-1}$, $\hat{\sigma}_{n-1}$, and $\hat{\alpha}_{2,n-1}$ be the empirical mean, standard deviation, and empirical (uncentered) second moment:

$$\hat{\mu}_{n-1} = \frac{1}{n-1}\sum_{1\leq t<n} x_t,$$

$$\hat{\sigma}_{n-1} = \sqrt{\frac{1}{n-2}\sum_{1\leq t<n}\left(x_t - \hat{\mu}_{n-1}\right)^2},$$

$$\hat{\alpha}_{2,n-1} = \frac{1}{n-1}\sum_{1\leq t<n} {x_t}^2.$$

Consider the effect of the next return, x_n.

1. Express the empirical mean $\hat{\mu}_n$ as a function of x_n, n, and $\hat{\mu}_{n-1}$.
2. Express the empirical second moment $\hat{\alpha}_{2,n}$ as a function of x_n, n, $\hat{\alpha}_{2,n-1}$ and $\hat{\mu}_{n-1}$.

3. Show that $\hat{\sigma}_n = \sqrt{\frac{n}{n-1}}\sqrt{\hat{\alpha}_{2,n} - \hat{\mu}_n^2}$.
4. Show that $\hat{\sigma}_n = \sqrt{\hat{\alpha}_{2,n-1} + \frac{1}{n}{x_n}^2 - \frac{n-1}{n}\hat{\mu}_{n-1}^2 - \frac{2}{n}\hat{\mu}_{n-1}x_n}$.
5. Compute the derivatives

$$\frac{\mathrm{d}\hat{\mu}_n}{\mathrm{d}x_n} \text{ and } \frac{\mathrm{d}\hat{\sigma}_n}{\mathrm{d}x_n},$$

again in terms of n, $\hat{\mu}_{n-1}$, and $\hat{\alpha}_{2,n-1}$.

6. Show that $\hat{\zeta}_n$, which is defined as $(\hat{\mu}_n - r_0)/\hat{\sigma}_n$, has an optimum precisely when

$$x_n = \frac{\hat{\alpha}_{2,n-1} - r_0\hat{\mu}_{n-1}}{\hat{\mu}_{n-1} - r_0} = \hat{\mu}_{n-1} + \frac{n-2}{n-1}\frac{\hat{\sigma}_{n-1}}{\hat{\zeta}_{n-1}}.$$

This can be expressed as

$$x_n - r_0 = \hat{\sigma}_{n-1}\left(\hat{\zeta}_{n-1} + \frac{n-2}{n-1}\frac{1}{\hat{\zeta}_{n-1}}\right).$$

Show that

$$\frac{x_n - r_0}{\hat{\sigma}_{n-1}} \geq 2\sqrt{\frac{n-2}{n-1}} \approx 2.$$

About how many (historical) standard deviations beyond the risk free rate should you seek to gain in the next period to maximize your achieved Sharpe ratio?

* 7. Supposing x_n satisfies the first-order optimality condition for $\hat{\zeta}_n$. Show that if it is a local *maximum*, *i.e.*, if the second-order optimality conditions hold then it must be the case that $0 \leq \hat{\zeta}_{n-1}$.

* 8. Suppose that $\hat{\zeta}_{n-1} > 0$, and that x_n maximizes the Sharpe ratio, $\hat{\zeta}_n$. Show that the optimal value it takes is

$$\hat{\zeta}_n = \sqrt{\left(\frac{n-1}{n-2}\right)^2 \hat{\zeta}_{n-1}^2 + \frac{1}{n}}.$$

* 9. Suppose that $\hat{\zeta}_{n-1} < 0$. Show that

$$\lim_{x_n\to\infty} \hat{\zeta}_n = \sqrt{\frac{1}{n}}.$$

Ex. 5.3 **Alice relevered** Consider Alice and Bob from Example 5.0.2.

1. What is the volatility of Bob's returns? Of Alice's returns?

2. Suppose that Alice holds L proportion of her wealth long when $f_{t-1} = -1$, and $L/3$ long otherwise. Find the value of L such that Alice's returns have the same volatility as Bob's.
3. When Alice relevers in this way, what is her expected monthly return?

4. Suppose that conditional f_{t-1}, the asset's returns are normally distributed with the given moments. Suppose that Alice relevers to achieve the same volatility as Bob. What is the probability that Alice sees higher returns than Bob in a given month?

Ex. 5.4 Optimality of the Markowitz portfolio Suppose $r_0 = 0$.

1. Defining

$$\zeta(\boldsymbol{\nu}) = \frac{\boldsymbol{\mu}^\top \boldsymbol{\nu}}{\sqrt{\boldsymbol{\nu}^\top \Sigma \boldsymbol{\nu}}},$$

prove that

$$H_{\boldsymbol{\nu}} \zeta(\boldsymbol{\nu})\big|_{\boldsymbol{\nu}=c\Sigma^{-1}\boldsymbol{\mu}} = \frac{-\Sigma}{c^2 \sqrt{\boldsymbol{\mu}^\top \Sigma^{-1} \boldsymbol{\mu}}} + \frac{\boldsymbol{\mu}\boldsymbol{\mu}^\top}{c^2 \left(\boldsymbol{\mu}^\top \Sigma^{-1} \boldsymbol{\mu}\right)^{3/2}}. \tag{5.58}$$

2. Prove that this Hessian is singular, and has null vector $\Sigma^{-1}\boldsymbol{\mu}$.

Ex. 5.5 Optimization under risk budget Prove that Equation 5.16 provides the unique solution to Equation 5.15 when $r_0 > 0, R > 0$ and Σ is positive semi-definite.

Ex. 5.6 Optimization under positivity constraint For $r_0 > 0$ try to solve the portfolio optimization problem

$$\max_{\substack{\boldsymbol{x}:\, \boldsymbol{\nu}=e^{\boldsymbol{x}} \\ \boldsymbol{\nu}^\top \Sigma_1 \boldsymbol{\nu} \le R^2}} \frac{\boldsymbol{\nu}^\top \boldsymbol{\mu} - r_0}{\sqrt{\boldsymbol{\nu}^\top \Sigma \boldsymbol{\nu}}}, \tag{5.59}$$

where $R > 0$ is the risk budget. Is this just a dead end?

Ex. 5.7 Optimization under dollar constraint Consider the portfolio optimization problem

$$\max_{\boldsymbol{\nu}:\mathbf{1}^\top \boldsymbol{\nu}=1} \frac{\boldsymbol{\nu}^\top \boldsymbol{\mu} - r_0}{\sqrt{\boldsymbol{\nu}^\top \Sigma \boldsymbol{\nu}}} \tag{5.60}$$

1. Suppose that $\mathbf{1}^\top \Sigma^{-1} \boldsymbol{\mu} > 0$ and $r_0 = 0$. Show that the problem is solved by

$$\frac{\Sigma^{-1}\boldsymbol{\mu}}{\mathbf{1}^\top \Sigma^{-1} \boldsymbol{\mu}}.$$

What happens if $\mathbf{1}^\top\Sigma^{-1}\boldsymbol{\mu} < 0$ and $r_0 = 0$?

2. Suppose that $\mathbf{1}^\top\Sigma^{-1}\boldsymbol{\mu} = 0$ and $r_0 > 0$. Show that the problem is *not* solved by $c\Sigma^{-1}\boldsymbol{\mu}$ for any c.
3. Write the first order necessary conditions for a solution to this problem. Can you see an easy route to an analytical solution?
4. Simplify to the two asset case, letting the portfolio take value $\left[x, 1-x\right]^\top$. Solve for x for general $\boldsymbol{\mu}$, Σ and r_0.

Ex. 5.8 Softmax portfolio Consider the *softmax function* which takes $\mathbb{R}^n$ to $\mathbb{R}^n$ defined as

$$f\mathrm{smax}\left(x\right)_i = \frac{e^{x_i}}{\sum_j e^{x_j}}. \tag{5.61}$$

For $r_0 > 0$ solve the portfolio optimization problem

$$\max_{\substack{\eta:\ \boldsymbol{\nu}=f\mathrm{smax}(\eta)\\ \boldsymbol{\nu}^\top\Sigma_1\boldsymbol{\nu}\le R^2}} \frac{\boldsymbol{\nu}^\top\boldsymbol{\mu} - r_0}{\sqrt{\boldsymbol{\nu}^\top\Sigma\boldsymbol{\nu}}}, \tag{5.62}$$

where $R > 0$ is the risk budget.

Ex. 5.9 Mean utility optimization Maximizing geometric mean return is often approximated by the following optimization problem:

$$\max_{\theta} \mu\left(\theta\right) - \frac{\lambda}{2}\sigma^2\left(\theta\right).$$

Suppose that θ_* is an optimum for this maximization problem, and also solves the signal-noise ratio optimization of Equation 5.1. How are $\zeta\left(\theta_*\right)$, $\sigma\left(\theta_*\right)$, and λ related?

Ex. 5.10 Optimization under hedging constraint Prove that Equation 5.22 provides the unique solution to Equation 5.21 when $r_0 > 0, R > 0$, Σ is positive semi-definite, and

$$\boldsymbol{\mu}^\top\Sigma^{-1}\boldsymbol{\mu} > \left(\mathsf{G}\boldsymbol{\mu}\right)^\top\left(\mathsf{G}\Sigma\mathsf{G}^\top\right)^{-1}\left(\mathsf{G}\boldsymbol{\mu}\right).$$

1. Under what conditions can $\boldsymbol{\mu}^\top\Sigma^{-1}\boldsymbol{\mu} = \left(\mathsf{G}\boldsymbol{\mu}\right)^\top\left(\mathsf{G}\Sigma\mathsf{G}^\top\right)^{-1}\left(\mathsf{G}\boldsymbol{\mu}\right)$?

Ex. 5.11 Tracking error Let Σ be the covariance of returns.

1. What is the plain-English interpretation of $\boldsymbol{\nu}_1{}^\top\Sigma\boldsymbol{\nu}_2 = 0$?
2. How do you express the "tracking error," defined as

$$\mathrm{E}\left[\left(\boldsymbol{\nu}_1{}^\top\boldsymbol{x} - \boldsymbol{\nu}_2{}^\top\boldsymbol{x}\right)^2\right]?$$

Ex. 5.12 Highly correlated assets Equation 5.14 suggests that the signal-noise ratio goes to ∞ as $|\rho| \rightarrow 1$. How do you interpret this?

Ex. 5.13 Maximum benefit of diversification Suppose that the smallest eigenvalue of the correlation matrix, R, is $\lambda_1 > 0$. Show that

$$\boldsymbol{\zeta}^{\top}\mathsf{R}^{-1}\boldsymbol{\zeta} \le \lambda_1^{-1}\left\|\boldsymbol{\zeta}\right\|_2^2.$$

How do you interpret this?

Ex. 5.14 Diversification benefit, correlated assets Consider k assets with mutual correlation, ρ. Suppose that the returns are re-scaled such that each asset has unit volatility. Then the covariance matrix equals the correlation matrix, which takes value

$$\Sigma = (1-\rho)\,\mathsf{I}_k + \rho\mathbf{1}_k{\mathbf{1}_k}^{\top}.$$

1. Confirm that

$$\Sigma^{-1} = \frac{1}{1-\rho}\left(\mathsf{I}_k - \frac{\rho\mathbf{1}_k{\mathbf{1}_k}^{\top}}{1+(k-1)\,\rho}\right),$$

   either via the Sherman-Morrison-Woodbury identity, or by simple matrix multiplication.
2. Suppose that the signal-noise ratios of the assets are collected in the vector $\boldsymbol{\zeta}$. Then the optimal squared signal-noise ratio is $\zeta_*^2 = \boldsymbol{\zeta}^{\top}\Sigma^{-1}\boldsymbol{\zeta}$. Show that

$$\zeta_*^2 = \frac{(k-1)\,\sigma_{\zeta}^2}{1-\rho} + \frac{k\bar{\zeta}^2}{1+(k-1)\,\rho},$$

   where $\bar{\zeta} := {\mathbf{1}_k}^{\top}\boldsymbol{\zeta}/k$, and

$$\sigma_{\zeta}^2 = \frac{\boldsymbol{\zeta}^{\top}\boldsymbol{\zeta} - k\bar{\zeta}^2}{k-1}.$$

3. Suppose that all assets have identical signal-noise ratio, ζ, and thus $\sigma_{\zeta}^2 = 0$. Show that

$$\lim_{k\rightarrow\infty}\zeta_*^2 = \frac{\zeta^2}{\rho}.$$

   Where did the diversification benefit go?
4. Suppose that the elements of $\boldsymbol{\zeta}$ are in the closed interval $[\zeta_l, \zeta_h)$, with $0 < \zeta_l \le \zeta_h$. Assuming k is even, show that there is some $\boldsymbol{\zeta}$ such that

$$\sigma_{\zeta}^2 = \frac{k}{k-1}\frac{\left(\zeta_h - \zeta_l\right)^2}{4}, \quad \text{and} \quad \bar{\zeta} = \frac{\zeta_l + \zeta_h}{2}.$$

Conclude that some diversification benefit is possible, with optimal signal-noise ratio growing as the square root of the number of assets:

$$\lim_{k\to\infty} \frac{\zeta_*^2}{k} > 0.$$

Ex. 5.15 **A bad example** Imagine a toy universe of 3 assets where

$$\boldsymbol{\mu} = \begin{bmatrix} 0.013 \\ -0.026 \\ 0.020 \end{bmatrix}, \qquad \Sigma = \begin{bmatrix} 0.1600 & 0.0480 & -0.0240 \\ 0.0480 & 0.1600 & 0.1080 \\ -0.0240 & 0.1080 & 0.0900 \end{bmatrix}.$$

1. Show that

$$\boldsymbol{\mu}^\top \Sigma^{-1} \boldsymbol{\mu} \approx -0.434.$$

2. This seems to suggest that ζ_*^2 is negative. How can that happen?
3. Suppose instead that

$$\boldsymbol{\mu} = \begin{bmatrix} 0.002 \\ 0.026 \\ 0.020 \end{bmatrix},$$

with the Σ above. Show now that $\boldsymbol{\mu}^\top \Sigma^{-1} \boldsymbol{\mu} > 0$. What has happened?

Ex. 5.16 **Effects of kurtosis** Repeat the computations of Example 5.3.2, but setting $\kappa = 1$ and $\kappa = 4$. What happens to the multi-period signal-noise ratio in these cases?

1. Repeat the computations of Example 5.3.3, but setting $\kappa = 1$. Confirm that you compute the same variance, passthrough, and signal-noise ratio for $\kappa = 1$ in the "unfettered" setting that you did above in the structured case.

Ex. 5.17 **Breakdown of diversification** Prove that diversification can lead to no improvement in signal-noise ratio:

1. Suppose that the Markowitz portfolio is concentrated on one asset, say $\Sigma^{-1}\boldsymbol{\mu} = \boldsymbol{e}_1$. How is this equivalent to "diversification has not improved the signal-noise ratio?"
2. Show that for any symmetric, positive-definite Σ, and any i, there is a $\boldsymbol{\mu}$ such that $\Sigma^{-1}\boldsymbol{\mu} = \boldsymbol{e}_i$.
3. * † Show that for any $\boldsymbol{\mu}$ not equal to zero, and i, there is a symmetric positive-definite Σ such that $\Sigma^{-1}\boldsymbol{\mu} = \boldsymbol{e}_i$.

Ex. 5.18 **Nonstop Kronecker madness** Derive the variance of Equation 5.29 from Question 8 of Exercise 3.2. It may be helpful to recall the property that for square matrices A and B, [Magnus and Neudecker, 1979]

$$\mathsf{K}(\mathsf{A} \otimes \mathsf{B}) = (\mathsf{B} \otimes \mathsf{A})\,\mathsf{K}.$$

Ex. 5.19 Ignoring conditioning information One can choose to ignore conditioning information in the allocation decision, while recognizing that it exists. This is equivalent to integrating out the conditioning information to arrive at unconditional estimates of the mean and covariance.

1. Denote the conditional mean and covariance as functions

$$\begin{aligned} \mathrm{E}\left[\boldsymbol{x}_t \,\middle|\, \boldsymbol{f}_{t-1}\right] &= \boldsymbol{\mu}\left(\boldsymbol{f}_{t-1}\right), \\ \mathrm{Var}\left(\boldsymbol{x}_t \,\middle|\, \boldsymbol{f}_{t-1}\right) &= \Sigma\left(\boldsymbol{f}_{t-1}\right). \end{aligned} \tag{5.63}$$

   Show that the unconditional mean and covariance are

$$\begin{aligned} \mathrm{E}\left[\boldsymbol{x}_t\right] &= \mathrm{E}_{\boldsymbol{f}}\left[\boldsymbol{\mu}\left(\boldsymbol{f}\right)\right], \\ \mathrm{Var}\left(\boldsymbol{x}_t\right) &= \mathrm{E}_{\boldsymbol{f}}\left[\Sigma\left(\boldsymbol{f}\right) + \boldsymbol{\mu}\left(\boldsymbol{f}\right)\boldsymbol{\mu}^{\top}\left(\boldsymbol{f}\right)\right] - \mathrm{E}_{\boldsymbol{f}}\left[\boldsymbol{\mu}\left(\boldsymbol{f}\right)\right]\mathrm{E}_{\boldsymbol{f}}\left[\boldsymbol{\mu}^{\top}\left(\boldsymbol{f}\right)\right]. \end{aligned}$$

   Supposing that the conditional mean function is constant and equal to $\boldsymbol{\mu}$, show that the covariance reduces to $\mathrm{E}_{\boldsymbol{f}}\left[\Sigma\left(\boldsymbol{f}\right)\right]$.

2. Express the unconditional mean and covariance when

$$\mathrm{E}\left[\boldsymbol{x}_t \,\middle|\, \boldsymbol{f}_{t-1}\right] = \boldsymbol{\mu}, \quad \mathrm{Var}\left(\boldsymbol{x}_t \,\middle|\, \boldsymbol{f}_{t-1}\right) = \left\|\boldsymbol{f}_{t-1}\right\|^2 \Sigma.$$

3. Express the unconditional mean and covariance when

$$\mathrm{E}\left[\boldsymbol{x}_t \,\middle|\, \boldsymbol{f}_{t-1}\right] = \boldsymbol{\mu}, \quad \mathrm{Var}\left(\boldsymbol{x}_t \,\middle|\, \boldsymbol{f}_{t-1}\right) = \Sigma_0 + \left\|\boldsymbol{f}_{t-1}\right\|^2 \Sigma_1.$$

* **Ex. 5.20 Curse of the insane risk manager** Suppose that the mean and covariance of returns are $\boldsymbol{\mu}$, Σ, but you are subject to the whims of an "insane" risk manager who believes that the risk of the returns are some symmetric positive-definite matrix Σ_1.

1. For $r_0 > 0$ solve the portfolio optimization problem

$$\max_{\boldsymbol{\nu}:\boldsymbol{\nu}^{\top}\Sigma_1\boldsymbol{\nu} \le R^2} \frac{\boldsymbol{\nu}^{\top}\boldsymbol{\mu} - r_0}{\sqrt{\boldsymbol{\nu}^{\top}\Sigma\boldsymbol{\nu}}}, \tag{5.64}$$

   where $R > 0$ is the "insane" risk budget.

2. Suppose you have L different insane risk managers each with a different view of covariance, $\Sigma_1, \Sigma_2, \ldots, \Sigma_L$ and a different risk budget $R_1, R_2, \ldots, R_L$. For $r_0 > 0$ solve the portfolio optimization problem

$$\max_{\boldsymbol{\nu}:\boldsymbol{\nu}^{\top}\Sigma_l\boldsymbol{\nu} \le {R_l}^2\,\forall l} \frac{\boldsymbol{\nu}^{\top}\boldsymbol{\mu} - r_0}{\sqrt{\boldsymbol{\nu}^{\top}\Sigma\boldsymbol{\nu}}}. \tag{5.65}$$

Ex. 5.21 Constrained passthroughs Assuming the conditional expectation model of Equation 5.26, create a passthrough matrix that maximizes the single-period signal-noise ratio subject to constraints.

1. Impose a subspace constraint of the form $\mathsf{G}^{C}\mathsf{N} = \mathbf{0}$.
2. Impose a hedging constraint of the form $\mathsf{G}\Sigma\mathsf{N} = \mathbf{0}$. Compute the sample analogue of the Hotelling-Lawley trace under this constraint; recall it is the expected squared single-period signal-noise ratio.

Ex. 5.22 Alternative Markowitz portfolio An alternative expression for the Markowitz portfolio is

$$\boldsymbol{\nu}_{*} = c\left(\Sigma + \boldsymbol{\mu}\boldsymbol{\mu}^{\top}\right)^{-1}\boldsymbol{\mu}$$

1. Using the Sherman-Morrison-Woodbury identity (*cf.* Equation A.3) prove that
$$\left(\Sigma + \boldsymbol{\mu}\boldsymbol{\mu}^{\top}\right)^{-1}\boldsymbol{\mu} = \frac{1}{1 + \boldsymbol{\mu}^{\top}\Sigma^{-1}\boldsymbol{\mu}}\Sigma^{-1}\boldsymbol{\mu}.$$
2. Compare this form of the Markowitz portfolio to the non-parametric optimum of Equation 5.40. Would it be fair to say that this form, with the inverted second moment matrix, is the more fundamental form of the Markowitz portfolio? [Černý, 2020]

Ex. 5.23 Alternative Markowitz portfolio II: projection Show that the portfolio optimization problem is equivalent to a projection problem. That is, show that the problem

$$\max_{\boldsymbol{\nu}:\mathsf{A}\boldsymbol{\nu}=\boldsymbol{b}} \frac{\boldsymbol{\nu}^{\top}\boldsymbol{\mu}}{\sqrt{\boldsymbol{\nu}^{\top}\Sigma\boldsymbol{\nu}}}, \tag{5.66}$$

and the problem

$$\min_{\boldsymbol{\nu}:\mathsf{A}\boldsymbol{\nu}=\boldsymbol{b}} \left(\boldsymbol{\nu} - \Sigma^{-1}\boldsymbol{\mu}\right)^{\top}\Sigma\left(\boldsymbol{\nu} - \Sigma^{-1}\boldsymbol{\mu}\right) \tag{5.67}$$

have the same solution. How do you interpret the latter problem?

Ex. 5.24 Alice is optimal Show that Alice in Example 5.0.2 has mazimized the multi-period signal-noise ratio, using the nonparametric results fom Section 5.3.2.

Ex. 5.25 Exponential heteroskedasticity Confirm the results of Example 5.3.5 empirically: draw 1 million values of exponential f_{t-1} with rate 1, then draw x_t with mean 0.1 and standard deviation $\sqrt{f_{t-1}}$. Allocate $\left(0.01 + f_{t-1}\right)^{-1}$ wealth proportionally to the strategy. Measure the empirical Sharpe ratio of the returns and of the re-levered strategy.

1. Repeat this experiment, but make the leverage look more like the Markowitz portfolio by taking

$$w_*\left(f_{t-1}\right)=\frac{c\mu\left(f_{t-1}\right)}{\sigma^2\left(f_{t-1}\right)}=\frac{c\mu}{f_{t-1}\sigma^2}.$$

Compute the theoretical signal-noise ratio for this leverage function and compare to your empirical estimate.

Ex. 5.26 Exponential heteroskedasticity II Suppose that f_{t-1} is a positive scalar and

$$\begin{aligned}\mathrm{E}\left[\boldsymbol{x}_t \,\middle|\, f_{t-1}\right] &= \boldsymbol{\mu},\\ \mathrm{Var}\left(\boldsymbol{x}_t \,\middle|\, f_{t-1}\right) &= f_{t-1}\Sigma.\end{aligned} \tag{5.68}$$

Consider the conditional portfolio

$$\boldsymbol{\nu} \mid f_{t-1} = \mathcal{P}\left(f_{t-1}\right)\Sigma^{-1}\boldsymbol{\mu},$$

for some polynomial $\mathcal{P}\left(x\right)$.

1. Show that the expected return of this portfolio is

$$\mathrm{E}_f\left[\mathcal{P}\left(f\right)\right]\boldsymbol{\mu}^\top\Sigma^{-1}\boldsymbol{\mu}.$$

2. Show that the variance of returns of this portfolio is

$$\mathrm{E}_f\left[f\mathcal{P}^2\left(f\right)\right]\boldsymbol{\mu}^\top\Sigma^{-1}\boldsymbol{\mu}+\mathrm{E}_f\left[\mathcal{P}^2\left(f\right)\right]\left(\boldsymbol{\mu}^\top\Sigma^{-1}\boldsymbol{\mu}\right)^2-\left(\mathrm{E}_f\left[\mathcal{P}\left(f\right)\right]\boldsymbol{\mu}^\top\Sigma^{-1}\boldsymbol{\mu}\right)^2.$$

3. Letting $\boldsymbol{\beta}$ be the vector of coefficients of the polynomial $\mathcal{P}\left(x\right)$, show that the expected return takes form $\boldsymbol{a}^\top\boldsymbol{\beta}$ and the variance takes the form $\boldsymbol{\beta}^\top\mathsf{M}\boldsymbol{\beta}$ for some vector $\boldsymbol{a}$ and matrix M defined in terms of the expected moments of f and $\boldsymbol{\mu}^\top\Sigma^{-1}\boldsymbol{\mu}$. Show how this leads to a necessary condition for optimality of the $\boldsymbol{\beta}$.
4. Suppose that f_{t-1} follows an exponential distribution with rate parameter λ. Thus in particular

$$\mathrm{E}_f\left[f^k\right]=\frac{k!}{\lambda^k}.$$

Show that the kth element of $\boldsymbol{a}$ (starting from index 0) is

$$a_k=\frac{k!}{\lambda^k}\boldsymbol{\mu}^\top\Sigma^{-1}\boldsymbol{\mu}.$$

Show that the i,jth element of M is

$$\mathsf{M}_{i,j}=\boldsymbol{\mu}^\top\Sigma^{-1}\boldsymbol{\mu}\left(\frac{(i+j+1)!}{\lambda^{i+j+1}}+\boldsymbol{\mu}^\top\Sigma^{-1}\boldsymbol{\mu}\left(\frac{(i+j)!}{\lambda^{i+j}}-\frac{i!}{\lambda^i}\frac{j!}{\lambda^j}\right)\right).$$

5. Suppose that $\lambda = 1$, $\boldsymbol{\mu}^{\top}\Sigma^{-1}\boldsymbol{\mu} = 0.01$. Show that the optimal 3-degree polynomial has coefficients (in increasing order):

$$\boldsymbol{\mu} = \begin{bmatrix} 3.954 & -2.950 & 0.655 & -0.041 \end{bmatrix}.$$

Show that the signal-noise ratio of this portfolio is 0.144.

6. Plot this polynomial. It should seem that for some values of f_{t-1}, we *short* the portfolio $\Sigma^{-1}\boldsymbol{\mu}$. Can the signal-noise ratio of the portfolio be improved by not shorting this portfolio? Also plot the optimal solution given by Example 5.3.5, namely $(x + 0.01)^{-1}$.
7. Repeat this analysis but vary the degree of the polynomial $\mathcal{P}(x)$ to be between 0 and 7. Confirm that the signal-noise ratio grows with polynomial degree. Compare the best signal-noise ratio to the theoretical maximum given in Example 5.3.5.

Ex. 5.27 Exponential heteroskedasticity hedged Prove that the optimal allocation under the constraint $\langle \mathbf{1}, \boldsymbol{\nu} \rangle = 0$ in Example 5.3.7 is the all-zero allocation.

Ex. 5.28 Exponential heteroskedasticity hedged II Confirm the results of Example 5.3.7 empirically: draw 1 million values of exponential f_{t-1} with rate 1, then draw x_t with mean 0.1 and standard deviation $\sqrt{f_{t-1}}$.

1. First test three different allocations: the buy-and-hold, the optimal unhedged allocation $(0.01 + f_{t-1})^{-1}$, and the hedged allocation $q - 1 + (0.01 + f_{t-1})^{-1}$. Here you should compute $q \approx 0.041$ via the exponential integral. Measure the empirical Sharpe ratio of each of these strategies. Confirm that the squared Sharpe ratios are nearly additive: hedged plus buy hold (nearly) equals the optimal.
2. Compute the correlation of the optimal unhedged returns to the buy-hold returns. Compute the correlation of the optimal hedged returns to the buy-hold returns.

Ex. 5.29 Beta heteroskedasticity Consider the market timing model of Equation 5.44, but where $f_{t-1} \sim \mathcal{B}(\alpha, \beta)$.

1. Compute the unconditional signal-noise ratio of the buy-and-hold strategy.
2. Compute the optimal leverage, and then compute the q.

Ex. 5.30 Parimutuel portfolio Consider a *parimutuel betting* situation where, for simplicity, one outcome out of k possible will occur. If vector $\boldsymbol{x}$ represents the dollar amounts the players have wagered, then all players who have wagered on the winning outcome, call it outcome i will receive $(1 - r)\,\mathbf{1}^{\top}\boldsymbol{x}/x_i$ for every dollar they wagered. Here $r > 0$ is an amount taken off the top by the management (and is sometimes rather large). Let $\boldsymbol{\pi}$ be the vector of probabilities of each outcome, summing to one. Consider the value of a portfolio $\boldsymbol{\nu}$, which is a vector of purchased wagers. Moreover, assume that the scale of $\boldsymbol{\nu}$ is so small it does not affect the allocations $\boldsymbol{x}$.

1. What is the expected value of the returns of portfolio $\boldsymbol{\nu}$?
2. What is the volatility of returns of $\boldsymbol{\nu}$?
3. Compute the Markowitz portfolio naïvely, without imposing a positivity constraint. Do you notice something odd?

Ex. 5.31 **Discrete state hedged market timing** Find the strategy that maximizes the signal-noise ratio under the discrete state market timing model of Example 5.3.6 subject to a hedge against the buy-and-hold strategy. That is, solve the λ_1 as in Equation 5.53, then write out the optimal allocation subject to $f_{t-1} = \mathcal{S}_j$.

1. Write out an expression for the signal-noise ratio of your allocation. How does it compare to the unconstrained maximal signal-noise ratio.

Ex. 5.32 **Look, ma! No returns!** Consider the hedged portfolio optimization problem where you hedge your portfolio against the returns of the Markowitz portfolio. What does the constraint simplify to? What does this imply?

Ex. 5.33 **Carole hedges against Bob** Under the model of Example 5.0.2, suppose Carole maximizes signal-noise ratio, conditional on zero correlation against the buy-and-hold strategy (*i.e.*, Bob). What is Carole's allocation and signal-noise ratio?

Ex. 5.34 **Nonparametric alternative formulations** Consider the nonparametric setting where $\boldsymbol{\mu}$ and A_2 are functions of some observable $\boldsymbol{f}$.

1. Solve the mean-variance optimization problem

$$\max_{\boldsymbol{\nu}} \langle \boldsymbol{\nu}, \boldsymbol{\mu} \rangle - \frac{k}{2} \left(\langle \boldsymbol{\nu}, \mathsf{A}_2 \boldsymbol{\nu} \rangle - \langle \boldsymbol{\nu}, \boldsymbol{\mu} \rangle^2 \right). \tag{5.69}$$

2. Solve the mean-variance optimization problem subject to the (expected) full investment constraint

$$\max_{\boldsymbol{\nu}:\langle \boldsymbol{\nu}, \mathbf{1} \rangle = 1} \langle \boldsymbol{\nu}, \boldsymbol{\mu} \rangle - \frac{k}{2} \left(\langle \boldsymbol{\nu}, \mathsf{A}_2 \boldsymbol{\nu} \rangle - \langle \boldsymbol{\nu}, \boldsymbol{\mu} \rangle^2 \right). \tag{5.70}$$

3. Solve the minimum-variance optimization problem

$$\min_{\boldsymbol{\nu}:\langle \boldsymbol{\nu}, \mathbf{1} \rangle = 1} \frac{1}{2} \left(\langle \boldsymbol{\nu}, \mathsf{A}_2 \boldsymbol{\nu} \rangle - \langle \boldsymbol{\nu}, \boldsymbol{\mu} \rangle^2 \right). \tag{5.71}$$

Ex. 5.35 Nonparametric to parametric Another way of turning the non-parametric problem into a parametric one is to restrict the search space to the linear combination of certain *basis functions*. So supposed that $\boldsymbol{H}_1, \ldots, \boldsymbol{H}_m$ are vector-valued functions of the feature, $\boldsymbol{f}$. Turn the problem:

$$\max_{\substack{\boldsymbol{\nu}=\sum_i \lambda_i \boldsymbol{H}_i \\ \langle\boldsymbol{\nu},\mathsf{A}_2\boldsymbol{\nu}\rangle - \langle\boldsymbol{\nu},\boldsymbol{\mu}\rangle^2 \le R^2}} \frac{\langle\boldsymbol{\nu},\boldsymbol{\mu}\rangle - r_0}{\sqrt{\langle\boldsymbol{\nu},\mathsf{A}_2\boldsymbol{\nu}\rangle - \langle\boldsymbol{\nu},\boldsymbol{\mu}\rangle^2}} \tag{5.72}$$

into a parametric problem on the λ_i.

* **Ex. 5.36 Research problem: spanning decomposition** ☨ Prove a spanning decomposition for the non-parametric conditional hedged setting. That is, consider the solution to Equation 5.48 where the constraints are hedging type constraints: $\boldsymbol{B}_j(\boldsymbol{f}) = \mathsf{A}_2(\boldsymbol{f})\,\boldsymbol{G}_j(\boldsymbol{f}) - \langle\boldsymbol{G}_j,\boldsymbol{\mu}\rangle\boldsymbol{\mu}(\boldsymbol{f})$, and the $b_j = 0$. Assume $r_0 = 0$. Show that the maximal squared signal-noise ratio under this constraint plus the maximal squared signal-noise ratio over linear combinations of the $\boldsymbol{G}_j(\boldsymbol{f})$ allocations equals the unconstrained maximal squared signal-noise ratio.

6

Portfolio Inference for Gaussian Returns

> You can't carve up the world.
> It's not a pie.
>
> PATTI SMITH

In Chapter 5, we considered the problem of maximizing the signal-noise ratio in the case where the population parameters are known. Since this kind of clairvoyance is rare, here we consider the case where the population parameters are unknown and must be inferred from data. In this chapter, we will focus on the case of *i.i.d.* Gaussian returns; deviations from this model will be considered in the sequel. For the most part, we will assume that one constructs sample estimates of the population parameters which define a strategy, then employ those estimates (or "plug them in") in place to define a feasible strategy. For example, if the theoretical optimal portfolio for a given situation is the Markowitz portfolio, $\boldsymbol{\Sigma}^{-1}\boldsymbol{\mu}$, then we assume one will deploy $\hat{\boldsymbol{\Sigma}}^{-1}\hat{\boldsymbol{\mu}}$ for sample estimates $\hat{\boldsymbol{\mu}}$ and $\hat{\boldsymbol{\Sigma}}$.

The main kinds of inference we will consider are:

1. Inference on the population maximal signal-noise ratio. Presumably if this is not sufficiently high, one would not be willing to deploy the strategy. We will often refer to this population maximal signal-noise ratio as the "effect size."
2. Similarly, inference on the difference in maximal signal-noise ratio for nested classes of strategies. That is, determining whether the addition of optionality to an existing strategy increases the signal-noise ratio by a sufficiently large amount. For example, does adding a new asset to one's universe increase the signal-noise ratio, or is the new asset "spanned" by the existing universe of assets?
3. Arguably of less use, inference on parameters of the optimal strategy. For example, creating confidence intervals on an individual element of the population Markowitz portfolio. It is less clear what one should do with this information.

DOI: 10.1201/9781003181057-6

6.1 Optimal sample portfolios

Let $\boldsymbol{x}_1, \boldsymbol{x}_2, \ldots, \boldsymbol{x}_n$ be independent identically distributed relative returns drawn from a k-variate Gaussian distribution with population mean $\boldsymbol{\mu}$ and population covariance Σ: $\boldsymbol{x}_t \sim \mathcal{N}\left(\boldsymbol{\mu}, \Sigma\right)$. If one held a portfolio $\hat{\boldsymbol{\nu}}$ on these assets, rebalancing every period, then one would have experienced returns of

$$\hat{\boldsymbol{\nu}}^{\top}\boldsymbol{x}_t \sim \mathcal{N}\left(\hat{\boldsymbol{\nu}}^{\top}\boldsymbol{\mu}, \hat{\boldsymbol{\nu}}^{\top}\Sigma\hat{\boldsymbol{\nu}}\right).$$

Let $\hat{\boldsymbol{\mu}}$ be the usual sample estimate of the mean and $\hat{\Sigma}$ the usual sample estimate of the covariance:

$$\begin{aligned} \hat{\boldsymbol{\mu}} &:= \sum_t \boldsymbol{x}_t/n, \\ \hat{\Sigma} &:= \frac{1}{n-1}\sum_t \left(\boldsymbol{x}_t - \hat{\boldsymbol{\mu}}\right)\left(\boldsymbol{x}_t - \hat{\boldsymbol{\mu}}\right)^{\top}. \end{aligned} \tag{6.1}$$

Again, if you had rebalanced into $\hat{\boldsymbol{\nu}}$, your Sharpe ratio would have been $\hat{\boldsymbol{\nu}}^{\top}\hat{\boldsymbol{\mu}}/\sqrt{\hat{\boldsymbol{\nu}}^{\top}\hat{\Sigma}\hat{\boldsymbol{\nu}}}$. Thus maximization of the Sharpe ratio is expressed by the optimization problem

$$\max_{\hat{\boldsymbol{\nu}}:\hat{\boldsymbol{\nu}}^{\top}\hat{\Sigma}\hat{\boldsymbol{\nu}} \le R^2} \frac{\hat{\boldsymbol{\nu}}^{\top}\hat{\boldsymbol{\mu}} - r_0}{\sqrt{\hat{\boldsymbol{\nu}}^{\top}\hat{\Sigma}\hat{\boldsymbol{\nu}}}}, \tag{6.2}$$

where r_0 is the risk-free rate, and $R > 0$ is a risk "budget." The population analogue of this problem was considered in Chapter 5, and the parallels should be very clear.

This problem is solved by the (sample) Markowitz portfolio

$$\hat{\boldsymbol{\nu}}_* := c\hat{\Sigma}^{-1}\hat{\boldsymbol{\mu}}, \tag{6.3}$$

where the constant c is chosen to maximize return under the given risk budget:

$$c = \frac{R}{\sqrt{\hat{\boldsymbol{\mu}}^{\top}\hat{\Sigma}^{-1}\hat{\boldsymbol{\mu}}}}.$$

The Sharpe ratio of this portfolio is

$$\hat{\zeta}_* := \frac{\hat{\boldsymbol{\nu}}_*{}^{\top}\hat{\boldsymbol{\mu}} - r_0}{\sqrt{\hat{\boldsymbol{\nu}}_*{}^{\top}\hat{\Sigma}\hat{\boldsymbol{\nu}}_*}} = \sqrt{\hat{\boldsymbol{\mu}}^{\top}\hat{\Sigma}^{-1}\hat{\boldsymbol{\mu}}} - \frac{r_0}{R}. \tag{6.4}$$

The population analogue of this quantity is defined as

$$\zeta_* := \sqrt{\boldsymbol{\mu}^{\top}\Sigma^{-1}\boldsymbol{\mu}} - \frac{r_0}{R}. \tag{6.5}$$

Remark (Drag Term). The "drag term," $\frac{r_0}{R}$ is an annoying additive constant that would have to be inserted and removed throughout the results that follow. For instance, rather than consider the distribution of $\hat{\zeta}_*^2$, we would have to discuss the distribution of

$$\left(\hat{\zeta}_* + \frac{r_0}{R}\right)^2.$$

To avoid this pain, we will mostly assume for the rest of the chapter that $r_0 = 0$.

Though the language is clunky, we will refer to $\hat{\zeta}_*^2$ as the "squared maximal Sharpe ratio," and the population analogue, ζ_*^2 as the "squared maximal signal-noise ratio." The quantity $\hat{\zeta}_*^2$ is a summary statistic; it provides some clue as to whether there is a "there there" in the population. If it were the case that $\zeta_*^2 = 0$, then any portfolio on the assets would have zero signal-noise ratio.

Example 6.1.1 (Portfolio on the Fama French Four Factors). We consider the returns of the Market, SMB, HML, and UMD portfolios downloaded from Ken French's website as introduced in Example 1.1.1. Using monthly data from Jan 1927 to Dec 2020, we compute

$$\hat{\boldsymbol{\mu}}^\top = \left[\begin{array}{cccc} Mkt & SMB & HML & UMD \\ \hline 0.949 & 0.208 & 0.322 & 0.648 \end{array}\right] \%\,\text{mo.}^{-1},$$

$$\hat{\Sigma} = \left[\begin{array}{c|cccc} & Mkt & SMB & HML & UMD \\ \hline Mkt & 28.669 & 5.513 & 4.503 & -8.619 \\ SMB & 5.513 & 10.154 & 1.432 & -2.377 \\ HML & 4.503 & 1.432 & 12.305 & -6.823 \\ UMD & -8.619 & -2.377 & -6.823 & 22.138 \end{array}\right] \%^2\,\text{mo.}^{-1}.$$

Based on a risk budget of $R = 1.5\,\%\,\text{mo.}^{-1/2}$, we compute the sample Markowitz portfolio as

$$\hat{\boldsymbol{\nu}}_*{}^\top = \left[\begin{array}{cccc} Mkt & SMB & HML & UMD \\ \hline 0.212 & 0.024 & 0.209 & 0.292 \end{array}\right] \%^{-1}.$$

Assuming a risk-free rate of $r_0 = 0.25\,\%\,\text{mo.}^{-1}$, we compute

$$\hat{\zeta}_* = 0.308\ \text{mo.}^{-1/2} - 0.167\ \text{mo.}^{-1/2} = 0.141\ \text{mo.}^{-1/2}.$$

Ignoring the drag term, this corresponds to $\left(\hat{\zeta}_* + \frac{r_0}{R}\right)^2 = 0.095\ \text{mo.}^{-1}$. ⊣

Example 6.1.2 (Portfolio on the five industries). The Sharpe ratio of the five baskets of industry stocks introduced in Example 1.2.5 were computed on the monthly relative returns from Jan 1927 to Dec 2020. Based on a risk budget of $R = 1.5\,\%\,\text{mo.}^{-1/2}$, we compute the sample Markowitz portfolio as

$$\hat{\boldsymbol{\nu}}_*{}^\top = \left[\begin{array}{ccccc} Consumer & Manufacturing & Technology & Healthcare & Other \\ \hline 0.211 & 0.077 & 0.085 & 0.146 & -0.215 \end{array}\right] \%^{-1}.$$

Note that despite the fact that all five industries have positive sample mean (*cf.* Example 1.2.6), the Markowitz portfolio holds some of the industries short. Setting $r_0 = 0$, we compute $\hat{\zeta}_* = 0.223\ \text{mo.}^{-1/2}$. ⊣

Subspace restriction

Suppose, as in Section 5.2.2, that G is an $k_g \times k$ matrix of rank $k_g \le k$. Let G^C be the matrix whose rows span the null space of the rows of G, *i.e.*, $\mathsf{G}^C\mathsf{G}^\top = 0$. Consider the constrained portfolio optimization problem

$$\max_{\substack{\hat{\boldsymbol{\nu}}:\ \mathsf{G}^C\hat{\boldsymbol{\nu}}=0,\\ \hat{\boldsymbol{\nu}}^\top\hat{\Sigma}\hat{\boldsymbol{\nu}}\le R^2}} \frac{\hat{\boldsymbol{\nu}}^\top\hat{\boldsymbol{\mu}} - r_0}{\sqrt{\hat{\boldsymbol{\nu}}^\top\hat{\Sigma}\hat{\boldsymbol{\nu}}}}. \tag{6.6}$$

This constraint forces feasible portfolios to be in the column space of $\mathsf{G}^\top$. That is, $\hat{\boldsymbol{\nu}} = \mathsf{G}^\top\hat{\boldsymbol{\nu}}_g$, for some unknown vector $\hat{\boldsymbol{\nu}}_g$. Thus this subspace restriction corresponds to the problem of optimizing the portfolio on k_g assets with sample mean $\mathsf{G}\hat{\boldsymbol{\mu}}$ and sample covariance $\mathsf{G}\hat{\Sigma}\mathsf{G}^\top$. This problem has unique solution

$$\hat{\boldsymbol{\nu}}_{*,\mathsf{G}} := \frac{R}{\sqrt{\left(\mathsf{G}\hat{\boldsymbol{\mu}}\right)^\top\left(\mathsf{G}\hat{\Sigma}\mathsf{G}^\top\right)^{-1}\left(\mathsf{G}\hat{\boldsymbol{\mu}}\right)}}\mathsf{G}^\top\left(\mathsf{G}\hat{\Sigma}\mathsf{G}^\top\right)^{-1}\mathsf{G}\hat{\boldsymbol{\mu}} \tag{6.7}$$

when $r_0 > 0$.

The Sharpe ratio of this portfolio is

$$\hat{\zeta}_{*,\mathsf{G}} := \sqrt{\left(\mathsf{G}\hat{\boldsymbol{\mu}}\right)^\top\left(\mathsf{G}\hat{\Sigma}\mathsf{G}^\top\right)^{-1}\left(\mathsf{G}\hat{\boldsymbol{\mu}}\right)} - \frac{r_0}{R}. \tag{6.8}$$

Example 6.1.3 (Portfolio on the five industries, dollar neutrality). Continuing Example 6.1.2, we consider optimizing a portfolio on the five industry baskets, subject to a dollar neutrality constraint. That is, we set

$$\mathsf{G}^C = \left[1.00 \quad 1.00 \quad 1.00 \quad 1.00 \quad 1.00\right].$$

With a risk budget of $R = 2\,\%\,\text{mo.}^{-1/2}$, we compute

$$\hat{\boldsymbol{\nu}}_{*,\mathsf{G}}{}^\top = \begin{bmatrix} \textit{Consum} & \textit{Manufa} & \textit{Techno} & \textit{Health} & \textit{Other} \\ \hline 0.356 & -0.270 & 0.010 & 0.342 & -0.439 \end{bmatrix} \%^{-1}.$$

We note that the elements of $\hat{\boldsymbol{\nu}}_{*,\mathsf{G}}$ effectively sum to zero. Setting $r_0 = 0$, we compute $\hat{\zeta}_{*,\mathsf{G}} = 0.044\ \text{mo.}^{-1/2}$. This is rather smaller than the value we computed in the unconstrained case, $0.223\ \text{mo.}^{-1/2}$. ⊣

Hedging constraint

Suppose, as in Section 5.2.3, that G is an $k_g \times k$ matrix G of rank $k_g \le k$. Consider the problem of portfolio optimization for portfolios "hedged against" the rows of G:

$$\max_{\substack{\hat{\boldsymbol{\nu}}:\ \mathsf{G}\hat{\Sigma}\hat{\boldsymbol{\nu}}=0,\\ \hat{\boldsymbol{\nu}}^\top\hat{\Sigma}\hat{\boldsymbol{\nu}}\le R^2}} \frac{\hat{\boldsymbol{\nu}}^\top\hat{\boldsymbol{\mu}} - r_0}{\sqrt{\hat{\boldsymbol{\nu}}^\top\hat{\Sigma}\hat{\boldsymbol{\nu}}}}. \tag{6.9}$$

This problem is solved by

$$\hat{\boldsymbol{\nu}}_{*,\mathsf{I}\backslash\mathsf{G}} = c\left(\hat{\Sigma}^{-1}\hat{\boldsymbol{\mu}} - \mathsf{G}^{\top}\left(\mathsf{G}\hat{\Sigma}\mathsf{G}^{\top}\right)^{-1}\mathsf{G}\hat{\boldsymbol{\mu}}\right) = c_1\hat{\boldsymbol{\nu}}_{*,\mathsf{I}} - c_2\hat{\boldsymbol{\nu}}_{*,\mathsf{G}}, \quad (6.10)$$

using the notation from Equation 6.7, for some constants c_i which are fixed by the risk budget. Note that, up to scaling, $\hat{\Sigma}^{-1}\hat{\boldsymbol{\mu}}$ is the unconstrained optimal portfolio, and thus the imposition of the G constraint only changes the unconstrained portfolio in assets corresponding to columns of G containing non-zero elements. In the garden variety application where G is a single row of the identity matrix, the imposition of the constraint only changes the holdings in the asset to be hedged (modulo changes in the leading constant to satisfy the risk budget).

The squared Sharpe ratio of the optimal portfolio is

$$\Delta_{\mathsf{I}\backslash\mathsf{G}}\hat{\zeta}_*^2 = \hat{\boldsymbol{\mu}}^{\top}\hat{\Sigma}^{-1}\hat{\boldsymbol{\mu}} - (\mathsf{G}\hat{\boldsymbol{\mu}})^{\top}\left(\mathsf{G}\hat{\Sigma}\mathsf{G}^{\top}\right)^{-1}(\mathsf{G}\hat{\boldsymbol{\mu}}) = \hat{\zeta}_{*,\mathsf{I}}^2 - \hat{\zeta}_{*,\mathsf{G}}^2, \quad (6.11)$$

using the notation from Equation 6.8, and setting $r_0 = 0$.

Example 6.1.4 (Portfolio on the Fama French Four Factors, hedged against the Market). We continue Example 6.1.1 but introduce the constraint that a portfolio must be hedged against the Market. With a risk budget of $R = 1.5\,\%\,\text{mo.}^{-1/2}$, we compute

$$\hat{\boldsymbol{\nu}}_{*,\mathsf{I}\backslash\mathsf{G}}^{\top} = \left[\frac{Mkt \quad SMB \quad HML \quad UMD}{0.062 \quad 0.029 \quad 0.256 \quad 0.357}\right]\%^{-1}.$$

Note that despite the hedging constraint, the portfolio still has nonzero allocation to the Market component. We note that this portfolio is proportional to the unconstrained optimal Markowitz portfolio of Example 6.1.1, except for the Market component.

We compute

$$\Delta_{\mathsf{I}\backslash\mathsf{G}}\hat{\zeta}_* = 0.252\ \text{mo.}^{-1/2}.$$

⊣

Example 6.1.5 (Portfolio on the five industries, hedged against the equal dollar allocation). Continuing Example 6.1.3 we consider optimizing a portfolio on the returns of the baskets of five industry stocks, subject to a hedge against

$$\mathsf{G}^{\top} = \left[\frac{Consumer \quad Manufacturing \quad Technology \quad Healthcare \quad Other}{1.00 \quad 1.00 \quad 1.00 \quad 1.00 \quad 1.00}\right].$$

That is, we hedge against the portfolio of equal investment in all five industries. With a risk budget of $R = 2\,\%\,\text{mo.}^{-1/2}$, we compute

$$\hat{\boldsymbol{\nu}}_{*,\mathsf{I}\backslash\mathsf{G}}^{\top} = \left[\frac{Consum \quad Manufa \quad Techno \quad Health \quad Other}{0.413 \quad 0.071 \quad 0.092 \quad 0.248 \quad -0.674}\right]\%^{-1}.$$

We compute

$$\Delta_{\mathsf{I}\backslash\mathsf{G}}\hat{\zeta}_*^2 = 0.014\ \text{mo.}^{-1} = 0.05\ \text{mo.}^{-1} - 0.036\ \text{mo.}^{-1}.$$

⊣

6.1.1 Optimal sample conditional portfolios

As in Section 5.3, suppose that $\boldsymbol{f}_{t-1}$ is some f-vector of factors observable before the time required to capture returns $\boldsymbol{x}_t$, the corresponding k-vector of asset returns. Recall the conditional expectation model, Equation 5.26:

$$\begin{aligned} \mathrm{E}\left[\boldsymbol{x}_t \,\middle|\, \boldsymbol{f}_{t-1}\right] &= \mathsf{B}\boldsymbol{f}_{t-1}, \\ \mathrm{Var}\left(\boldsymbol{x}_t \,\middle|\, \boldsymbol{f}_{t-1}\right) &= \Sigma. \end{aligned} \tag{6.12}$$

Here B is some $k \times f$ regression coefficient matrix. In this chapter we expand this model to assume that, conditional on $\boldsymbol{f}_{t-1}$, the returns take a normal distribution:

$$\boldsymbol{x}_t \,\middle|\, \boldsymbol{f}_{t-1} \sim \mathcal{N}\left(\mathsf{B}\boldsymbol{f}_{t-1}, \Sigma\right). \tag{6.13}$$

First we describe how to estimate B and Σ under this model. [Rencher, 2002; Reinsel and Velu, 1998] Suppose you collect n paired observations of $\boldsymbol{f}_{t-1}$ and $\boldsymbol{x}_t$. Place these vectors, respectively, into corresponding rows of the $n \times f$ matrix F, and the $n \times k$ matrix X. Under the normal model of returns,

$$\mathsf{X} = \mathsf{F}\mathsf{B}^{\top} + \mathsf{E}, \quad \mathrm{vec}\left(\mathsf{E}\right) \sim \mathcal{N}\left(\mathbf{0}, \Sigma \otimes \mathsf{I}\right).$$

The estimates from multivariate multiple regression are

$$\begin{aligned} \hat{\mathsf{B}}^{\top} &= \left(\mathsf{F}^{\top}\mathsf{F}\right)^{-1}\mathsf{F}^{\top}\mathsf{X}, \\ \hat{\Sigma} &= \left(\hat{\mathsf{E}}^{\top}\hat{\mathsf{E}}\right) / \left(n - f\right), \quad \text{where } \hat{\mathsf{E}} = \mathsf{X} - \mathsf{F}\hat{\mathsf{B}}^{\top}. \end{aligned} \tag{6.14}$$

Under the the assumptions of normality, exogeneity, independence of errors, and conditional on the observed features, the sample estimate is unbiased and normally distributed: [Reinsel and Velu, 1998]

$$\mathrm{vec}\left(\hat{\mathsf{B}}\right) | \mathsf{F} \sim \mathcal{N}\left(\mathsf{B}, \Sigma \otimes \left[\mathsf{F}^{\top}\mathsf{F}\right]^{-1}\right). \tag{6.15}$$

Traditionally in multivariate statistics it is assumed that the factors are deterministic, though in our case they are almost always random (or out of the control of the experimenter). In the sequel we will use the second moment of the factors,

$$\Gamma_f := \mathrm{E}\left[\boldsymbol{f}\boldsymbol{f}^{\top}\right].$$

We estimate this from this sample by

$$\hat{\Gamma}_f = \frac{1}{n} \sum_{0 \le t < n} \boldsymbol{f}_t \boldsymbol{f}_t^{\top}. \tag{6.16}$$

Maximizing single-period Sharpe ratio

Now consider the problem of maximizing the single-period Sharpe ratio, conditional on observing $\boldsymbol{f}_{t-1}$:

$$\max_{\hat{\mathsf{N}}:\left(\hat{\mathsf{N}}\boldsymbol{f}_{t-1}\right)^{\top}\hat{\Sigma}\left(\hat{\mathsf{N}}\boldsymbol{f}_{t-1}\right)\le R^2} \frac{\left(\hat{\mathsf{N}}\boldsymbol{f}_{t-1}\right)^{\top}\hat{\mathsf{B}}\boldsymbol{f}_{t-1} - r_0}{\sqrt{\left(\hat{\mathsf{N}}\boldsymbol{f}_{t-1}\right)^{\top}\hat{\Sigma}\left(\hat{\mathsf{N}}\boldsymbol{f}_{t-1}\right)}}. \tag{6.17}$$

This is merely Equation 6.2 with $\hat{\mathsf{B}}\boldsymbol{f}_{t-1}$ plugged in for $\hat{\boldsymbol{\mu}}$, and assuming a portfolio linear in $\boldsymbol{f}_{t-1}$. This problem has solution

$$\hat{\mathsf{N}}_* := c\hat{\Sigma}^{-1}\hat{\mathsf{B}}, \tag{6.18}$$

where the constant c is chosen to maximize return under the given risk budget:

$$c = \frac{R}{\sqrt{\left(\hat{\mathsf{B}}\boldsymbol{f}_{t-1}\right)^{\top}\hat{\Sigma}^{-1}\left(\hat{\mathsf{B}}\boldsymbol{f}_{t-1}\right)}}.$$

Conditional on $\boldsymbol{f}_{t-1}$, the one-period Sharpe ratio of this portfolio is

$$\hat{\zeta}_* := \frac{\left(\hat{\mathsf{N}}_*\boldsymbol{f}_{t-1}\right)^{\top}\hat{\mathsf{B}}\boldsymbol{f}_{t-1} - r_0}{\sqrt{\left(\hat{\mathsf{N}}_*\boldsymbol{f}_{t-1}\right)^{\top}\hat{\Sigma}\left(\hat{\mathsf{N}}_*\boldsymbol{f}_{t-1}\right)}} = \sqrt{\left(\hat{\mathsf{B}}\boldsymbol{f}_{t-1}\right)^{\top}\hat{\Sigma}^{-1}\left(\hat{\mathsf{B}}\boldsymbol{f}_{t-1}\right)} - \frac{r_0}{R}. \tag{6.19}$$

We can estimate a kind of average (over $\boldsymbol{f}_{t-1}$) of the squared single-period Sharpe ratio of this strategy by computing

$$\frac{1}{n}\sum_{t=1}^{n}\operatorname{tr}\left(\hat{\mathsf{B}}^{\top}\hat{\Sigma}^{-1}\hat{\mathsf{B}}\boldsymbol{f}_{t-1}{\boldsymbol{f}_{t-1}}^{\top}\right) = \operatorname{tr}\left(\hat{\mathsf{B}}^{\top}\hat{\Sigma}^{-1}\hat{\mathsf{B}}\hat{\Gamma}_f\right), \tag{6.20}$$

assuming $r_0 = 0$. This quantity is the Hotelling Lawley trace, a well-known statistic in testing of multivariate multiple linear regression, as we will see below.

Example 6.1.6 (Conditional Portfolio on the Fama French Four Factors). We consider the returns of the Market, SMB, HML, and UMD portfolios introduced in Example 1.1.1. We compute the 12-month running mean of the Mkt returns, as well as the 12-month running standard deviation of Mkt returns, upon which we apply a log transformation. Together with a constant 1 for an intercept term, we use these variables as the features $\boldsymbol{f}_{t-1}$. We delay these by one month, then align with the following month's returns. That is we compute the mean of Mkt returns on the 12-month period ending in a January; we then use these to forecast the returns in March. This gives us one month minus an epsilon to compute the features before making the investment decision, and avoids any "bounce effect." To get a complete sample we restrict our attention to the returns from Apr 1928 to Dec 2020.

Thus we have $n = 1,113$ mo., $k = 4$ and $f = 3$. Using this data we estimate the regression coefficient and covariance as

$$\hat{\mathsf{B}}^{\top} = \left[\begin{array}{c|cccc} & Mkt & SMB & HML & UMD \\ \hline intercept & 0.717 & -0.474 & 0.568 & 0.809 \\ mean & -0.007 & -0.089 & -0.098 & 0.383 \\ log\_sd & 0.155 & 0.544 & -0.104 & -0.374 \end{array}\right] \%\,\text{mo.}^{-1},$$

$$\hat{\Sigma} = \left[\begin{array}{c|cccc} & Mkt & SMB & HML & UMD \\ \hline Mkt & 28.884 & 5.602 & 4.548 & -8.787 \\ SMB & 5.602 & 10.146 & 1.428 & -2.204 \\ HML & 4.548 & 1.428 & 12.370 & -6.827 \\ UMD & -8.787 & -2.204 & -6.827 & 21.924 \end{array}\right] \%^2\,\text{mo.}^{-1}.$$

Compared to the covariance estimated in Example 6.1.1, the covariance here has changed very little, though the intercept term differs from the unconditional mean seen in that previous example. We compute the passthrough matrix as

$$\hat{\mathsf{N}}_*^{\top} = \left[\begin{array}{c|cccc} & Mkt & SMB & HML & UMD \\ \hline intercept & 0.048 & -0.068 & 0.076 & 0.073 \\ mean & 0.007 & -0.009 & 0.001 & 0.020 \\ log\_sd & -0.009 & 0.057 & -0.024 & -0.022 \end{array}\right].$$

We compute the expected squared maximal Sharpe ratio of Equation 6.20 as $0.129\,\text{mo.}^{-1}$. The square root of this quantity is $0.359\,\text{mo.}^{-1/2}$, which is only modestly larger than the $\hat{\zeta}_*^2$ computed in Example 6.1.1 (when you ignore the drag term there). Moreover, as noted in Section 5.3, this expected single-period squared maximal Sharpe ratio can be quite a bit larger than the multi-period variant one computes from the flattening trick. ⊣

Flattening

As introduced in Section 5.3, the flattening trick is a way of turning conditional portfolio problems into unconditional ones by working with a vectorized, or "flattened," product of the returns and feature vectors. That is, use unconditional techniques on the vector $\operatorname{vec}\left(\boldsymbol{x}_t \boldsymbol{f}_{t-1}^{\top}\right) = \boldsymbol{f}_{t-1} \otimes \boldsymbol{x}_t$ to construct the (flattened) optimal passthrough matrix and perform inference. This trick allows one to easily optimize the multi-period Sharpe ratio.

In the spirit of this chapter, one can consider flattening under the assumption that $\operatorname{vec}\left(\boldsymbol{x}_t \boldsymbol{f}_{t-1}^{\top}\right)$ is normally distributed,

$$\boldsymbol{f}_{t-1} \otimes \boldsymbol{x}_t \sim \mathcal{N}\left(\boldsymbol{\xi}, \Omega\right). \tag{6.21}$$

Inference on the multi-period signal-noise ratio is then via the unconditional inference on $\boldsymbol{\xi}$, via Hotelling's test, *etc.*

Practical experience suggests the extra degrees of freedom associated with this approach can lead to poor performance. Alternatively one can impose additional structure on the problem. Staying within the assumption of Gaussian returns we have imposed on this chapter, one can imagine a multivariate normal product:

$$\boldsymbol{x}_t \boldsymbol{f}_{t-1}^{\top} \sim \mathcal{MN}_{k,f}\left(\mathsf{M}, \Psi, \Xi\right). \tag{6.22}$$

The extra structure does not seem to add much, since the matrix normal distribution reduces to

$$\boldsymbol{f}_{t-1} \otimes \boldsymbol{x}_t \sim \mathcal{N}\left(\operatorname{vec}\left(\mathsf{M}\right), \Psi \otimes \Xi\right).$$

Moreover, the estimation of the Ψ and Ξ require an iterative method, and the resultant summary statistic has unknown distribution.

Example 6.1.7 (Conditional Portfolio on the Fama French Four Factors, Flattening). We continue Example 6.1.6, with the same features and returns, but use the flattening trick. Without imposing a structure on the covariance, we compute the passthrough matrix

$$\hat{\mathsf{N}}^{\top} = \left[\begin{array}{c|cccc} & Mkt & SMB & HML & UMD \\ \hline intercept & 0.099 & -0.041 & 0.135 & 0.159 \\ mean & -0.001 & -0.011 & 0.005 & 0.005 \\ log\_sd & -0.035 & 0.035 & -0.057 & -0.058 \end{array}\right].$$

We compute the squared maximal Sharpe ratio as $\hat{\zeta}_*^2 = 0.136\,\text{mo.}^{-1}$, slightly larger than the effect size seen in Example 6.1.6.

We also consider the structured approach with $\boldsymbol{x}_t \boldsymbol{f}_{t-1}^{\top}$ following a matrix normal distribution. We estimate Ψ and Ξ via Dutilleul's iterative method, *cf.* Section 6.4. [Dutilleul, 1999] This yields the passthrough

$$\hat{\mathsf{N}}^{\top} = \left[\begin{array}{c|cccc} & Mkt & SMB & HML & UMD \\ \hline intercept & 0.105 & -0.052 & 0.119 & 0.176 \\ mean & 0.004 & -0.001 & 0.002 & 0.009 \\ log\_sd & -0.038 & 0.035 & -0.050 & -0.068 \end{array}\right],$$

and corresponding estimate $\hat{\zeta}_*^2 = 0.148\,\text{mo.}^{-1}$. ⊣

Hedging out a signal

The flattening trick easily allows one to impose certain constraints to the portfolio problem. Previously we considered maximizing the Sharpe ratio under the constraint of zero (sample) covariance to a space of fixed portfolios. This is simple with the flattening trick. We can also entertain the problem of hedging out certain features, or even interactions between features and returns.

So consider some $f_m \times f$ matrix M, the rows of which span the space of features we wish to hedge out. Under the unfettered flattening model of Equation 6.21, the optimization problem can be expressed as

$$\max_{\hat{\mathsf{N}}:\ \substack{(\mathsf{M}^{\top}\otimes\mathsf{I})\hat{\Omega}\,\mathrm{vec}(\mathsf{D})=\mathbf{0},\\ \mathrm{vec}(\hat{\mathsf{N}})^{\top}\hat{\Omega}\,\mathrm{vec}(\hat{\mathsf{N}})\le R^2}} \frac{\hat{\boldsymbol{\xi}}^{\top}\mathrm{vec}\left(\hat{\mathsf{N}}\right) - r_0}{\sqrt{\mathrm{vec}\left(\hat{\mathsf{N}}\right)^{\top}\hat{\Omega}\,\mathrm{vec}\left(\hat{\mathsf{N}}\right)}}. \tag{6.23}$$

Under Equation 6.22 the problem is quoted with the constrained estimate of $\hat{\Omega}$. This problem is, up to the dreadful bookkeeping of which dimension corresponds to feature or asset, simply a realization of the problem of Equation 6.9, and the analysis considered for hedging constraint applies here under the unfettered flattening model.

Example 6.1.8 (Conditional Portfolio on the Fama French Four Factors, Feature Hedged). We continue Example 6.1.7, with the same features and returns. Here we consider the problem of hedging out the intercept term. That is we take M to be the row of the identity matrix corresponding to the constant "feature." Under the unfettered flattening model, we compute the hedged passthrough matrix as

$$\hat{\mathsf{N}}_{\mathsf{I}\setminus\mathsf{M}}^{\top} = \left[\begin{array}{c|cccc} & Mkt & SMB & HML & UMD \\ \hline intercept & 0.056 & -0.046 & 0.091 & 0.100 \\ mean & -0.001 & -0.011 & 0.005 & 0.005 \\ log\_sd & -0.035 & 0.035 & -0.057 & -0.058 \end{array}\right].$$

Compared to the passthrough computed in Example 6.1.7, only the row corresponding to the intercept term has changed. We compute the hedged squared maximal Sharpe ratio as $\Delta_{\mathsf{I}\setminus\mathsf{M}}\hat{\zeta}_*^2 = 0.043\,\mathrm{mo.}^{-1}$, much smaller than the effect size seen in Example 6.1.7. We note that this value is computed as

$$\Delta_{\mathsf{I}\setminus\mathsf{M}}\hat{\zeta}_*^2 = 0.136\,\mathrm{mo.}^{-1} - 0.092\,\mathrm{mo.}^{-1}.$$

The first number (the minuend) is as we computed in Example 6.1.7 via flattening; the second number (the subtrahend) is almost, but not quite equal to the value for the unconditional portfolio we computed in Example 6.1.1. It does not match exactly because we have discarded some of the rows of returns to allow "burn-in" of the features. ⊣

Conditional expectation and heteroskedasticity models

Now consider the case where you observe features that affect both the expected value and the volatility of returns. That is, suppose you observe the scalar s_{t-1} and f-vector $\boldsymbol{f}_{t-1}$ prior to the time required to capture $\boldsymbol{x}_t$. The variables s and $\boldsymbol{f}$ need not be independent. To forecast volatility, s must be positive. The conditional expectation model of Equation 5.33 can be expressed as

$$\begin{aligned} \mathrm{E}\left[s_{t-1}\boldsymbol{x}_t \,\middle|\, s_{t-1}, \boldsymbol{f}_{t-1}\right] &= \mathsf{B}\boldsymbol{f}_{t-1}, \\ \mathrm{Var}\left(s_{t-1}\boldsymbol{x}_t \,\middle|\, s_{t-1}, \boldsymbol{f}_{t-1}\right) &= \Sigma. \end{aligned} \tag{6.24}$$

As we are assuming normally distributed returns in this chapter, we model this situation as

$$s_{t-1}\boldsymbol{f}_{t-1} \otimes \boldsymbol{x}_t \sim \mathcal{N}(\boldsymbol{\xi}, \Omega). \quad (6.25)$$

One can maximize the single-period signal-noise ratio via the conditional estimation procedure, or maximize the multi-period signal-noise ratio via the flattening trick.

Example 6.1.9 (Conditional Heteroskedasticity Portfolio on the Fama French Four Factors, Flattening). We continue Example 6.1.6, with the same features and returns, however we use only the running 12-month running mean of the Mkt returns to predict expected returns. We take one divided by the 12-month running variance of Mkt returns as s_{t-1}. The s_{t-1} are normalized to have mean one for better interpretation. We delay both features by one month as we did in Example 6.1.6.

Without imposing a structure on the covariance, we compute the passthrough matrix

$$\hat{\mathsf{N}}^{\top} = \begin{bmatrix} & Mkt & SMB & HML & UMD \\ \hline intercept & 0.052 & 0.030 & 0.052 & 0.070 \\ mean & -0.011 & -0.017 & -0.007 & -0.004 \end{bmatrix}.$$

This has to be multiplied by $s_{t-1}\boldsymbol{f}_{t-1}$ at every time to find the portfolio, as this is effectively the new feature. We compute the multi-period squared maximal Sharpe ratio as $\hat{\zeta}_*^2 = 0.099\,\text{mo.}^{-1}$, somewhat smaller than the effect size seen in Example 6.1.6.

We also consider the structured approach with $s_{t-1}\boldsymbol{x}_t{\boldsymbol{f}_{t-1}}^{\top}$ following a matrix normal distribution. We estimate Ψ and Ξ via Dutilleul's iterative method, *cf.* Section 6.4. [Dutilleul, 1999] This yields the passthrough

$$\hat{\mathsf{N}}^{\top} = \begin{bmatrix} & Mkt & SMB & HML & UMD \\ \hline intercept & 0.043 & 0.027 & 0.028 & 0.060 \\ mean & -0.001 & -0.015 & 0.002 & -0.001 \end{bmatrix},$$

and corresponding estimate $\hat{\zeta}_*^2 = 0.097\,\text{mo.}^{-1}$. ⊣

6.1.2 Hotelling's T^2

In Chapter 2 we noted that the Sharpe ratio and the t-statistic were identical up to scaling by $\sqrt{n}$. This fact allowed us to turn known results about the t-statistic into results on the Sharpe ratio. This "natural arb" was quite fruitful because the t-statistic is one of the most widely studied estimators in univariate statistics. We will apply similar translation here in the multivariate setting, converting known results about Hotelling's T^2, the multivariate analogue of Student's t, into results on the squared Sharpe ratio. These analogues are outlined in Table 6.1.

	inference problem	
Response variable	Inference on mean	Inference on regression coefficient
Univariate	t-statistic	t-statistic
Multivariate	Hotelling's T^2	MGLH tests, including Hotelling-Lawley trace

TABLE 6.1
The classical frequentist statistical procedures are indicated for univariate and multivariate responses and for the problem form. Each of these have analogues in quantitative finance: as indicated above, the t-statistic is the Sharpe ratio up to scaling by $\sqrt{n}$; in the sequel we will explore the connections in the multivariate case.

The statistic $n\left(\hat{\boldsymbol{\mu}}^{\top}\hat{\Sigma}^{-1}\hat{\boldsymbol{\mu}}\right) = n\left(\hat{\zeta}_{*} + \frac{r_0}{R}\right)^2$ is Hotelling's T^2, which up to scaling, follows a non-central F distribution:

$$\frac{n}{n-1}\frac{n-k}{k}\left(\hat{\zeta}_{*} + \frac{r_0}{R}\right)^2 \sim F\left(k, n-k, n\left(\zeta_{*} + \frac{r_0}{R}\right)^2\right), \tag{6.26}$$

where $F(v_1, v_2, \delta)$ is the non-central F-distribution with v_1, v_2 degrees of freedom and non-centrality parameter δ. This allows us to make inference about ζ_{*}. By using the "biased" covariance estimate, defined as

$$\tilde{\Sigma} := \frac{n-1}{n}\hat{\Sigma} = \frac{1}{n}\sum_{t}\left(\boldsymbol{x}_t - \hat{\boldsymbol{\mu}}\right)\left(\boldsymbol{x}_t - \hat{\boldsymbol{\mu}}\right)^{\top},$$

the above expression can be simplified slightly as

$$\frac{n-k}{k}\hat{\boldsymbol{\mu}}^{\top}\tilde{\Sigma}^{-1}\hat{\boldsymbol{\mu}} \sim F\left(k, n-k, n\left(\zeta_{*} + \frac{r_0}{R}\right)^2\right).$$

Again, the drag term is visually noisy; if we assume $r_0 = 0$, then we can compactly rewrite Equation 6.26 as

$$\frac{n}{n-1}\frac{n-k}{k}\hat{\zeta}_{*}^2 \sim F\left(k, n-k, n\zeta_{*}^2\right).$$

Caution (Expect a Haircut). We will use the connection to Hotelling's T^2 to perform inference on ζ_{*}^2, which is the *maximal* effect size in the population. If $\zeta_{*}^2 = 0$, then any portfolio on the assets has zero signal-noise ratio. However, if $\zeta_{*}^2 > 0$, the sample Markowitz portfolio can have *negative* signal-noise ratio due to misestimation error. We call this loss in achieved signal-noise ratio due to estimation error the *haircut*, and will consider it in greater detail in Chapter 8. In the meantime, one should keep in mind that a large ζ_{*}^2 might be necessary for a profitable portfolio, *but it may not be sufficient.*

Summary Statistics for Constrained Problems

The summary statistic under the subspace restriction immediately follows from the above, since if $\boldsymbol{x}_{t-1} \sim \mathcal{N}(\boldsymbol{\mu}, \Sigma)$, then $\mathsf{G}\boldsymbol{x}_{t-1} \sim \mathcal{N}(\mathsf{G}\boldsymbol{\mu}, \mathsf{G}\Sigma\mathsf{G}^{\top})$. Then Equation 6.26 becomes

$$\frac{n}{n-1}\frac{n-k_g}{k_g}\left(\hat{\zeta}_{*,\mathsf{G}} + \frac{r_0}{R}\right)^2 \sim F\left(k_g, n-k_g, n\left(\zeta_{*,\mathsf{G}} + \frac{r_0}{R}\right)^2\right). \tag{6.27}$$

The hedging constraint has a less obvious derivation. Rao [1952] showed that the test statistic

$$F_{\mathsf{G}} = \frac{n-k}{k-k_g}\frac{\hat{\zeta}^2_{*,\mathsf{I}} - \hat{\zeta}^2_{*,\mathsf{G}}}{\frac{n-1}{n} + \hat{\zeta}^2_{*,\mathsf{G}}} = \frac{n}{n-1}\frac{n-k}{k-k_g}\frac{\Delta_{\mathsf{I}\setminus\mathsf{G}}\hat{\zeta}^2_*}{1 + \frac{n}{n-1}\hat{\zeta}^2_{*,\mathsf{G}}} \tag{6.28}$$

follows a (central) F distribution under the null hypothesis of $\Delta_{\mathsf{I}\setminus\mathsf{G}}\zeta^2_* = 0$. Giri [1964] showed that, under the alternative, and conditional on observing $\hat{\zeta}^2_{*,\mathsf{G}}$,

$$F_{\mathsf{G}} \,\Big|\, \hat{\zeta}^2_{*,\mathsf{G}} \sim F\left(k-k_g, n-k, \frac{n}{1+\frac{n}{n-1}\hat{\zeta}^2_{*,\mathsf{G}}}\Delta_{\mathsf{I}\setminus\mathsf{G}}\zeta^2_*\right), \tag{6.29}$$

where $F(v_1, v_2, \delta)$ is the non-central F-distribution with v_1, v_2 degrees of freedom and non-centrality parameter δ. Gibbons *et al.* [1989] describe this test for the case where $k_g = 1$. Britten-Jones [1999] proved that for general k_g, this statistic follows a central F-distribution under the null, but does not discuss the distribution under the alternative.

Distribution of portfolio weights

There is a connection between the spanning test (which we view as a hedging constraint) and portfolio weights, namely that when $\Delta_{\mathsf{I}\setminus\mathsf{G}}\zeta^2_* = 0$, then the Markowitz portfolio is a linear combination of the rows of G. In the case where G consists of $k-1$ rows of the identity matrix, all but the jth row, if $\Delta_{\mathsf{I}\setminus\mathsf{G}}\zeta^2_* = 0$, then the jth element of the Markowitz portfolio is zero. Thus the distribution of $\Delta_{\mathsf{I}\setminus\mathsf{G}}\,\hat{\zeta}^2_*$ can be re-expressed as a fact about portfolio weights.

Given vector $\boldsymbol{g}$, Britten-Jones showed that

$$t_g := \frac{\boldsymbol{g}^{\top}\hat{\Sigma}^{-1}\hat{\boldsymbol{\mu}}\sqrt{n-k}}{\sqrt{\left(\frac{n-1}{n} + \hat{\zeta}^2_*\right)\boldsymbol{g}^{\top}\hat{\Sigma}^{-1}\boldsymbol{g} - \left(\boldsymbol{g}^{\top}\hat{\Sigma}^{-1}\hat{\boldsymbol{\mu}}\right)^2}} \tag{6.30}$$

follows a t-distribution with $n-k$ degrees of freedom under the null hypothesis $\boldsymbol{g}^{\top}\Sigma^{-1}\boldsymbol{\mu} = 0$. However, under the alternative hypothesis, the distribution of this statistic is conditional on some non-tested elements of $\hat{\Sigma}^{-1}\hat{\boldsymbol{\mu}}$.

In this form, the test statistic is unpleasant to look at, but it can be simply computed as the t-statistic of an unusual linear regression problem. The regression problem has the constant 1 as the dependent variable, and the vector of returns as the independent variable, that is one regresses $\mathbf{1} \sim \mathsf{X}\boldsymbol{\beta}$, without an intercept term, then performs the usual regression t test on contrasts of the form $\boldsymbol{g}^{\top}\hat{\boldsymbol{\beta}}$. [Britten-Jones, 1999]

While the relationship to regression testing is convenient, it is not the only way to view this statistic. Rather one can write

$$t_{\boldsymbol{g}} = \sqrt{\frac{n}{n-1}} \frac{\boldsymbol{g}^{\top}\hat{\Sigma}^{-1}\hat{\boldsymbol{\mu}}\sqrt{n-k}}{\sqrt{\boldsymbol{g}^{\top}\hat{\Sigma}^{-1}\boldsymbol{g}\left(1 + \frac{n}{n-1}\hat{\zeta}^{2}_{*,\mathsf{G}}\right)}}, \tag{6.31}$$

where G is the $k-1 \times k$ matrix of rank $k-1$ whose rows are orthogonal to $\boldsymbol{g}$. In this form, one can easily see that $t_{\boldsymbol{g}}^{2}$ is equal to the statistic F_{G} from Equation 6.28. Thus it takes the conditional distribution

$$t_{\boldsymbol{g}} \Big| \hat{\zeta}^{2}_{*,\mathsf{G}} \sim t\left(n-k, \sqrt{\frac{n}{1 + \frac{n}{n-1}\hat{\zeta}^{2}_{*,\mathsf{G}}}} \frac{\boldsymbol{g}^{\top}\Sigma^{-1}\boldsymbol{\mu}}{\sqrt{\boldsymbol{g}^{\top}\hat{\Sigma}^{-1}\boldsymbol{g}}}\right), \tag{6.32}$$

Bodnar and Okhrin [2011] describe the unconditional density of the elements of the sample Markowitz portfolio, $\hat{\Sigma}^{-1}\hat{\boldsymbol{\mu}}$. [Kotsiuba and Mazur, 2015] The density is too complicated for practical use and the author knows of no implementation of it. Thankfully, in their Corollary 1, Bodnar and Okhrin give the following stochastic form for a contrast of the sample Markowitz portfolio:

$$\boldsymbol{g}^{\top}\hat{\Sigma}^{-1}\hat{\boldsymbol{\mu}} \sim \frac{n-1}{u_1}\left[\boldsymbol{g}^{\top}\Sigma^{-1}\boldsymbol{\mu} + \frac{u_2}{\sqrt{n}}\sqrt{\left(1 + \frac{k-1}{n-k+1}u_3\right)\boldsymbol{g}^{\top}\Sigma^{-1}\boldsymbol{g}}\right], \tag{6.33}$$

where u_1, u_2, u_3 are independent and $u_1 \sim \chi^2\left(n-k\right)$, $u_2 \sim \mathcal{N}\left(0,1\right)$ and $u_3 \sim F\left(\frac{k-1}{2}, \frac{n-k+1}{2}, n\delta\right)$, where $\delta = \boldsymbol{\mu}^{\top}\Sigma^{-1}\boldsymbol{\mu} - \frac{\left(\boldsymbol{g}^{\top}\Sigma^{-1}\boldsymbol{\mu}\right)^2}{\boldsymbol{g}^{\top}\Sigma^{-1}\boldsymbol{g}}$. Thus to perform unconditional tests on $\boldsymbol{g}^{\top}\hat{\Sigma}^{-1}\hat{\boldsymbol{\mu}}$ one would draw many random variates from the right hand side under the null hypothesis and compare some suitable sample quantile to the observed value. One can also see from this formulation that elements of the sample Markowitz portfolio are unbiased up to scaling: [Kotsiuba and Mazur, 2015]

$$\mathrm{E}\left[\boldsymbol{g}^{\top}\hat{\Sigma}^{-1}\hat{\boldsymbol{\mu}}\right] = \frac{n-1}{n-k-2}\boldsymbol{g}^{\top}\Sigma^{-1}\boldsymbol{\mu}. \tag{6.34}$$

More generally, we have

$$\mathrm{E}\left[\hat{\Sigma}^{-1}\hat{\boldsymbol{\mu}}\right] = \frac{n-1}{n-k-2}\Sigma^{-1}\boldsymbol{\mu}, \tag{6.35}$$

$$\mathrm{Var}\left(\hat{\Sigma}^{-1}\hat{\boldsymbol{\mu}}\right) = c\left(\frac{n-k}{n-k-2}\Sigma^{-1}\boldsymbol{\mu}\boldsymbol{\mu}^{\top}\Sigma + \left(1 + \zeta_*^2 - \frac{2}{n}\right)\Sigma^{-1}\right), \tag{6.36}$$

$$\text{where} \quad c = \frac{\left(n-1\right)^2}{\left(n-k-1\right)\left(n-k-4\right)\left(n-k-2\right)}.$$

Example 6.1.10 (Markowitz portfolio covariance, Fama French Four Factors). We consider the 1,128 months of returns of the Market, SMB, HML, and UMD portfolios introduced in Example 1.1.1, spanning from Jan 1927 to Dec 2020. We estimate the covariance of the Markowitz portfolio on the factor by plugging in $\hat{\boldsymbol{\mu}}$, $\hat{\Sigma}$ and $\hat{\zeta}_*^2$ for $\boldsymbol{\mu}$, Σ and ζ_*^2 in Equation 6.36. As the units are somewhat arbitrary, we convert these into a correlation matrix, which takes the value

	Mkt	*SMB*	*HML*	*UMD*
Mkt	1.000	−0.276	−0.071	0.286
SMB	−0.276	1.000	−0.036	0.039
HML	−0.071	−0.036	1.000	0.378
UMD	0.286	0.039	0.378	1.000

.

⊣

6.1.3 Multivariate general linear hypothesis

Under the conditional expectation model of Section 6.1.1, the summary statistic does not reduce to an F-statistic. Rather we turn to techniques from testing multivariate linear regressions. Testing for significance of the elements of B is via the *Multivariate General Linear Hypothesis* (MGLH). [Muller and Peterson, 1984; Shieh, 2003, 2005; O'Brien and Shieh, 1999; Timm, 2002; Yanagihara, 2001] The MGLH can be posed as

$$H_0 : \mathsf{ABC} = \Theta, \tag{6.37}$$

for $a \times k$ matrix A, $f \times c$ matrix C, and $a \times c$ matrix Θ. We require A and C to have full rank, and $a \leq k$ and $c \leq f$. In the garden-variety application, one tests whether B is all zero by letting A and C be identity matrices, and Θ a matrix of all zeros.

Testing the MGLH proceeds by one of four test statistics, each defined in terms of two matrices, the model variance matrix, $\hat{\mathsf{H}}$, and the error variance matrix, $\hat{\mathsf{E}}$, defined as

$$\hat{\mathsf{H}} := n\left(\mathsf{A}\hat{\mathsf{B}}\mathsf{C} - \Theta\right)\left(\mathsf{C}^\top\hat{\Gamma}_f^{-1}\mathsf{C}\right)^{-1}\left(\mathsf{A}\hat{\mathsf{B}}\mathsf{C} - \Theta\right)^\top, \quad \hat{\mathsf{E}} := n\mathsf{A}^\top\hat{\Sigma}\mathsf{A}, \tag{6.38}$$

where $\hat{\Gamma}_f$, as defined in Equation 6.16 is $\hat{\Gamma}_f = \frac{1}{n}\sum_t \boldsymbol{f}_{t-1}\boldsymbol{f}_{t-1}^\top$. Note that typically in non-finance applications, the regressors are deterministic and controlled by the experimenter (giving rise to the term "design matrix"). In this case, it is assumed that $\hat{\Gamma}_f$ estimates the population analogue, Γ_f, without error, though some work has been done for the case of "random explanatory variables." [Shieh, 2005]

The four test statistics for the MGLH are:

$$\text{Hotelling-Lawley trace:} \quad \hat{T} := \operatorname{tr}\left(\hat{\mathsf{E}}^{-1}\hat{\mathsf{H}}\right), \tag{6.39}$$

$$\text{Pillai-Bartlett trace:} \quad \hat{P} := \operatorname{tr}\left(\left(\hat{\mathsf{E}}+\hat{\mathsf{H}}\right)^{-1}\hat{\mathsf{H}}\right), \tag{6.40}$$

$$\text{Wilk's LRT:} \quad \hat{U} := \left|\left(\hat{\mathsf{E}}+\hat{\mathsf{H}}\right)^{-1}\hat{\mathsf{H}}\right|, \tag{6.41}$$

$$\text{Roy's largest root:} \quad \hat{R} := max\left(eig\left(\hat{\mathsf{E}}^{-1}\hat{\mathsf{H}}\right)\right). \tag{6.42}$$

These are the classical representation of the four statistics, though they can all be expressed as functions of the eigenvalues of the matrix $\mathsf{I}_a + \hat{\mathsf{E}}^{-1}\hat{\mathsf{H}}$. Under the null hypothesis the matrix $\hat{\mathsf{H}}$ "should" be all zeros, and the four MGLH statistics "should" be zero, or close enough. We will focus almost entirely on the Hotelling Lawley trace because of the interpretation as the expected squared maximal Sharpe ratio of the linear passthrough under the conditional model. We should note that the distributions of these statistics, under the null and alternative, are still not fully understood.

Approximate distribution of test statistics

The exact distributions of the four MGLH test statistics are unknown even under the simplifying assumptions of normality, deterministic design matrix, and under the null hypothesis, except in the case where $min\,(a, c) = 1$, when the problem essentially reduces to the "univariate" case. Instead we have a number of approximations, typically to a non-central F-distribution or non-central χ^2-distribution. The approximations are:

1. The Pillai approximation for the Pillai Bartlett trace is [Pillai, 1956; Muller and Peterson, 1984; Shieh, 2005]

$$\frac{\hat{P}}{s-\hat{P}}\frac{s\,(n+s-c)}{ca} \approx F\left(ca, s\,(n+s-c), n\frac{P}{s-P}\,(s\,(n+s-c))\right),$$

with $s = min\,(a, c)$, and $P = \operatorname{tr}\left((\mathsf{E}+\mathsf{H})^{-1}\mathsf{H}\right)$. (We note that Shieh attributes a different approximation to Pillai.)

2. The Pillai approximation for the Hotelling Lawley trace is [Pillai, 1956; Muller and Peterson, 1984]

$$\frac{s\,(n-c-1)+2}{sca}\hat{T} \approx F\left(ca, s\,(n-c-1)+2, n\operatorname{tr}\left(\mathsf{E}^{-1}\mathsf{H}\right)\right),$$

with $s = min\,(a, c)$.

3. The Pillai-Samson approximation for the Hotelling Lawley trace is [Pillai and Samson, 1959; Shieh, 2005]

$$\frac{s\,(n-k-a-1)+2}{sca}\hat{T} \approx F\left(ca, s\,(n-k-a-1)+2, n\operatorname{tr}\left(\mathsf{E}^{-1}\mathsf{H}\right)\right),$$

with $s = min\,(a, c)$.

4. The McKeon approximation for the Hotelling Lawley trace is [McKeon, 1974; Shieh, 2005]

$$\frac{g\left(ca+2\right)+4}{\left(g\left(ca+2\right)+2\right)ca/\left(n-k-a-1\right)}\hat{T}\approx F\left(ca,g\left(ca+2\right)+4,n\operatorname{tr}\left(\mathsf{E}^{-1}\mathsf{H}\right)\right),$$

with

$$g=\frac{\left(n-k\right)^{2}-\left(n-k\right)\left(2a+3\right)+a\left(a+3\right)}{\left(n-k\right)\left(c+a+1\right)-\left(c+2a+a^{2}-1\right)}.$$

Testing for nonzero B

Perhaps the most common variant of the MGLH is testing

$$H_0:\mathsf{B}=\mathsf{0},\tag{6.43}$$

which corresponds to the above analysis with A and C equal to conformal identity matrices (with $a=k$ and $c=f$), and Θ equal to the zero matrix. In this case we have

$$\hat{\mathsf{H}}=n\hat{\mathsf{B}}\hat{\Gamma}_f\hat{\mathsf{B}}^{\top},\quad \hat{\mathsf{E}}=n\hat{\Sigma},$$

and thus the test statistics have the form

$$\hat{T}=\operatorname{tr}\left(\hat{\Sigma}^{-1}\hat{\mathsf{B}}\hat{\Gamma}_f\hat{\mathsf{B}}^{\top}\right),\tag{6.44}$$

$$\hat{P}=\operatorname{tr}\left(\left(\hat{\Sigma}+\hat{\mathsf{B}}\hat{\Gamma}_f\hat{\mathsf{B}}^{\top}\right)^{-1}\hat{\mathsf{B}}\hat{\Gamma}_f\hat{\mathsf{B}}^{\top}\right).\tag{6.45}$$

6.1.4 † Distribution of summary statistics

Using the connection to the F-distribution (Equation 6.26), we can express the density, distribution and quantile functions of $\hat{\zeta}_*^2$. The usual change of variables formula tells us that these are, respectively

$$f_{SR*}\left(x;n,k,\zeta_*^2\right)=\frac{n}{n-1}\frac{n-k}{k}f_f\left(\frac{n}{n-1}\frac{n-k}{k}x;k,n-k,n\zeta_*^2\right),\tag{6.46}$$

$$F_{SR*}\left(x;n,k,\zeta_*^2\right)=F_f\left(\frac{n}{n-1}\frac{n-k}{k}x;k,n-k,n\zeta_*^2\right),\tag{6.47}$$

$$SR*_p\left(n,k,\zeta_*^2\right)=\frac{n-1}{n}\frac{n-k}{k}f_p\left(k,n-k,n\zeta_*^2\right),\tag{6.48}$$

where $f_f\left(x;\nu_1,\nu_2,\delta\right)$, $F_f\left(x;\nu_1,\nu_2,\delta\right)$, and $f_p\left(\nu_1,\nu_2,\delta\right)$ are the density, cumulative distribution and quantile functions of the non-central F-distribution with non-centrality parameter δ and ν_1,ν_2 degrees of freedom. Here we are setting $r_0=0$. Note that the density of the non-central F-distribution is zero for negative arguments.

The PDF, CDF, and quantile functions of the non-central F-distribution have commonly available implementations. For example, in **R**, these are available as `df`, `pf`, and `qf`. As a practical matter, these should be used instead of writing your own. *If your interests are practical, rather than theoretical, you can safely skip the rest of this section.*

The density and distribution functions of the *central* F-distribution can be expressed as [Walck, 1996]

$$f_f(x; \nu_1, \nu_2) = \left(\frac{\nu_1}{\nu_2}\right)^{\frac{\nu_1}{2}} \frac{1}{B\left(\frac{\nu_1}{2}, \frac{\nu_2}{2}\right)} \frac{x^{\frac{\nu_1}{2}-1}}{\left(1 + \frac{\nu_1 x}{\nu_2}\right)^{\frac{\nu_1+\nu_2}{2}}}, \tag{6.49}$$

$$F_f(x; \nu_1, \nu_2) = \mathrm{I}_{(\nu_1 x)/(\nu_2+\nu_1 x)}\left(\frac{\nu_1}{2}, \frac{\nu_2}{2}\right), \tag{6.50}$$

where $B(a, b)$ is the beta function and $\mathrm{I}_x(a, b)$ is the regularized incomplete beta function [Abramowitz and Stegun, 1964, 6.6.2].

The density function of the non-central F-distribution can then be expressed as a Poisson mixture of the central density [Walck, 1996]

$$f_f(x; \nu_1, \nu_2, \delta) = e^{-\frac{\delta}{2}} \sum_{r=0}^{\infty} \frac{1}{r!}\left(\frac{\delta}{2}\right)^r f_f(x; \nu_1 + 2r, \nu_2). \tag{6.51}$$

The cumulative distribution function can be written as

$$F_f(x; \nu_1, \nu_2, \delta) = e^{-\frac{\delta}{2}} \sum_{r=0}^{\infty} \frac{1}{r!}\left(\frac{\delta}{2}\right)^r \mathrm{I}_{(\nu_1 x)/(\nu_2+\nu_1 x)}\left(\frac{\nu_1}{2} + r, \frac{\nu_2}{2}\right). \tag{6.52}$$

6.2 Moments of summary statistics

The moments of the non-central F-distribution are known, and can easily be translated into those of the squared maximal Sharpe ratio. [Walck, 1996] Suppose that $f \sim F(v_1, v_2, \delta)$. Then for $0 \le i < v_2/2$,

$$\mathrm{E}\left[f^i\right] = e^{-\delta/2}\left(\frac{v_2}{v_1}\right)^i \frac{\Gamma\left(\frac{v_2-2i}{2}\right)}{\Gamma\left(\frac{v_2}{2}\right)} \sum_{r=0}^{\infty} \frac{1}{r!}\left(\frac{\delta}{2}\right)^r \frac{\Gamma\left(\frac{v_1+2r+2i}{2}\right)}{\Gamma\left(\frac{v_1+2r}{2}\right)}. \tag{6.53}$$

When $\delta = 0$ this simplifies to

$$\mathrm{E}\left[f^i\right] = \left(\frac{v_2}{v_1}\right)^i \frac{\Gamma\left(\frac{v_2-2i}{2}\right)}{\Gamma\left(\frac{v_2}{2}\right)} \frac{\Gamma\left(\frac{v_1+2i}{2}\right)}{\Gamma\left(\frac{v_1}{2}\right)}. \tag{6.54}$$

Because of the connection to the non-central F-distribution under normal returns (Equation 6.26), the moments of the squared maximal Sharpe ratio can be expressed. Again, to reduce visual noise, we assume $r_0 = 0$, so the drag term disappears. We then have:

$$\mathrm{E}\left[\hat{\zeta}_*^{2i}\right] = \exp\left(-\frac{n}{2}\zeta_*^2\right)\left(\frac{n-1}{n}\right)^i \frac{\Gamma\left(\frac{n-k-2i}{2}\right)}{\Gamma\left(\frac{n-k}{2}\right)} \sum_{r=0}^{\infty} \frac{1}{r!}\left(\frac{n\zeta_*^2}{2}\right)^r \frac{\Gamma\left(\frac{k+2r+2i}{2}\right)}{\Gamma\left(\frac{k+2r}{2}\right)}, \quad (6.55)$$

when $0 \leq 2i < n - k$.

Simplified versions without infinite expansion for small i are

$$\begin{aligned}
\alpha_1\left(\hat{\zeta}_*^2\right) &= \left(\frac{n-1}{n}\right)\frac{\delta + k}{(n-k-2)},\\
\alpha_2\left(\hat{\zeta}_*^2\right) &= \left(\frac{n-1}{n}\right)^2\frac{\delta^2 + (2\delta + k)(k+2)}{(n-k-2)(n-k-4)},\\
\alpha_3\left(\hat{\zeta}_*^2\right) &= \left(\frac{n-1}{n}\right)^3\frac{\delta^3 + 3\delta^2(k+4) + (3\delta + k)(k+4)(k+2)}{(n-k-2)(n-k-4)(n-k-6)},\\
\alpha_4\left(\hat{\zeta}_*^2\right) &= \left(\frac{n-1}{n}\right)^4\frac{\delta^4 + 4\delta^3(k+6) + 6\delta^2(k+6)(k+4) + (4\delta + k)(k+6)(k+4)(k+2)}{(n-k-2)(n-k-4)(n-k-6)(n-k-8)},
\end{aligned} \quad (6.56)$$

where we plug in

$$\delta = n\zeta_*^2.$$

Expressed explicitly the expected value of $\hat{\zeta}_*^2$ is then

$$\mathrm{E}\left[\hat{\zeta}_*^2\right] = \left(\frac{n-1}{n}\right)\frac{n\zeta_*^2 + k}{(n-k-2)}. \quad (6.57)$$

The variance, skewness, and excess kurtosis of $\hat{\zeta}_*^2$ can then be computed from the first two uncentered moments:

$$\begin{aligned}
\mathrm{Var}\left(\hat{\zeta}_*^2\right) &= \alpha_2\left(\hat{\zeta}_*^2\right) - \left(\alpha_1\left(\hat{\zeta}_*^2\right)\right)^2,\\
&= 2\left(\frac{n-1}{n}\right)^2\frac{n^2\zeta_*^4 + 2n^2\zeta_*^2 - 4n\zeta_*^2 + kn - 2k}{(n-k-2)^2(n-k-4)}.\\
\mathrm{skew}\left(\hat{\zeta}_*^2\right) &= \frac{\alpha_3\left(\hat{\zeta}_*^2\right) - 3\alpha_2\left(\hat{\zeta}_*^2\right)\alpha_1\left(\hat{\zeta}_*^2\right) + 2\alpha_1\left(\hat{\zeta}_*^2\right)^3}{\left(\alpha_2\left(\hat{\zeta}_*^2\right) - \alpha_1\left(\hat{\zeta}_*^2\right)^2\right)^{3/2}}.\\
\mathrm{ex\,kurtosis}\left(\hat{\zeta}_*^2\right) &= \frac{\alpha_4\left(\hat{\zeta}_*^2\right) - 4\alpha_3\left(\hat{\zeta}_*^2\right)\alpha_1\left(\hat{\zeta}_*^2\right) + 6\alpha_2\left(\hat{\zeta}_*^2\right)\alpha_1\left(\hat{\zeta}_*^2\right)^2 - 3\alpha_1\left(\hat{\zeta}_*^2\right)^4}{\left(\alpha_2\left(\hat{\zeta}_*^2\right) - \alpha_1\left(\hat{\zeta}_*^2\right)^2\right)^2} - 3.
\end{aligned} \quad (6.58)$$

We note that the variance of $\hat{\zeta}_*^2$ is approximately

$$\mathrm{Var}\left(\hat{\zeta}_*^2\right) \approx \frac{2\zeta_*^4 + 4\zeta_*^2}{n} + \frac{2k}{n^2}. \quad (6.59)$$

As the last term is $\mathcal{O}\left(n^{-2}\right)$, it is ignored in the typical asymptotic analysis. However, for the n, k, and ζ_*^2 one is likely to encounter in practice, a large proportion of the variance of $\hat{\zeta}_*^2$ can come from the $\mathcal{O}\left(n^{-2}\right)$ term. For example, if $\zeta_*^2 = 1\ \mathrm{yr}^{-1} = 0.004\ \mathrm{day}^{-1}$, $n = 1,260$ days, and $k = 10$, then the terms $4\zeta_*^2$ and $2k/n$ are equal to each other, both taking value $0.016\ \mathrm{day}^{-1}$. Thus the typically ignorable $\mathcal{O}\left(n^{-2}\right)$ term is significant in this case.

Moments of ζ_*

Via Taylor expansion, we can find the approximate expected value and approximate variance of ζ_*. The former is

$$\mathrm{E}\left[\hat{\zeta}_*\right] \approx \sqrt{\alpha_1\left(\hat{\zeta}_*^2\right)} - \frac{1}{8}\left(\frac{\alpha_2\left(\hat{\zeta}_*^2\right) - \left(\alpha_1\left(\hat{\zeta}_*^2\right)\right)^2}{\left(\alpha_1\left(\hat{\zeta}_*^2\right)\right)^{3/2}}\right) + \ldots,$$

$$= \frac{9}{8}\sqrt{\frac{n-1}{n}}\sqrt{\frac{\delta+k}{(n-k-2)}}\left(1 - \frac{1}{9}\frac{\delta^2 + (2\delta+k)(k+2)}{(\delta+k)^2}\frac{(n-k-2)}{(n-k-4)} + \ldots\right). \quad (6.60)$$

Letting $k = c_a n$, and taking the limit as $n \to \infty$, we have

$$\mathrm{E}\left[\hat{\zeta}_*\right] \to \sqrt{\frac{\zeta_*^2 + c_a}{1 - c_a}},$$

which is only approximately equal to ζ_*. Note that if c_a becomes arbitrarily small (k is fixed while n grows without bound), then $\hat{\zeta}_*$ is asymptotically unbiased.

The variance we can write as

$$\mathrm{Var}\left(\hat{\zeta}_*\right) = \alpha_1\left(\hat{\zeta}_*^2\right) - \left(\mathrm{E}\left[\hat{\zeta}_*\right]\right)^2,$$

$$\approx \frac{1}{4}\left(\frac{\alpha_2\left(\hat{\zeta}_*^2\right) - \left(\alpha_1\left(\hat{\zeta}_*^2\right)\right)^2}{\alpha_1\left(\hat{\zeta}_*^2\right)}\right) + \frac{1}{64}\left(\frac{\alpha_2\left(\hat{\zeta}_*^2\right) - \left(\alpha_1\left(\hat{\zeta}_*^2\right)\right)^2}{\left(\alpha_1\left(\hat{\zeta}_*^2\right)\right)^{3/2}}\right)^2 + \ldots,$$

$$\approx \frac{1}{4}\left(\frac{\alpha_2\left(\hat{\zeta}_*^2\right)}{\alpha_1\left(\hat{\zeta}_*^2\right)} - \alpha_1\left(\hat{\zeta}_*^2\right)\right), \quad (6.61)$$

$$= \frac{n}{4(n-1)}\left(\frac{n^2\zeta_*^4 + (k+2)\left(2n\zeta_*^2 + k\right)}{(n-k-4)\left(n\zeta_*^2 + k\right)} - \frac{n\zeta_*^2 + k}{n-k-2}\right). \quad (6.62)$$

These results can trivially be translated into equivalent results for $\hat{\zeta}_{*,\mathsf{G}}$, and on $\Delta_{\mathsf{I}\setminus\mathsf{G}}\,\hat{\zeta}_*^2$ under the null $\Delta_{\mathsf{I}\setminus\mathsf{G}}\zeta_*^2 = 0$.

Example 6.2.1 (Simulated Moments). We simulate $n = 504$ days of normal returns (2 years) on $k = 5$ assets with $\zeta_* = 1.2\,\mathrm{yr}^{-1/2}$, where we assume 252 trading days per year. We compute $\hat{\zeta}_*^2$. We repeat this simulation 100,000 times, and compute the empirical raw first through fourth moments of $\hat{\zeta}_*^2$. We also compute the theoretical moments from Equation 6.56. We compute the empirical and theoretical means and variances of $\hat{\zeta}_*$, from Equation 6.60 and Equation 6.62, and the empirical and theoretical variance of $\hat{\zeta}_*^2$. These are all given in Table 6.2, and we see good agreement between theoretical and empirical moments.

Note that we compute the expected value of $\hat{\zeta}_*$ to be around $1.91\,\mathrm{yr}^{-1/2}$. This is quite a bit larger than $\zeta_* = 1.2\,\mathrm{yr}^{-1/2}$. The squared maximal Sharpe ratio is a severely biased estimator of the squared maximal signal-noise ratio. ⊣

Moment	Theoretical	Empirical
$\alpha_1\left(\hat{\zeta}_*^2\right)$	3.988	3.98
$\alpha_2\left(\hat{\zeta}_*^2\right)$	21.498	21.40
$\alpha_3\left(\hat{\zeta}_*^2\right)$	145.216	144.00
$\alpha_4\left(\hat{\zeta}_*^2\right)$	1175.380	1170.00
$\mathrm{Var}\left(\hat{\zeta}_*^2\right)$	5.597	5.57
$\mathrm{skew}\left(\hat{\zeta}_*^2\right)$	1.122	1.12
$\mathrm{ex\,kurtosis}\left(\hat{\zeta}_*^2\right)$	1.840	1.85
$\mathrm{E}\left[\hat{\zeta}_*\right]$	1.909	1.91
$\mathrm{Var}\left(\hat{\zeta}_*\right)$	0.351	0.34

TABLE 6.2
Theoretical and empirical (over 100,000 simulations) moments, variance, and cumulants of $\hat{\zeta}_*^2$ are given for $n = 504$ and $k = 5$. We quote $\hat{\zeta}_*^2$ and its moments in annualized terms, though skewness and kurtosis are unitless. We round the empirical moments to three significant digits. The theoretical moments of $\hat{\zeta}_*$ are approximations.

Example 6.2.2 (Simulated Moments, Hedging Constraint). We repeat Example 6.2.1, but impose a hedge constraint. We simulate $n = 504$ days of normal returns (2 years) on $k = 5$ assets with $\zeta_* = 1.2\,\mathrm{yr}^{-1/2}$, where we assume 252 trading days per year. We let $\boldsymbol{\mu} = \sqrt{\zeta_*^2/k}\mathbf{1}$ and $\Sigma = \mathsf{I}$, and thus $\boldsymbol{\nu}_{*,\propto}\mathbf{1}$. We take $\mathsf{G} = \mathbf{1}^\top$, and thus have hedged out all the available signal-noise ratio. Thus we are under the null, $\Delta_{\mathsf{I}\setminus\mathsf{G}}\zeta_*^2 = 0$, and the sample statistic $\Delta_{\mathsf{I}\setminus\mathsf{G}}\hat{\zeta}_*^2$ follows a central F-distribution, up to scaling.

We compute $\Delta_{\mathsf{I}\setminus\mathsf{G}}\,\hat{\zeta}_*^2$, repeating the simulation 100,000 times. We compute the empirical and theoretical raw first through fourth moments of $\Delta_{\mathsf{I}\setminus\mathsf{G}}\,\hat{\zeta}_*^2$, the latter via Equation 6.56, as well as the empirical and theoretical means and variances of $\Delta_{\mathsf{I}\setminus\mathsf{G}}\,\hat{\zeta}_*$ and the variance of $\Delta_{\mathsf{I}\setminus\mathsf{G}}\,\hat{\zeta}_*^2$. These are all given in Table 6.3, and again we see good agreement between theoretical and empirical moments. Again, the sample statistic, $\Delta_{\mathsf{I}\setminus\mathsf{G}}\,\hat{\zeta}_*$ is a severely biased estimate of the population analogue, which is zero in this case. ⊣

Moment	Theoretical	Empirical
$\alpha_1\left(\Delta_{\mathsf{I}\setminus\mathsf{G}}\hat{\zeta}_*^2\right)$	2.024	2.04
$\alpha_2\left(\Delta_{\mathsf{I}\setminus\mathsf{G}}\hat{\zeta}_*^2\right)$	6.171	6.29
$\alpha_3\left(\Delta_{\mathsf{I}\setminus\mathsf{G}}\hat{\zeta}_*^2\right)$	25.183	25.90
$\alpha_4\left(\Delta_{\mathsf{I}\setminus\mathsf{G}}\hat{\zeta}_*^2\right)$	128.992	134.00
$\operatorname{Var}\left(\Delta_{\mathsf{I}\setminus\mathsf{G}}\hat{\zeta}_*^2\right)$	2.073	2.12
$\operatorname{skew}\left(\Delta_{\mathsf{I}\setminus\mathsf{G}}\hat{\zeta}_*^2\right)$	1.440	1.44
$\operatorname{ex\,kurtosis}\left(\Delta_{\mathsf{I}\setminus\mathsf{G}}\hat{\zeta}_*^2\right)$	3.147	3.14
$\operatorname{E}\left[\Delta_{\mathsf{I}\setminus\mathsf{G}}\hat{\zeta}_*\right]$	1.333	1.34
$\operatorname{Var}\left(\Delta_{\mathsf{I}\setminus\mathsf{G}}\hat{\zeta}_*\right)$	0.256	0.24

TABLE 6.3
Theoretical and empirical (over 100,000 simulations) moments, variance, and cumulants of $\Delta_{\mathsf{I}\setminus\mathsf{G}}\,\hat{\zeta}_*^2$ are given for $n = 504$ and $k = 5$. We quote $\Delta_{\mathsf{I}\setminus\mathsf{G}}\,\hat{\zeta}_*^2$ and its moments in annualized terms, though skewness and kurtosis are unitless. Simulations are under the null, so the sample statistic follows a central F-distribution up to scaling. We round the empirical moments to three significant digits. The theoretical moments of $\Delta_{\mathsf{I}\setminus\mathsf{G}}\,\hat{\zeta}_*$ are approximations.

Example 6.2.3 (The Fundamental Flaw of Backtesting). We consider the case where one adds assets to one's universe without an increase in ζ_*. We plot the approximate expected value of $\hat{\zeta}_*$ from Equation 6.60 versus k for the case where $n = 208$ weeks, and ζ_* is fixed at $1\,\mathrm{yr}^{-1/2}$ in Figure 6.1. We see that $\hat{\zeta}_*$ seems to grow at a rate slightly greater than $c\sqrt{k}$. ⊣

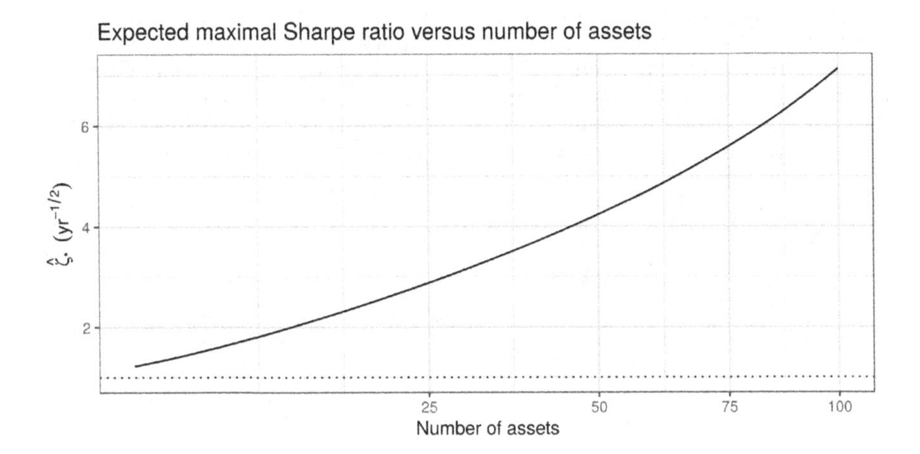

FIGURE 6.1
The approximate expected value of $\hat{\zeta}_*$ is plotted versus k for the case where ζ_* remains constant. Expectation is based on $n = 208$ weeks of data with $\zeta_* = 1\,\mathrm{yr}^{-1/2}$. We plot a horizontal line at ζ_*. The y-axis is in annual units, and the x-axis is in square root scale.

6.2.1 Asymptotic moments

First consider the case where as $n \to \infty$, the ratio $k/n \to c_a < 1$, and ζ_* is fixed. In this case, the asymptotic moments are:

$$\begin{aligned}
\alpha_1\left(\hat{\zeta}_*^2\right) &\to \frac{\zeta_*^2 + c_a}{1 - c_a},\\
\alpha_2\left(\hat{\zeta}_*^2\right) &\to \left(\frac{\zeta_*^2 + c_a}{1 - c_a}\right)^2,\\
\alpha_3\left(\hat{\zeta}_*^2\right) &\to \left(\frac{\zeta_*^2 + c_a}{1 - c_a}\right)^3.
\end{aligned} \tag{6.63}$$

The asymptotic variance of $\hat{\zeta}_*^2$ in this case is

$$\operatorname{Var}\left(\hat{\zeta}_*^2\right) \to \frac{2}{n}\frac{2\zeta_*^2 + c_a}{\left(1 - c_a\right)^2}. \tag{6.64}$$

The asymptotic approximate variance of ζ_*, from Equation 6.62, is

$$\mathrm{Var}\left(\hat{\zeta}_*\right) \rightarrow \frac{1}{2n} \frac{\zeta_*^4 + 2\zeta_*^2 + c_a}{\left(1 - c_a\right)^2 \left(\zeta_*^2 + c_a\right)} \approx \frac{1 + 2c_a}{2n} \left(1 + \frac{1}{1 + c_a/\zeta_*^2}\right).$$

Moments under up-sampling

Consider now the case of "up-sampling" the returns. That is, suppose that each return period is divided into m subintervals, and one considers the portfolio that rebalances at each period. The sample size is replaced with mn, and the signal-noise ratio is replaced with $\zeta_*/\sqrt{m}$. We then consider the case where $m \rightarrow \infty$, while n, k and ζ_* remain constant. Because of how the units work, we are concerned with the moments of $m\hat{\zeta}_*^2$. Asymptotically these are:

$$\begin{aligned}
\alpha_1\left(m\hat{\zeta}_*^2\right) &\rightarrow \zeta_*^2 + \frac{k}{n}, \\
\alpha_2\left(m\hat{\zeta}_*^2\right) &\rightarrow \zeta_*^4 + \left(2\zeta_*^2 + \frac{k}{n}\right)\left(\frac{k+2}{n}\right), \\
\alpha_3\left(m\hat{\zeta}_*^2\right) &\rightarrow \zeta_*^6 + 3\zeta_*^4\frac{k+4}{n} + \left(3\zeta_*^2 + \frac{k}{n}\right)\left(\frac{k+4}{n}\right)\left(\frac{k+2}{n}\right), \\
\alpha_4\left(m\hat{\zeta}_*^2\right) &\rightarrow \zeta_*^8 + 4\zeta_*^6\frac{k+6}{n} + 6\zeta_*^4\left(\frac{k+6}{n}\right)\left(\frac{k+4}{n}\right) \\
&\quad + \left(4\zeta_*^2 + \frac{k}{n}\right)\left(\frac{k+6}{n}\right)\left(\frac{k+4}{n}\right)\left(\frac{k+2}{n}\right).
\end{aligned} \tag{6.65}$$

6.2.2 Unbiased estimation

Equation 6.56 suggests a simple unbiased estimator for ζ_*^2, namely

$$\mathrm{E}\left[\frac{(n-k-2)}{n-1}\hat{\zeta}_*^2 - \frac{k}{n}\right] = \zeta_*^2. \tag{6.66}$$

However, this estimator may be *negative.* In general, there is no non-negative function of $\hat{\zeta}_*^2, n$ and k that is monotonic in $\hat{\zeta}_*^2$ and which is an unbiased estimate of ζ_*.

Kubokawa, Robert, and Saleh consider improved methods for estimating the non-centrality parameter given an observation of a non-central F statistic. [Kubokawa *et al.*, 1993]. They note that the estimator of Equation 6.66, which they call δ_0, is the Uniform Minimum Variance Unbiased Estimator (UMVUE). These "KRS estimators" are computed as

$$\begin{aligned}
\delta_0 &= \frac{(n-k-2)}{n-1}\hat{\zeta}_*^2 - \frac{k}{n}, \\
\delta_1 &= max\left(\delta_0, 0\right), \\
\delta_2 &= max\left(\delta_0, \frac{2}{k+2}\left(\delta_0 + \frac{k}{n}\right)\right).
\end{aligned} \tag{6.67}$$

The estimators δ_1, δ_2 are non-negative, and dominate δ_0 in having lower expected squared error.

Example 6.2.4 (Simulated moments, estimation). We consider the 100,000 simulated values of $\hat{\zeta}_*^2$ from Example 6.2.1. We compute the KRS δ_i estimators for each simulation. The empirical mean value of δ_0 is $1.432\,\text{yr}^{-1}$, which is indeed close to the population value $\zeta_*^2 = 1.44\,\text{yr}^{-1}$. However, around 30.7% of the δ_0 are negative. We compute the root mean squared error of the δ_0 estimator as $2.33\,\text{yr}^{-1}$; that of δ_1 to be $2.08\,\text{yr}^{-1}$, and that of δ_2 to be $1.97\,\text{yr}^{-1}$. For this choice of n, k and ζ_*^2, the δ_2 estimator dominates δ_1, and both do exhibit lower squared error than the unbiased estimator, δ_0. ⊣

Example 6.2.5 (Estimation on the Fama French Four Factors). We consider the returns of the Fama French four factors considered in Example 6.1.1. In that example, we computed (ignoring the drag term) $\hat{\zeta}_* = 0.308\,\text{mo.}^{-1/2}$, corresponding to $\hat{\zeta}_*^2 = 0.095\,\text{mo.}^{-1}$. With $n = 1{,}128, k = 4$, we compute the unbiased estimator and two KRS estimators of ζ_*^2 as $\delta_0 = \delta_1 = \delta_2 = 0.091\,\text{mo.}^{-1}$. They take the same value because δ_0 is positive and larger than the alternative calculation. Note that $\delta_0^{1/2}$ corresponds to a signal-noise ratio of around $1.044\,\text{yr}^{-1/2}$, which is not much smaller than $\hat{\zeta}_*$. ⊣

Constrained problems

The KRS estimators can be applied to the subspace constraint, simply by adjusting the k to represent the effective universe size of the subspace. Some work is required to adapt estimation to hedged portfolios. This proceeds by using the same reasoning on the F-statistic implied by Equation 6.29. The estimators are:

$$\begin{aligned}
\delta_0 &= \frac{(n-k-2)}{n-1}\Delta_{\mathsf{I}\setminus\mathsf{G}}\hat{\zeta}_*^2 - \frac{k-k_g}{n}\left(1 + \frac{n}{n-1}\hat{\zeta}_{*,\mathsf{G}}^2\right), \\
\delta_1 &= max\left(\delta_0, 0\right), \\
\delta_2 &= max\left(\delta_0, \frac{2}{k-k_g+2}\,\frac{n-k-2}{n-1}\Delta_{\mathsf{I}\setminus\mathsf{G}}\hat{\zeta}_*^2\right).
\end{aligned} \tag{6.68}$$

Example 6.2.6 (Estimation on the Fama French Four Factors, hedged against Market). We continue Example 6.1.4 where we considered the Fama French four factors with a hedge against the Market. Based on $n = 1{,}128, k = 4$, we compute the unbiased estimator and two KRS estimators of $\Delta_{\mathsf{I}\setminus\mathsf{G}}\,\zeta_*^2$ as $\delta_0 = \delta_1 = \delta_2 = 0.06\,\text{mo.}^{-1}$. Note that $\delta_0^{1/2}$ corresponds to a signal-noise ratio of around $0.246\,\text{mo.}^{-1/2}$, which is around the same value as $\Delta_{\mathsf{I}\setminus\mathsf{G}}\hat{\zeta}_* = 0.252\,\text{mo.}^{-1/2}$, as computed in the previous example. ⊣

6.3 Frequentist inference on squared maximal signal-noise ratio

6.3.1 Confidence intervals

One can find confidence intervals on ζ_* implicitly via the connection to the non-central F distribution. So to find confidence interval $[\zeta_l, \zeta_u)$ with coverage $1-\alpha$, solve the implicit equations

$$\begin{aligned} 1-\alpha/2 &= F_f\left(\left(\frac{n(n-k)}{k(n-1)}\right)\hat{\zeta}_*^2; k, n-k, n\zeta_l^2\right), \\ \alpha/2 &= F_f\left(\left(\frac{n(n-k)}{k(n-1)}\right)\hat{\zeta}_*^2; k, n-k, n\zeta_u^2\right), \end{aligned} \tag{6.69}$$

where $F_f\left(x; \nu_1, \nu_2, \delta\right)$ is the CDF of the non-central F-distribution with non-centrality parameter δ and ν_1 and ν_2 degrees of freedom. This method requires computational inversion of the CDF function.

However, there may not be such ζ_l, ζ_u such that the above hold with equality, and in practice one should use the limiting forms:

$$\begin{aligned} \zeta_l &= \min\left\{z \,\middle|\, z \ge 0,\ 1-\alpha/2 \ge F_f\left(\left(\frac{n(n-k)}{k(n-1)}\right)\hat{\zeta}_*^2; k, n-k, nz^2\right)\right\}, \\ \zeta_u &= \min\left\{z \,\middle|\, z \ge 0,\ \alpha/2 \ge F_f\left(\left(\frac{n(n-k)}{k(n-1)}\right)\hat{\zeta}_*^2; k, n-k, nz^2\right)\right\}. \end{aligned} \tag{6.70}$$

Since $F_f\left(\cdot; k, n-k, nz^2\right)$ is a decreasing function of z^2, and approaches zero in the limit, the above confidence interval is well defined.

To construct a confidence interval for $\Delta_{\mathsf{I}\setminus\mathsf{G}}\zeta_*$, for the hedging constrained problem of Equation 6.9, use the conditional distribution of Equation 6.29. So to find confidence interval $[\zeta_l, \zeta_u)$ with coverage $1-\alpha$, solve the limiting forms:

$$\begin{aligned} \zeta_l &= \min\left\{z \ge 0 \,\middle|\, 1-\alpha/2 \ge F_f\left(\frac{n(n-k)}{(k-k_g)(n-1)}\frac{\Delta_{\mathsf{I}\setminus\mathsf{G}}\hat{\zeta}_*^2}{c_g}; k-k_g, n-k, \frac{nz^2}{c_g}\right)\right\}, \\ \zeta_u &= \min\left\{z \ge 0 \,\middle|\, \alpha/2 \ge F_f\left(\frac{n(n-k)}{(k-k_g)(n-1)}\frac{\Delta_{\mathsf{I}\setminus\mathsf{G}}\hat{\zeta}_*^2}{c_g}; k-k_g, n-k, \frac{nz^2}{c_g}\right)\right\}, \\ c_g &= 1 + \frac{n}{n-1}\hat{\zeta}_{*,\mathsf{G}}^2. \end{aligned} \tag{6.71}$$

Example 6.3.1 (Fama French Four Factors, confidence interval). We consider the 1,128 months of returns of the Market, SMB, HML, and UMD portfolios introduced in Example 1.1.1, spanning from Jan 1927 to Dec 2020. Setting $r_0 = 0$, we compute $\hat{\zeta}_* = 0.308\ \text{mo.}^{-1/2}$. We compute a 95% confidence interval on ζ_* as $[0.243, 0.363)\ \text{mo.}^{-1/2}$. ⊣

Example 6.3.2 (Portfolio on the Fama French Four Factors, Market hedged confidence interval). We continue Example 6.1.4, where we consider the optimal portfolio on the Fama French four factors with a hedge against the Market. As in Example 6.1.4, we compute

$$\hat{\zeta}_*^2 = 0.095\,\text{mo.}^{-1}, \quad \hat{\zeta}_{*,\mathsf{G}}^2 = 0.031\,\text{mo.}^{-1}, \quad \Delta_{\mathsf{I}\backslash\mathsf{G}}\hat{\zeta}_*^2 = 0.063\,\text{mo.}^{-1}.$$

We compute a one-sided 95% confidence interval on $\Delta_{\mathsf{I}\backslash\mathsf{G}}\ \zeta_*$ as $[0.204, \infty)$ mo.$^{-1/2}$. ⊣

Example 6.3.3 (Portfolio on the five industries, hedged against all). Continuing Example 6.1.5 we consider the optimal portfolio on the five baskets of industry stocks, subject to a hedge against the equal dollar allocation. As in Example 6.1.5, we compute

$$\hat{\zeta}_*^2 = 0.05\,\text{mo.}^{-1}, \quad \hat{\zeta}_{*,\mathsf{G}}^2 = 0.036\,\text{mo.}^{-1}, \quad \Delta_{\mathsf{I}\backslash\mathsf{G}}\hat{\zeta}_*^2 = 0.014\,\text{mo.}^{-1}.$$

We compute two-sided 99% confidence interval on $\Delta_{\mathsf{I}\backslash\mathsf{G}}\ \zeta_*$ as $[0.009, 0.191)$ mo.$^{-1/2}$. ⊣

Example 6.3.4 (Conditional Portfolio on the Fama French Four Factors, HLT). We consider the returns of the Market, SMB, HML, and UMD portfolios introduced in Example 1.1.1. In Example 6.1.6, we computed the Hotelling Lawley trace as $\hat{T} = 0.129$ mo.$^{-1}$. Using the Pillai-Samson approximation, we compute a two-sided 95% confidence interval on $T = \operatorname{tr}\left(\mathsf{E}^{-1}\mathsf{H}\right)$ as $[0.08, 0.163)$ mo.$^{-1}$. ⊣

We can construct very approximate $1 - \alpha$ confidence interval on ζ_*^2 as

$$\frac{(n-k-2)}{n-1}\hat{\zeta}_*^2 - \frac{k}{n} \pm z_\alpha \sqrt{2\frac{n^2\zeta_*^4 + 2n^2\hat{\zeta}_*^2 - 4n\hat{\zeta}_*^2 + kn - 2k}{n^2\,(n-k-4)}}. \tag{6.72}$$

These confidence intervals are based on a normal approximation, and are centered around the δ_0 estimator of Equation 6.67. Square roots of these endpoints give approximate confidence intervals on ζ_*.

Example 6.3.5 (Fama French Four Factors, rough confidence interval). Continuing Example 6.3.1, we consider returns of the Market, SMB, HML, and UMD portfolios. We compute the unbiased estimator of ζ_*^2 as $\delta_0 = 0.091\,\text{mo.}^{-1}$, and we compute the standard error as $0.019\,\text{mo.}^{-1}$. Based on these we compute an approximate 95% confidence interval on ζ_* as $[0.232, 0.358)$ mo.$^{-1/2}$. ⊣

6.3.2 Hypothesis tests

There are a few statistical tests for hypotheses involving the squared maximal signal-noise ratio. The classical Hotelling's T^2-test for the vector mean can be considered a hypothesis test on the squared maximal signal-noise ratio. Many of these tests can be considered multivariate analogues to tests described in Section 2.5.3:

1 sample vector mean test

The classical one-sample test for mean of a multivariate random variable uses Hotelling's T^2-statistic, just as the univariate test uses the t-statistic. Unlike the univariate case, we cannot easily perform a one-sided test (because $k > 1$ makes one-sidedness an odd concept), and thus we have the two-sided test. To test

$$H_0 : \boldsymbol{\mu} = \boldsymbol{\mu}_0 \quad \text{versus} \quad H_1 : \boldsymbol{\mu} \neq \boldsymbol{\mu}_0,$$

reject at the α level if

$$T_0^2 = n(\hat{\boldsymbol{\mu}} - \boldsymbol{\mu}_0)^\top \hat{\Sigma}^{-1} (\hat{\boldsymbol{\mu}} - \boldsymbol{\mu}_0) \geq \frac{k(n-1)}{n-k} f_{1-\alpha}(k, n-k),$$

where $f_{1-\alpha}(k, n-k)$ is the $1-\alpha$ quantile of the (central) F-distribution with k and $n-k$ degrees of freedom.

If $\boldsymbol{\mu} = \boldsymbol{\mu}_1 \neq \boldsymbol{\mu}_0$, then the test statistic T_0^2 is distributed as a non-central F-distribution with non-centrality parameter

$$\delta_1 = n(\boldsymbol{\mu}_1 - \boldsymbol{\mu}_0)^\top \Sigma^{-1} (\boldsymbol{\mu}_1 - \boldsymbol{\mu}_0),$$

$k, n-k$ degrees of freedom. The power of this test under this alternative is

$$1 - \beta = 1 - F_f\left(f_{1-\alpha}(k, n-k); k, n-k, \delta_1\right),$$

where $F_f(x; \nu_1, \nu_2, \delta)$ is the cumulative distribution function of the non-central F-distribution with non-centrality parameter δ and ν_1, ν_2 degrees of freedom. [Bilodeau and Brenner, 1999] Note that the non-centrality parameter is equal to the population analogue of the T^2 statistic itself.

2 sample vector mean test

Given two independent samples of sizes n_1, n_2, from populations assumed to have the same covariance, the test for the hypothesis of equal vector means also proceeds by a Hotelling's T^2-statistic. To test

$$H_0 : \boldsymbol{\mu}_1 = \boldsymbol{\mu}_2 \quad \text{versus} \quad H_1 : \boldsymbol{\mu}_1 \neq \boldsymbol{\mu}_2,$$

compute the statistic

$$T^2 = \frac{n_1 n_2}{n_1 + n_2} (\hat{\boldsymbol{\mu}}_1 - \hat{\boldsymbol{\mu}}_2) \left(\frac{1}{n_1 + n_2 - 2}\left((n_1 - 1)\hat{\Sigma}_1 + (n_2 - 1)\hat{\Sigma}_2\right)\right)^{-1} (\hat{\boldsymbol{\mu}}_1 - \hat{\boldsymbol{\mu}}_2)^\top,$$

then reject at the α level if

$$T^2 \geq \frac{k(n_1 + n_2 - 2)}{n_1 + n_2 - 1 - k} f_{1-\alpha}(k, n_1 + n_2 - 1 - k),$$

where $f_{1-\alpha}(\nu_1, \nu_2)$ is the $1-\alpha$ quantile of the (central) F-distribution with ν_1 and ν_2 degrees of freedom.

1 sample maximal signal-noise ratio test

A one-sample test for maximal signal-noise ratio involves the T^2-statistic as follows. To test

$$H_0 : \zeta_* = \zeta_0 \quad \text{versus} \quad H_1 : \zeta_* > \zeta_0,$$

reject if

$$\hat{\zeta}_*^2 > \frac{n-1}{n}\frac{k}{n-k} f_{1-\alpha}\left(k, n-k, n\zeta_0^2\right),$$

where $f_q\left(\nu_1, \nu_2, \delta\right)$ is the q quantile of the non-central F-distribution with non-centrality parameter δ and degrees of freedom ν_1, ν_2. Here $\hat{\zeta}_*^2$ is measured with $r_0 = 0$ to avoid the visual noise of the drag term.

If $\zeta_* > \zeta_0$, then the power of this test is

$$1 - \beta = 1 - F_f\left(f_{1-\alpha}\left(k, n-k, n\zeta_0^2\right); k, n-k, n\zeta_*^2\right),$$

where $F_f\left(x; \nu_1, \nu_2, \delta\right)$ is the the cumulative distribution function of the non-central F-distribution with non-centrality parameter δ and ν_1, ν_2 degrees of freedom.

1 sample maximal signal-noise ratio test, hedged portfolio

Let G be a $k_g \times k$ matrix of rank $k_g \leq k$. The one-sample test for maximal signal-noise ratio for a portfolio hedged against G involves the delta T^2-statistic as follows. To test

$$H_0 : \Delta_{\mathsf{I}\setminus\mathsf{G}}\zeta_* = \zeta_0 \quad \text{versus} \quad H_1 : \Delta_{\mathsf{I}\setminus\mathsf{G}}\zeta_* > \zeta_0,$$

reject if

$$\Delta_{\mathsf{I}\setminus\mathsf{G}}\hat{\zeta}_*^2 > \frac{n-1}{n}\frac{k-k_g}{n-k}\left(1 + \frac{n}{n-1}\hat{\zeta}_{*,\mathsf{G}}^2\right) f_{1-\alpha}\left(k-k_g, n-k, \frac{n\zeta_0^2}{1 + \frac{n}{n-1}\hat{\zeta}_{*,\mathsf{G}}^2}\right),$$

where $f_q\left(\nu_1, \nu_2, \delta\right)$ is the q quantile of the non-central F-distribution with non-centrality parameter δ and degrees of freedom ν_1, ν_2. Because the connection to the F-distribution is only conditional on $\hat{\zeta}_{*,\mathsf{G}}^2$, we cannot easily describe the power of this test under the alternative. This test is also the test for portfolio spanning.

1 sample one-sided portfolio weight test

To test a single contrast on portfolio weights against a two-sided alternative

$$H_0 : \boldsymbol{v}^\top \Sigma^{-1}\boldsymbol{\mu} = c \quad \text{versus} \quad H_1 : \boldsymbol{v}^\top \Sigma^{-1}\boldsymbol{\mu} \neq c$$

one can use the 1 sample test for hedged maximal signal-noise ratio above, with G equal to the $k-1 \times k$ matrix of rank $k-1$ whose rows are orthogonal to $\boldsymbol{v}$.

The test against a one-sided alternative is via a t-statistic. To test

$$H_0 : \boldsymbol{v}^\top \Sigma^{-1} \boldsymbol{\mu} = c \quad \text{versus} \quad H_1 : \boldsymbol{v}^\top \Sigma^{-1} \boldsymbol{\mu} > c,$$

for k-vector $\boldsymbol{v}$ and constant c, compute

$$t = \sqrt{\frac{n}{n-1}} \frac{\boldsymbol{v}^\top \hat{\Sigma}^{-1} \hat{\boldsymbol{\mu}} \sqrt{n-k}}{\sqrt{\boldsymbol{v}^\top \hat{\Sigma}^{-1} \boldsymbol{v} \left(1 + \frac{n}{n-1} \hat{\zeta}^2_{*,\mathsf{G}}\right)}}.$$

Conditional on $\hat{\zeta}^2_{*,\mathsf{G}}$, this statistic is distributed as a non-central t-distribution with non-centrality parameter

$$\delta = \sqrt{\frac{n}{1 + \frac{n}{n-1} \hat{\zeta}^2_{*,\mathsf{G}}}} \frac{c}{\sqrt{\boldsymbol{g}^\top \hat{\Sigma}^{-1} \boldsymbol{g}}},$$

and $n-k$ degrees of freedom. Here G is the $k-1 \times k$ matrix of rank $k-1$ whose rows are orthogonal to $\boldsymbol{v}$. Thus we reject at the α level if t is greater than $t_{1-\alpha}(\delta, n-1)$, the $1-\alpha$ quantile of the non-central t-distribution with $n-1$ degrees of freedom and non-centrality parameter δ.

1 sample multivariate regression coefficient test

Under the conditional expectation model of Equation 6.12 with Gaussian returns (Equation 6.13), to test

$$H_0 : \mathsf{B} = 0 \quad \text{versus} \quad H_1 : \mathsf{B} \neq 0,$$

compute the sample Hotelling Lawley trace, $\hat{T}$, and compare to an F-distribution using one of the approximations of Section 6.1.3, *e.g.*, McKeon's approximation. That is, reject if $\hat{T}$ exceeds some cutoff from a transformed upper $1-\alpha$ quantile of an appropriate F distribution.

1 sample expected squared maximal signal-noise ratio test

Under the conditional expectation model of Equation 6.12 with Gaussian returns (Equation 6.13), to test

$$H_0 : \operatorname{tr}\left(\Sigma^{-1} \mathsf{B} \Gamma_f \mathsf{B}^\top\right) = \zeta_0^2 \quad \text{versus} \quad H_1 : \operatorname{tr}\left(\Sigma^{-1} \mathsf{B} \Gamma_f \mathsf{B}^\top\right) > \zeta_0^2,$$

compute the sample Hotelling Lawley trace, $\hat{T}$, and compare to a non-central F-distribution using one of the approximations of Section 6.1.3, *e.g.*, McKeon's approximation. That is, reject if $\hat{T}$ exceeds some cutoff from a transformed upper $1 - \alpha$ quantile of an appropriate F distribution.

Example 6.3.6 (Fama French Four Factors, testing the mean). We consider the returns of the Market, SMB, HML, and UMD portfolios introduced in Example 1.1.1. Using monthly data from Jan 1927 to Dec 2020, we test the hypothesis

$$H_0 : \boldsymbol{\mu} = \begin{bmatrix} Mkt & SMB & HML & UMD \\ \hline 0.50 & 0.50 & 0.50 & 0.50 \end{bmatrix} \%\,\text{mo.}^{-1}.$$

In Example 6.1.1 we list the sample mean and covariance of returns. We compute

$$T^2 = 32.045.$$

The cutoff value at the 0.05 significance level is computed as

$$\frac{k(n-1)}{n-k} f_{1-\alpha}\left(k, n-k\right) = 9.545,$$

and we reject the null hypothesis. ⊣

Example 6.3.7 (Fama French Four Factors, January effect). We consider the returns of the Market, SMB, HML, and UMD portfolios introduced in Example 1.1.1. We consider the hypothesis that the vector mean of returns is the same in January as the rest of the year. We split the monthly returns into two groups: the $n_1 = 94$ Januaries, and the $n_2 = 1,034$ non-Januaries. We compute the sample means and pooled covariance matrix as

$$\hat{\boldsymbol{\mu}}_1^{\top} = \begin{bmatrix} Mkt & SMB & HML & UMD \\ \hline 1.506 & 2.022 & 1.790 & -1.553 \end{bmatrix} \%\,\text{mo.}^{-1},$$

$$\hat{\boldsymbol{\mu}}_2^{\top} = \begin{bmatrix} Mkt & SMB & HML & UMD \\ \hline 0.898 & 0.043 & 0.189 & 0.848 \end{bmatrix} \%\,\text{mo.}^{-1},$$

$$\frac{(n_1-1)\,\hat{\boldsymbol{\Sigma}}_1 + (n_2-1)\,\hat{\boldsymbol{\Sigma}}_2}{n_1+n_2-2} = \begin{bmatrix} & Mkt & SMB & HML & UMD \\ \hline Mkt & 28.666 & 5.426 & 4.433 & -8.515 \\ SMB & 5.426 & 9.864 & 1.191 & -2.016 \\ HML & 4.433 & 1.191 & 12.120 & -6.535 \\ UMD & -8.515 & -2.016 & -6.535 & 21.717 \end{bmatrix} \%^2\,\text{mo.}^{-1}.$$

We compute

$$T^2 = 61.756.$$

The cutoff value at the 0.05 significance level is computed as

$$\frac{k(n_1+n_2-2)}{n_1+n_2-1-k} f_{1-\alpha}\left(k, n_1+n_2-1-k\right) = 9.545,$$

and we reject the null hypothesis. ⊣

Example 6.3.8 (Fama French Four Factors, testing the maximal signal-noise ratio). We consider the returns of the Market, SMB, HML, and UMD portfolios introduced in Example 1.1.1. Using monthly data from Jan 1927 to Dec 2020, we test the hypothesis

$$H_0 : \zeta_* = 0.1 \text{ mo.}^{-1/2}.$$

Setting $r_0 = 0$, we compute $\hat{\zeta}_*^2 = 0.095 \text{ mo.}^{-1}$. The cutoff value at the 0.05 significance level is computed as

$$\frac{k(n-1)}{n(n-k)} f_{1-\alpha}\left(k, n-k, n\zeta_0^2\right) = 0.026 \text{ mo.}^{-1},$$

and we reject the null hypothesis at the 0.05 level. ⊣

Example 6.3.9 (Portfolio on the Fama French Four Factors, testing hedged portfolio). We continue Example 6.1.4, where we consider the optimal portfolio on the Fama French four factors with a hedge against the Market. We test

$$H_0 : \Delta_{\mathsf{I}\setminus\mathsf{G}}\zeta_* = 0.25 \text{ mo.}^{-1/2}.$$

As in Example 6.1.4, we compute

$$\hat{\zeta}_*^2 = 0.095 \,\text{mo.}^{-1}, \quad \hat{\zeta}_{*,\mathsf{G}}^2 = 0.031 \,\text{mo.}^{-1}, \quad \Delta_{\mathsf{I}\setminus\mathsf{G}}\hat{\zeta}_*^2 = 0.063 \,\text{mo.}^{-1}.$$

We compute the cutoff

$$\frac{n-1}{n}\frac{k-k_g}{n-k}\left(1 + \frac{n}{n-1}\hat{\zeta}_{*,\mathsf{G}}^2\right) f_{1-\alpha}\left(k-k_g, n-k, \frac{n\zeta_0^2}{1+\frac{n}{n-1}\hat{\zeta}_{*,\mathsf{G}}^2}\right) = 0.093 \text{ mo.}^{-1},$$

and we fail to reject at the 0.05 level. ⊣

Example 6.3.10 (Portfolio on the five industries, hedged against all). Continuing Example 6.1.5 we consider the optimal portfolio on the five industry baskets, subject to a hedge against the equal dollar allocation. We test

$$H_0 : \Delta_{\mathsf{I}\setminus\mathsf{G}}\zeta_* = 0 \text{ mo.}^{-1/2}.$$

As in Example 6.1.5, we compute

$$\hat{\zeta}_*^2 = 0.05 \,\text{mo.}^{-1}, \quad \hat{\zeta}_{*,\mathsf{G}}^2 = 0.036 \,\text{mo.}^{-1}, \quad \Delta_{\mathsf{I}\setminus\mathsf{G}}\hat{\zeta}_*^2 = 0.014 \,\text{mo.}^{-1}.$$

We compute the cutoff

$$\frac{n-1}{n}\frac{k}{n-k}\left(1 + \frac{n}{n-1}\hat{\zeta}_{*,\mathsf{G}}^2\right) f_{1-\alpha}\left(k-k_g, n-k, \frac{n\zeta_0^2}{1+\frac{n}{n-1}\hat{\zeta}_{*,\mathsf{G}}^2}\right) = 0.011 \text{ mo.}^{-1},$$

and narrowly reject the null hypothesis at the 0.05 level. ⊣

Example 6.3.11 (Portfolio on the Fama French Four Factors, value of SMB). We consider the returns of the Market, SMB, HML, and UMD portfolios introduced in Example 1.1.1. We consider the null hypothesis that the size factor, SMB, has zero weight in the population Markowitz portfolio. Using monthly data from Jan 1927 to Dec 2020, we compute $\hat{\zeta}_{*,\mathsf{G}}^2 = 0.095 \text{ mo.}^{-1}$. We compute the t statistic as $t = 0.47$. Compared to a central t-distribution with 1,124 degrees of freedom we fail to reject the null hypothesis at the 0.05 level. ⊣

Example 6.3.12 (Conditional Portfolio on the Fama French Four Factors, HLT). We consider the returns of the Market, SMB, HML, and UMD portfolios introduced in Example 1.1.1. In Example 6.1.6, we computed the Hotelling Lawley trace as $\hat{T} = 0.129\,\text{mo.}^{-1}$. To test the hypothesis $H_0 : \mathsf{B} = 0$, we compute cutoffs based on the approximations of Section 6.1.3. The Pillai approximation gives a cutoff of $0.019\,\text{mo.}^{-1}$; the Pillai-Samson approximation yields $0.019\,\text{mo.}^{-1}$; the McKeon approximation cutoff is $0.019\,\text{mo.}^{-1}$. While these are only approximations, a Frequentist would likely reject the null hypothesis, at the approximate 0.05 level. ⊣

Example 6.3.13 (Conditional Portfolio on the Fama French Four Factors, Flattening, Hotelling). In Example 6.1.7, we used the flattening trick to compute the passthrough matrix that maximizes the multi-period Sharpe ratio. There we found $\hat{\zeta}_*^2 = 0.136\,\text{mo.}^{-1}$. While there is no reason to believe that the product of features and returns takes a Gaussian distribution, if we assume that they do, we can apply the test of ζ_*^2 based on the Hotelling T^2-statistic. To test the null hypothesis $H_0 : \zeta_*^2 = 0.289\,\text{mo.}^{-1}$ at the 0.05 level, we compute the cutoff value as $0.127\,\text{mo.}^{-1}$, based on $n = 1,113$ and $k = 12$ (the number of assets times the number of features). We reject the null at the 0.05 level. ⊣

6.3.3 Power and sample size

In Section 2.5.8 we examined the relationship between sample size and effect size for the one-sample t-test, or equivalently the one-sample test for signal-noise ratio. Here we consider the analogous results for Hotelling's test, or the test for zero maximal signal-noise ratio. As noted in Section 6.3.2, the power of this test is

$$1 - \beta = 1 - F_f\left(f_{1-\alpha}\left(k, n-k, 0\right); k, n-k, n\zeta_*^2\right). \tag{6.73}$$

This equation implicitly defines a sample size, n given α, β, k, and ζ_*^2; however, it is not enlightening.

As it happens, for fixed values of α, β, and k, the sample size relationship is similar to that found for the t-test in Equation 2.37, namely

$$n \approx \frac{c}{\zeta_*^2}, \tag{6.74}$$

where the constant c depends on α, β, and k. For $\alpha = 0.05$, an approximate value of the numerator c is given in Table 6.4 for a few different values of the power. Note that for $k = 1$, we should recover the same sample-size relationship as shown in Table 2.2 for the two-sided test. This is because Hotelling's T^2 for $k = 1$ is simply Student's t-statistic squared (and thus directional information is lost).

As a test we computed the n required to achieve a power of 0.8 for different values of k and ζ_*, then plotted $n\zeta_*^2$ vesus ζ_* in Figure 6.2. Under the approximation of Equation 6.74, these points should fall on horizontal lines

power	numerator
0.25	$1.66k^{0.438+0.006 \log k}$
0.50	$3.86k^{0.351+0.012 \log k}$
0.80	$7.87k^{0.277+0.017 \log k}$
0.95	$13.03k^{0.228+0.020 \log k}$

TABLE 6.4
The approximate numerator, c, in the sample size relationship $n = c\zeta_*^{-2}$ required to achieve a fixed power for the one-sample Hotelling test is given as a function of k. The type I rate is 0.05.

for different values of k. We also plot the fit from Table 6.4, and find that the approximation is fairly good for the ζ_* one is likely to encounter in practice, and the relatively modest k tested. We also plot lines corresponding to $n = 2\,\text{yr}$ and $5\,\text{yr}$ for reference.

As the $k = 1$ case corresponds to the power relationship of the test for zero signal-noise ratio outlined in Section 2.5.8, we see that as you move to larger universe sizes, the n required to achieve a given power grows for fixed ζ_*. This should not be surprising as the possibility of committing a type I error presumably increases in the number of assets. However, one typically adds assets to one's investment universe to chase higher signal-noise ratio[1]. The fundamental law of active management posits that $\log \zeta_*^2 \approx c_0 + \log k$. This represents a growth rate in k that mostly outpaces the power relationship, which takes the form $\log \zeta_*^2 \approx b_0 + b_1 \log k + b_2 (\log k)^2$, with $b_1 < \frac{1}{2}$ and b_2 very small. However, if the growth of ζ_*^2 in k is slower than the fundamental law suggests, the addition of assets can require an increase in n to achieve the same power and type I rate.

As in the $k = 1$ case, the power rule is rather harsh. For example, suppose you believed that for your universe of $k = 10$ assets that $\zeta_* = 0.9\ \text{yr}^{-1/2}$. If you were to rely on Hotelling's test with $\alpha = 0.05$, then to achieve a power of $\frac{1}{2}$, you would have to observe at least $n = 2,869$ days of daily returns, or around 11.4 yr. In fact, in this case the power of Hotelling's test when one observes only one year of daily data is around 0.074, barely above the type I rate.

[1]Or to increase capacity.

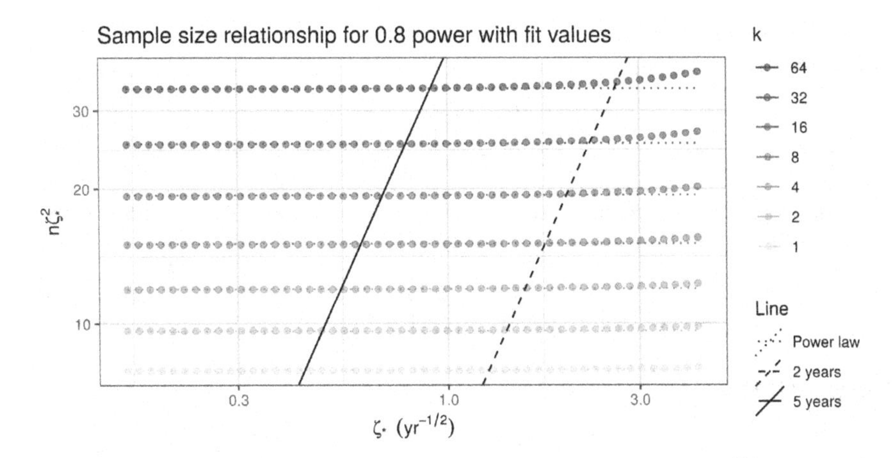

FIGURE 6.2
The n required to achieve a power of 0.8 and a type I rate of 0.05 is computed for different k and ζ_*. The product $n\zeta_*^2$ is plotted as points versus ζ_* for varying k. The fit given by $7.87k^{0.277+0.017\log k}$ is plotted as lines. The fits are reasonably good except for large (perhaps unrealistically so) values of ζ_*. Lines corresponding to $n = 2\,\text{yr}$ and $n = 5\,\text{yr}$ are also drawn. Fits are based on daily data assuming 252 days per year. Both x and y axes are in log space.

6.4 † Likelihoodist inference on squared maximal signal-noise ratio

The likelihood of the squared maximal signal-noise ratio is, as likelihoods are, just a different view of the density of $\hat{\zeta}_*^2$:

$$\mathcal{L}_{SR,*}\left(\zeta_*^2 \,\middle|\, \hat{\zeta}_*^2, k, n\right) := f_{SR*}\left(\hat{\zeta}_*^2; n, k, \zeta_*^2\right), \tag{6.75}$$

with the latter density given in Equation 6.46. As the likelihood consists of an infinite sum of central F densities, themselves being hard to evaluate, we are unlikely to find analytic solutions to common likelihood problems.

Indeed to compute the MLE of ζ_*^2, it is simplest to employ a numerical procedure to maximize the likelihood. It appears that the MLE will typically be smaller than $\hat{\zeta}_*$ (and is non-negative), and thus a simple golden section search can be employed.

There is, however, one simple analytic result regarding the MLE: Spruill [1986] gives a sufficient condition for the MLE of the non-centrality parameter of a non-central F distribution to be zero, given multiple observations. For the case of a single observation, which corresponds to our case, the condition is particularly simple: if the random variable is no greater than one, the MLE of the non-centrality parameter is equal to zero. The equivalent fact about the optimal Sharpe ratio is that if

$$\hat{\zeta}_*^2 \leq \left(\frac{n-1}{n}\right)\frac{c_a}{1-c_a}, \tag{6.76}$$

then the MLE of ζ_* is zero, where $c_a = k/n$ is the "aspect ratio." The equivalent condition for the case of spanned portfolios is if

$$\Delta_{\mathsf{I}\setminus\mathsf{G}}\hat{\zeta}_*^2 \leq \left(\frac{n-1}{n}\right)\frac{k-k_g}{n-k}\left(1+\frac{n}{n-1}\hat{\zeta}_{*,\mathsf{G}}^2\right), \tag{6.77}$$

then the MLE for $\Delta_{\mathsf{I}\setminus\mathsf{G}}\,\zeta_*^2$ is zero.

Spruill's condition

If $\hat{\zeta}_*$ is smaller than around $\sqrt{c_a/\left(1-c_a\right)}$, then the MLE of ζ_* is zero.

Example 6.4.1 (MLE and Spruill's Condition). We consider the case of $n = 1,008$ days, $k = 10$. We spawn hypothetical $\hat{\zeta}_*$, then compute the MLE for ζ_* using a numerical search and the density of Equation 6.46. We repeat this for a range of hypothetical $\hat{\zeta}_*$. We plot the MLE versus $\hat{\zeta}_*$, with a vertical line at Spruill's cutoff of $\sqrt{\frac{n-1}{n}\frac{c_a}{1-c_a}}$ in Figure 6.3. We see that Spruill's condition is indeed respected in this case. Moreover, we see that when $\hat{\zeta}_*$ exceeds Spruill's cutoff, it appears to approach $y = x$ asymptotically. ⊣

Example 6.4.2 (Fama French Four Factors, MLE). We consider the returns of the Fama French four factors considered in Example 6.1.1. In that example, we computed (ignoring the drag term) $\hat{\zeta}_* = 0.308\,\text{mo.}^{-1/2}$. Here we compute the numerical MLE of ζ_* to be $0.303\ \text{mo.}^{-1/2}$. Compare to the improved risk estimates, which in Example 6.2.5 we computed to be $\sqrt{\delta_2} = 0.301\,\text{mo.}^{-1/2}$. ⊣

Example 6.4.3 (Five industry portfolios, MLE). We consider the returns of the five baskets of stocks categorized by major industry introduced in Example 1.2.5. We compute $\hat{\zeta}_* = 0.223\ \text{mo.}^{-1/2}$. Based on this and $n = 1,128$ mo., we compute the MLE of ζ_* to be $0.215\ \text{mo.}^{-1/2}$. ⊣

Example 6.4.4 (Fama French Four Factors, hedged against Market, MLE). We continue Example 6.1.4 where we considered the optimal portfolio on the Fama French four factors, hedged against the Market. There we computed $\Delta_{\mathsf{I}\setminus\mathsf{G}}\hat{\zeta}_* = 0.252\ \text{mo.}^{-1/2}$. Here we compute the MLE of $\Delta_{\mathsf{I}\setminus\mathsf{G}}\zeta_*$ to be $0.248\ \text{mo.}^{-1/2}$. ⊣

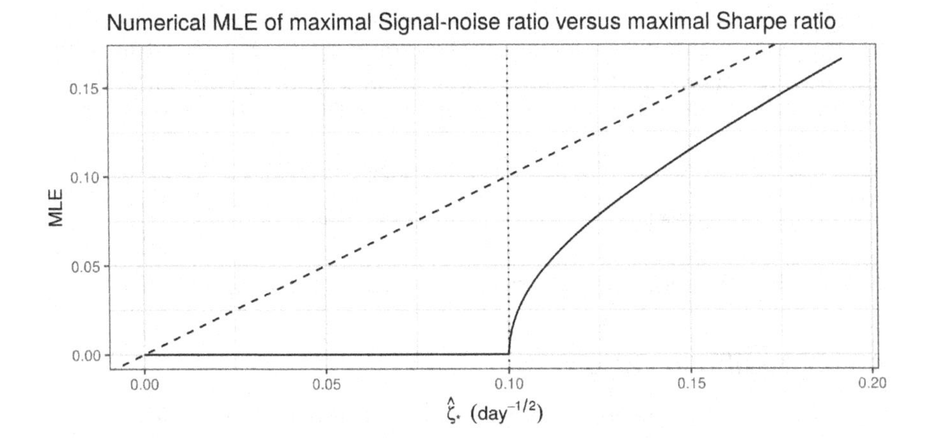

FIGURE 6.3
The MLE, found numerically, of ζ_* is plotted versus hypothetical $\hat{\zeta}_*$ for the case where $k = 10$ and $n = 1,008$ days. Both axes are in daily units, $\text{day}^{-1/2}$. A vertical line is plotted at the cutoff from Spruill's condition on the MLE, which is confirmed here. The line $y = x$ is also plotted.

MLE for the conditional expectation model

Now consider the conditional expectation model, where the statistic of interest is the Hotelling Lawley trace. To find the approximate MLE of the population analogue of the HLT, one can use one of the approximations via the F distribution given in Section 6.1.3, *e.g.*, McKeon's approximation. As with inference on ζ_*, the MLE must be found numerically.

Example 6.4.5 (Conditional Portfolio on the Fama French Four Factors, MLE). In Example 6.1.6, we computed the sample Hotelling Lawley trace of the the Market, SMB, HML, and UMD portfolios introduced in Example 1.1.1 as $\hat{T} = 0.129\ \text{mo.}^{-1}$. Via the Pillai approximation we estimate the MLE of T as $0.118\ \text{mo.}^{-1}$; via Pillai approximation we have $0.118\ \text{mo.}^{-1}$; McKeon's approximation yields $0.118\ \text{mo.}^{-1}$. ⊣

† MLE for the matrix normal model

Consider the model where the product of returns and features follow a matrix normal law

$$\boldsymbol{x}_t \boldsymbol{f}_{t-1}^{\top} \sim \mathcal{MN}_{k,f}\left(\mathsf{M}, \Psi, \Xi\right). \tag{6.78}$$

First let $\hat{\mathsf{M}}$ be the MLE for M, which is simply expressed as

$$\hat{\mathsf{M}} = \frac{1}{n} \mathsf{X}^{\top} \mathsf{F}.$$

Now suppose you wish to maximize the likelihood conditional on $\mathsf{M} = \mathsf{M}_0$.

Typically one uses $\mathsf{M}_0 = \hat{\mathsf{M}}$. Dutilleul's method is an iterative scheme for estimating the Ψ and Ξ that maximize the profile likelihood. [Dutilleul, 1999] It proceeds as follows: first estimate $\hat{\Psi}_0$ in some way, by letting $\hat{\Psi}_0 = \hat{\Sigma}$. Set $i = 0$ and repeat the following iterative update until the estimates converge

$$\begin{aligned}\hat{\Xi}_{i+1} &\leftarrow \frac{1}{nk}\mathsf{F}^{\top}\operatorname{Diag}\left(\operatorname{diag}\left(\mathsf{X}\hat{\Psi}_i^{-1}\mathsf{X}^{\top}\right)\right)\mathsf{F} - \frac{1}{k}\left(\mathsf{M}_0{}^{\top}\hat{\Psi}_i^{-1}\hat{\mathsf{M}} + \hat{\mathsf{M}}^{\top}\hat{\Psi}_i^{-1}\mathsf{M}_0 - \hat{\mathsf{M}}^{\top}\hat{\Psi}_i^{-1}\hat{\mathsf{M}}\right),\\ \hat{\Psi}_{i+1} &\leftarrow \frac{1}{nf}\mathsf{X}^{\top}\operatorname{Diag}\left(\operatorname{diag}\left(\mathsf{F}\hat{\Xi}_{i+1}^{-1}\mathsf{F}^{\top}\right)\right)\mathsf{X} - \frac{1}{f}\left(\mathsf{M}_0\hat{\Xi}_i^{-1}\hat{\mathsf{M}}^{\top} + \hat{\mathsf{M}}\hat{\Xi}_i^{-1}\mathsf{M}_0{}^{\top} - \hat{\mathsf{M}}\hat{\Xi}_i^{-1}\hat{\mathsf{M}}^{\top}\right).\end{aligned} \tag{6.79}$$

When $\mathsf{M}_0 = \hat{\mathsf{M}}$ the terms in parentheses simplify by some cancellation.

6.4.1 Likelihood ratio test

To test $H_0 : \zeta_* = \zeta_0$ versus an unrestricted alternative, find the MLE, then compute the test statistic

$$D = -2\log\frac{\mathcal{L}_{SR,*}\left(\zeta_0^2 \,\middle|\, \hat{\zeta}_*^2, n, k\right)}{\mathcal{L}_{SR,*}\left(\zeta_{\mathrm{MLE}}^2 \,\middle|\, \hat{\zeta}_*^2, n, k\right)}. \tag{6.80}$$

It appears that the unrestricted model has only one extra degree of freedom, namely that ζ_*^2 can take any value. This is certainly the case under the null hypothesis when $\zeta_*^2 > 0$; in this case, by Wilk's theorem we expect D to converge (in n) to a χ^2 random variable with one degree of freedom. [Pawitan, 2001] However, when $\zeta_*^2 = 0$ the parameter is on the boundary of acceptable values of ζ_*^2, and one sees a rather more complicated asymptotic distribution of D. [Andrews, 2001] Indeed, the hypothesis $\zeta_*^2 = 0$ implies that $\boldsymbol{\mu} = \mathbf{0}$, which would seem to consume k degrees of freedom. It turns out that the likelihood ratio test for $\boldsymbol{\mu} = \mathbf{0}$ is actually equivalent to Hotelling's T^2 test, *cf.* Exercise 6.25, suggesting that the LRT to test $H_0 : \zeta_*^2 = 0$ should not be used, rather the Frequentist test based on the f distribution should be used instead.

When ζ_*^2 is merely "near" zero, the quality of the asymptotic approximation will depend on the sample size, n. Moreover, it appears that deviations from the asymptotic χ^2 distribution would mostly affect tests against the alternative $\zeta_* < \zeta_0$, while achieving an upper bound on the type I rate against alternatives of the form $\zeta_* > \zeta_0$. In summary, we urge caution in using the LRT for tests of this hypothesis.

Example 6.4.6 (Likelihood Ratio Test). We draw 120 months of normal returns for 5 assets for $\zeta_* = \zeta_0$. From $\hat{\zeta}_*$, we compute the MLE of ζ_* numerically, then compute the LRT against the null hypothesis $H_0 : \zeta_* = \zeta_0$. We convert the D to a p-value under the $\chi^2(1)$ approximation. We repeat this 1,000 times. We perform this experiment for $\zeta_0 = 0\ \text{mo.}^{-1/2}$ and $\zeta_0 = 0.3\ \text{mo.}^{-1/2}$. In Figure 6.4, we P-P plot the putative p-values against a uniform law.

For $\zeta_0 = 0\ \text{mo.}^{-1/2}$ we see that a large proportion of the computed p-values are zero, as the MLE is equal to zero, and thus $D = 0$. We plot a vertical dotted line at the theoretical probability that $\hat{\zeta}_*$ does not exceed

Spruill's cutoff, which we compute as 0.579. This theoretical cutoff coincides with the empirical rate of zero D values.

For $\zeta_0 = 0.3$ mo.$^{-1/2}$, while there is a non-zero probability that $\hat{\zeta}_*$ does not exceed Spruill's cutoff, this does not correspond to $D = 0$. Instead one can witness a slight deviation from the $\chi^2(1)$ approximation at larger p-values. $\dashv$

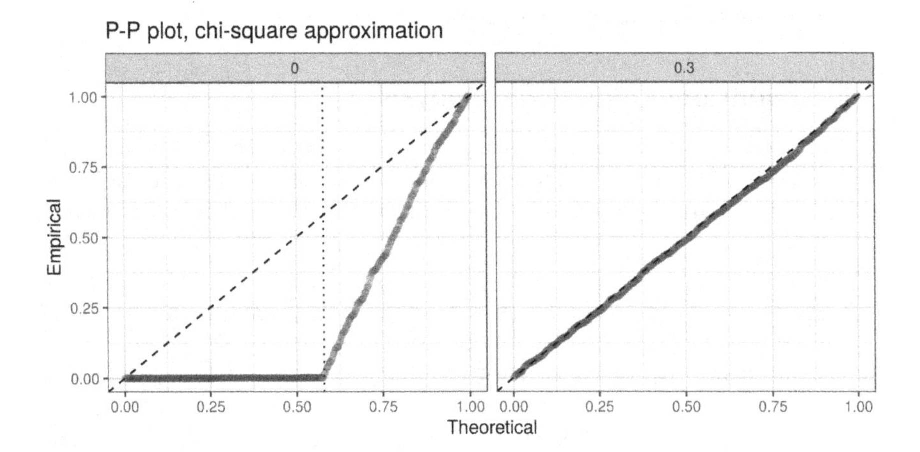

FIGURE 6.4
We P-P plot p-values under the $\chi^2(1)$ approximation to D for simulations under the null hypothesis. The two facets are differentiated by the value of ζ_*. Under the boundary value $\zeta_* = 0$ the $\chi^2(1)$ approximation is clearly poor.

Example 6.4.7 (Fama French Four Factors, testing the maximal signal-noise ratio). In Example 6.3.8 we tested the hypothesis

$$H_0 : \zeta_* = 0.1 \text{ mo.}^{-1/2}.$$

on the returns of the Market, SMB, HML, and UMD portfolios introduced in Example 1.1.1. We compute the MLE of ζ_* as 0.303 mo.$^{-1/2}$. We compute the likelihood statistic $D = 43.194$, which is much larger than the 0.95 quantile of the χ^2-distribution with four degrees of freedom, and we reject the null hypothesis at the 0.05 level. $\dashv$

To test $H_0 : \zeta = \zeta_0$ versus $H_1 : \zeta = \zeta_1$, compute the test statistic

$$\Lambda = \frac{\mathcal{L}_{SR,*}\left(\zeta_0^2 \middle| \hat{\zeta}_*^2, n, k\right)}{\mathcal{L}_{SR,*}\left(\zeta_1^2 \middle| \hat{\zeta}_*^2, n, k\right)}. \tag{6.81}$$

When Λ is small, smaller than some cutoff, we prefer H_1; when it is large, we prefer H_0. To use this test statistic in a null hypothesis test, one must find the cutoff value to reject H_0 in favor of H_1, with the cutoff value chosen to achieve the desired type I rate. That is, the Likelihood Ratio Test is performed by rejecting H_0 when $\Lambda < c$, where c is chosen to give the test the requisite type I error rate. However, it is unknown whether one can easily compute such a c for given ζ_0, ζ_1. If the likelihood ratio Λ was monotonic in $\hat{\zeta}_*^2$ (recall we saw an analogous result for testing the signal-noise ratio in Section 2.6.1), then the LRT would be equivalent to the Frequentist F-test. We conjecture that Λ is indeed monotonic, but know of no proof.

6.4.2 Traditional likelihood analysis

The likelihood analysis above focused only on ζ_*^2 and $\hat{\zeta}_*^2$. For comparison of different models for the same data, the traditional multivariate normal likelihood analysis should be used. The log likelihood of the multivariate normal is

$$\log \mathcal{L}_{\mathcal{N}}\left(\boldsymbol{\mu}, \Sigma \,|\mathsf{X}\right) = -\frac{nk}{2} \log 2\pi - \frac{n}{2} \log |\Sigma| - \frac{1}{2} \operatorname{tr}\left(\Sigma^{-1} \sum_t \left(\boldsymbol{x}_t - \boldsymbol{\mu}\right) \left(\boldsymbol{x}_t - \boldsymbol{\mu}\right)^\top\right). \quad (6.82)$$

Conditional on $\boldsymbol{\mu} = \boldsymbol{\mu}_0$ the maximum likelihood estimate of Σ is [Anderson, 2003]

$$\Sigma_0 = \frac{1}{n} \sum_t \left(\boldsymbol{x}_t - \boldsymbol{\mu}_0\right) \left(\boldsymbol{x}_t - \boldsymbol{\mu}_0\right)^\top. \quad (6.83)$$

Thus the log likelihood for a given choice of $\boldsymbol{\mu}_0$ is

$$\log \mathcal{L}_{\mathcal{N}}\left(\boldsymbol{\mu}_0, \Sigma_0 \,|\mathsf{X}\right) = -\frac{nk}{2} \log 2\pi - \frac{n}{2} \log |\Sigma_0| - \frac{nk}{2}. \quad (6.84)$$

Note that a very similar analysis holds for the conditional expectation case where the features are deterministic. Though in this case the maximum likelihood estimate of the covariance, conditional on $\mathsf{B} = \mathsf{B}_0$, is

$$\Sigma_0 = \frac{1}{n} \sum_t \left(\boldsymbol{x}_t - {\mathsf{B}_0}^\top \boldsymbol{f}_{t-1}\right) \left(\boldsymbol{x}_t - {\mathsf{B}_0}^\top \boldsymbol{f}_{t-1}\right)^\top. \quad (6.85)$$

Moreover, in the more common case where the features are considered random, unless we are performing inference on them, we can marginalize them out and treat them as fixed. Thus we can generally use Equation 6.84 to compute the maximized likelihood to compare models on the same data. See also Exercise 6.25.

Comparison of models should proceed by computing and comparing an Information Criterion, such as the *Akaike Information Criterion* or *Bayesian Information Criterion.* [Pawitan, 2001] If m is the number of estimated parameters in $\boldsymbol{\mu}_0$ or in B_0, then

$$AIC = 2m + \left(nk \log 2\pi + n \log |\Sigma_0| + nk\right), \quad (6.86)$$

$$BIC = \log nm + \left(nk \log 2\pi + n \log |\Sigma_0| + nk\right). \quad (6.87)$$

The term in the parenthesis is negative two times the log likelihood at the MLE. One selects the model which *minimizes* the chosen information criterion.

Example 6.4.8 (Information Criteria, Fama French Four Factors). We consider the returns of the Market, SMB, HML, and UMD portfolios introduced in Example 1.1.1. We choose every nonempty subset of the four assets to fit the $\boldsymbol{\mu}_0$ letting any non-selected asset have zero mean. We compute the corresponding Σ_0, then the log likelihood, AIC and BIC. We find that the full model, which fits a non-zero mean to all four assets, has the lowest AIC, with value 25037.4. (This value is not interpretable in any way.) The model with the lowest BIC fits non-zero mean to Mkt, HML, UMD only. ⊣

6.5 † Bayesian inference on squared maximal signal-noise ratio

In analogy to Bayesian inference on the signal-noise ratio, we first consider prior distributions on the mean and covariance, $\boldsymbol{\mu}$ and Σ, and only consider ζ_*^2 as a function of these parameters. This is directly analogous to the Normal-Inverse-Gamma prior of Equation 2.45. In the *Normal-Inverse-Wishart* prior, one has an unconditional inverse Wishart prior distribution on Σ, and conditional on Σ a normal prior on $\boldsymbol{\mu}$. [Gelman *et al.*, 2013; Murphy, 2007] These can be stated as

$$\begin{aligned} \Sigma &\propto \mathcal{IW}\left(m_0\Omega_0, m_0\right), \\ \boldsymbol{\mu}\,|\Sigma &\propto \mathcal{N}\left(\boldsymbol{\mu}_0, \frac{1}{n_0}\Sigma\right), \end{aligned} \tag{6.88}$$

where $\Omega_0, m_0, \boldsymbol{\mu}_0$, and n_0 are the hyper-parameters. Note that this formulation differs from the traditional multivariate prior by a factor of m_0 in the inverse Wishart parameter. However, in this formulation, Ω_0 is a kind of estimate of Σ. The density function of the inverse Wishart is (*cf.* Equation A.14)

$$f_{\mathcal{IW}}\left(\mathsf{Y}; \Psi, n\right) = \frac{\left|\mathsf{Y}\right|^{n+p+1/2}\left|\Psi\right|^{n/2}}{2^{np/2}\Gamma\left(n/2\right)} e^{-\frac{1}{2}\operatorname{tr}(\Psi\mathsf{Y})}. \tag{6.89}$$

Under this formulation, an non-informative prior corresponds to $m_0 = 0 = n_0$.

After observing n *i.i.d.* draws from a normal distribution, $\mathcal{N}\left(\boldsymbol{\mu}, \Sigma\right)$, let $\hat{\boldsymbol{\mu}}$ and $\hat{\Sigma}$ be the sample estimates from Equation 6.1. The posterior is then

$$\begin{aligned} \Sigma &\propto \mathcal{IW}\left(m_1\Omega_1, m_1\right), \\ \boldsymbol{\mu}\,|\Sigma &\propto \mathcal{N}\left(\boldsymbol{\mu}_1, \frac{1}{n_1}\Sigma\right), \end{aligned} \tag{6.90}$$

where

$$n_1 = n_0 + n, \quad \boldsymbol{\mu}_1 = \frac{n_0 \boldsymbol{\mu}_0 + n\hat{\boldsymbol{\mu}}}{n_1}, \tag{6.91}$$

$$m_1 = m_0 + n, \quad \Omega_1 = \frac{1}{m_1}\left(m_0\Omega_0 + (n-1)\hat{\Sigma} + \frac{n_0 n}{n_1}\left(\hat{\boldsymbol{\mu}} - \boldsymbol{\mu}_0\right)\left(\hat{\boldsymbol{\mu}} - \boldsymbol{\mu}_0\right)^{\top}\right). \tag{6.92}$$

This model can be modified to one on the covariance matrix and the maximal signal-noise ratio, where the former is a nuisance parameter. Transforming Equation 6.90, we arrive at

$$\begin{aligned} \Sigma &\propto \mathcal{IW}\left(m_1\Omega_1, m_1\right), \\ n_1\zeta_*^2 \left|\Sigma\right. &\propto \chi^2\left(k; n_1 {\boldsymbol{\mu}_1}^{\top}\Sigma^{-1}\boldsymbol{\mu}_1\right). \end{aligned} \tag{6.93}$$

The marginal posterior can be expressed in terms of Lecoutre's *lambda-square distribution*:

$$n_1\zeta_*^2 \propto \Lambda_{k,m_1}^2\left(n_1 {\boldsymbol{\mu}_1}^{\top}\Omega_1^{-1}\boldsymbol{\mu}_1\right), \tag{6.94}$$

where $\Lambda_{m,q}^2\left(a^2\right)$ is the lambda-square distribution with m and q degrees of freedom and "eccentricity" a^2, *cf.* Section A.2.2.

Example 6.5.1 (Bayesian update, Fama French Four Factors). We consider the returns of the Market, SMB, HML, and UMD portfolios. We assume prior parameters of $n_0 = 24$ mo., $m_0 = 36$ mo.,

$${\boldsymbol{\mu}_0}^{\top} = \begin{bmatrix} Mkt & SMB & HML & UMD \\ \hline 0.150 & 0.150 & 0.150 & 0.150 \end{bmatrix} \%\,\text{mo.}^{-1},$$

$$\Omega_0 = \begin{bmatrix} & Mkt & SMB & HML & UMD \\ \hline Mkt & 6.000 & 0.000 & 0.000 & 0.000 \\ SMB & 0.000 & 6.000 & 0.000 & 0.000 \\ HML & 0.000 & 0.000 & 6.000 & 0.000 \\ UMD & 0.000 & 0.000 & 0.000 & 6.000 \end{bmatrix} \%^2\,\text{mo.}^{-1}.$$

In Example 6.1.1, we list the sample mean and covariance of returns, based on $n = 1,128$ mo. Integrating these we arrive at the posterior $n_1 = 1,152$ mo., $m_1 = 1,164$ mo.,

$${\boldsymbol{\mu}_1}^{\top} = \begin{bmatrix} Mkt & SMB & HML & UMD \\ \hline 0.932 & 0.207 & 0.319 & 0.637 \end{bmatrix} \%\,\text{mo.}^{-1},$$

$$\Omega_1 = \begin{bmatrix} & Mkt & SMB & HML & UMD \\ \hline Mkt & 27.956 & 5.339 & 4.363 & -8.337 \\ SMB & 5.339 & 10.017 & 1.387 & -2.301 \\ HML & 4.363 & 1.387 & 12.100 & -6.604 \\ UMD & -8.337 & -2.301 & -6.604 & 21.625 \end{bmatrix} \%^2\,\text{mo.}^{-1}.$$

The marginal posterior for ζ_*^2 is then

$$1152\zeta_*^2 \propto \Lambda_{4,1164}^2\left(107.571\,\text{mo.}^{-1}\right).$$

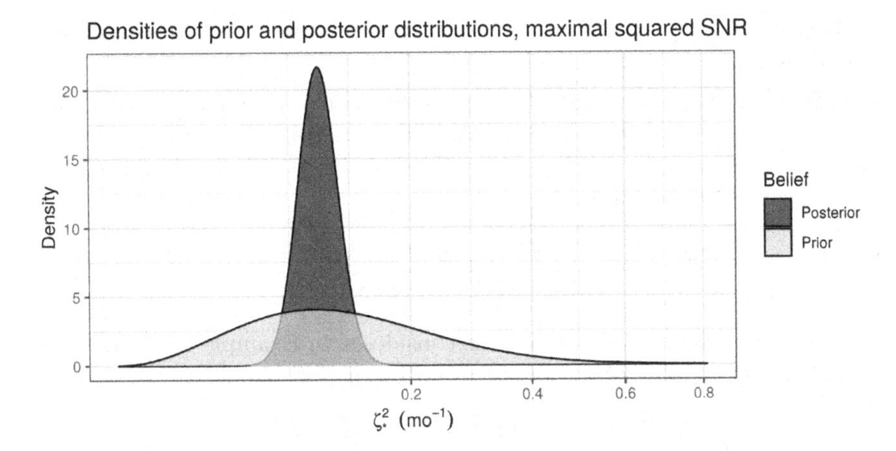

FIGURE 6.5
Prior and posterior density of ζ_*^2 are plotted. The x-axis is in square root scale.

In Figure 6.5, we plot the prior and posterior marginal density of belief on ζ_*^2. We note that the mode of the posterior is approximately 0.094 mo.$^{-1}$ which is indeed very close to the squared maximal Sharpe ratio, $\hat{\zeta}_*^2 = 0.095$ mo.$^{-1}$. ⊣

Example 6.5.2 (Bayesian update, non-informative prior, five industries). In Example 6.1.2, we considered a portfolio on the five industry baskets. Based on a non-informative prior, $m_0 = 0 = n_0$, we perform the Bayesian update. We compute the mode of the posterior marginal distribution on ζ_*^2 to be around 0.052 mo.$^{-1}$, the square root of which is 0.227 mo.$^{-1/2}$, very close to $\hat{\zeta}_* = 0.223$ mo.$^{-1/2}$. ⊣

Bayesian Portfolios

We have so far only focused on Bayesian inference on ζ_*^2. To be true to their beliefs, the Bayesian should not allocate into the plug-in Markowitz portfolio, as a Frequentist might. Rather, they should pull a portfolio out of their posterior. We can rewrite the posterior of Equation 6.90 as

$$\begin{aligned} \Sigma &\propto \mathcal{IW}\left(m_1 \Omega_1, m_1\right), \\ \Sigma^{-1}\boldsymbol{\mu} | \Sigma &\propto \mathcal{N}\left(\Sigma^{-1}\boldsymbol{\mu}_1, \frac{1}{n_1}\Sigma^{-1}\right). \end{aligned} \tag{6.95}$$

The posterior marginal distribution on $\Sigma^{-1}\boldsymbol{\mu}$ does not take a well-known form. However, the expected value of this distribution is

$$\frac{1}{m_1\left(m_1 - k - 1\right)} {\Omega_1}^{-1} \boldsymbol{\mu}_1. \tag{6.96}$$

Alternatively, one could take draws from this distribution and estimate the median value.

Example 6.5.3 (Bayesian Portfolio on the Fama French Four Factors). We consider the returns of the Market, SMB, HML, and UMD portfolios. In Example 6.1.1, based on a risk budget of $R = 1.5\,\%\,\text{mo.}^{-1/2}$, we computed the sample Markowitz portfolio as

$$\hat{\boldsymbol{\nu}}_*{}^{\top} = \frac{\begin{bmatrix} Mkt & SMB & HML & UMD \end{bmatrix}}{\begin{bmatrix} 0.212 & 0.024 & 0.209 & 0.292 \end{bmatrix}} \,\%^{-1}.$$

From the Bayesian posterior on returns considered in Example 6.5.1, we compute the expected value of the Bayesian posterior portfolio as

$$\hat{\boldsymbol{\nu}}_{B,*}{}^{\top} = \frac{\begin{bmatrix} Mkt & SMB & HML & UMD \end{bmatrix}}{\begin{bmatrix} 0.212 & 0.026 & 0.208 & 0.292 \end{bmatrix}} \,\%^{-1}.$$

Here we have rescaled the Bayesian portfolio to have the same ℓ_2 norm as the plug-in Markowitz portfolio. Drawing 10,000 simulations from the posterior distribution, and rescaling them to have the same the ℓ_2 norm as the plug-in portfolio, we estimate the median from the posterior marginal distribution as

$$\hat{\boldsymbol{\nu}}_{B,*}{}^{\top} = \frac{\begin{bmatrix} Mkt & SMB & HML & UMD \end{bmatrix}}{\begin{bmatrix} 0.210 & 0.025 & 0.205 & 0.289 \end{bmatrix}} \,\%^{-1}.$$

We conclude that the very weak prior is overwhelmed by the evidence, and thus the posterior belief portfolio is almost identical to the plug-in Markowitz portfolio. ⊣

Black Litterman portfolio

The *Black Litterman* portfolio is a Bayesian portfolio construction that combines simple "views" on the returns with likelihood supplied by inference on the market portfolio to approximate the likelihood part. [Satchell and Scowcroft, 2000; Bertsimas *et al.*, 2012] We assume that the covariance Σ is known, then follow the classical Bayesian conjugate analysis for the multivariate Gaussian case with unknown mean and known covariance. [Murphy, 2007] The Black Litterman prior can be expressed as

$$\mathsf{P}\boldsymbol{\mu} \propto \mathcal{N}\left(\boldsymbol{q}, \operatorname{Diag}\left(\boldsymbol{\tau}\right)\right), \tag{6.97}$$

where P is the matrix of views, each of which is a portfolio, $\boldsymbol{q}$ is the prior belief on the expected returns of the view portfolios, and $\boldsymbol{\tau}$ are the uncertainties in belief.

Observing $\hat{\boldsymbol{\mu}}$ over n samples, we assume $\hat{\boldsymbol{\mu}} \sim \mathcal{N}\left(\boldsymbol{\mu}, \frac{1}{n}\Sigma\right)$. The posterior belief should then be

$$\boldsymbol{\mu} \propto \mathcal{N}\left(\boldsymbol{\mu}_1, \Omega_1\right),$$

where

$$\Omega_1 = \left(\mathsf{P}\mathrm{Diag}\left(\boldsymbol{\tau}\right)^{-1}\mathsf{P}^\top + n\Sigma^{-1}\right)^{-1}, \tag{6.98}$$

$$\boldsymbol{\mu}_1 = \Omega_1\left(\mathsf{P}^\top\mathrm{Diag}\left(\boldsymbol{\tau}\right)^{-1}\boldsymbol{q} + n\Sigma^{-1}\hat{\boldsymbol{\mu}}\right). \tag{6.99}$$

But rather than compute $\hat{\boldsymbol{\mu}}$, it is inferred from the portfolio held by the market as a whole, which is assumed to be proportional to $\Sigma^{-1}\hat{\boldsymbol{\mu}}$. Thus one only need to know (estimate, really), Σ and n. The posterior of the Markowitz portfolio then has expected value $\Sigma^{-1}\boldsymbol{\mu}_1$.

6.5.1 Credible intervals on the squared maximal signal-noise ratio

One can construct credible intervals on the squared maximal signal-noise ratio based on the posterior, so-called *posterior intervals*, via quantiles of the lambda-square distribution. [Gelman *et al.*, 2013] For example, a $(1-\alpha)$ credible interval on ζ_*^2 is given by

$$\frac{1}{n_1}\left[{\Lambda^2}_{\alpha/2;\,k,m_1}\left(n_1{\boldsymbol{\mu}_1}^\top{\Omega_1}^{-1}\boldsymbol{\mu}_1\right), {\Lambda^2}_{(2-\alpha)/2;\,k,m_1}\left(n_1{\boldsymbol{\mu}_1}^\top{\Omega_1}^{-1}\boldsymbol{\mu}_1\right)\right). \tag{6.100}$$

Here ${\Lambda^2}_{p;\,m,q}\left(a^2\right)$ is the p-quantile of the lambda-square distribution with m and q degrees of freedom and "eccentricity" a^2.

Example 6.5.4 (Fama French Four Factors, credible interval). In Example 6.3.1, we computed the 95% confidence interval on ζ_* as $[0.243, 0.363)$ mo.$^{-1/2}$. Here we compute 95% *credible interval* via Equation 6.100, using a non-informative prior $m_0 = 0 = n_0$. Rather than compute the quantiles of the lambda-square distribution directly (no good implementation is available), we simulate draws from it. We compute the credible interval as $[0.253, 0.372)$ mo.$^{-1/2}$. ⊣

6.5.2 Bayesian inference, conditional expectation

We now turn our attention to the Bayesian update for the conditional expectation model of Equation 6.13. First, we consider the case where the features are assumed to be deterministic, rather than random. The Normal-Inverse-Wishart prior can be generalized to this case. We start with the prior

$$\begin{aligned} \Sigma &\propto \mathcal{IW}\left(m_0\Omega_0, m_0\right), \\ \mathsf{B}\,|\,\Sigma &\propto \mathcal{MN}\left(\mathsf{B}_0, \Sigma, \frac{1}{n_0}{\mathsf{\Gamma}_0}^{-1}\right), \end{aligned} \tag{6.101}$$

where $\Omega_0, n_0, m_0, \mathsf{B}_0$, and $\mathsf{\Gamma}_0$ are the hyper-parameters.

The posterior is then

$$\begin{aligned} \Sigma &\propto \mathcal{IW}\left(m_1 \Omega_1, m_1\right), \\ \mathsf{B} \,|\Sigma &\propto \mathcal{MN}\left(\mathsf{B}_1, \Sigma, \frac{1}{n_1}\mathsf{\Gamma}_1^{-1}\right), \end{aligned} \tag{6.102}$$

where

$$n_1 = n_0 + n, \tag{6.103}$$

$$m_1 = m_0 + n, \tag{6.104}$$

$$\mathsf{\Gamma}_1 = \frac{1}{n_1}\left(n_0 \mathsf{\Gamma}_0 + \mathsf{F}^\top \mathsf{F}\right), \tag{6.105}$$

$$\mathsf{B}_1 = \frac{1}{n_1}\left(\mathsf{X}^\top \mathsf{F} + n_0 \mathsf{B}_0 \mathsf{\Gamma}_0\right) {\mathsf{\Gamma}_1}^{-1}, \tag{6.106}$$

$$\Omega_1 = \frac{1}{m_1}\left(m_0 \Omega_0 + \left(\mathsf{X} - \mathsf{F}{\mathsf{B}_1}^\top\right)^\top \left(\mathsf{X} - \mathsf{F}{\mathsf{B}_1}^\top\right) + n_0 \left(\mathsf{B}_1 - \mathsf{B}_0\right) \mathsf{\Gamma}_0 (\mathsf{B}_1 - \mathsf{B}_0)^\top\right). \tag{6.107}$$

As above where we tweaked the posterior to reflect the distribution of ζ_*^2, we can modify the posterior here to give the conditional belief on the Hotelling Lawley trace:

$$\begin{aligned} \Sigma &\propto \mathcal{IW}\left(m_1 \Omega_1, m_1\right), \\ n_1 \operatorname{tr}\left(\mathsf{B}^\top \Sigma^{-1} \mathsf{B} \mathsf{\Gamma}_1\right) |\Sigma &\propto \chi^2\left(kf; n_1 \operatorname{tr}\left({\mathsf{B}_1}^\top \Sigma^{-1} \mathsf{B}_1 \mathsf{\Gamma}_1\right)\right). \end{aligned} \tag{6.108}$$

One is tempted to try to express the marginal posterior in terms of Lecoutre's lambda-square distribution. However, this would involve the trace of a Wishart matrix, which does not reduce to a chi-square distribution except in the case where $f = 1$. [Jensen, 1970] The marginal posterior distribution can simulated by drawing an inverse Wishart, then a chi-square.

Example 6.5.5 (Conditional Portfolio on the Fama French Four Factors, Bayesian Inference). Consider the returns of the Market, SMB, HML, and UMD portfolios introduced in Example 1.1.1. In Example 6.1.6, we described two technical features of lagged returns, then computed the Hotelling Lawley trace as $\hat{T} = 0.129 \text{ mo.}^{-1}$.

We assume prior parameters of $n_0 = 72$ mo., $m_0 = 36$ mo.,

$$\mathsf{\Gamma}_0 = \left[\begin{array}{c|ccc} & \textit{intercept} & \textit{mean} & \textit{log\_sd} \\ \hline \textit{intercept} & 2.0 & 1.0 & 1.0 \\ \textit{mean} & 1.0 & 2.0 & 1.0 \\ \textit{log\_sd} & 1.0 & 1.0 & 2.0 \end{array}\right],$$

$$\mathsf{B}_0 = \left[\begin{array}{c|ccc} & \textit{intercept} & \textit{mean} & \textit{log\_sd} \\ \hline Mkt & 0.0 & 0.0 & 0.0 \\ SMB & 0.0 & 0.0 & 0.0 \\ HML & 0.0 & 0.0 & 0.0 \\ UMD & 0.0 & 0.0 & 0.0 \end{array}\right],$$

$$\mathsf{\Omega}_0 = \left[\begin{array}{c|cccc} & Mkt & SMB & HML & UMD \\ \hline Mkt & 10.0 & 0.0 & 0.0 & 0.0 \\ SMB & 0.0 & 10.0 & 0.0 & 0.0 \\ HML & 0.0 & 0.0 & 10.0 & 0.0 \\ UMD & 0.0 & 0.0 & 0.0 & 10.0 \end{array}\right] \%^2\,\text{mo.}^{-1}.$$

Integrating these with the observed F and X, we arrive at the posterior $n_1 = 1,185$ mo., $m_1 = 1,149$ mo.,

$$\mathsf{\Gamma}_1 = \left[\begin{array}{c|ccc} & \textit{intercept} & \textit{mean} & \textit{log\_sd} \\ \hline \textit{intercept} & 1.061 & 0.928 & 1.384 \\ \textit{mean} & 0.928 & 3.436 & 1.088 \\ \textit{log\_sd} & 1.384 & 1.088 & 2.188 \end{array}\right],$$

$$\mathsf{B}_1 = \left[\begin{array}{c|ccc} & \textit{intercept} & \textit{mean} & \textit{log\_sd} \\ \hline Mkt & 0.383 & 0.013 & 0.328 \\ SMB & -0.182 & -0.106 & 0.353 \\ HML & 0.290 & -0.078 & 0.054 \\ UMD & 0.344 & 0.401 & -0.101 \end{array}\right],$$

$$\mathsf{\Omega}_1 = \left[\begin{array}{c|cccc} & Mkt & SMB & HML & UMD \\ \hline Mkt & 28.277 & 5.411 & 4.421 & -8.440 \\ SMB & 5.411 & 10.133 & 1.373 & -2.145 \\ HML & 4.421 & 1.373 & 12.280 & -6.570 \\ UMD & -8.440 & -2.145 & -6.570 & 21.555 \end{array}\right] \%^2\,\text{mo.}^{-1}.$$

We draw 100,000 simulations from the posterior distribution on T. The median of these is $0.118\,\text{mo.}^{-1}$. Based on the same simulations we construct a 95% credible interval on the Hotelling Lawley trace as $[0.082, 0.161)$ mo.$^{-1}$. Recall that in Example 6.3.4, we computed approximate two-sided 95% confidence interval on T as $[0.08, 0.163)$ mo.$^{-1}$. ⊣

6.6 Exercises

Ex. 6.1 Optimal backtested functional Sharpe ratio Suppose you have a quantitative strategy which is parametrized by some vector $\boldsymbol{\theta}$. Let $\boldsymbol{x}(\boldsymbol{\theta})$

be the vector of (backtested) returns associated with the strategy captured by $\boldsymbol{\theta}$. Suppose you wish to find the $\boldsymbol{\theta}$ that maximizes the Sharpe ratio of the backtested returns,

$$\boldsymbol{\theta}_* = \operatorname*{argmax}_{\boldsymbol{\theta}} \frac{\hat{\mu}(\boldsymbol{\theta}) - r_0}{\hat{\sigma}(\boldsymbol{\theta})},$$

where

$$\hat{\mu}(\boldsymbol{\theta}) = \frac{\mathbf{1}^\top \boldsymbol{x}(\boldsymbol{\theta})}{n},$$
$$\hat{\sigma}(\boldsymbol{\theta}) = \sqrt{\frac{\boldsymbol{x}(\boldsymbol{\theta})^\top \boldsymbol{x}(\boldsymbol{\theta}) - n\hat{\mu}^2(\boldsymbol{\theta})}{n-1}},$$
$$n = \mathbf{1}^\top \mathbf{1}.$$

1. Show that

$$\left[\left(\frac{\hat{\mu}(\boldsymbol{\theta}) - r_0}{\hat{\sigma}^2(\boldsymbol{\theta})}\right)(\boldsymbol{x}(\boldsymbol{\theta}) - \hat{\mu}(\boldsymbol{\theta})\mathbf{1}) - \frac{n-1}{n}\mathbf{1}\right]^\top \left.\frac{\mathrm{d}\boldsymbol{x}(\boldsymbol{\theta})}{\mathrm{d}\boldsymbol{\theta}}\right|_{\boldsymbol{\theta}=\boldsymbol{\theta}_*} = \mathbf{0}.$$

(Here, as elsewhere, we take the derivative to be in *numerator layout*, with a row for each element of $\boldsymbol{x}$ and a column for each element of $\boldsymbol{\theta}$.)

2. Show, therefore, that the optimal value of the Sharpe ratio, achieved at $\boldsymbol{\theta}_*$, satisifies

$$\left.\frac{\hat{\mu}(\boldsymbol{\theta}) - r_0}{\hat{\sigma}(\boldsymbol{\theta})}\right|_{\boldsymbol{\theta}=\boldsymbol{\theta}_*} = \sqrt{\frac{n-1}{n}} \left.\frac{\frac{1}{\sqrt{n}}\mathbf{1}^\top \frac{\mathrm{d}\boldsymbol{x}(\boldsymbol{\theta})}{\mathrm{d}\boldsymbol{\theta}}}{\left(\frac{\boldsymbol{x}(\boldsymbol{\theta}) - \hat{\mu}(\boldsymbol{\theta})\mathbf{1}}{\sqrt{n-1}\hat{\sigma}(\boldsymbol{\theta})}\right)^\top \frac{\mathrm{d}\boldsymbol{x}(\boldsymbol{\theta})}{\mathrm{d}\boldsymbol{\theta}}}\right|_{\boldsymbol{\theta}=\boldsymbol{\theta}_*}.$$

Ex. 6.2 **Root F** Using Taylor expansion, prove Equation 6.60.

Ex. 6.3 **Log squared maximal Sharpe ratio** Using Taylor expansion, find the approximate expected value of $\log\left(\hat{\zeta}_*^2\right)$.

Ex. 6.4 **No unbiased estimator** Show that there is no non-negative function of $\hat{\zeta}_*^2, n$, and k that is monotonic in $\hat{\zeta}_*^2$ that is an unbiased estimate of ζ_*. *Hint:* considering the case $\zeta_* = 0$, show that the function would have to be identically zero.

Ex. 6.5 **Alternative sample Markowitz portfolio** Define $\tilde{\Sigma} := \frac{n-1}{n}\hat{\Sigma}$.

1. Show that

$$\left(\tilde{\Sigma} + \hat{\boldsymbol{\mu}}\hat{\boldsymbol{\mu}}^\top\right)^{-1}\hat{\boldsymbol{\mu}} = \left(\mathsf{X}^\top\mathsf{X}\right)^{-1}\mathsf{X}^\top\mathbf{1},$$

where X is the $n \times k$ matrix of returns.

2. Interpret the right-hand side of the equation above as the result of a regression problem.
3. In Exercise 5.22, you were asked to prove that

$$\left(\Sigma + \boldsymbol{\mu}\boldsymbol{\mu}^{\top}\right)^{-1}\boldsymbol{\mu} = \frac{1}{1 + \boldsymbol{\mu}^{\top}\Sigma^{-1}\boldsymbol{\mu}}\Sigma^{-1}\boldsymbol{\mu}.$$

A similar relation holds, of course, for $\hat{\boldsymbol{\mu}}$ and $\tilde{\Sigma}$. Show that, up to scaling, the sample Markowitz portfolio $\hat{\Sigma}^{-1}\hat{\boldsymbol{\mu}}$ can be interpreted as the regression coefficients of a certain regression problem. [Britten-Jones, 1999]

Ex. 6.6 Portfolio weight test via regression Let X be the $n \times k$ matrix of returns.

1. Show that

$$\left(\mathsf{X}^{\top}\mathsf{X}\right)^{-1} = \frac{1}{n-1}\left(\hat{\Sigma}^{-1} - \frac{\hat{\Sigma}^{-1}\hat{\boldsymbol{\mu}}\hat{\boldsymbol{\mu}}^{\top}\hat{\Sigma}^{-1}}{\frac{n-1}{n} + \hat{\zeta}_*^2}\right).$$

2. Show that

$$\hat{\boldsymbol{\beta}} = \left(\mathsf{X}^{\top}\mathsf{X}\right)^{-1}\mathsf{X}^{\top}\mathbf{1} = \frac{n}{n-1}\frac{\hat{\Sigma}^{-1}\hat{\boldsymbol{\mu}}}{1 + \frac{n}{n-1}\hat{\zeta}_*^2}.$$

3. Show that

$$SSR_u = \frac{n}{1 + \frac{n}{n-1}\hat{\zeta}_*^2}.$$

4. Recall that usual regression t statistic takes the form

$$t_{\boldsymbol{g}} = \frac{\boldsymbol{g}^{\top}\hat{\boldsymbol{\beta}}}{s\sqrt{a_{ii}}},$$

where $s^2 = SSR_u / (n-k)$ and $a_{ii} = \boldsymbol{g}^{\top}\left(\mathsf{X}^{\top}\mathsf{X}\right)^{-1}\boldsymbol{g}$. Show that the usual regression t takes the form given in Equation 6.30. [Britten-Jones, 1999]

Ex. 6.7 Spanning tests, Fama French factors Perform the frequentist hypothesis test of spanning (equivalently, testing for $H_0 : \Delta_{\mathsf{I}\setminus\mathsf{G}}\zeta_* = 0$) for subsets of the Fama French factors.

1. Test the spanning of the four factors by the three factors Market, SMB, and HML.
2. Test the spanning of the six factors by the four factors Market, SMB, HML, and UMD. The six factors are available via the **tsrsa** package as the dataset `mff6`:

```
library(tsrsa)
data(mff6)
```

Ex. 6.8 Hotelling and Hotelling Lawley Show that the Hotelling Lawley trace in the case where the feature is identically 1 reduces to Hotelling's T^2. Which of the HLT F approximations, if any, is exact?

Ex. 6.9 Parametric market timing Consider the conditional expectation model with $k = f = 1$:

$$x_t = \mu + \beta f_{t-1} + \epsilon_t,$$

where the error ϵ_t has zero mean, variance σ^2, and is independent of f_{t-1}. Describe how to perform a two-sided hypothesis test on β via the MGLH. Describe how to do it via the univariate tests of Section 2.5.5. Do these give the same answer? Will it depend on which F approximation you use?

Ex. 6.10 Meta momentum Consider a market timing model on the Fama French UMD portfolio returns using the previous period's UMD returns as a feature. Perform inference using the Hotelling-Lawley trace.

Ex. 6.11 Moments of the Markowitz portfolio For *i.i.d.* Gaussian returns, $\hat{\boldsymbol{\mu}}$ and $\hat{\Sigma}$ are independent, with $\hat{\boldsymbol{\mu}} \sim \mathcal{N}\left(\boldsymbol{\mu}, \frac{1}{n}\Sigma\right)$ and $(n-1)\,\hat{\Sigma} \sim \mathcal{W}\left(\Sigma, n-1\right)$. Assume the following facts about moments of $\hat{\Sigma}^{-1}$, which follows a rescaled inverse Wishart distribution:

$$\begin{aligned} \mathrm{E}\left[\hat{\Sigma}^{-1}\right] &= \frac{n-1}{n-k-2}\Sigma^{-1}, \\ \mathrm{E}\left[\hat{\Sigma}^{-1} \otimes \hat{\Sigma}^{-1}\right] &= c_1 \Sigma^{-1} \otimes \Sigma^{-1} + c_2 \operatorname{vec}\left(\Sigma^{-1}\right) \operatorname{vec}\left(\Sigma^{-1}\right)^\top + c_2 \mathsf{K}\left(\Sigma^{-1} \otimes \Sigma^{-1}\right) \end{aligned}$$

where

$$\begin{aligned} c_2 &= \frac{(n-1)^2}{(n-k-1)(n-k-2)(n-k-4)} \text{ and} \\ c_1 &= (n-k-3)\, c_2, \end{aligned}$$

and K is the commutation matrix.

1. Derive Equation 6.35.
2. Derive Equation 6.36.
3. Find the expectation of the hedged portfolio $\hat{\boldsymbol{\nu}}_{*,\mathsf{I}\setminus\mathsf{G}}$ defined in Equation 6.10.

Ex. 6.12 Simulation of distributions under the null Via Monte Carlo simulation verify the connection between summary statistics and the posited distributional forms by drawing samples under the null. Perform 10^5 simulations for each of the following:

1. Confirm the distribution of $\hat{\zeta}_*^2$. Draw $n = 1,024$ days of daily returns from a normal distribution with $k = 8$ and $\zeta_* = 0.125\,\text{day}^{-1/2}$. Confirm Equation 6.26. One way to perform this task is to compute p-values and confirm they are uniformly distributed by Q-Q plotting them against a uniform law. Another way would be to Q-Q plot $\hat{\zeta}_*^2$ against the quantile given by the law.
2. Confirm the distribution of $\Delta_{\mathsf{I}\backslash\mathsf{G}}\,\hat{\zeta}_*^2$. Draw $n = 1,024$ days of daily returns from a normal distribution with $k = 8$ and $\zeta_* = 0.125\,\text{day}^{-1/2}$. Let $\boldsymbol{\mu} = c\mathbf{1}$. Let G be the first four rows of the identity matrix. Confirm the distribution of $\Delta_{\mathsf{I}\backslash\mathsf{G}}\,\hat{\zeta}_*^2$ given by Equation 6.28.
3. Confirm the distribution of $\hat{\boldsymbol{\nu}}_*$. Draw $n = 1,024$ days of daily returns from a normal distribution with $k = 8$ and $\zeta_* = 0.125\,\text{day}^{-1/2}$. Confirm that $\hat{\boldsymbol{\nu}}_*$ follows the distribution of Equation 6.30.

Ex. 6.13 **Empirical power computation** Via Monte Carlo simulation compute the approximate power of some of the Frequentist hypothesis tests of Section 6.3.2:

1. Compute the power of the 1 sample test for the maximal signal-noise ratio. Draw $n = 1,024$ days of daily returns from a normal distribution with $k = 8$ and test the hypothesis $\zeta_* = 0$ at the 0.05 level. Let ζ_* vary from 0 to $0.25\,\text{day}^{-1/2}$. Plot the power (the rejection probability) versus ζ_*.
2. Compute the power of the 1 sample test for the maximal signal-noise ratio under a hedge constraint. Draw $n = 1,024$ days of daily returns from a normal distribution with $k = 8$ and G is the first four rows of the identity matrix. Let $\boldsymbol{\mu} = c\mathbf{1}$. Let ζ_* vary from 0 to $0.25\,\text{day}^{-1/2}$. Test the hypothesis $\Delta_{\mathsf{I}\backslash\mathsf{G}}\zeta_*^2 = 0$. Plot the power versus $\Delta_{\mathsf{I}\backslash\mathsf{G}}\,\hat{\zeta}_*^2$.
3. Compute the power of the 1 sample portfolio weight test. Draw $n = 1,024$ days of daily returns from a normal distribution with $k = 8$ and test the hypothesis $\mathbf{e}_1^\top\zeta_* = 0$ at the 0.05 level. Let $\boldsymbol{\mu} = c\mathbf{1}$, and let ζ_* vary from 0 to $0.25\,\text{day}^{-1/2}$. Plot the power versus the actual $\mathbf{e}_1^\top\zeta_*$.
4. Compute the power of the test on the HLT. Draw $n = 1,024$ days of daily returns from a normal distribution with $k = 8$ under the conditional expectation model with $f = 2$. Let $\mathsf{B} = c\mathbf{1}_k\mathbf{1}_f{}^\top$, and let T vary from 0 to $0.25\,\text{day}^{-1/2}$. Test the hypothesis $T = 0$ at the 0.05 level. Plot the power versus T.

Ex. 6.14 **Boring hypothesis testing** Perform some hypothesis tests on the Fama French Six factor data.

1. Test $H_0 : \zeta_* = 0.25\,\text{mo.}^{-1/2}$.
2. Hedging out the Market component, test $H_0 : \Delta_{\mathsf{I}\backslash\mathsf{G}}\zeta_* = 0.15\,\text{mo.}^{-1/2}$.
3. Using previous period's Market returns and intercept as features, consider the conditional expectation model. Test $H_0 : \mathsf{B} = 0$.

4. Using previous period's Market returns and intercept as features, consider the conditional expectation model. Test $H_0 : T = 0.0625\,\text{mo.}^{-1}$.

* **Ex. 6.15** **Testing hedged portfolios** ☨ How would you perform the hypothesis test

$$H_0 : \boldsymbol{v}^\top \boldsymbol{\nu}_{*,\mathsf{I}\setminus\mathsf{G}} = c \quad \text{versus} \quad H_1 : \boldsymbol{v}^\top \boldsymbol{\nu}_{*,\mathsf{I}\setminus\mathsf{G}} > c,$$

for given G, $\boldsymbol{v}$ and c?

Ex. 6.16 **Alternative CIs, maximal signal-noise ratio** Suppose you observe $\hat{\zeta}_* = 0.2\,\text{day}^{-1/2}$ for $k = 5$, $n = 2{,}000\,\text{days}$. Compute a 95% confidence interval on ζ_* in the usual way, via Equation 6.70.

1. Find other confidence intervals with 95% coverage, but with different type I rates for the lower and upper bounds. Numerically compute the narrowest interval with 95% coverage. Is it very different than the usual CI?
2. Compute confidence interval with 95% coverage that is symmetric around $\hat{\zeta}_*$. Are they very different?

Ex. 6.17 **Two sample test for maximal signal-noise ratio** ☨ Suppose you observe $\hat{\zeta}^2_{*,i}$ for $i = 1, 2$ on independent returns, and wish to test the null hypothesis $H_0 : \zeta^2_{*,1} = \zeta^2_{*,2}$. How would you do this?

Ex. 6.18 **Chow test** Describe how to perform a Chow test on ζ^2_*. That is, given data from two time periods or under two conditions, describe how to test the null hypothesis that ζ^2_* is the same in the two periods, via the MGLH.

1. Test your hypothesis test empirically via simulation to see if it has the advertised type I rate.

Ex. 6.19 **Testing Value** A recent complaint among factor investors is that Value underperformed in the 2010s. Investigate this via the HML returns from the Fama French factors.

1. Perform the hypothesis test that the optimal portfolio weight on HML in the Markowitz portfolio on the four factors is zero.
2. Create an indicator variable that is one only for months after December, 2009. Using the flattening trick with an intercept and this indicator, test the hypothesis that the optimal weight on HML is zero after December, 2009.

Ex. 6.20 **Hedging out a signal** Consider the conditional expectation model of Equation 6.12. Suppose an adversary observes not the entire feature vector $\boldsymbol{f}_{t-1}$, rather a restricted version, $\mathsf{M}\boldsymbol{f}_{t-1}$ for some $f_m \times f$ matrix M. The adversary maximizes the single-period Sharpe ratio conditional on $\mathsf{M}\boldsymbol{f}_{t-1}$ That is, they compute

$$\begin{aligned}
\hat{\mathsf{B}}_m^\top &= \left(\mathsf{M}\mathsf{F}^\top\mathsf{F}\mathsf{M}^\top\right)^{-1}\mathsf{M}\mathsf{F}^\top\mathsf{X},\\
\hat{\Sigma}_m &= \left(\mathsf{X}^\top\mathsf{X} - \mathsf{X}^\top\mathsf{F}\mathsf{M}^\top\left(\mathsf{M}\mathsf{F}^\top\mathsf{F}\mathsf{M}^\top\right)^{-1}\mathsf{M}\mathsf{F}^\top\mathsf{X}\right)/(n-f_m).
\end{aligned} \tag{6.109}$$

Upon observing $\mathsf{M}\boldsymbol{f}_{t-1}$, they allocate $c\hat{\Sigma}_m^{-1}\hat{\mathsf{B}}_m\mathsf{M}\boldsymbol{f}_{t-1}$.

1. Conditional on observing $\boldsymbol{f}_{t-1}$, solve the following optimization problem:

$$\max_{\substack{\hat{\boldsymbol{\nu}}:\boldsymbol{f}_{t-1}^\top\mathsf{M}^\top\hat{\mathsf{B}}_m^\top\hat{\Sigma}_m^{-1}\hat{\Sigma}\hat{\boldsymbol{\nu}}=0,\\ \hat{\boldsymbol{\nu}}^\top\hat{\Sigma}\hat{\boldsymbol{\nu}}\le R^2}} \frac{\hat{\boldsymbol{\nu}}^\top\hat{\mathsf{B}}\boldsymbol{f}_{t-1} - r_0}{\sqrt{\hat{\boldsymbol{\nu}}^\top\hat{\Sigma}\hat{\boldsymbol{\nu}}}}. \tag{6.110}$$

   Your solution should be homogeneous of order zero in $\boldsymbol{f}_{t-1}$, but not linear.

Ex. 6.21 **Boring KRS computations** Compute the unbiased and two KRS estimators of ζ_*^2 for the case of five industry baskets from Example 6.1.2.

Ex. 6.22 **MGLH KRS** Adapt the KRS estimators of Equation 6.67 to the Hotelling Lawley statistic via the various MGLH approximations of Section 6.1.3.

1. Compute the KRS estimators for the HLT computed in Example 6.1.6.

Ex. 6.23 **Variance stabilization** Consider variance stabilizing transform on $\hat{\zeta}_*^2$.

1. Starting from Equation 6.56, rewrite the second moment $\alpha_2\left(\hat{\zeta}_*^2\right)$ in the form

$$\alpha_2\left(\hat{\zeta}_*^2\right) = c_0 + c_1\alpha_1\left(\hat{\zeta}_*^2\right) + c_2\left(\alpha_1\left(\hat{\zeta}_*^2\right)\right)^2,$$

   where the constants c_i have no dependence on ζ_*^2.
2. Express the variance of $\hat{\zeta}_*^2$ as

$$\operatorname{Var}\left(\hat{\zeta}_*^2\right) = \frac{2}{n-k-4}\left[\left(\alpha_1\left(\hat{\zeta}_*^2\right) + \frac{n-1}{n}\right)^2 - \left(\frac{n-1}{n}\right)^2\left(\frac{n-2}{n-k-2}\right)\right].$$

3. Integrate

$$\int \frac{1}{\sqrt{(x+a)^2 - b}}\,\mathrm{d}x$$

to find a variance stabilizing transform like

$$f\left(\hat{\zeta}_*^2\right) = \sqrt{\frac{n-k-4}{2}}\tanh^{-1}\left(\frac{\hat{\zeta}_*^2 + \frac{n-1}{n}}{\sqrt{\left(\hat{\zeta}_*^2 + \frac{n-1}{n}\right)^2 - c}}\right).$$

Is this transform defined for all $\hat{\zeta}_*^2$? *cf.* Laubscher [1960].

Ex. 6.24 ***F* likelihood derivative** Consider the density of the non-central F-distribution given in Equation 6.51.

1. Show that
$$\frac{\mathrm{d}f_f\left(x;\nu_1,\nu_2,\delta\right)}{\mathrm{d}\delta} = \frac{1}{2}\left(f_f\left(x;\nu_1+2,\nu_2,\delta\right) - f_f\left(x;\nu_1,\nu_2,\delta\right)\right).$$

* **Ex. 6.25** **Equivalence of LRT and Hotelling's T^2** Given $\boldsymbol{x}_1, \boldsymbol{x}_2, \ldots, \boldsymbol{x}_n$ *i.i.d.* normal with mean $\boldsymbol{\mu}$ and covariance Σ, treating the parameters as unknown, the likelihood is

$$\mathcal{L}_{\mathcal{N}}\left(\boldsymbol{\mu},\Sigma\,|\boldsymbol{x}_t\right) = \frac{1}{\sqrt{\left(2\pi\right)^k|\Sigma|}}\exp\left(-\frac{1}{2}\left(\boldsymbol{x}_t-\boldsymbol{\mu}\right)^{\top}\Sigma^{-1}\left(\boldsymbol{x}_t-\boldsymbol{\mu}\right)\right).$$

1. Conditional on $\boldsymbol{\mu} = \boldsymbol{\mu}_0$, show that the MLE for Σ is
$$\Sigma_0 = \frac{1}{n}\sum_{t=1}^{n}\left(\boldsymbol{x}_t-\boldsymbol{\mu}_0\right)\left(\boldsymbol{x}_t-\boldsymbol{\mu}_0\right)^{\top}.$$
2. Conditional on $\boldsymbol{\mu} = \boldsymbol{\mu}_0$, show that the maximal value of the likelihood is
$$\max\prod_{t=1}^{n}\mathcal{L}_{\mathcal{N}}\left(\boldsymbol{\mu}_0,\Sigma\,|\boldsymbol{x}_t\right) = \left(2\pi\right)^{-nk/2}\left|\Sigma_0\right|^{-n/2}e^{-nk/2}.$$
3. Show that the likelihood ratio test for the hypothesis $\boldsymbol{\mu} = \boldsymbol{\mu}_0$ against the unrestricted alternative takes value:
$$\lambda = \frac{\prod_{t=1}^{n}\mathcal{L}_{\mathcal{N}}\left(\boldsymbol{\mu}_0,\Sigma_0\,|\boldsymbol{x}_t\right)}{\prod_{t=1}^{n}\mathcal{L}_{\mathcal{N}}\left(\hat{\boldsymbol{\mu}},\hat{\Sigma}\,|\boldsymbol{x}_t\right)} = \left(\frac{\left|\Sigma_0\right|}{\left|\hat{\Sigma}\right|}\right)^{-n/2} = \left|\hat{\Sigma}^{-1}\Sigma_0\right|^{-n/2}.$$
4. The key observation now is that $\hat{\Sigma}$ and Σ_0 differ by a rank one update. Show that
$$\begin{aligned}\Sigma_0 &= \frac{1}{n}\sum_{t=1}^{n}\left(\boldsymbol{x}_t-\boldsymbol{\mu}_0\right)\left(\boldsymbol{x}_t-\boldsymbol{\mu}_0\right)^{\top},\\ &= \frac{n-1}{n}\hat{\Sigma} + \left(\boldsymbol{\mu}_0-\hat{\boldsymbol{\mu}}\right)\left(\boldsymbol{\mu}_0-\hat{\boldsymbol{\mu}}\right)^{\top}.\end{aligned}$$

5. Conclude that

$$\hat{\Sigma}^{-1}\Sigma_0 = \frac{n-1}{n}\mathsf{I} + \hat{\Sigma}^{-1}\left(\boldsymbol{\mu}_0 - \hat{\boldsymbol{\mu}}\right)\left(\boldsymbol{\mu}_0 - \hat{\boldsymbol{\mu}}\right)^\top.$$

Using Sylvester's determinant theorem conclude that λ is a monotonic function of $T^2 = n(\boldsymbol{\mu}_0 - \hat{\boldsymbol{\mu}})^\top \hat{\Sigma}^{-1}\left(\boldsymbol{\mu}_0 - \hat{\boldsymbol{\mu}}\right)$.

Ex. 6.26 **Likelihood alternative form** Conditional on $\boldsymbol{\mu} = \boldsymbol{\mu}_0$, the MLE for Σ is

$$\Sigma_0 = \frac{1}{n}\sum_{t=1}^{n}\left(\boldsymbol{x}_t - \boldsymbol{\mu}_0\right)\left(\boldsymbol{x}_t - \boldsymbol{\mu}_0\right)^\top.$$

In Exercise 6.25 you proved essentially that

$$\Sigma_0 = \hat{\Sigma}\left(\frac{n-1}{n}\mathsf{I} + \hat{\Sigma}^{-1}\left(\boldsymbol{\mu}_0 - \hat{\boldsymbol{\mu}}\right)\left(\boldsymbol{\mu}_0 - \hat{\boldsymbol{\mu}}\right)^\top\right).$$

1. Re-express the normal likelihood of Equation 6.84 as

$$\begin{aligned}\log \mathcal{L}_{\mathcal{N}}\left(\boldsymbol{\mu}_0, \Sigma_0 \left|\mathsf{X}\right.\right) = &-\frac{nk}{2}\log 2\pi - \frac{n}{2}\log\left|\hat{\Sigma}\right| \\ &-\frac{n}{2}\log\left|\frac{n-1}{n} + \left(\boldsymbol{\mu}_0 - \hat{\boldsymbol{\mu}}\right)^\top\hat{\Sigma}^{-1}\left(\boldsymbol{\mu}_0 - \hat{\boldsymbol{\mu}}\right)\right| - \frac{nk}{2}.\end{aligned}$$

Ex. 6.27 **LRT for matrix normal** Suppose you wished to test the null hypothesis

$$H_0 : \boldsymbol{x}_t\boldsymbol{f}_{t-1}^\top \sim \mathcal{MN}_{k,f}\left(\mathsf{M}, \Psi, \Xi\right) \quad \text{vs.} \quad H_1 : \boldsymbol{f}_{t-1}\otimes\boldsymbol{x}_t \sim \mathcal{N}\left(\boldsymbol{\xi}, \Omega\right),$$

for unstructured Ω.

1. Compute the likelihood of the matrix normal distribution for the maximum likelihood estimators, $\hat{\mathsf{M}}$, and $\hat{\Psi}_\infty$, $\hat{\Xi}_\infty$, with the latter assumed to be the converged values of Dutilleul's method.
2. Compute the likelihood of the flattened normal distribution for the maximum likelihood estimators $\hat{\boldsymbol{\xi}}$ and $\hat{\Omega}$.
3. Show that the likelihood ratio takes the form

$$\left(\frac{\left|\hat{\Omega}\right|}{\left|\hat{\Psi}_\infty \otimes \hat{\Xi}_\infty\right|}\right)^{\frac{n}{2}}.$$

Rather than rely on the asymptotic χ^2 distribution of twice the log likelihood difference, Gerard and Hoff [2014] describe a procedure to find the rejection cutoff via Monte Carlo simulations.

4. Compute the likelihood ratio for the returns times features from Example 6.1.7.

Ex. 6.28 **LRT near zero** Repeat the simulations of Example 6.4.6 under the null for $\zeta_* = 0.01\,\text{mo.}^{-1/2}$.

Ex. 6.29 **Choosing momentum** Test conditional expectation models on the Fama French 4 factors, using the previous period's returns on all four factors as the feature. Including an intercept term, your full model would have 20 fit coefficients.

1. Compute the log likelihood for all $2^{20} - 1$ subsets of the possible coefficients being fit. Which minimizes the Bayesian Information Criterion?

* **Ex. 6.30** **MLE monotonicity** ✝ Can you prove that the MLE for ζ_* is monotonically non-decreasing in $\hat{\zeta}_*^2$?

Ex. 6.31 **Likelihood intervals** A *p likelihood interval* for ζ_* is defined as a set

$$\left\{ \zeta_*^2 \;\middle|\; \frac{\mathcal{L}_{SR,*}\left(\zeta_*^2 \middle| \hat{\zeta}_*^2, n, k\right)}{\mathcal{L}_{SR,*}\left(\zeta_{*,\text{MLE}}^2 \middle| \hat{\zeta}_*^2, n, k\right)} \geq p \right\}$$

It is said that a 14.7% likelihood interval will be equivalent to the 95% confidence interval in certain cases. Continue Example 6.3.1 by finding the 14.7% likelihood interval for ζ_*^2 given $\hat{\zeta}_*^2 = 0.095\,\text{mo.}^{-1/2}$ for $n = 1,128\,\text{mo.}$ and $k = 4$.

Ex. 6.32 **Likelihood test, flattened conditional portfolio** Test the hypothesis H_0 of Example 6.3.13 via the Likelihood Ratio Test.

Ex. 6.33 **Uninformative prior** How should you set the hyperparameters in Equation 6.88 to express an "uninformative prior"?

Ex. 6.34 **Boring credible interval computation** Repeat the analysis of Example 6.5.4, but on the Fama French six factor data, computing 95% credible interval on ζ_*

Ex. 6.35 **Checking the Bayesian update** Show that the prior and posterior of Section 6.5.2 (*e.g.*, Equation 6.102) reduce to those of the unconditional case (*e.g.*, Equation 6.90) for the case where the "feature," $\boldsymbol{f}_{t-1}$ is identically the scalar 1. That is, where $\mathsf{F} = \mathbf{1}_n$.

Ex. 6.36 **Bayesian update, stochastic features** Generalize the prior and posterior of Section 6.5.2 to deal with the case where the features vector $\boldsymbol{f}_t$ is random with unknown mean and covariance. Your model should have hyperparameters expressing belief on those parameters.

Ex. 6.37 **Hedged Bayesian** Consider, for example, the belief that one should hold only the Market portfolio because there are no other premia to be harvested. How would a Bayesian express such a belief in a prior? That is, given $k_g \times k$ matrix G of rank $k_g \leq k$, how would you adapt Equation 6.88 to reflect the belief that $\mathsf{G}\hat{\Sigma}\boldsymbol{\nu}_* = \mathbf{0}$?

1. Given the posterior of Equation 6.90, and the constraint $\mathsf{G}\hat{\Sigma}\boldsymbol{\nu}_* = \mathbf{0}$, what is the hedged Bayesian portfolio?

Ex. 6.38 Kalman filter, maximal signal-noise ratio Consider the Kalman Filter on multivariate estimation. Suppose the prediction step is of the form $\boldsymbol{\mu} \mapsto \mathsf{A}\boldsymbol{\mu} + \boldsymbol{b} + \boldsymbol{z}$, where $\boldsymbol{z} \sim \mathcal{N}(\mathbf{0}, \Omega)$.

1. From the prior of Equation 6.88, write the prediction step as

$$\begin{aligned} \Sigma &\propto \mathcal{IW}\left(m_0\Omega_0, m_0\right), \\ \boldsymbol{\mu}^- \,|\Sigma &\propto \mathcal{N}\left(\mathsf{A}\boldsymbol{\mu}_0 + \boldsymbol{b}, \Omega + \frac{1}{n_0}\mathsf{A}\Sigma\mathsf{A}^\top\right), \end{aligned} \tag{6.111}$$

2. Suppose that after the prediction step, you observe n *i.i.d.* draws of returns drawn from $\mathcal{N}(\boldsymbol{\mu}^-, \Sigma)$. Express the posterior belief of Equation 6.90 by writing out the equations for the posterior hyperparameters.
3. What is the marginal belief on ζ_*?
4. How might you estimate Ω?

Ex. 6.39 Crisis investing Define "post-crisis" as the Market level being 10% or more below its highest historical value. Using the flattening trick, create a conditional portfolio on the Fama French four factor returns using an intercept term and the binary post-crisis indicator.

1. Compute the passthrough matrix $\hat{\mathsf{N}}_*$.
2. Compute the multi-period $\hat{\zeta}_*^2$ for this conditional portfolio.
3. Consider hedging out the Market term. Compute the hedged passthrough matrix.
4. Compute the $\Delta_{\mathsf{I}\setminus\mathsf{G}}\,\hat{\zeta}_*^2$ for the conditional portfolio with the Market term hedged out.
5. Compute a confidence interval on the $\Delta_{\mathsf{I}\setminus\mathsf{G}}\,\zeta_*^2$ for this hedged portfolio.
6. Consider hedging out the intercept feature. Compute the hedged passthrough matrix.
7. Compute the $\Delta_{\mathsf{I}\setminus\mathsf{G}}\,\hat{\zeta}_*^2$ for the conditional portfolio with the intercept feature hedged out.
8. Compute a confidence interval on the $\Delta_{\mathsf{I}\setminus\mathsf{G}}\,\zeta_*^2$ for this hedged portfolio.

Ex. 6.40 **Black Litterman portfolio** Consider the Black Litterman portfolio.

1. How would you infer the holdings of the Market as a whole for equities?

2. What does a non-informative prior look like under Black Litterman?
3. Empirically compare the Black Litterman portfolio to the Bayesian portfolio of Equation 6.96 for the case where $\mathsf{P} = \mathsf{I}$, $\boldsymbol{q} = \boldsymbol{\mu}_0$, $\Omega_0 = \operatorname{Diag}(\boldsymbol{\tau})$, $k = 8$, and letting n vary.

* **Ex. 6.41** **Credible intervals vs confidence intervals** ↑ Can you show that the Bayesian credible interval under an uninformative prior is nearly the same as the Frequentist confidence interval?

1. Can you demonstrate it empirically?

* **Ex. 6.42** **Research problem: LRT and f-test** ↑ Show that the Λ of Equation 6.81 is monotonic in $\hat{\zeta}_*^2$ for $0 \le \zeta_0 < \zeta_1$.

* **Ex. 6.43** **Research problem: paired portfolio test** Devise a test for the squared maximal signal-noise ratio of two portfolios on possibly correlated asset universes given paired observations. That is, suppose you have two matrices B_i for $i = 1, 2$, which are $l_i \times n$, of rank $l_i < n$. Recall the standard nomenclature

$$\hat{\zeta}_{*,\mathsf{B}_i}^2 := \left(\mathsf{B}_i\hat{\boldsymbol{\mu}}\right)^\top \left(\mathsf{B}_i\hat{\Sigma}\mathsf{B}_i^\top\right)^{-1} \left(\mathsf{B}_i\hat{\boldsymbol{\mu}}\right),$$

and similarly define ζ_{*,B_i} as the population analogue.

1. Find the distribution of $\hat{\zeta}_{*,\mathsf{B}_1}^2 - \hat{\zeta}_{*,\mathsf{B}_2}^2$.
2. Can this statistic be used to perform inference on $\zeta_{*,\mathsf{B}_1}^2 - \zeta_{*,\mathsf{B}_2}^2$?

7

Portfolio Inference for Other Returns

let go of everything you're
holding onto
now let go of everything else

PAUL WILLIAMS
Das Energi

In the previous chapter we considered portfolio inference under the assumption that returns were *i.i.d.* Gaussian. We now consider what can be said when those assumptions are violated. Regrettably, this is still an area of research, and there are many unknowns. Unlike in the univariate case the asymptotic results that can easily be derived are of little use for the sample sizes one is likely to encounter in practice. We will see that *i.i.d.* elliptical returns will not have much effect on the statistics previously described. However, autocorrelation and heteroskedasticity of returns can cause significant problems. The solution we offer is of perhaps limited utility.

Before proceeding we consider a simple diagnostic device that can be used to detect and describe deviations from the *i.i.d.* assumption. Under the assumption that returns are *i.i.d.* multivariate normal, the observation level *Mahalanobis distances*, defined as

$$d_t^2 := \left(\boldsymbol{x}_t - \hat{\boldsymbol{\mu}}\right)^{\top} \hat{\Sigma}^{-1} \left(\boldsymbol{x}_t - \hat{\boldsymbol{\mu}}\right) \tag{7.1}$$

follow a beta distribution up to scaling: [Wilks, 1962; Hardin and Rocke, 2005]

$$\frac{n}{\left(n-1\right)^2} d_t^2 \sim \mathcal{B}\left(\frac{k}{2}, \frac{n-k-1}{2}\right). \tag{7.2}$$

While the individual Mahalanobis distances are not independent of each other (*cf.* Exercise 7.1), they show only weak dependence for large n. The Mahalanobis distances give a way to graphically diagnose the assumptions of independence and heteroskedasticity. For this purpose, one only need plot the distances over time to detect conditional heteroskedasticity, or Q-Q plot them against the beta law to confirm deviance from the *i.i.d.* multivariate normality assumption.

Example 7.0.1 (Fama French Four Factors, Mahalanobis Distances). Consider the returns of the Market, SMB, HML, and UMD portfolios introduced in Example 1.1.1. We compute the Mahalanobis distances on the monthly returns

DOI: 10.1201/9781003181057-7

from Jan 1927 to Dec 2020 ($n = 1,128$ mo.), then compare to a Beta distribution with parameters 2 and 561.5.

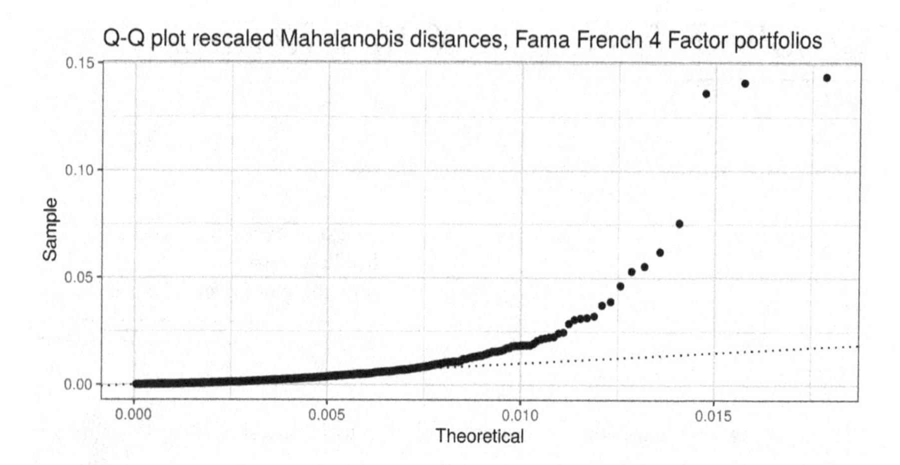

FIGURE 7.1
Q-Q plot of rescaled Mahalanobis distances of monthly returns of the Fama French four factor portfolios against the theoretical beta law.

In Figure 7.1, we Q-Q plot the distances, transformed to a putative beta variable, against quantiles of $\mathcal{B}(2, 561.5)$. In Figure 7.2, we plot the rescaled distances over time, along with a horizontal line at the 0.95 quantile of the corresponding Beta distribution. There is clear clustering of the extremely large values, along with periods of relative quiet. We see periods of larger Mahalanobis distances during the Great Depression, the Dot-Com Bubble, as well as April 2009 and October 1987. In Figure 7.3, we plot the autocorrelation of the Mahalanobis distances at various lags and see a very high 1-period autocorrelation as well as some longer term dependence. This plot is a multivariate analogue of the absolute correlation plot in Figure 3.1, though the effect here is perhaps stronger than in the univariate case. ⊣

For real assets' returns series, it is usually a foregone conclusion that these tests (and other tests for multivariate normality [Mardia, 1980; Zhou and Shao, 2014]) will indicate deviation from assumptions. In this chapter we will consider the distribution of $\hat{\zeta}_*^2$ and the Hotelling-Lawley statistic when returns are autocorrelated, heteroskedastic or non-normal.

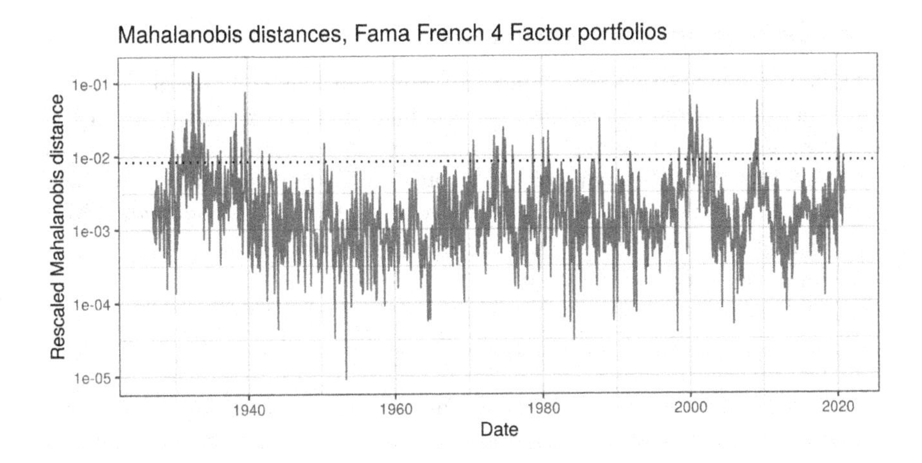

FIGURE 7.2
Rescaled Mahalanobis distances of monthly returns of the Fama French four factor portfolios are plotted over time. The rescaled distances should follow a Beta distribution. A horizontal line is plotted at the 0.95 quantile of that distribution. The y-axis is in log scale.

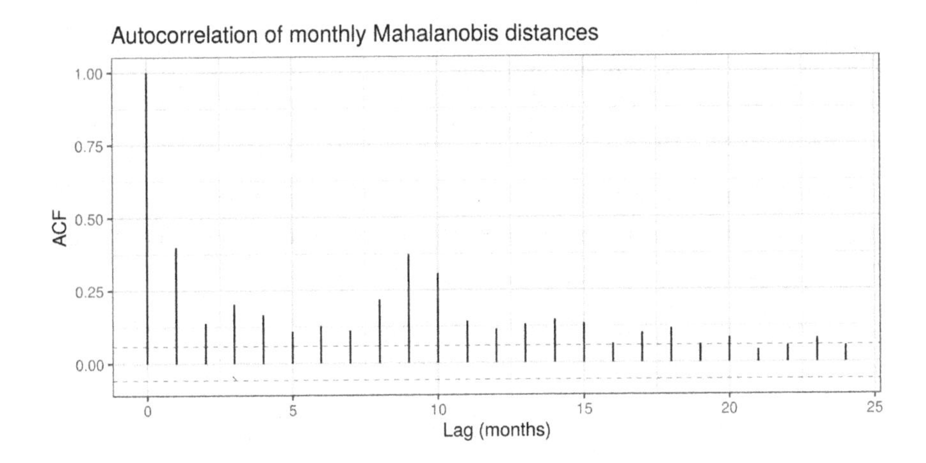

FIGURE 7.3
Autocorrelation of the Mahalanobis distances on the four Fama French portfolios' returns are plotted versus lag.

7.1 The second moment matrix

To study the distribution of $\hat{\zeta}_*^2$ and the Hotelling-Lawley statistic, we first consider a trick for collecting the first two moments of a multivariate distribution into a single parameter[1]. So as usual, let $\boldsymbol{x}_t$ be the vector of returns of k assets at time t, with mean $\boldsymbol{\mu}$, and covariance Σ. For the moment we will not assume a distributional form to the returns vector. The *augmented form* of $\boldsymbol{x}_t$ is simply $\boldsymbol{x}_t$ prepended with a 1, which we write as $\tilde{\boldsymbol{x}}_t = \left[1, {\boldsymbol{x}_t}^\top\right]^\top$. The second moment of $\tilde{\boldsymbol{x}}_t$ has the form

$$\Theta := \mathrm{E}\left[\tilde{\boldsymbol{x}}_t {\tilde{\boldsymbol{x}}_t}^\top\right] = \left[\begin{array}{cc} 1 & \boldsymbol{\mu}^\top \\ \boldsymbol{\mu} & \Sigma + \boldsymbol{\mu}\boldsymbol{\mu}^\top \end{array}\right]. \tag{7.3}$$

The matrix Θ contains the first and second moments, and in fact is bijective with them, meaning that Θ completely describes the first two moments and vice-versa. The matrix Θ contains the first and second moment of $\boldsymbol{x}$, but is also the uncentered second moment of $\tilde{\boldsymbol{x}}_t$, a fact which makes it amenable to analysis via the central limit theorem.

By inspection one can confirm that the inverse of Θ is

$$\Theta^{-1} = \left[\begin{array}{cc} 1 + \boldsymbol{\mu}^\top\Sigma^{-1}\boldsymbol{\mu} & -\boldsymbol{\mu}^\top\Sigma^{-1} \\ -\Sigma^{-1}\boldsymbol{\mu} & \Sigma^{-1} \end{array}\right] = \left[\begin{array}{cc} 1 + \zeta_*^2 & -{\boldsymbol{\nu}_*}^\top \\ -\boldsymbol{\nu}_* & \Sigma^{-1} \end{array}\right], \tag{7.4}$$

where $\boldsymbol{\nu}_* = \Sigma^{-1}\boldsymbol{\mu}$ is the Markowitz portfolio, and $\zeta_* = \sqrt{\boldsymbol{\mu}^\top\Sigma^{-1}\boldsymbol{\mu}}$ is the Sharpe ratio of that portfolio. In the lower right corner, we have the inverse covariance Σ^{-1} which is called the *precision matrix.*

Given n *i.i.d.* observations $\boldsymbol{x}_t$, let $\tilde{\mathsf{X}}$ be the matrix whose rows are the vectors ${\tilde{\boldsymbol{x}}_t}^\top$. The naïve sample estimator

$$\hat{\Theta} := \frac{1}{n}\tilde{\mathsf{X}}^\top\tilde{\mathsf{X}} \tag{7.5}$$

is an unbiased estimator since $\Theta = \mathrm{E}\left[\tilde{\boldsymbol{x}}^\top\tilde{\boldsymbol{x}}\right]$. We note that in this form $\hat{\Theta}$ essentially uses a different denominator for estimating covariance than we had used previously:

$$\hat{\Theta} = \left[\begin{array}{cc} 1 & \hat{\boldsymbol{\mu}}^\top \\ \hat{\boldsymbol{\mu}} & \frac{n-1}{n}\hat{\Sigma} + \hat{\boldsymbol{\mu}}\hat{\boldsymbol{\mu}}^\top \end{array}\right].$$

In the inverse we have

$$\hat{\Theta}^{-1} = \left[\begin{array}{cc} 1 + \frac{n}{n-1}\hat{\zeta}_*^2 & -\frac{n}{n-1}{\hat{\boldsymbol{\nu}}_*}^\top \\ -\frac{n}{n-1}\hat{\boldsymbol{\nu}}_* & \frac{n}{n-1}\hat{\Sigma}^{-1} \end{array}\right] = \frac{n}{n-1}\left[\begin{array}{cc} \frac{n-1}{n} + \hat{\zeta}_*^2 & -{\hat{\boldsymbol{\nu}}_*}^\top \\ -\hat{\boldsymbol{\nu}}_* & \hat{\Sigma}^{-1} \end{array}\right].$$

Thus one can use $\hat{\Theta}^{-1}$ to perform inference on ζ_*^2, $\boldsymbol{\nu}_*$, Σ^{-1}, and combinations thereof.

[1]This material is abridged from the author's paper on the asymptotic distribution of the Markowitz portfolio. [Pav, 2013]

7.1.1 Distribution under Gaussian returns

We now consider the distribution of $\hat{\Theta}$ when returns are *i.i.d.* Gaussian. Let $\boldsymbol{x} \sim \mathcal{N}(\boldsymbol{\mu}, \Sigma)$, $\tilde{\boldsymbol{x}} = \left[1, \boldsymbol{x}^\top\right]^\top$, and $\Theta = \mathrm{E}\left[\tilde{\boldsymbol{x}}\tilde{\boldsymbol{x}}^\top\right]$. Given n *i.i.d.* samples $\boldsymbol{x}_t$, defined $\hat{\Theta}$ as in Equation 7.5. Then the density of $\hat{\Theta}$ is

$$f\left(\hat{\Theta}; \Theta\right) = \exp\left(c'_{n,k}\right) \frac{\left|\hat{\Theta}\right|^{\frac{n-k-2}{2}}}{\left|\Theta\right|^{\frac{n}{2}}} \exp\left(-\frac{n}{2} \operatorname{tr}\left(\Theta^{-1}\hat{\Theta}\right)\right), \tag{7.6}$$

for some $c'_{n,k}$. [Pav, 2013]

The derivatives of the log likelihood are given by

$$\begin{aligned} \frac{\mathrm{d}\log f\left(\hat{\Theta}; \Theta\right)}{\mathrm{d}\operatorname{vec}\left(\Theta\right)} &= -\frac{n}{2}\left[\operatorname{vec}\left(\Theta^{-1} - \Theta^{-1}\hat{\Theta}\Theta^{-1}\right)\right]^\top, \\ \frac{\mathrm{d}\log f\left(\hat{\Theta}; \Theta\right)}{\mathrm{d}\operatorname{vec}\left(\Theta^{-1}\right)} &= -\frac{n}{2}\left[\operatorname{vec}\left(\Theta - \hat{\Theta}\right)\right]^\top. \end{aligned} \tag{7.7}$$

As a consequence $\hat{\Theta}$ is the maximum likelihood estimator of Θ.

7.2 Asymptotic distribution of the second moment

Collecting the mean and covariance into the second moment matrix gives the asymptotic distribution of the sample Markowitz portfolio without much work. Let $\hat{\Theta}$ be the unbiased sample estimate of Θ, based on n *i.i.d.* samples of $\boldsymbol{x}$. Let Ω be the variance of $\operatorname{vech}\left(\tilde{\boldsymbol{x}}\tilde{\boldsymbol{x}}^\top\right)$. Then, asymptotically in n,

$$\sqrt{n}\left(\operatorname{vech}\left(\hat{\Theta}^{-1}\right) - \operatorname{vech}\left(\Theta^{-1}\right)\right) \rightsquigarrow \mathcal{N}\left(0, \mathsf{H}\Omega\mathsf{H}^\top\right), \tag{7.8}$$

where

$$\mathsf{H} = -\mathsf{L}\left(\Theta^{-1} \otimes \Theta^{-1}\right)\mathsf{D}. \tag{7.9}$$

Here L is the elimination matrix, and D is the duplication matrix, *cf.* Definition A.1.3. Furthermore, we may replace Ω in this equation with an asymptotically consistent estimator, $\hat{\Omega}$. This follows from the central limit theorem and multivariate delta method, where the expression of H follows from the matrix derivative of the inverse, Equation A.25.

To estimate the covariance of $\operatorname{vech}\left(\hat{\Theta}^{-1}\right)$, plug in $\hat{\Theta}$ for Θ in the covariance computation, and use some consistent estimator for Ω, call it $\hat{\Omega}$. One way to compute $\hat{\Omega}$ is to via the sample covariance of the vectors $\operatorname{vech}\left(\tilde{\boldsymbol{x}}_t\tilde{\boldsymbol{x}}_t^\top\right) =$

$\left[1, \boldsymbol{x}_t^{\top}, \operatorname{vech}\left(\boldsymbol{x}_t \boldsymbol{x}_t^{\top}\right)^{\top}\right]^{\top}$. An analogous procedure was used in Section 3.2. More elaborate covariance estimators can be used, for example, to deal with violations of the *i.i.d.* assumptions. [Zeileis, 2004] Note that because the first element of vech $\left(\tilde{\boldsymbol{x}}_t \tilde{\boldsymbol{x}}_t^{\top}\right)$ is a deterministic 1, the first row and column of Ω is all zeros, and we need not estimate it.

The main use of the relationship in Equation 7.8 is to estimate the standard error on $\hat{\zeta}_*^2$, though it can also be used to estimate the covariance of $\hat{\boldsymbol{\nu}}_*$ and of vech $\left(\hat{\Sigma}^{-1}\right)$. We can also use this relationship to find the standard error of $\hat{\zeta}_*$, since if v is the asymptotic variance of $\hat{\zeta}_*^2$, then $v/\left(4\zeta_*^2\right)$ is the asymptotic variance of $\hat{\zeta}_*$, by the delta method.

To be explicit, the asymptotic results for $\hat{\zeta}_*^2$ and $\hat{\zeta}_*$ are

$$\sqrt{n}\left(\hat{\zeta}_*^2 - \zeta_*^2\right) \rightsquigarrow \mathcal{N}\left(0, \operatorname{tr}\left(\mathsf{D}\Omega\mathsf{D}^{\top}\left(\Theta^{-1}\boldsymbol{e}_1\boldsymbol{e}_1^{\top}\Theta^{-1} \otimes \Theta^{-1}\boldsymbol{e}_1\boldsymbol{e}_1^{\top}\Theta^{-1}\right)\right)\right), \tag{7.10}$$

$$\sqrt{n}\left(\hat{\zeta}_* - \zeta_*\right) \rightsquigarrow \mathcal{N}\left(0, \frac{1}{4\zeta_*^2}\operatorname{tr}\left(\mathsf{D}\Omega\mathsf{D}^{\top}\left(\Theta^{-1}\boldsymbol{e}_1\boldsymbol{e}_1^{\top}\Theta^{-1} \otimes \Theta^{-1}\boldsymbol{e}_1\boldsymbol{e}_1^{\top}\Theta^{-1}\right)\right)\right). \tag{7.11}$$

We will use the asymptotic distribution of $\hat{\Theta}$ to explore the robustness of $\hat{\zeta}_*^2$ to the assumption of *i.i.d.* Gaussian returns that was used in Chapter 6. As in Chapter 3, we will consider returns drawn from elliptical distribution, heteroskedastic independent returns, and autocorrelated returns. Unfortunately, the asymptotic results have a fundamental limitation, and alternative analysis will have to be applied.

Caution (Limitations of asymptotics). The asymptotic results we present in this chapter typically contain an error term on the order of $\mathcal{O}\left(n^{-2}\right)$. While this would seem sufficient for most applications, the coefficient of the n^{-2} term can be quite large, often depending on k. Recall, a similar effect was observed in the Gaussian case in Section 6.2. The results given here will be misleading for values of k/n one is likely to encounter in practice. Any result in this chapter which appears to have no dependence on k should be viewed with some suspicion.

7.2.1 Asymptotic distribution, *i.i.d.* elliptical returns

We now consider the case where returns follow an elliptical distribution, which includes the *i.i.d.* Gaussian returns case. So let $\boldsymbol{x}_t$ follow an elliptical distribution with mean and covariance $\boldsymbol{\mu}$ and Σ and kurtosis factor κ. Let $\hat{\Theta}$ be the unbiased sample estimate of Θ, based on n *i.i.d.* samples of $\boldsymbol{x}$. In analogue to how $\tilde{\boldsymbol{x}}$ are built from $\boldsymbol{x}$, define

$$\tilde{\boldsymbol{\mu}} := \left[1, \boldsymbol{\mu}^{\top}\right]^{\top}, \quad \text{and} \quad \tilde{\Sigma} := \left[\begin{array}{cc} 0 & \boldsymbol{0}^{\top} \\ \boldsymbol{0} & \Sigma \end{array}\right]. \tag{7.12}$$

Note that $\Theta = \tilde{\Sigma} + \tilde{\mu}\tilde{\mu}^{\top}$. Then it can be shown that [Pav, 2013]

$$\Omega_0 = n \operatorname{Var}\left(\operatorname{vec}\left(\hat{\Theta}\right)\right) = (\kappa - 1)\left[\operatorname{vec}\left(\tilde{\Sigma}\right)\operatorname{vec}\left(\tilde{\Sigma}\right)^{\top} + (\mathsf{I} + \mathsf{K})\,\tilde{\Sigma} \otimes \tilde{\Sigma}\right] + (\mathsf{I} + \mathsf{K})\left[\Theta \otimes \Theta - \tilde{\mu}\tilde{\mu}^{\top} \otimes \tilde{\mu}\tilde{\mu}^{\top}\right]. \quad (7.13)$$

Here K is the "commutation matrix," *cf.* Definition A.1.3. Below we will use $\mathsf{N} = \frac{1}{2}(\mathsf{I} + \mathsf{K})$ as the "symmetrizing matrix." We can plug this Ω_0 into Equation 7.8 to get the asymptotic distribution of $\hat{\Theta}^{-1}$ in the elliptical case. By working with the L, D, and N matrices, the covariance in this case can be further reduced to:

$$\sqrt{n}\left(\operatorname{vech}\left(\hat{\Theta}^{-1}\right) - \operatorname{vech}\left(\Theta^{-1}\right)\right) \rightsquigarrow \mathcal{N}\left(0, \mathsf{B}\right), \quad (7.14)$$

where

$$\begin{aligned}\mathsf{B} = {} & (\kappa - 1)\,\mathsf{L}\left[\operatorname{vec}\left(\Theta^{-1} - e_1 e_1^{\top}\right)\operatorname{vec}\left(\Theta^{-1} - e_1 e_1^{\top}\right)^{\top}\right]\mathsf{L}^{\top} \\ & + 2(\kappa - 1)\,\mathsf{L}\mathsf{N}\left[\left(\Theta^{-1} - e_1 e_1^{\top}\right) \otimes \left(\Theta^{-1} - e_1 e_1^{\top}\right)\right]\mathsf{N}^{\top}\mathsf{L}^{\top} \\ & + 2\mathsf{L}\mathsf{N}\left[\Theta^{-1} \otimes \Theta^{-1} - e_1 e_1^{\top} \otimes e_1 e_1^{\top}\right]\mathsf{N}^{\top}\mathsf{L}^{\top}.\end{aligned}$$

As we are mostly concerned with the signal-noise ratio and sample Markowitz portfolio, we can select their joint asymptotic distribution out from this expression to arrive at:

$$\sqrt{n}\left(\left[\begin{array}{c}\frac{n}{n-1}\hat{\zeta}_*^2 \\ \frac{n}{n-1}\hat{\nu}_*\end{array}\right] - \left[\begin{array}{c}\zeta_*^2 \\ \nu_*\end{array}\right]\right) \rightsquigarrow \mathcal{N}\left(0, \mathsf{C}\right), \quad (7.15)$$

where

$$\mathsf{C} = \left[\begin{array}{cc} 2\zeta_*^2\left(2 + \zeta_*^2\right) + 3(\kappa - 1)\zeta_*^4 & \left(2\left(1 + \zeta_*^2\right) + 3(\kappa - 1)\zeta_*^2\right)\nu_*^{\top} \\ \left(2\left(1 + \zeta_*^2\right) + 3(\kappa - 1)\zeta_*^2\right)\nu_* & \left(1 + \kappa\zeta_*^2\right)\Sigma^{-1} + (2\kappa - 1)\nu_*\nu_*^{\top}\end{array}\right]. \quad (7.16)$$

This can be proved by the cental limit theorem and multivariate method. [Pav, 2013] This normal asymptotic distribution of $\hat{\nu}_*$ was found in the Gaussian case by Kotsiuba and Mazur[2]. [Kotsiuba and Mazur, 2015; Jobson and Korkie, 1980]

The asymptotic variance of $\hat{\nu}_*$ given here is comparable to the exact value under Gaussian returns in Equation 6.36. Compare the variance of $\frac{n}{n-1}\hat{\zeta}_*^2$ given in $C_{1,1}$, plugging in $\kappa = 1$, to the value from the F distribution in Equation 6.64.

The asymptotic variance of $\hat{\zeta}_*^2$ in this aproximation is a quadratic function of ζ_*^2, and omits a $\mathcal{O}\left(n^{-2}\right)$ term. When $\zeta_*^2 = 0$, it is useless. Instead by

[2]The covariance published in Kotsiuba and Mazur [2015] deviates from Equation 7.16 by an erroneous sign flip. [Mazur, 2020]

performing one more transformation and using the delta method once again on the map $x \mapsto \sqrt{x}$ we arrive at the asymptotic variance

$$\sqrt{n}\left(\sqrt{\frac{n}{n-1}}\hat{\zeta}_{*} - \zeta_{*}\right) \rightsquigarrow \mathcal{N}\left(0, 1 + \frac{2 + 3(\kappa - 1)}{4}\zeta_{*}^{2}\right). \tag{7.17}$$

Note that the variance for ζ_* given here is the square of Mertens' form of the standard error of the Sharpe ratio (*cf.* Equation 2.28), since elliptical distributions have zero skew and excess kurtosis of $3(\kappa - 1)$.

Higher order asymptotics

Iwashita [1997] described a higher order asymptotic distribution of Hotelling's T^2 under independent elliptical returns, both under the null and the alternative. [Kano, 1995] Translating into the equivalent result for $\hat{\zeta}_{*}^{2}$, it was shown that

$$\frac{f_{SR*}\left(x; n, k, \zeta_{*}^{2}\right)}{n} = f_{\chi^2}\left(nx; k, n\zeta_{*}^{2}\right) + \frac{1}{8n}\left\{\sum_{j=0}^{4} b_j f_{\chi^2}\left(nx; k + 2j, n\zeta_{*}^{2}\right)\right\} + o\left(n^{-1}\right),$$

where the constants b_j depend on k, κ and ζ_{*}^{2}, and $f_{\chi^2}(x; v, \delta)$ is the density of the non-central chi-square distribution with v degrees of freedom and non-centrality parameter δ. From this approximation one can easily compute the approximate first and second moments of $\hat{\zeta}_{*}^{2}$, namely

$$\mathrm{E}\left[\hat{\zeta}_{*}^{2}\right] = \zeta_{*}^{2} + \frac{1}{n}\left[\kappa\zeta_{*}^{2}(k+2) - \zeta_{*}^{2} + k\right] + \frac{1}{n^2}\left(k(k+1)\right) + o\left(n^{-2}\right), \tag{7.18}$$

$$\begin{aligned}\mathrm{Var}\left(\hat{\zeta}_{*}^{2}\right) = {} & \frac{\zeta_{*}^{2}}{n}\left[4 + 2\zeta_{*}^{2} + 3(\kappa - 1)\zeta_{*}^{2}\right] + \frac{2k}{n^2} \\ & + \frac{\zeta_{*}^{2}}{n^2}\left[8(1+k) - \zeta_{*}^{2} - \kappa^2\zeta_{*}^{2}(k+2)^2 + \kappa\left(2\zeta_{*}^{2} + 4\right)(k+2)\right] + o\left(n^{-2}\right),\end{aligned} \tag{7.19}$$

Based on these higher order formulae we claim:

The squared maximal Sharpe ratio for elliptical returns

For modest κ, the mean and variance of $\hat{\zeta}_{*}^{2}$ for elliptical returns are very close to those for Gaussian returns.

Example 7.2.1 (Elliptical returns and squared maximal Sharpe ratio standard error). We draw returns from an elliptical distribution with a fixed value of κ: for $\kappa = 1$ we take the multivariate normal, for $\kappa > 1$ we take a multivariate t distribution with the appropriate degrees of freedom. We draw a fixed number of returns with a fixed value of $\zeta_* = 0.07\ \mathrm{day}^{-1/2}$, then compute $\hat{\zeta}_{*}^{2}$. We let number of assets k vary from 1 to 16; we let the sample size n vary from 1,260 days to 2,520 days; we let the kurtosis factor κ take values of 1 and

10. For each Cartesian corner of the parameter space, we perform 10,000 simulations.

In Figure 7.4, we plot the empirical mean value of $\hat{\zeta}_*$ versus k, with facets for the different values of n. We plot a horizontal line at the population value, ζ_*. We also plot a single line based on the formula for the Gaussian *i.i.d.* case, as given in Equation 6.60. We note that the kurtosis factor has negligible effect on the empirical mean value of $\hat{\zeta}_*$, so much so that the points are essentially overlapping each other. Because of this, the mean value from the Gaussian *i.i.d.* case is fairly accurate across the range of κ tested.

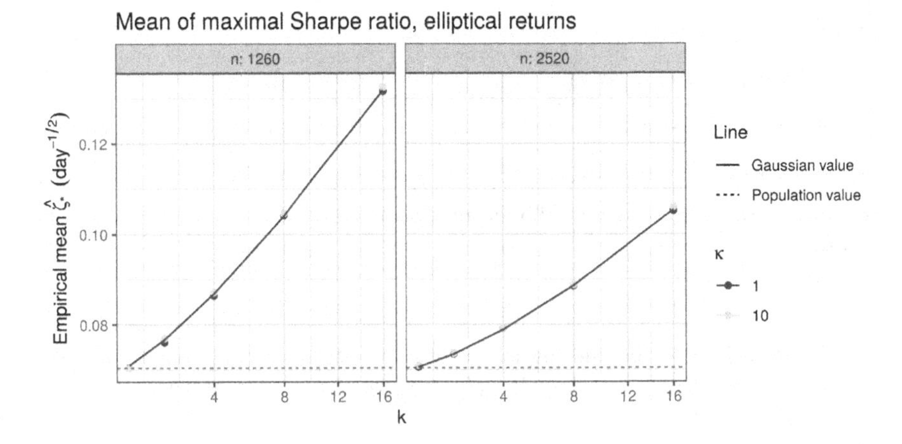

FIGURE 7.4
The empirical mean over 10,000 simulations of $\hat{\zeta}_*$ is plotted versus k, with facets for different n. Returns are drawn from elliptical distributions with varying κ, but we see only weak dependence of the mean on this kurtosis factor. A horizontal line is plotted at ζ_*, along with a line from the law under the Gaussian *i.i.d.* case.

In Figure 7.5, we plot the empirical variance of $\hat{\zeta}_*$ versus k, with facets for the different values of n. We plot a line for the approximate variance in the Gaussian *i.i.d.* case from Equation 6.62. We also plot lines for the asymptotic variance, from Equation 7.17. The latter have no dependence on k and so are horizontal.

In Figure 7.6, we plot the empirical mean value of $\hat{\zeta}_*^2$ versus k, with facets for the different values of n. We plot a horizontal line at the population value ζ_*^2. We plot lines for the higher order approximate mean from Equation 7.18, and a line for the approximate mean in the Gaussian *i.i.d.* case from Equation 6.57. Each of these theoretical values are very close to the empirical values in this plot, and they essentially overlap each other.

In Figure 7.7, we plot the empirical variance of $\hat{\zeta}_*^2$ versus k. We plot lines for the asymptotic approximate variance from Equation 7.16, the higher order asymptotic variance from Equation 7.19, and the value for the Gaussian *i.i.d.* case from Equation 6.58. The asymptotic value has no dependence on k and

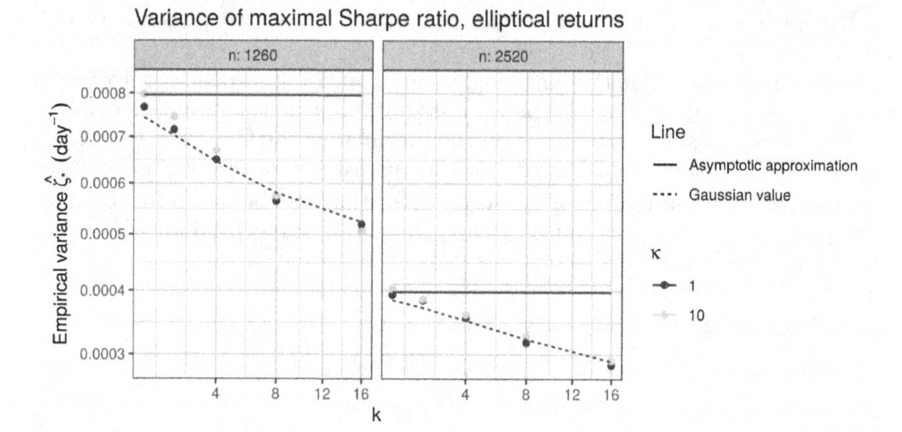

FIGURE 7.5
The empirical variance over 10,000 simulations of $\hat{\zeta}_*$ are plotted versus k. We also plot the asymptotic approximate variance and the variance for the Gaussian *i.i.d.* case.

is a horizontal line, while the higher order asymptotic and Gaussian theoretical values vary with k and describe the empirical variance quite well in this experiment. ⊣

7.2.2 Asymptotic distribution, matrix normal returns

In Chapter 3, we considered the distribution of the Sharpe ratio under various deviations from *i.i.d.* normality: autocorrelation, heteroskedasticity and so on. We wish to perform similar analysis on $\hat{\zeta}_*^2$, to check whether its distribution is robust to violations of these assumptions[3]. To do so we model the matrix of returns by a matrix normal distribution.

So suppose that the augmented returns[4] follow a matrix normal distribution:

$$\tilde{\mathsf{X}} \sim \mathcal{MN}\left(\tilde{\mathsf{M}}, \Psi, \Xi\right),$$

for $n \times (1+p)$ matrix $\tilde{\mathsf{M}}$ (whose first column is all ones) and symmetric positive semi-definite matrices Ξ and Ψ, respectively of size $(1+p)\times(1+p)$ and $n \times n$. There is an ambiguity in the scaling of the two covariance matrices, which we resolve by assuming that $\operatorname{tr}(\Psi) = n$. Again we let

$$\hat{\Theta} = \frac{1}{n}\tilde{\mathsf{X}}^\top\tilde{\mathsf{X}}.$$

[3]Spoiler alert: the bias and variance of $\hat{\zeta}_*^2$ are affected by autocorrelation and heteroskedasticity in a way that $\hat{\zeta}$ of a single asset is not.

[4]Recall that the first column of $\tilde{\mathsf{X}}$ is identically one.

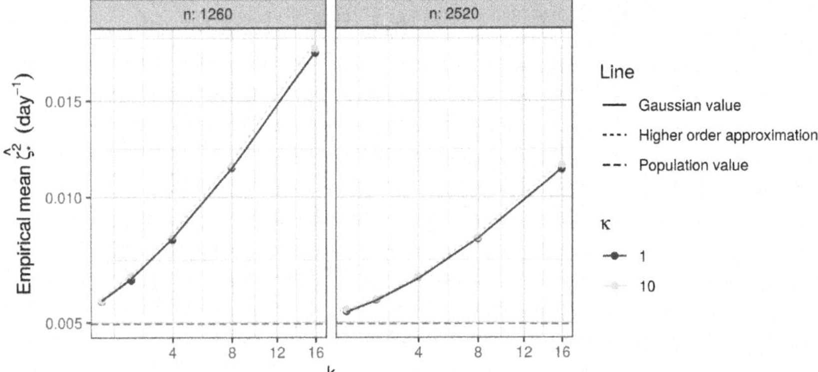

FIGURE 7.6
The empirical mean over 10,000 simulations of $\hat{\zeta}_*^2$ are plotted versus k. A horizontal line is plotted at ζ_*^2, along with lines for the higher order asymptotics and the law under the Gaussian *i.i.d.* case.

The mean and covariance of $\hat{\Theta}$ are then: [Pav, 2013]

$$\mathrm{E}\left[\hat{\Theta}\right] = \frac{1}{n}\tilde{\mathsf{M}}^\top\tilde{\mathsf{M}} + \frac{\operatorname{tr}\left(\Psi\right)}{n}\Xi = \frac{1}{n}\tilde{\mathsf{M}}^\top\tilde{\mathsf{M}} + \Xi, \tag{7.20}$$

and

$$\operatorname{Var}\left(\operatorname{vec}\left(\hat{\Theta}\right)\right) = \frac{1}{n^2}\left(\mathsf{I}+\mathsf{K}\right)\left[\Xi\otimes\Xi\operatorname{tr}\left(\Psi^2\right) + \left(\tilde{\mathsf{M}}^\top\Psi\tilde{\mathsf{M}}\right)\otimes\Xi + \Xi\otimes\left(\tilde{\mathsf{M}}^\top\Psi\tilde{\mathsf{M}}\right)\right]. \tag{7.21}$$

We note that in the *i.i.d.* normal case, $\tilde{\mathsf{M}} = \mathbf{1}\tilde{\boldsymbol{\mu}}^\top$, $\Psi = \mathsf{I}$, $\Xi = \tilde{\Sigma}$, where $\tilde{\boldsymbol{\mu}}$ and $\tilde{\Sigma}$ are defined in Equation 7.12. Then the variance in Equation 7.21 reduces to

$$\frac{1}{n}\left(\mathsf{I}+\mathsf{K}\right)\left[\tilde{\Sigma}\otimes\tilde{\Sigma} + \left(\tilde{\boldsymbol{\mu}}\tilde{\boldsymbol{\mu}}^\top\right)\otimes\tilde{\Sigma} + \tilde{\Sigma}\otimes\left(\tilde{\boldsymbol{\mu}}\tilde{\boldsymbol{\mu}}^\top\right)\right].$$

This is equivalent to Equation 7.13 with $\kappa = 1$.

Letting

$$\Theta = \lim_{n\to\infty}\frac{1}{n}\tilde{\mathsf{M}}^\top\tilde{\mathsf{M}} + \Xi,$$

then asymptotically in n,

$$\sqrt{n}\left(\hat{\Theta}^{-1} - \Theta^{-1}\right) \rightsquigarrow \mathcal{N}\left(0, \frac{1}{n}\mathsf{G}\right), \tag{7.22}$$

with

$$\begin{aligned}\mathsf{G} = {} & 2\mathsf{N}\left[\left(\Theta^{-1}\Xi\Theta^{-1}\right)\otimes\left(\Theta^{-1}\Xi\Theta^{-1}\right)\operatorname{tr}\left(\Psi^2\right)\right] \\ & + 2\mathsf{N}\left[\left(\Theta^{-1}\tilde{\mathsf{M}}^\top\Psi\tilde{\mathsf{M}}\Theta^{-1}\right)\otimes\left(\Theta^{-1}\Xi\Theta^{-1}\right) + \left(\Theta^{-1}\Xi\Theta^{-1}\right)\otimes\left(\Theta^{-1}\tilde{\mathsf{M}}^\top\Psi\tilde{\mathsf{M}}\Theta^{-1}\right)\right].\end{aligned} \tag{7.23}$$

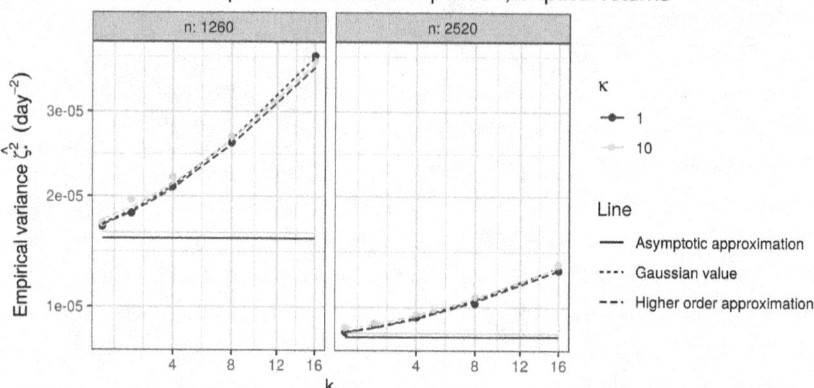

FIGURE 7.7
The empirical variance over 10,000 simulations of $\hat{\zeta}_*^2$ is plotted versus k. We also plot the asymptotic approximate variance, which has no dependence on k, and the higher order approximation and and the variance for the Gaussian *i.i.d.* case.

To find the asymptotic variance of $\hat{\zeta}_*^2$, which is the $1,1$ element of $\hat{\Theta}^{-1}$ minus one, simply compute $\left(\boldsymbol{e}_1^\top \otimes \boldsymbol{e}_1^\top\right) \mathsf{G} \left(\boldsymbol{e}_1 \otimes \boldsymbol{e}_1\right)$. This is hard to express compactly in the general case, but in specific cases it reduces nicely.

Higher order approximate moments

A higher order approximation for the expected value of $\hat{\zeta}_*^2$ can be found, from which an approximate asymptotic variance can be derived, *cf.* Exercise 7.24. Let U_{-1} be the "remove first" matrix, whose size should be inferred in context: it is a matrix of all rows but the first of the identity matrix. Let $c_1 = \mathbf{1}^\top \Psi \mathbf{1}/n$, let $\Sigma = \mathsf{U}_{-1} \Xi \mathsf{U}_{-1}^\top$, $\mathsf{M} = \tilde{\mathsf{M}} \mathsf{U}_{-1}^\top$, and define $\zeta_{*,m}^2 = n^{-2} \mathbf{1}^\top \mathsf{M} \Sigma^{-1} \mathsf{M}^\top \mathbf{1}$. Then

$$\mathrm{E}\left[\hat{\zeta}_*^2\right] \approx \frac{n-1}{n} \left(\frac{c_1 k}{n} + \zeta_{*,m}^2\right), \tag{7.24}$$

$$\mathrm{Var}\left(\hat{\zeta}_*^2\right) \approx \left(\frac{n-1}{n}\right)^2 \frac{2c_1}{n} \left(\frac{c_1 k}{n} + 2\zeta_{*,m}^2\right). \tag{7.25}$$

Compare these to the formulæ for the Gaussian *i.i.d.* case, given in Equation 6.57 and Equation 6.58. Note that the order of the error terms of these approximations is not known, see Exercise 7.25.

By a similar analysis one has the approximate expected value and variance of the Markowitz portfolio. Define $\boldsymbol{\nu}_* = n^{-1} \Sigma^{-1} \mathsf{M}^\top \mathbf{1}$. Then

$$\mathrm{E}\left[\hat{\boldsymbol{\nu}}_*\right] \approx (n-1) \left[\left(\frac{c_1}{n} + \zeta_{*,m}^2 + 1\right) \boldsymbol{\nu}_* - \frac{1}{n} \Sigma^{-1/2} \mathsf{M}^\top \Sigma^{-1} \mathsf{M} \Sigma^{-1/2} \mathsf{M}^\top \mathbf{1} + \ldots\right], \tag{7.26}$$

$$\mathrm{Var}\left(\hat{\boldsymbol{\nu}}_*\right) \approx n c_1 \left(\frac{n-1}{n}\right)^2 \Sigma^{-1} \tag{7.27}$$

See Exercise 7.26.

7.2.3 Asymptotic squared maximal Sharpe ratio, heteroskedastic returns

In Section 3.1.3, we considered the distribution of $\hat{\zeta}$ when returns are heteroskedastic. Here we consider matrix normal returns under a similar model. Suppose that $\boldsymbol{l}$ is a strictly positive random n-vector such that $\mathsf{M} = \boldsymbol{l}^{\lambda}\boldsymbol{\mu}^{\top}$, for some λ, $\Psi = \operatorname{Diag}(\boldsymbol{l})$, and $\Xi = \tilde{\Sigma}$, where $\tilde{\boldsymbol{\mu}}$ and $\tilde{\Sigma}$ are defined in Equation 7.12. Recall that the 'optimistic model' of market heteroskedasticity, where changes in volatility are due to an acceleration of price discovery, corresponds to $\lambda = \frac{1}{2}$, and the 'pessimistic model' has $\lambda = 0$, under which extra volatility only contributes noise. Under a random leverage model one has $\lambda = \frac{1}{2}$; when one observes returns from jumbled periodicities one has $\lambda = 1$.

Define M_k as the empirical raw kth moment of $\boldsymbol{l}$:

$$M_k := \frac{1}{n} \sum_{1 \le t \le n} {l_t}^k. \tag{7.28}$$

The assumption on the trace of Ψ implies that $\boldsymbol{l}$ has been rescaled such that $M_1 = 1$. We note then that $c_1 = 1$, $\operatorname{tr}\left(\Psi^2\right) = nM_2$ and $\mathsf{M}^{\top}\mathsf{M} = nM_{2\lambda}\boldsymbol{\mu}\boldsymbol{\mu}^{\top}$. Moreover, $\zeta_{*,m}^2 = M_{\lambda}^2\boldsymbol{\mu}^{\top}\Sigma^{-1}\boldsymbol{\mu}$. When $\lambda = 0$ or $\lambda = 1$ we have that $\zeta_{*,m}^2 = \zeta_*^2$; for modest values of λ this should nearly hold. Thus the higher order approximations of Equation 7.24 and Equation 7.25 suggest that the mean and variance of $\hat{\zeta}_*^2$ will only slightly deviate from those under Gaussian *i.i.d.* returns.

Heteroskedasticity and maximal Sharpe ratio

Modest heteroskedasticity only causes small bias and inflation to standard error of $\hat{\zeta}_*^2$, as compared to those under *i.i.d.* Gaussian returns.

Asymptotic results

We note again that the asymptotic results have no dependence on k, as this is considered fixed as $n \to \infty$. And thus we will see these results are of limited use for finite sample sizes. For heteroskedastic returns Θ takes value

$$\begin{aligned}
\Theta = \mathrm{E}\left[\hat{\Theta}\right] &= \begin{bmatrix} 1 & M_{\lambda}\boldsymbol{\mu}^{\top} \\ M_{\lambda}\boldsymbol{\mu} & M_{2\lambda}\boldsymbol{\mu}\boldsymbol{\mu}^{\top} + \Sigma \end{bmatrix}, \\
&= \begin{bmatrix} 1 & M_{\lambda}\boldsymbol{\mu}^{\top} \\ M_{\lambda}\boldsymbol{\mu} & M_{\lambda}^2\boldsymbol{\mu}\boldsymbol{\mu}^{\top} + \left(\left(M_{2\lambda} - M_{\lambda}^2\right)\boldsymbol{\mu}\boldsymbol{\mu}^{\top} + \Sigma\right) \end{bmatrix}.
\end{aligned}$$

The inverse of this is

$$\Theta^{-1} = \begin{bmatrix} 1 + \frac{\zeta_{*,m}^2}{d_2} & -\frac{M_\lambda}{d_2}\boldsymbol{\mu}^\top\Sigma^{-1} \\ -\frac{M_\lambda}{d_2}\Sigma^{-1}\boldsymbol{\mu} & \Sigma^{-1} - \frac{d_1\Sigma^{-1}\boldsymbol{\mu}\boldsymbol{\mu}^\top\Sigma^{-1}}{d_2} \end{bmatrix},$$

where $d_1 = M_{2\lambda} - M_\lambda^2$, $d_2 = 1 + d_1\boldsymbol{\mu}^\top\Sigma^{-1}\boldsymbol{\mu}$. Note that the term d_1 is like the variance of $\boldsymbol{l}^\lambda$. When $\lambda = 0$ it equals zero and d_2 equals 1, as there is no variation in $\boldsymbol{l}^\lambda$.

Then asymptotically in n,

$$\sqrt{n}\left(\hat{\zeta}_*^2 - \frac{\zeta_{*,m}^2}{d_2}\right) \rightsquigarrow \mathcal{N}\left(0, v^2\right), \tag{7.29}$$

where

$$v^2 = 4\frac{\zeta_{*,m}^2}{d_2^4}\left(\frac{M_2\zeta_{*,m}^2}{2} + \left(d_2 + \zeta_{*,m}^2\left(1 - \frac{M_{\lambda+1}}{M_\lambda}\right)\right)^2 + \frac{\left(M_{2\lambda+1} - M_{\lambda+1}^2\right)\zeta_{*,m}^4}{M_\lambda^2}\right), \tag{7.30}$$

$$\approx 4\frac{\zeta_{*,m}^2}{d_2^4}\left(d_2^2 + \frac{M_2\zeta_{*,m}^2}{2}\right), \tag{7.31}$$

where the approximation holds when $\boldsymbol{\mu}^\top\Sigma^{-1}\boldsymbol{\mu}$ is small, as we expect it to be in real applications. Note that when d_1 is nonzero then ζ_*^2 is bounded and can take values no larger than $M_\lambda^2 d_1^{-1}$.

Setting $\zeta_*^2 = \boldsymbol{\mu}^\top\Sigma^{-1}\boldsymbol{\mu}$ as usual, for a few special cases we have the approximations:

$$(\lambda = 0)\quad \hat{\zeta}_*^2 \rightsquigarrow \mathcal{N}\left(\zeta_*^2, \frac{1}{n}2\zeta_*^2\left(2 + M_2\zeta_*^2\right)\right),$$

$$(\lambda = 1/2)\quad \hat{\zeta}_*^2 \rightsquigarrow \mathcal{N}\left(\frac{M_{1/2}^2\zeta_*^2}{1 + \left(1 - M_{1/2}^2\right)\zeta_*^2}, \frac{1}{n}v^2\right),$$

$$v^2 \approx \frac{4M_{1/2}^2\zeta_*^2}{\left(1 + \left(1 - M_{1/2}^2\right)\zeta_*^2\right)^2} + \frac{2M_2M_{1/2}^4\zeta_*^4}{\left(1 + \left(1 - M_{1/2}^2\right)\zeta_*^2\right)^4}, \tag{7.32}$$

$$(\lambda = 1)\quad \hat{\zeta}_*^2 \rightsquigarrow \mathcal{N}\left(\frac{\zeta_*^2}{1 + \left(M_2 - 1\right)\zeta_*^2}, \frac{1}{n}v^2\right),$$

$$v^2 \approx \frac{4\zeta_*^2}{\left(1 + \left(M_2 - 1\right)\zeta_*^2\right)^2} + \frac{2M_2\zeta_*^4}{\left(1 + \left(M_2 - 1\right)\zeta_*^2\right)^4}.$$

In the $\lambda = 0$ case, $\hat{\zeta}_*^2$ is asymptotically unbiased for ζ_*^2, and the variance of $\hat{\zeta}_*^2$ is barely affected. For the other cases, one should see only modest changes to the expectation and variance of $\hat{\zeta}_*^2$ as compared to the Gaussian *i.i.d.* case.

Example 7.2.2 (Heteroskedastic returns and squared maximal Sharpe ratio standard error). We draw returns from a heteroskedastic normal distribution, with $\boldsymbol{l}$ sampled from 1,260 values of the VIX level squared, then rescaled to have mean one. We draw a fixed number of returns with a fixed value of $\zeta_* = 0.07\ \text{day}^{-1/2}$, then compute $\hat{\zeta}_*^2$. We let number of assets k vary from 1 to 16. We take the sample size n to be either $1,260$ days or twice that value, recycling $\boldsymbol{l}$ for the longer series, which avoids variation in the moments of $\boldsymbol{l}$

among different simulations. We let λ take values between 0 and 2. For each Cartesian corner of the parameter space, we perform 10,000 simulations.

In Figure 7.8, we plot the empirical mean value of $\hat{\zeta}_*^2$ versus k, along with lines for the higher order approximation from Equation 7.24 noting that $c_1 = 1$ and $\zeta_{*,m}^2 = M_\lambda^2 \zeta_*^2$. We also plot the limiting asymptotic value $\zeta_{*,m}^2 d_2^{-2}$, and the expected value for the Gaussian *i.i.d.* case, from Equation 6.57. The higher order approximate expected value matches the empirical mean fairly well, but the asymptotic value does not.

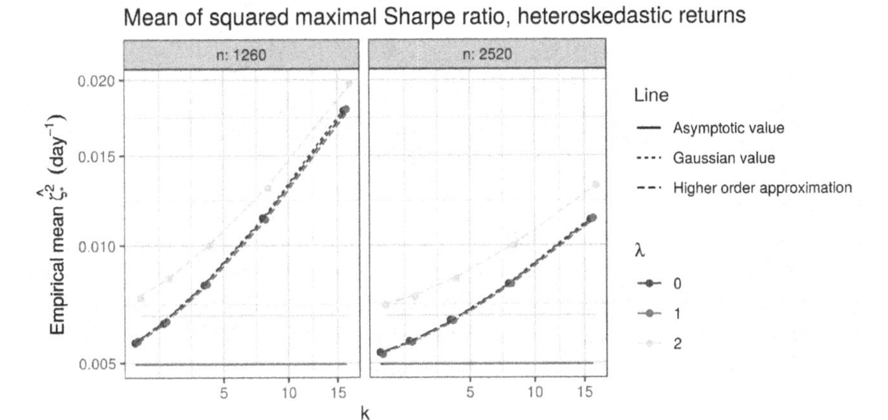

FIGURE 7.8
The empirical mean over 10,000 simulations of $\hat{\zeta}_*^2$ is plotted versus k, for different values of λ and facets for different n. Returns are drawn from heteroskedastic distribution with samples of rescaled VIX squared for $\boldsymbol{l}$. Lines are plotted for the higher order approximate expected value, the expected value in the Gaussian case, and the asymptotic value, which has no dependence on k. The higher order expected line overlaps the Gaussian line. Points are slightly jittered in the x direction to better show them.

In Figure 7.9, we plot the empirical variance of $\hat{\zeta}_*^2$ versus k, along with lines for the higher order approximation from Equation 7.25, and $\zeta_{*,m}^2 = M_\lambda^2 \zeta_*^2$. We also plot the limiting asymptotic value of v^2 from Equation 7.31, and the variance from the Gaussian *i.i.d.* case in Equation 6.58. The higher order approximate variance appears to be reasonably close to the empirical variance, but the asymptotic variance is not. ⊣

Example 7.2.3 (VIX and heteroskedastic bias). In Example 3.1.1, we considered the rescaled squared VIX index as $\boldsymbol{l}$. Using data from 1990-01-02 through 2020-12-31, for values of j between 0 and 3, empirical estimates of M_j range (non-monotonically) from around 0.923 to 11.3. In particular, $M_2 = 2.19$, and $M_3 = 11.3$.

For values of λ tested between 0 and 1, the geometric "bias" of $\hat{\zeta}_{*,m}^2$, (*i.e.*,

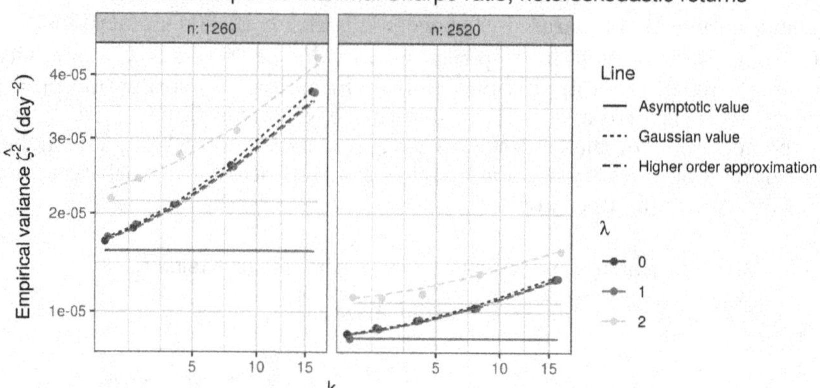

FIGURE 7.9
The empirical variance over 10,000 simulations of $\hat{\zeta}^2_*$ is plotted versus k. Lines are plotted for the higher order approximate variance, the variance in the Gaussian case, and the asymptotic value, which has no dependence on k. The Gaussian and higher order expected values are nearly overlapping. Points are slightly jittered in the x direction to better show them.

M^2_λ) ranged from from around 0.852 to 1. The relative decrease in the variance of $\hat{\zeta}^2_*$ from heteroskedasticity is similar, as evident from Equation 7.25. ⊣

Example 7.2.4 (Weekly and daily returns). In Example 3.1.3, we considered the effect of jumbling together weekly and daily returns when computing $\hat{\zeta}$, which we here extend to the multivariate case. Suppose you have observed 520 weeks of hypothetical weekly returns and 520 days of daily returns for a strategy, with the returns independent. Because of how expectation and variance scale, we have $\lambda = 1$. To get $M_1 = 1$, we define $\boldsymbol{l}$ to be 520 values of 5/3 and 520 values of 1/3. This means that if we compute $\hat{\zeta}^2_*$ naïvely, jumbling together daily and weekly returns, we are effectively estimating ζ^2_* at the scale of 3 days, instead of at the daily frequency.

Because $\lambda = 1$, then $M^2_\lambda = 1$, and so $\zeta^2_{*,m} = \zeta^2_*$. Thus, the expected value and variance of $\hat{\zeta}^2_*$, as approximated in Equation 7.24 and Equation 7.25 are effectively the same as one would observe in the Gaussian *i.i.d.* case, and so one could perform inference using the parametric results from Chapter 6, simply adjusting for the different time scale.

Supposing that $k = 8$ and $\zeta^2_* = 0.009\ \text{day}^{-1}$, the expected value and standard error of $\hat{\zeta}^2_*$ from the higher order approximations are, respectively, $0.012\ \text{day}^{-1}$ and $0.004\ \text{day}^{-1}$, where we have adjusted the units to daily time scale. If we had, instead, observed 3,120 days of *daily* returns, the expected value and standard error of $\hat{\zeta}^2_*$ would be $0.012\ \text{day}^{-1}$ and $0.004\ \text{day}^{-1}$, effectively the same values.

As a check, we perform 10,000 Monte Carlo simulations, where $k = 8$, Σ is diagonal, and $\boldsymbol{\mu}$ is $\mathbf{1}$ rescaled to achieve $\zeta_*^2 = 0.009\ \text{day}^{-1}$. We compute $\hat{\zeta}_*^2$ and convert it back to daily units by dividing by 3. The empirical mean of the $\hat{\zeta}_*^2$ is $0.012\ \text{day}^{-1}$, The standard deviation of the rescaled $\hat{\zeta}_*^2$ over these 10,000 realizations is around $0.004\,\text{day}^{-1}$, confirming the theoretical results. ⊣

7.2.4 Asymptotic squared maximal Sharpe ratio, autocorrelated returns

In Section 3.1.4, we considered univariate returns drawn from an AR(1) process. Analogously, here we assume $\tilde{\mathsf{M}} = \mathbf{1}\tilde{\boldsymbol{\mu}}^\top$ and $\Xi = \tilde{\Sigma}$, where $\tilde{\boldsymbol{\mu}}$ and $\tilde{\Sigma}$ are defined in Equation 7.12, and Ψ is an AR(1) matrix defined by $\Psi_{i,j} = \rho^{|i-j|}$. We will assume that $|\rho|$ is bounded away from one. We note that asymptotically $\hat{\Theta}$ goes to $\Theta = \tilde{\boldsymbol{\mu}}\tilde{\boldsymbol{\mu}}^\top + \tilde{\Sigma}$, which is independent of ρ. Thus $\hat{\zeta}_*$ is *asymptotically* unbiased for ζ_*. Regrettably for reasonable sample sizes, $\hat{\zeta}_*^2$ is affected by ρ.

Because mean returns do not vary, $\zeta_{*,m}^2 = \zeta_*^2$. We had used the following approximations in the univariate case (*cf.* Exercise 3.7), and they will be of use in our analysis here:

$$c_1 = \mathbf{1}^\top \Psi \mathbf{1}/n \approx \left(\frac{1+\rho}{1-\rho} + \frac{1}{n}\frac{2\rho}{(1-\rho)^2}\right),$$

$$c_2 = \operatorname{tr}\left(\Psi^2\right)/n \approx \left(\frac{1+\rho^2}{1-\rho^2} + \frac{1}{n}\frac{2\rho^2}{(1-\rho^2)^2}\right).$$

Plugging in c_1 and $\zeta_{*,m}^2$ into the higher order approximations of Equation 7.24 and Equation 7.25, we then have

$$\operatorname{E}\left[\hat{\zeta}_*^2\right] \approx \frac{n-1}{n}\left(\frac{k}{n}\frac{1+\rho}{1-\rho} + \zeta_*^2\right), \tag{7.33}$$

$$\operatorname{Var}\left(\hat{\zeta}_*^2\right) \approx 2\frac{1+\rho}{1-\rho}\left(\frac{n-1}{n}\right)^2\left(\frac{k}{n}\frac{1+\rho}{1-\rho} + 2\zeta_*^2\right). \tag{7.34}$$

Thus the expected value and standard error of $\hat{\zeta}_*^2$ are both affected by ρ. Contrast this to the univariate case where you will recall only the standard error of $\hat{\zeta}$ was affected by ρ.

Autocorrelation and squared maximal Sharpe ratio

A small autocorrelation of ρ introduces a geometric bias to $\hat{\zeta}_*^2$ of around $100\rho\%$, and inflates its standard error by the same amount.

Turning to the asymptotic analysis, because $\tilde{\mathsf{M}} = \mathbf{1}\tilde{\boldsymbol{\mu}}^\top$ we have $\tilde{\mathsf{M}}^\top \Psi \tilde{\mathsf{M}} = nc_1 \tilde{\boldsymbol{\mu}}\tilde{\boldsymbol{\mu}}^\top$. Then the G of Equation 7.23 becomes

$$\begin{aligned} \mathsf{G} = & 2nc_2 \mathsf{N} \left[\left(\Theta^{-1}\tilde{\Sigma}\Theta^{-1}\right) \otimes \left(\Theta^{-1}\tilde{\Sigma}\Theta^{-1}\right) \right] \\ & + 2nc_1 \mathsf{N} \left[\left(\Theta^{-1}\tilde{\boldsymbol{\mu}}\tilde{\boldsymbol{\mu}}^\top\Theta^{-1}\right) \otimes \left(\Theta^{-1}\tilde{\Sigma}\Theta^{-1}\right) + \left(\Theta^{-1}\tilde{\Sigma}\Theta^{-1}\right) \otimes \left(\Theta^{-1}\tilde{\boldsymbol{\mu}}\tilde{\boldsymbol{\mu}}^\top\Theta^{-1}\right) \right]. \end{aligned} \tag{7.35}$$

Then asymptotically in n,

$$\sqrt{n}\left(\hat{\zeta}_*^2 - \zeta_*^2\right) \rightsquigarrow \mathcal{N}\left(0, \frac{4c_1}{n}\zeta_*^2 + \frac{2c_2}{n}\zeta_*^4\right). \tag{7.36}$$

Again, the variance is independent of k, and is of limited practical use.

Example 7.2.5 (AR(1) returns and squared maximal Sharpe ratio standard error). We draw a fixed number of returns from an autocorrelated Gaussian distribution, with a fixed value of $\zeta_* = 0.07\,\text{day}^{-1/2}$, then compute $\hat{\zeta}_*^2$. We let number of assets k vary from 3 to 10. We fix the sample size as $n = 1,260$ days. We let ρ vary from -0.2 to 0.2. For each Cartesian corner of the parameter space, we perform 5,000 simulations.

In Figure 7.10, we plot the empirical mean value of $\hat{\zeta}_*^2$ versus ρ along with a line for the higher order approximation from Equation 7.24. We also plot the limiting asymptotic value ζ_*^2, and the expected value for the Gaussian *i.i.d.* case, from Equation 6.57. The higher order approximate expected value matches the empirical mean fairly well.

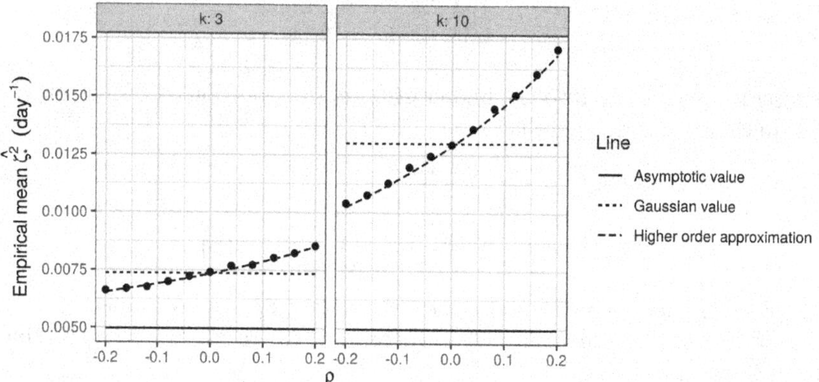

FIGURE 7.10
The empirical mean over 5,000 simulations of $\hat{\zeta}_*^2$ is plotted versus ρ, with facets for different n. Returns are drawn from autocorrelated Gaussian distribution. Lines are plotted for the higher order approximate expected value, the expected value in the uncorrelated Gaussian case, and the asymptotic value. The latter two have no dependence on ρ.

In Figure 7.11, we plot the empirical variance of $\hat{\zeta}_*^2$ versus ρ, along with lines for the higher order approximation from Equation 7.25. We also plot the asymptotic variance from Equation 7.36 and the variance from the Gaussian *i.i.d.* case in Equation 6.58. The higher order approximate variance appears to be reasonably close to the empirical variance. ⊣

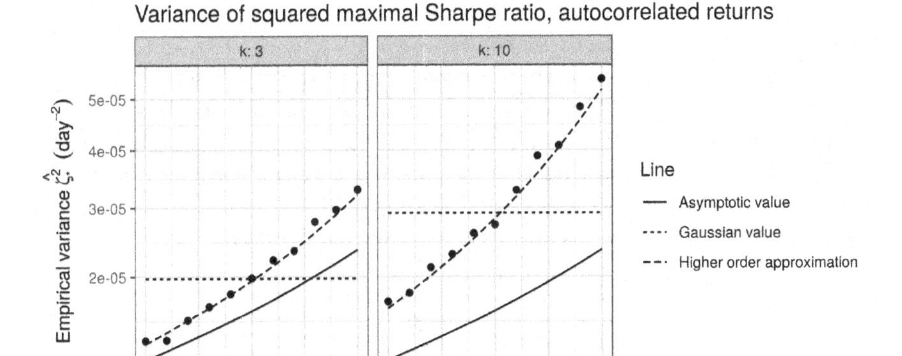

FIGURE 7.11
The empirical variance over 5,000 simulations of $\hat{\zeta}^2_*$ is plotted versus ρ. Lines are plotted for the higher order approximate variance, the variance in the uncorrelated Gaussian case, and the asymptotic value, which has no dependence on ρ. The asymptotic value depends on ρ but not on k.

7.3 † Constrained problems

We now adapt the second moment matrix approach to many of the generalizations of the portfolio optimization problem we introduced in Chapter 5. We will consider subspace constraints, hedging, conditional expectation and so on. For each of these extensions, we will link back to the problem formulation, then show how inference can be performed via some transformation of $\hat{\Theta}^{-1}$ using the central limit theorem and delta method.

7.3.1 † Subspace constraint

Consider first the portfolio optimization problem with basic subspace constraint introduced in Equation 5.18, *viz.*

$$\max_{\substack{\boldsymbol{\nu}:\mathsf{G}^C\boldsymbol{\nu}=\mathbf{0},\\ \boldsymbol{\nu}^\top\Sigma\boldsymbol{\nu}\le R^2}} \frac{\boldsymbol{\nu}^\top\boldsymbol{\mu}-r_0}{\sqrt{\boldsymbol{\nu}^\top\Sigma\boldsymbol{\nu}}}, \tag{7.37}$$

where G^C is a $(k-k_j)\times k$ matrix of rank $k-k_j$, r_0 is the disastrous rate, and $R>0$ is the risk budget. Again, the rows of G span the null space of the rows of G^C; that is, $\mathsf{G}^C\mathsf{G}^\top=0$, and $\mathsf{G}\mathsf{G}^\top=\mathsf{I}$. We can interpret the orthogonality constraint $\mathsf{G}^C\boldsymbol{\nu}=\mathbf{0}$ as stating that $\boldsymbol{\nu}$ must be a linear combination of the columns of $\mathsf{G}^\top$, thus $\boldsymbol{\nu}=\mathsf{G}^\top\boldsymbol{\xi}$.

The solution to this problem, given in Equation 5.19, is

$$\boldsymbol{\nu}_{R,\mathsf{G},*} := c\mathsf{G}^\top\left(\mathsf{G}\Sigma\mathsf{G}^\top\right)^{-1}\mathsf{G}\boldsymbol{\mu},$$

$$c=\frac{R}{\sqrt{\boldsymbol{\mu}^\top\mathsf{G}^\top\left(\mathsf{G}\Sigma\mathsf{G}^\top\right)^{-1}\mathsf{G}\boldsymbol{\mu}}}.$$

As in Equation 5.20, this portfolio has signal-noise ratio

$$\zeta_{*,\mathsf{G}} := \zeta\left(\boldsymbol{\nu}_{*,\mathsf{G}}\right) = \sqrt{\left(\mathsf{G}\boldsymbol{\mu}\right)^\top\left(\mathsf{G}\Sigma\mathsf{G}^\top\right)^{-1}\left(\mathsf{G}\boldsymbol{\mu}\right)} - \frac{r_0}{R}. \tag{7.38}$$

We can easily find the asymptotic distribution of $\hat{\boldsymbol{\nu}}_{R,\mathsf{G},*}$ and $\left(\zeta_{*,\mathsf{G}}+\frac{r_0}{R}\right)^2$ in the inverse second moment matrix by means of a transform. To do so, let $\tilde{\mathsf{G}}$ be the $(1+k_j)\times(k+1)$ matrix,

$$\tilde{\mathsf{G}} := \begin{bmatrix} 1 & 0\\ 0 & \mathsf{G}\end{bmatrix}.$$

Then $\tilde{\mathsf{G}}^\top\left(\tilde{\mathsf{G}}\Theta\tilde{\mathsf{G}}^\top\right)^{-1}\tilde{\mathsf{G}}$ takes the form

$$\tilde{\mathsf{G}}^\top\left(\tilde{\mathsf{G}}\Theta\tilde{\mathsf{G}}^\top\right)^{-1}\tilde{\mathsf{G}} = \begin{bmatrix} 1+\boldsymbol{\mu}^\top\mathsf{G}^\top\left(\mathsf{G}\Sigma\mathsf{G}^\top\right)^{-1}\mathsf{G}\boldsymbol{\mu} & -\boldsymbol{\mu}^\top\mathsf{G}^\top\left(\mathsf{G}\Sigma\mathsf{G}^\top\right)^{-1}\mathsf{G}\\ -\mathsf{G}^\top\left(\mathsf{G}\Sigma\mathsf{G}^\top\right)^{-1}\mathsf{G}\boldsymbol{\mu} & \mathsf{G}^\top\left(\mathsf{G}\Sigma\mathsf{G}^\top\right)^{-1}\mathsf{G}\end{bmatrix}. \tag{7.39}$$

In particular, elements 2 through $k+1$ of $-\operatorname{vech}\left(\tilde{\mathsf{G}}^\top\left(\tilde{\mathsf{G}}\Theta\tilde{\mathsf{G}}^\top\right)^{-1}\tilde{\mathsf{G}}\right)$ are the portfolio $\hat{\boldsymbol{\nu}}_{R,\mathsf{G},*}$, up to the scaling constant c which is the ratio of R to the square root of the first element of $\operatorname{vech}\left(\tilde{\mathsf{G}}^\top\left(\tilde{\mathsf{G}}\Theta\tilde{\mathsf{G}}^\top\right)^{-1}\tilde{\mathsf{G}}\right)$ minus one.

Analogous to Equation 7.8, asymptotically in n,

$$\sqrt{n}\left(\operatorname{vech}\left(\tilde{\mathsf{G}}^\top\left(\tilde{\mathsf{G}}\hat{\Theta}\tilde{\mathsf{G}}^\top\right)^{-1}\tilde{\mathsf{G}}\right) - \operatorname{vech}\left(\tilde{\mathsf{G}}^\top\left(\tilde{\mathsf{G}}\Theta\tilde{\mathsf{G}}^\top\right)^{-1}\tilde{\mathsf{G}}\right)\right) \rightsquigarrow \mathcal{N}\left(0,\mathsf{H}\Omega\mathsf{H}^\top\right), \tag{7.40}$$

where

$$\mathsf{H} = -\mathsf{L}\left(\tilde{\mathsf{G}}^\top\otimes\tilde{\mathsf{G}}^\top\right)\left(\left(\tilde{\mathsf{G}}\Theta\tilde{\mathsf{G}}^\top\right)^{-1}\otimes\left(\tilde{\mathsf{G}}\Theta\tilde{\mathsf{G}}^\top\right)^{-1}\right)\left(\tilde{\mathsf{G}}\otimes\tilde{\mathsf{G}}\right)\mathsf{D}.$$

7.3.2 † Hedging constraint

Consider now the portfolio optimization problem with hedging constraint given in Equation 5.21, namely.

$$\max_{\substack{\boldsymbol{\nu}:\mathsf{G}\Sigma\boldsymbol{\nu}=\mathbf{0},\\ \boldsymbol{\nu}^\top\Sigma\boldsymbol{\nu}\le R^2}} \frac{\boldsymbol{\nu}^\top\boldsymbol{\mu} - r_0}{\sqrt{\boldsymbol{\nu}^\top\Sigma\boldsymbol{\nu}}}, \tag{7.41}$$

where G is a $k_g \times k$ matrix of rank k_g. As given in Equation 5.22 the solution to this problem is the portfolio

$$\boldsymbol{\nu}_{*,\mathsf{I}\backslash\mathsf{G}} := c\left(\Sigma^{-1}\boldsymbol{\mu} - \mathsf{G}^\top\left(\mathsf{G}\Sigma\mathsf{G}^\top\right)^{-1}\mathsf{G}\boldsymbol{\mu}\right),$$
$$c = \frac{R}{\sqrt{\boldsymbol{\mu}^\top\Sigma^{-1}\boldsymbol{\mu} - \boldsymbol{\mu}^\top\mathsf{G}^\top\left(\mathsf{G}\Sigma\mathsf{G}^\top\right)^{-1}\mathsf{G}\boldsymbol{\mu}}}.$$

As noted previously this is, up to scaling, the difference of the unconstrained optimal portfolio and the subspace-constrained portfolio.

With $\tilde{\mathsf{G}} := \begin{bmatrix} 1 & 0 \\ 0 & \mathsf{G} \end{bmatrix}$, define the "delta inverse second moment" as

$$\Delta_{\mathsf{I}\backslash\mathsf{G}}\Theta^{-1} := \Theta^{-1} - \tilde{\mathsf{G}}^\top\left(\tilde{\mathsf{G}}\Theta\tilde{\mathsf{G}}^\top\right)^{-1}\tilde{\mathsf{G}}.$$

We then have

$$\Delta_{\mathsf{I}\backslash\mathsf{G}}\Theta^{-1} = \begin{bmatrix} \boldsymbol{\mu}^\top\Sigma^{-1}\boldsymbol{\mu} - \boldsymbol{\mu}^\top\mathsf{G}^\top\left(\mathsf{G}\Sigma\mathsf{G}^\top\right)^{-1}\mathsf{G}\boldsymbol{\mu} & -\boldsymbol{\mu}^\top\Sigma^{-1} + \boldsymbol{\mu}^\top\mathsf{G}^\top\left(\mathsf{G}\Sigma\mathsf{G}^\top\right)^{-1}\mathsf{G} \\ -\Sigma^{-1}\boldsymbol{\mu} + \mathsf{G}^\top\left(\mathsf{G}\Sigma\mathsf{G}^\top\right)^{-1}\mathsf{G}\boldsymbol{\mu} & \Sigma^{-1} - \mathsf{G}^\top\left(\mathsf{G}\Sigma\mathsf{G}^\top\right)^{-1}\mathsf{G} \end{bmatrix}.$$

So again the first column of this matrix has the total effect size as well as (negative) the optimal portfolio.

By the delta method, asymptotically in n,

$$\sqrt{n}\left(\operatorname{vech}\left(\Delta_{\mathsf{I}\backslash\mathsf{G}}\hat{\Theta}^{-1}\right) - \operatorname{vech}\left(\Delta_{\mathsf{I}\backslash\mathsf{G}}\Theta^{-1}\right)\right) \rightsquigarrow \mathcal{N}\left(0, \mathsf{H}\Omega\mathsf{H}^\top\right), \tag{7.42}$$

where

$$\mathsf{H} = -\mathsf{L}\left[\Theta^{-1}\otimes\Theta^{-1} - \left(\tilde{\mathsf{G}}^\top\otimes\tilde{\mathsf{G}}^\top\right)\left(\left(\tilde{\mathsf{G}}\Theta\tilde{\mathsf{G}}^\top\right)^{-1}\otimes\left(\tilde{\mathsf{G}}\Theta\tilde{\mathsf{G}}^\top\right)^{-1}\right)\left(\tilde{\mathsf{G}}\otimes\tilde{\mathsf{G}}\right)\right]\mathsf{D}.$$

7.3.3 † Conditional heteroskedasticity

In Section 5.3.1 we considered conditional heteroskedasticity models, where one observes a strictly positive scalar random variable, s_{t-1}, prior to the time the investment decision is required to capture $\boldsymbol{x}_t$. The two simple models for conditional heteroskedasticity given in Equation 5.30 and Equation 5.31 are

$$\begin{aligned}
&\text{(constant):} \quad \mathrm{E}\left[\boldsymbol{x}_t \,|\, s_{t-1}\right] = s_{t-1}^{-1}\boldsymbol{\mu}, \qquad \operatorname{Var}\left(\boldsymbol{x}_t \,|\, s_{t-1}\right) = s_{t-1}^{-2}\Sigma,\\
&\text{(floating):} \quad \mathrm{E}\left[\boldsymbol{x}_t \,|\, s_{t-1}\right] = \boldsymbol{\mu}, \qquad \operatorname{Var}\left(\boldsymbol{x}_t \,|\, s_{t-1}\right) = s_{t-1}^{-2}\Sigma.
\end{aligned}$$

Under the "constant" model the maximal single-period signal-noise ratio is $\sqrt{\boldsymbol{\mu}^\top \Sigma^{-1} \boldsymbol{\mu}}$, independent of s_{t-1}; under the "floating" model, it is is $s_{t-1}\sqrt{\boldsymbol{\mu}^\top \Sigma^{-1} \boldsymbol{\mu}}$. Thus the names reflect whether or not the maximal signal-noise ratio varies conditional on s_{t-1}.

The portfolio which maximizes the single-period signal-noise ratio is the same under both models, namely

$$\boldsymbol{\nu}_* = \frac{s_{t-1} R}{\sqrt{\boldsymbol{\mu}^\top \Sigma^{-1} \boldsymbol{\mu}}} \Sigma^{-1} \boldsymbol{\mu}. \tag{7.43}$$

Here again R is the risk budget. To perform inference on the portfolio $\boldsymbol{\nu}_*$ from Equation 7.43 under the "constant" model, apply the unconditional techniques to the sample second moment of $s_{t-1}\tilde{\boldsymbol{x}}_t$. For the "floating" model, however, some adjustment to the technique is required. Define $\tilde{\tilde{\boldsymbol{x}}}_t := s_{t-1}\tilde{\boldsymbol{x}}_t$; that is, $\tilde{\tilde{\boldsymbol{x}}}_t = \left[s_{t-1}, s_{t-1}\boldsymbol{x}_t{}^\top\right]^\top$. Consider the second moment of $\tilde{\tilde{\boldsymbol{x}}}$:

$$\Theta_s := \mathrm{E}\left[\tilde{\tilde{\boldsymbol{x}}}\tilde{\tilde{\boldsymbol{x}}}^\top\right] = \left[\begin{array}{cc} \gamma^2 & \gamma^2 \boldsymbol{\mu}^\top \\ \gamma^2 \boldsymbol{\mu} & \Sigma + \boldsymbol{\mu}\gamma^2\boldsymbol{\mu}^\top \end{array}\right], \tag{7.44}$$

where $\gamma^2 := \mathrm{E}\left[s^2\right]$. The inverse of Θ_s is

$$\Theta_s{}^{-1} = \left[\begin{array}{cc} \gamma^{-2} + \boldsymbol{\mu}^\top\Sigma^{-1}\boldsymbol{\mu} & -\boldsymbol{\mu}^\top\Sigma^{-1} \\ -\Sigma^{-1}\boldsymbol{\mu} & \Sigma^{-1} \end{array}\right] \tag{7.45}$$

Once again, the optimal portfolio (up to scaling and sign), appears in $\mathrm{vech}\left(\Theta_s{}^{-1}\right)$. Similarly, define the sample analogue:

$$\hat{\Theta}_s := \frac{1}{n}\sum_t \tilde{\tilde{\boldsymbol{x}}}_t \tilde{\tilde{\boldsymbol{x}}}_t{}^\top. \tag{7.46}$$

Again by the central limit theorem and multivariate delta method, asymptotically in n,

$$\sqrt{n}\left(\mathrm{vech}\left(\hat{\Theta}_s^{-1}\right) - \mathrm{vech}\left(\Theta_s{}^{-1}\right)\right) \rightsquigarrow \mathcal{N}\left(0, \mathsf{H}\Omega\mathsf{H}^\top\right), \tag{7.47}$$

where Ω is the variance of $\mathrm{vech}\left(\tilde{\tilde{\boldsymbol{x}}}\tilde{\tilde{\boldsymbol{x}}}^\top\right)$, and

$$\mathsf{H} = -\mathsf{L}\left(\Theta_s{}^{-1} \otimes \Theta_s{}^{-1}\right)\mathsf{D}. \tag{7.48}$$

The only real difference from the unconditional case is that we cannot automatically assume that the first row and column of Ω is zero (unless s is actually constant, which misses the point).

7.3.4 † Conditional expectation and heteroskedasticity

Suppose you observe random variables $s_{t-1} > 0$, and f-vector $\boldsymbol{f}_{t-1}$ prior to the decision required to capture $\boldsymbol{x}_t$. The s and $\boldsymbol{f}$ need not be independent. The general model introduced in Equation 5.33 is

$$\text{(bi-conditional):}\quad \mathrm{E}\left[\boldsymbol{x}_t \,\middle|\, s_{t-1}, \boldsymbol{f}_{t-1}\right] = \mathsf{B}\boldsymbol{f}_{t-1}, \qquad \mathrm{Var}\left(\boldsymbol{x}_t \,\middle|\, s_{t-1}, \boldsymbol{f}_{t-1}\right) = s_{t-1}^{-2}\Sigma,$$

where B is some $k \times f$ matrix. The portfolio which maximizes the single-period signal-noise ratio is

$$\boldsymbol{\nu}_* = \frac{s_{t-1} R}{\sqrt{{\boldsymbol{f}_{t-1}}^\top \mathsf{B}^\top \Sigma^{-1} \mathsf{B} \boldsymbol{f}_{t-1}}} \Sigma^{-1} \mathsf{B} \boldsymbol{f}_{t-1}. \tag{7.49}$$

Conditional on observing $\boldsymbol{f}_{t-1}$ and s_{t-1}, the squared maximal signal-noise ratio is

$$\zeta_*^2 \,\big|\, s_{t-1}, \boldsymbol{f}_{t-1} := \left(\frac{\mathrm{E}\left[{\boldsymbol{\nu}_*}^\top \boldsymbol{x}_t \,\middle|\, s_{t-1}, \boldsymbol{f}_{t-1} \right]}{\sqrt{\mathrm{Var}\left({\boldsymbol{\nu}_*}^\top \boldsymbol{x}_t \,\middle|\, s_{t-1}, \boldsymbol{f}_{t-1} \right)}} \right)^2 = {\boldsymbol{f}_{t-1}}^\top \mathsf{B}^\top \Sigma^{-1} \mathsf{B} \boldsymbol{f}_{t-1}.$$

This is independent of s_{t-1}, but depends on $\boldsymbol{f}_{t-1}$. The unconditional expected value of ζ_*^2 is

$$\mathrm{E}_{\boldsymbol{f}}\left[\zeta_*^2 \right] = \mathrm{tr}\left(\left(\mathsf{B}^\top \Sigma^{-1} \mathsf{B} \right) \Gamma_f \right),$$

the population Hotelling-Lawley trace.

To perform inference on the Markowitz coefficient via the second moment matrix, let

$$\tilde{\tilde{\boldsymbol{x}}}_t := \left[s_{t-1} \boldsymbol{f}_{t-1}^\top, s_{t-1} {\boldsymbol{x}_t}^\top \right]^\top. \tag{7.50}$$

The second moment of $\tilde{\tilde{\boldsymbol{x}}}$ is

$$\Theta_f := \mathrm{E}\left[\tilde{\tilde{\boldsymbol{x}}} \tilde{\tilde{\boldsymbol{x}}}^\top \right] = \left[\begin{array}{cc} \Gamma_f & \Gamma_f \mathsf{B}^\top \\ \mathsf{B} \Gamma_f & \Sigma + \mathsf{B} \Gamma_f \mathsf{B}^\top \end{array} \right], \tag{7.51}$$

where $\Gamma_f := \mathrm{E}\left[s^2 \boldsymbol{f} \boldsymbol{f}^\top \right]$. The inverse of Θ_f is

$$\Theta_f^{-1} = \left[\begin{array}{cc} {\Gamma_f}^{-1} + \mathsf{B}^\top \Sigma^{-1} \mathsf{B} & -\mathsf{B}^\top \Sigma^{-1} \\ -\Sigma^{-1} \mathsf{B} & \Sigma^{-1} \end{array} \right] \tag{7.52}$$

Once again, the Markowitz coefficient (up to scaling and sign), appears in $\mathrm{vech}\left({\Theta_f}^{-1} \right)$.

Define the sample estimate as

$$\hat{\Theta}_f := \frac{1}{n} \sum_t \tilde{\tilde{\boldsymbol{x}}}_t {\tilde{\tilde{\boldsymbol{x}}}_t}^\top. \tag{7.53}$$

Then asymptotically in n,

$$\sqrt{n} \left(\mathrm{vech}\left(\hat{\Theta}_f^{-1} \right) - \mathrm{vech}\left({\Theta_f}^{-1} \right) \right) \rightsquigarrow \mathcal{N}\left(0, \mathsf{H} \Omega \mathsf{H}^\top \right), \tag{7.54}$$

where

$$\mathsf{H} = -\mathsf{L}\left({\Theta_f}^{-1} \otimes {\Theta_f}^{-1} \right) \mathsf{D}. \tag{7.55}$$

Unlike for other problems, however, the Hotelling-Lawley trace does not exactly appear in the upper left corner of Θ^{-1}, rather it must be multiplied by Γ_f, then the trace must be taken.

7.4 Frequentist inference via the second moment

We now turn to using the second moment matrix $\hat{\Theta}$, or rather its inverse $\hat{\Theta}^{-1}$, to perform inference on elements of Θ^{-1}, for the case where returns deviate significantly from the assumption of *i.i.d.* Gaussian returns. Of primary interest is the squared maximal signal-noise ratio ζ_*^2. We will also consider inference on the population Markowitz portfolio, $\boldsymbol{\nu}_*$, although this is of questionable utility except perhaps in the context of spanning tests. One could also perform inference on the precision matrix.

In theory, asymptotic inference on ζ_*^2 would essentially be via Equation 7.10. To practically perform inference one must estimate Ω and substitute in values for Θ^{-1}. To estimate Ω, one can assume a distribution on returns, as in Equation 7.13 or Equation 7.21. Alternatively one can estimate $\mathsf{D}\Omega\mathsf{D}^\top$ directly as introduced in the univariate case in Section 3.2. This would proceed by constructing the vectors $\operatorname{vech}\left(\tilde{\boldsymbol{x}}_t\tilde{\boldsymbol{x}}_t{}^\top\right)$, then computing the sample covariance. One could deal with autocorrelation of heteroskedasticity by using alternative covariance estimators. [Zeileis, 2004]

However, we note again that $\hat{\zeta}_*^2$ is a squared quantity and will be fairly biased for ζ_*^2 for finite sample sizes. Inference based on the asymptotic results for $\hat{\Theta}^{-1}$ entirely ignore this bias. Moreover, the standard error estimates we get from the multivariate delta method ignore terms in n^{-2} which can be significant for real sample sizes. Thus for the task of performing inference on ζ_*^2, one should apply these results with great caution. Empirically it seems that the asymptotic distribution of ζ_* is much closer in shape to a Gaussian, but again the asymptotic results ignore significant bias, and are not to be trusted. This is unfortunate since the non-parametric estimation of Ω seems like a promising approach to autocorrelation, heteroskedasticity and deviations from normality. However, it seems that there are fewer such issues for the asymptotic distribution of $\hat{\boldsymbol{\nu}}_*$, and the Gaussian approximation hold fairly well.

Example 7.4.1 (Asymptotic normality (?), squared maximal Sharpe ratio). We perform 1,000 simulations of drawing $n = 1,260$ days of *i.i.d.* Gaussian data on $k = 8$ assets with $\zeta_*^2 = 1.5\ \mathrm{yr}^{-1}$. In Figure 7.12, we Q-Q plot $\hat{\zeta}_*^2$ against a normal distribution with mean ζ_*^2 and variance given by Equation 7.16. We also Q-Q plot $\hat{\zeta}_*$ against a normal distribution with mean ζ_* and variance from the multivariate delta method and Equation 7.16. Though the distributional "shape" of $\hat{\zeta}_*$ is closer to normal, it still is significantly biased. ⊣

Example 7.4.2 (Asymptotic Normality, Markowitz portfolio). We perform 2,000 simulations of drawing $n = 1,260$ days of *i.i.d.* Gaussian data on $k = 8$ assets with $\zeta_*^2 = 1.5\ \mathrm{yr}^{-1}$, where Σ is the identity matrix, and $\boldsymbol{\mu} = \sqrt{\zeta_*^2/k}\mathbf{1}$. We compute the Markowitz portfolio, $\hat{\boldsymbol{\nu}}_*$. In Figure 7.13, we Q-Q plot the first element of $\hat{\boldsymbol{\nu}}_*$ against a normal distribution with mean $\boldsymbol{e}_1^\top\boldsymbol{\nu}_*$ and variance given by Equation 7.16. We see that the distribution of $\boldsymbol{e}_1^\top\hat{\boldsymbol{\nu}}_*$ is accurately

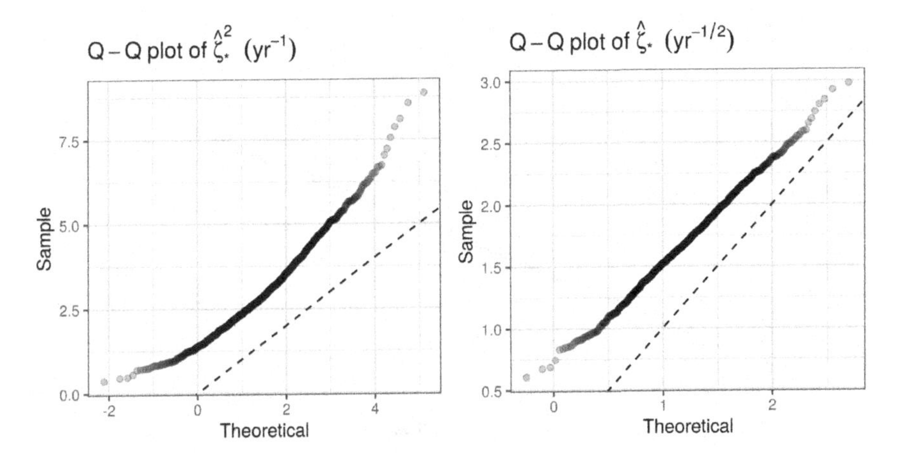

FIGURE 7.12
Q-Q plots of $\hat{\zeta}_*^2$ (left) and $\hat{\zeta}_*$ (right) are shown over 1,000 simulations of *i.i.d.* Gaussian data with $n = 1,260$ and $k = 8$. The theoretical distribution is Gaussian with mean ζ_*^2 and ζ_*, respectively. Clearly there is significant uncaptured bias.

captured by the normal approximation in this case, without any apparent bias issues, although perhaps with fatter tails. ⊣

We note that for testing the inference that $\zeta_*^2 = 0$, it suffices to test the hypothesis $\boldsymbol{\nu}_* = \mathbf{0}$. Since $\hat{\boldsymbol{\nu}}_*$ appears to be normally distributed around $\boldsymbol{\nu}_*$ with a covariance that can be estimated, we can test $\boldsymbol{\nu}_* = \mathbf{0}$ via a (rescaled) chi-squared approximation to $\hat{\boldsymbol{\nu}}_*{}^\top \hat{\boldsymbol{\nu}}_*$. Along the same lines we note that $\zeta_*^2 = \left(\Sigma^{-1/2}\boldsymbol{\mu}\right)^\top \left(\Sigma^{-1/2}\boldsymbol{\mu}\right)$, where here $\Sigma^{-1/2}$ is a matrix square root of the inverse of Σ. Thus if one could find a normal approximation $\hat{\Sigma}^{-1/2}\hat{\boldsymbol{\mu}} \approx \mathcal{N}\left(\Sigma^{-1/2}\boldsymbol{\mu}, d^2\mathsf{I}\right)$, then one would approximate $\hat{\zeta}_*^2$ as d^2 times a non-central chi-square distribution with k degrees of freedom and non-centrality parameter $\zeta_*^2 d^{-2}$. If, however, the asymptotic covariance of $\hat{\Sigma}^{-1/2}\hat{\boldsymbol{\mu}}$ were some general symmetric positive definite matrix, the problem is more complicated. [Baldessari, 1967; Tan, 1977] If again one were testing $\zeta_*^2 = 0$, then one could rescale to a central chi-square approximation for $\hat{\zeta}_*^2$. If not and the covariance were "close" to $d^2\mathsf{I}$ one could take the matrix trace of the covariance divided by k as an approximate d^2 and hope for the best.

Cholesky inverse second moment

To find a normal approximation $\hat{\Sigma}^{-1/2}\hat{\boldsymbol{\mu}} \approx \mathcal{N}\left(\Sigma^{-1/2}\boldsymbol{\mu}, d^2\mathsf{I}\right)$, we appeal to the *Cholesky trick*, by which the asymptotic distribution of $\hat{\Sigma}^{-1/2}\hat{\boldsymbol{\mu}}$ is found by a transform of $\hat{\Theta}$. This requires a rearrangement of the augmented vector and

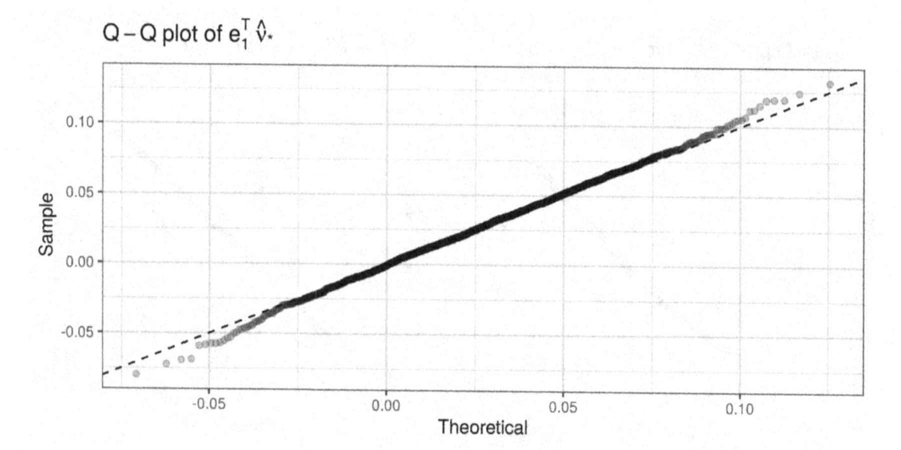

FIGURE 7.13
A Q-Q plot of $\boldsymbol{e}_1^\top \hat{\boldsymbol{\nu}}_*$ over 2,000 simulations of *i.i.d.* Gaussian data with $n = 1,260$ and $k = 8$ is shown. The theoretical distribution is Gaussian with mean $\boldsymbol{e}_1^\top \boldsymbol{\nu}_*$.

$\hat{\Theta}$. If you defined $\tilde{\boldsymbol{x}}_t = \left[\boldsymbol{x}_t{}^\top, 1\right]^\top$, then the second moment matrix has value

$$\Theta = \mathrm{E}\left[\tilde{\boldsymbol{x}}_t \tilde{\boldsymbol{x}}_t{}^\top\right] = \left[\begin{array}{cc} \Sigma + \boldsymbol{\mu}\boldsymbol{\mu}^\top & \boldsymbol{\mu} \\ \boldsymbol{\mu}^\top & 1 \end{array}\right].$$

The inverse is rearranged as well:

$$\Theta^{-1} = \left[\begin{array}{cc} \Sigma^{-1} & -\Sigma^{-1}\boldsymbol{\mu} \\ -\boldsymbol{\mu}^\top\Sigma^{-1} & 1 + \zeta_*^2 \end{array}\right].$$

The (lower) Cholesky factor of Θ^{-1} takes value

$$\Theta^{-1/2} = \left[\begin{array}{cc} \Sigma^{-1/2} & \boldsymbol{0} \\ -\boldsymbol{\mu}^\top\Sigma^{-1/2} & 1 \end{array}\right]. \tag{7.56}$$

Because $-\boldsymbol{\mu}^\top\Sigma^{-1/2}$ appears on the bottom row, one could use the multivariate delta method to find an asymptotic normal approximation for $\Sigma^{-1/2}\boldsymbol{\mu}$. The derivative is given in Equation A.33. Thus we have asymptotically in n,

$$\sqrt{n}\left(\hat{\Sigma}^{-1/2}\hat{\boldsymbol{\mu}} - \Sigma^{-1/2}\boldsymbol{\mu}\right) \rightsquigarrow \mathcal{N}\left(0, \mathsf{H}\Omega\mathsf{H}^\top\right), \tag{7.57}$$

where

$$\mathsf{H} = \mathsf{M}\left(\mathsf{L}\left(\mathsf{I} + \mathsf{K}\right)\left(\left(\Theta^{-\top/2}\right)^{-1} \otimes \Theta\right)\mathsf{L}^\top\right)^{-1}, \tag{7.58}$$

for some M which picks out the first k elements of the bottom row of $\Sigma^{-1/2}\boldsymbol{\mu}$ from vech $\left(\Sigma^{-1/2}\boldsymbol{\mu}\right)$. Here Ω is the covariance of the (rearranged) vech $\left(\tilde{\boldsymbol{x}}\tilde{\boldsymbol{x}}^\top\right)$, and, up to the rearrangement, is given by Equation 7.13 in the elliptical case. It can be consistently estimated under deviations from the *i.i.d.* Gaussian assumption, as discussed above. It has not been established symbolically, but simple to check numerically, that the covariance in Equation 7.57 is diagonal for general symmetric positive definite Σ and general $\boldsymbol{\mu}$, *cf.* Exercise 7.7. Thus the Cholesky trick consists of performing this cholesky inverse computation, then estimating the factor d as the average trace: $\hat{d^2} = \operatorname{tr}\left(\frac{1}{n}\hat{\mathsf{H}}\hat{\Omega}\hat{\mathsf{H}}^\top\right)/k$. We note that to carry out these computations for inference requires a lot of bookkeeping, which can be radically simplified via the use of automatic differentiation. [Pav, 2016b]

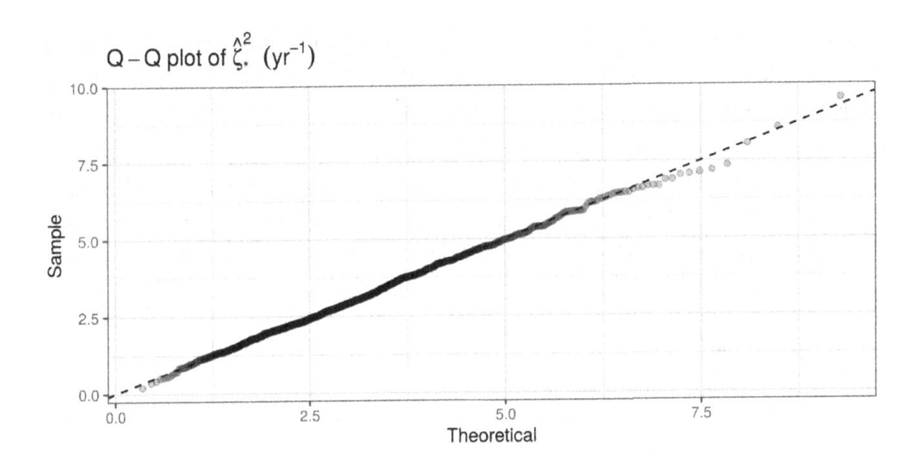

FIGURE 7.14
Annualized $\hat{\zeta}_*^2$ over 1,000 simulations of *i.i.d.* Gaussian data with $n = 1,260$ and $k = 8$ is Q-Q plotted. The theoretical distribution is a rescaled non-central chi-square with k degrees of freedom and non-centrality parameter $\hat{\zeta}_*^2 d^{-2}$, where we compute d^2 as the mean of the diagonal of the variance-covariance matrix.

Example 7.4.3 (Asymptotic Chi-square, squared maximal Sharpe ratio). We perform 1,000 simulations of drawing $n = 1,260$ days of *i.i.d.* Gaussian data on $k = 8$ assets with $\zeta_*^2 = 1.5\ \mathrm{yr}^{-1}$. We set Σ to be diagonal but with the elements $1, 2, \ldots, k$ on the diagonal, and let $\boldsymbol{\mu}$ be $\mu\mathbf{1}$ for some μ. We compute the actual Ω and H to get the covariance matrix of Equation 7.57, then compute d^2 as the trace of that covariance divided by k. Thus d^2 is computed with clairvoyance of the population parameters, and not estimated in each simulation. In Figure 7.14, we Q-Q plot $\hat{\zeta}_*^2$ against a rescaled (by d^2) non-central chi-square distribution with k degrees of freedom and non-centrality parameter $\zeta_*^2 d^{-2}$. We see good agreement with the theoretical distribution. ⊣

† Cholesky Inference, Constrained problems

We now consider extending the Cholesky inference to the constrained portfolio problems. The technique is trivially adapted to the subspace-constrained problem. Consider the hedged portfolio problem of Equation 6.9. Recall

$$\begin{aligned}
\Delta_{\mathsf{I}\backslash\mathsf{G}}\hat{\zeta}_{*}^{2} &= \hat{\mu}^{\top}\hat{\Sigma}^{-1}\hat{\mu} - \left(\mathsf{G}\hat{\mu}\right)^{\top}\left(\mathsf{G}\hat{\Sigma}\mathsf{G}^{\top}\right)^{-1}\left(\mathsf{G}\hat{\mu}\right),\\
&= \left(\hat{\Sigma}^{-1/2}\hat{\mu}\right)^{\top}\left(\mathsf{I} - \hat{\Sigma}^{\top/2}\mathsf{G}^{\top}\left(\mathsf{G}\hat{\Sigma}\mathsf{G}^{\top}\right)^{-1}\mathsf{G}\hat{\Sigma}^{1/2}\right)\left(\hat{\Sigma}^{-1/2}\hat{\mu}\right),\\
&= \left(\hat{\Sigma}^{-1/2}\hat{\mu}\right)^{\top}\mathsf{A}\left(\hat{\Sigma}^{-1/2}\hat{\mu}\right),
\end{aligned}$$

for matrix A, which is symmetric and idempotent with rank $k - k_g$. If $\hat{\Sigma}^{-1/2}\hat{\mu} \approx \mathcal{N}\left(\Sigma^{-1/2}\mu, d^2\mathsf{I}\right)$, then conditional on A being observed, $d^{-2}\Delta_{\mathsf{I}\backslash\mathsf{G}}\hat{\zeta}_{*}^{2}$ will follow a non-central χ^2 distribution with $k - k_g$ degrees of freedom and non-centrality parameter equal to $d^{-2}\left(\Sigma^{-1/2}\mu\right)^{\top}\mathsf{A}\left(\Sigma^{-1/2}\mu\right)$. [Graybill and Marsaglia, 1957] However, the matrix A includes the random $\hat{\Sigma}$, rather than Σ. Even conditioning on $\hat{\zeta}_{*,\mathsf{G}}^{2}$, it is unlikely that $\Delta_{\mathsf{I}\backslash\mathsf{G}}\hat{\zeta}_{*}^{2}$ will follow the rescaled non-central chi-square law. Using the Cholesky trick for inference on $\Delta_{\mathsf{I}\backslash\mathsf{G}}\zeta_{*}$ should be done with caution.

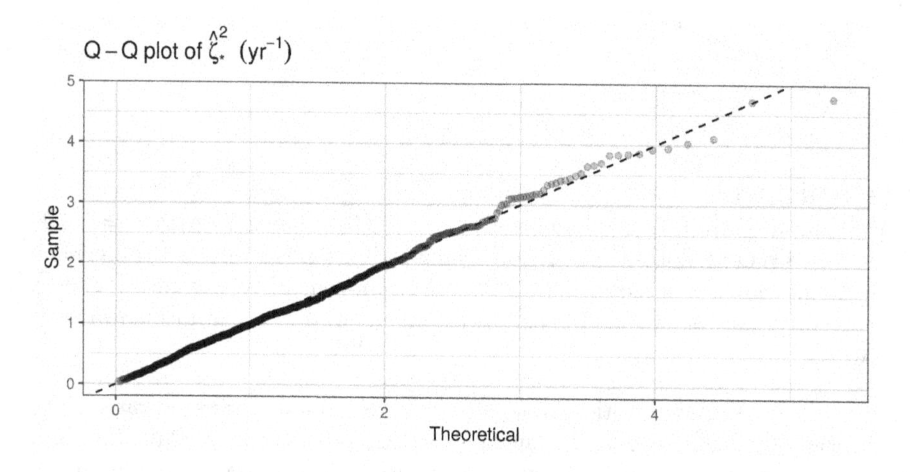

FIGURE 7.15
Annualized $\Delta_{\mathsf{I}\backslash\mathsf{G}}\ \hat{\zeta}_{*}^{2}$ over 1,000 simulations of *i.i.d.* Gaussian data with $n = 1,260$ and $k = 8$ and $k_g = 4$ is Q-Q plotted. The theoretical distribution is a rescaled non-central chi-square with $k - k_g$ degrees of freedom and non-centrality parameter $\Delta_{\mathsf{I}\backslash\mathsf{G}}\zeta_{*}^{2}d^{-2}$, where we compute d^2 as the mean of the diagonal of the variance-covariance matrix.

Example 7.4.4 (Asymptotic Chi-square, hedged portfolio squared maximal Sharpe ratio). We perform 1,000 simulations of drawing $n = 1,260$ days of *i.i.d.* Gaussian data on $k = 8$ assets, then hedge out the first $k_g = 4$. We set $\Delta_{\mathsf{I}\backslash\mathsf{G}}\zeta_*^2 = 0.35\ \mathrm{yr}^{-1}$. We set Σ to be diagonal but with the elements $1, 2, \ldots, k$ on the diagonal, and let $\boldsymbol{\mu}$ be $\mu\mathbf{1}$ for some μ. We compute the actual Ω and H to get the covariance matrix of Equation 7.57, then estimate d^2 as the trace of that covariance divided by k. In Figure 7.15, we Q-Q plot $\Delta_{\mathsf{I}\backslash\mathsf{G}}\hat{\zeta}_*^2$ against a rescaled (by d^2) non-central chi-square distribution with $k - k_g$ degrees of freedom and non-centrality parameter $\Delta_{\mathsf{I}\backslash\mathsf{G}}\zeta_*^2 d^{-2}$. We see good agreement with the theoretical distribution. Again we stress that the d^2 here was computed with knowledge of the population parameters, whereas in actual inference it would have to be estimated. ⊣

7.4.1 Confidence intervals

With all the caveats noted above, we consider confidence intervals on ζ_*^2 using the Cholesky trick, appealing to asymptotic results. We assume one has computed $\hat{\Theta}$ and has some estimate of Ω, call it $\hat{\Omega}$. Then one should compute an estimate of the matrix H from Equation 7.58, then compute $\hat{\mathsf{H}}\hat{\Omega}\hat{\mathsf{H}}^{\top}$. Let $d^2 = \mathrm{tr}\left(\frac{1}{n}\hat{\mathsf{H}}\hat{\Omega}\hat{\mathsf{H}}^{\top}\right)/k$. Then an approximate $1-\alpha$ confidence interval on ζ_*^2 has endpoints $\left[\zeta_{*,l}^2, \zeta_{*,u}^2\right]$ defined implicitly by

$$\begin{aligned} \zeta_{*,l}^2 &= \min\left\{z \,\middle|\, z \geq 0,\ 1-\alpha/2 \geq F_{\chi^2}\left(\hat{\zeta}_*^2 d^{-2}; k, z d^{-2}\right)\right\}, \\ \zeta_{*,u}^2 &= \min\left\{z \,\middle|\, z \geq 0,\ \alpha/2 \geq F_{\chi^2}\left(\hat{\zeta}_*^2 d^{-2}; k, z d^{-2}\right)\right\}, \end{aligned} \tag{7.59}$$

where $F_{\chi^2}(x; k, \delta)$ is the CDF of the non-central chi-square distribution with non-centrality parameter δ and k degrees of freedom.

Example 7.4.5 (Cholesky confidence intervals on ζ_*^2). We draw daily data from *i.i.d.* Gaussian returns distribution with k varying from 2 to 16, and n varying from 504 to 1,260 days. We let ζ_*^2 vary from $0\ \mathrm{yr}^{-1}$ to $1.5\ \mathrm{yr}^{-1}$. For each Cartesian corner of the parameter space, we simulate Gaussian returns, then estimate Ω, and estimate the covariance matrix of $\hat{\Sigma}^{-1/2}\hat{\boldsymbol{\mu}}$ using forward differentiation. From this we estimate the constant d^2 as the mean diagonal element of the estimated covariance of $\hat{\Sigma}^{-1/2}\hat{\boldsymbol{\mu}}$. We then construct a 95 % confidence interval as in Equation 7.59. We compute the empirical type I rate at a nominal 0.05 level of the confidence intervals.

In Figure 7.16, we plot the empirical type I rate against k with lines for different values of ζ_*^2, and facets for different values of n. We see that the type I rate of the confidence intervals is very close to the nominal type I rate. For comparison, in Figure 7.17, we plot the empirical type I rate of the confidence interval based on the normal approximation of Equation 7.15. These are clearly very poor and do not give near-nominal coverage for such "small" sample sizes. Moreover, the type I rate depends on k. ⊣

Example 7.4.6 (Cholesky confidence interval on ζ_*^2, heteroskedastic returns).

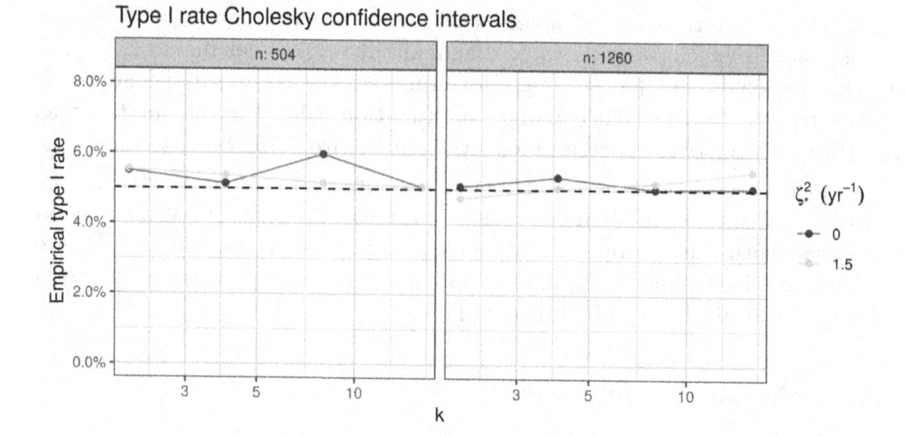

FIGURE 7.16
The empirical type I rate over 5,000 simulations of the Cholesky based confidence intervals of Equation 7.59 is shown. Returns are drawn from *i.i.d.* Gaussian distribution with identity covariance matrix, and varying n and ζ_*^2. A horizontal line is drawn at the nominal type I rate of 5%.

We repeat the experiment of Example 7.4.5, but draw returns from heteroskedastic matrix normal distribution, as in Example 7.2.2. As in that example, we sample $\boldsymbol{l}$ from 1,260 values of the VIX level squared, then rescaled to mean one. We draw returns with a fixed value of $\zeta_* = 0.07\,\text{day}^{-1/2}$; we k vary from 1 to 8. We take the sample size n to be either 1,260 days or twice that value, recycling $\boldsymbol{l}$ for the longer series, which avoids variation in the moments of $\boldsymbol{l}$ among different simulations. We let λ take values between 0 and 2. We then construct a 95 % confidence interval as in Equation 7.59 and compute the empirical type I rate of the confidence intervals over 2,000 simulations. The $\hat{\Omega}$ is estimated by a heteroskedasticity and autocorrelation consistent covariance estimator. [Zeileis, 2004]

In Figure 7.18, we plot the empirical type I rate of the confidence interval versus k, with lines for λ and facets for n. We see that the nominal type I rate is nearly conserved across the parameter space considered here, but the intervals seem too narrow for the larger n and λ case. ⊣

Example 7.4.7 (Cholesky confidence interval on ζ_*^2, autocorrelated returns). We repeat the experiment of Example 7.4.5, but draw returns from autocorrelated matrix normal distribution, as in Example 7.2.5. As in that example, we let number of assets k vary from 3 to 10; sample size is fixed at $n = 1,260$ days, and we let ρ vary from -0.4 to 0.4. For simulated returns, we construct a 95 % confidence interval as in Equation 7.59 and compute the empirical type I rate of the confidence intervals over 2,000 simulations. The $\hat{\Omega}$

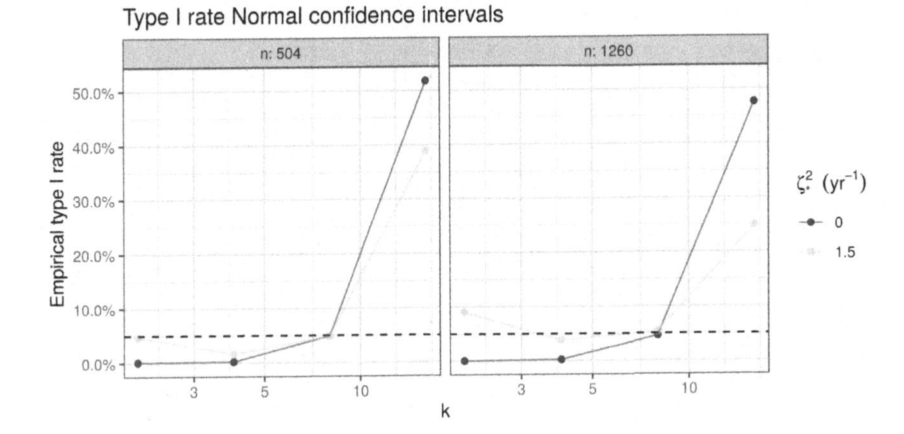

FIGURE 7.17
The empirical type I rate over 5,000 simulations of the normal confidence intervals, based on the multivariate delta method is shown. Returns are drawn from *i.i.d.* Gaussian distribution with identity covariance matrix, and varying n and ζ_*^2. A horizontal line is drawn at the nominal type I rate of 5%. This method for computing confidence intervals is clearly deficient.

is estimated by a heteroskedasticity and autocorrelation consistent covariance estimator. [Zeileis, 2004]

In Figure 7.19, we plot the empirical type I rate of the confidence intervals versus ρ, with facets for n. We see that the nominal type I rate is nearly conserved across the parameter space considered here. It is not clear if there is a small amount of bias in the confidence interval for large k, but this is unlikely to be an issue for the scale of ρ one is likely to encounter.

For comparison, we repeat this experiment, but test the confidence intervals based on Gaussian *i.i.d.* returns and the F distribution, as given in Equation 6.70. In Figure 7.20, we plot the empirical type I rate of these confidence intervals and find that the nominal type I rate is not maintained, except when $\rho = 0$, as we expect. $\dashv$

Example 7.4.8 (Fama French Four Factors, asymptotic confidence intervals). In Example 6.3.1, we computed a confidence interval on ζ_* over the Market, SMB, HML, and UMD portfolios. Using the Cholesky method with heteroskedasticity and autocorrelation consistent covariance estimators, we estimate $d^2 \approx 0.001$. From this we estimate a 95% confidence interval on ζ_* as $[0.232, 0.372)$ mo.$^{-1/2}$. Recall that we previously computed the confidence interval, via the F distribution, to be $[0.243, 0.363)$ mo.$^{-1/2}$, which is about the same though slightly narrower. $\dashv$

Example 7.4.9 (Fama French Four Factors, Market hedged asymptotic confidence interval). In Example 6.3.2, we computed a confidence interval on $\Delta_{\mathsf{I}\setminus\mathsf{G}}$

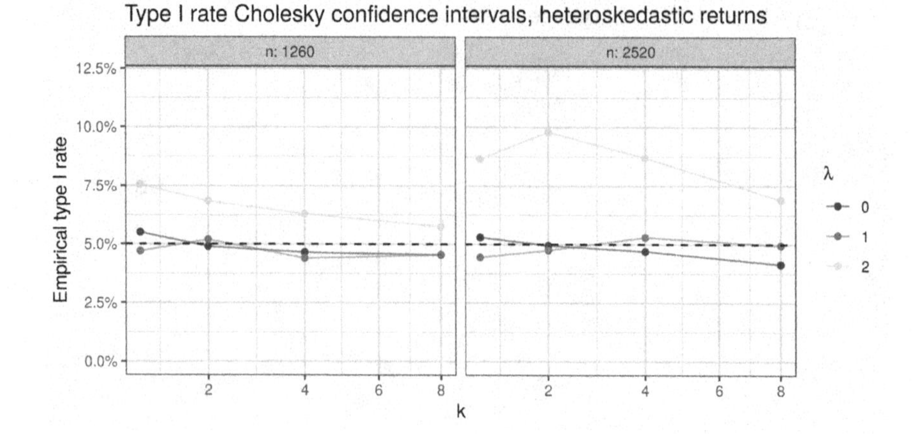

FIGURE 7.18
The empirical type I rate over 2,000 simulations of the Cholesky based confidence intervals of Equation 7.59 is shown. Returns are heteroskedastic with varying λ, using squared VIX as the $\boldsymbol{l}$. A horizontal line is drawn at the nominal type I rate of 5%. This method is anticonservative for large λ, but otherwise delivers near-nominal coverage.

ζ_* over the Market, SMB, HML, and UMD portfolios with a hedge against the Market. Using the Cholesky method with heteroskedasticity and autocorrelation consistent covariance estimators, we estimate $d^2 \approx 0.001$. From this we estimate the 95% one-sided confidence interval on $\Delta_{\mathsf{I}\backslash\mathsf{G}}\, \zeta_*$ as $[0.188, \infty)$ mo.$^{-1/2}$. Recall that we previously computed the confidence interval, via the F distribution, to be $[0.204, \infty)$ mo.$^{-1/2}$, which is slightly narrower. ⊣

The Markowitz portfolio

Confidence intervals on the Markowitz portfolio proceed by the normal approximation. To construct a confidence interval on $\boldsymbol{v}^\top\Sigma^{-1}\boldsymbol{\mu}$ for fixed vector $\boldsymbol{v}$, start by estimating Ω, then estimate the matrix H in Equation 7.8. By subsetting vech $\left(\hat{\Theta}^{-1}\right)$ one gets an estimate of the covariance of $-\hat{\Sigma}^{-1}\hat{\boldsymbol{\mu}}$. Taking the quadratic form of $\boldsymbol{v}$ then gives the estimated variance of $\boldsymbol{v}^\top\Sigma^{-1}\boldsymbol{\mu}$, the square root of which is the estimated standard error, $\hat{s}$. Then an approximate $1-\alpha$ confidence interval on $\boldsymbol{v}^\top\Sigma^{-1}\boldsymbol{\mu}$ is

$$\boldsymbol{v}^\top\hat{\Sigma}^{-1}\hat{\boldsymbol{\mu}} \pm \hat{s}z_{1-\alpha/2}, \tag{7.60}$$

where z_q is the q quantile of the standard normal distribution. The same procedure can be adapted to a compute confidence interval on elements of the hedged portfolio by computing an approximate covariance of vech $\left(\Delta_{\mathsf{I}\backslash\mathsf{G}}\hat{\Theta}^{-1}\right)$ and subsetting elements of it.

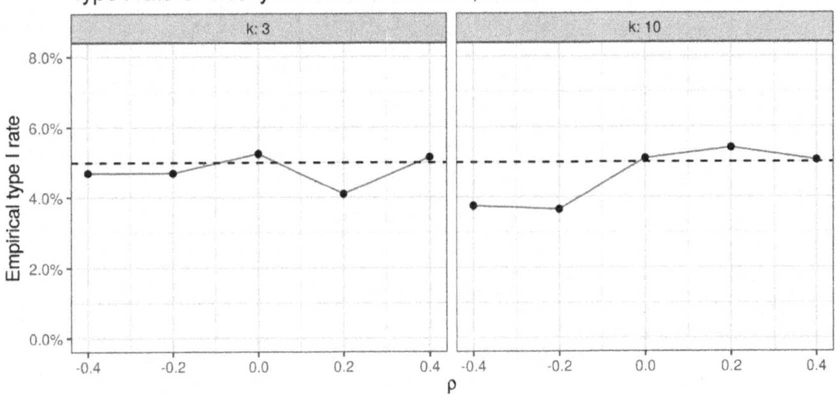

FIGURE 7.19
The empirical type I rate over 2,000 simulations of the Cholesky-based confidence intervals of Equation 7.59 is plotted versus ρ. Returns are autocorrelated Gaussians with ρ varying. A horizontal line is drawn at the nominal type I rate of 5%.

Example 7.4.10 (Markowitz portfolio covariance, Fama French Four Factors). We consider the 1,128 months of returns of the Market, SMB, HML, and UMD portfolios introduced in Example 1.1.1, spanning from Jan 1927 to Dec 2020. Mirroring Example 6.1.10, we estimate $\hat{\Omega}$ using the vanilla covariance estimator, then estimate the covariance of the Markowitz portfolio on the four factors. As the units are arbitrary, we convert these into a correlation matrix, which takes the value

	Mkt	SMB	HML	UMD
Mkt	1.000	−0.163	0.024	0.326
SMB	−0.163	1.000	−0.127	−0.026
HML	0.024	−0.127	1.000	0.493
UMD	0.326	−0.026	0.493	1.000

.

⊣

Example 7.4.11 (Fama French Four Factors, Size CI). We consider the 1,128 months of returns of the Market, SMB, HML, and UMD portfolios introduced in Example 1.1.1, spanning from Jan 1927 to Dec 2020. We estimate $\hat{\Omega}$ via the heteroskedasticity and autocorrelation consistent covariance estimator. We compute a 95% confidence interval on the weighting of SMB in the Markowitz portfolio as $[-0.015, 0.025)\ \%^{-1}$. These findings are consistent with the failure to reject the null of zero weighting found in Example 6.3.11. ⊣

Example 7.4.12 (Markowitz portfolio covariance, Fama French Four Factors, Hedged against Market). We consider a portfolio on the the Market, SMB,

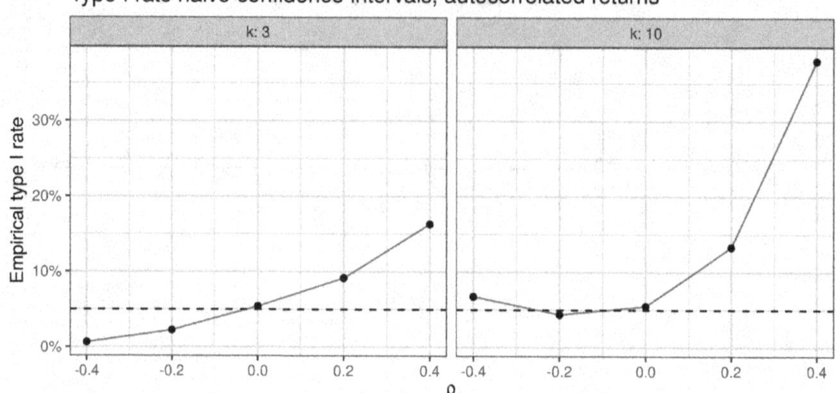

FIGURE 7.20
The empirical type I rate over 2,000 simulations of the confidence intervals of Equation 6.70 is plotted versus ρ. Returns are autocorrelated Gaussians with ρ varying. A horizontal line is drawn at the nominal type I rate of 5%.

HML, and UMD returns, hedged against the Market returns. Again we estimate $\hat{\Omega}$ using the vanilla covariance estimator, compute the covariance, then convert to the following correlation matrix:

	Mkt	SMB	HML	UMD
Mkt	1.000	−0.369	0.146	0.532
SMB	−0.369	1.000	−0.127	−0.026
HML	0.146	−0.127	1.000	0.493
UMD	0.532	−0.026	0.493	1.000

The correlations to the Market weight have changed compared to Example 7.4.10, but the other correlations appear unchanged. ⊣

7.4.2 Hypothesis tests

Hypothesis testing proceeds in a fashion similar to the construction of confidence intervals.

1 sample maximal signal-noise ratio test

To test

$$H_0 : \zeta_* = \zeta_0 \quad \text{versus} \quad H_1 : \zeta_* > \zeta_0,$$

estimate $\hat{\Omega}$ and H in Equation 7.58, compute $d^2 = \mathrm{tr}\left(\frac{1}{n}\hat{\mathsf{H}}\hat{\Omega}\hat{\mathsf{H}}^\top\right)/k$, then reject if

$$\hat{\zeta}_*^2 > d^2{\chi^2}_{1-\alpha}\left(k, \zeta_0^2 d^{-2}\right),$$

where ${\chi^2}_q\left(\nu, \delta\right)$ is the q quantile of the non-central chi-square distribution with ν degrees of freedom and non-centrality parameter δ.

1 sample maximal signal-noise ratio test, hedged portfolio

To test

$$H_0 : \Delta_{\mathsf{I}\setminus\mathsf{G}}\zeta_* = \zeta_0 \quad \text{versus} \quad H_1 : \Delta_{\mathsf{I}\setminus\mathsf{G}}\zeta_* > \zeta_0,$$

estimate $\hat{\Omega}$ and H in Equation 7.58, compute $d^2 = \mathrm{tr}\left(\frac{1}{n}\hat{\mathsf{H}}\hat{\Omega}\hat{\mathsf{H}}^\top\right)/k$, then reject approximately at the α level if

$$\Delta_{\mathsf{I}\setminus\mathsf{G}}\hat{\zeta}_*^2 > d^2{\chi^2}_{1-\alpha}\left(k - k_g, \zeta_0^2 d^{-2}\right),$$

where ${\chi^2}_q\left(\nu, \delta\right)$ is the q quantile of the non-central chi-square distribution with ν degrees of freedom and non-centrality parameter δ.

1 sample portfolio weight test

To test hypotheses on a single contrast on portfolio weights for contrast vector $\boldsymbol{v}$, estimate Ω and H in Equation 7.8, then use the multivariate delta method to estimate the variance of $\left[0, -\boldsymbol{v}^\top\right]\Theta^{-1}\boldsymbol{e}_1$. Let the square root of that estimated variance be $\hat{s}$. Then to test the contrast against a two-sided alternative

$$H_0 : \boldsymbol{v}^\top\Sigma^{-1}\boldsymbol{\mu} = c \quad \text{versus} \quad H_1 : \boldsymbol{v}^\top\Sigma^{-1}\boldsymbol{\mu} \neq c,$$

reject if $\left|\boldsymbol{v}^\top\hat{\Sigma}^{-1}\hat{\boldsymbol{\mu}} - c\right| > \hat{s}z_{1-\alpha/2}$, where z_q is the q quantile of the standard normal distribution. This test has approximately α type I error rate.

To test against a one-sided alternative,

$$H_0 : \boldsymbol{v}^\top\Sigma^{-1}\boldsymbol{\mu} = c \quad \text{versus} \quad H_1 : \boldsymbol{v}^\top\Sigma^{-1}\boldsymbol{\mu} > c,$$

reject if $\boldsymbol{v}^\top\hat{\Sigma}^{-1}\hat{\boldsymbol{\mu}} > c + \hat{s}z_{1-\alpha}$.

To test multiple conditions, use the appropriate χ^2 test. That is, given $\boldsymbol{v_1}, \boldsymbol{v_2}, \ldots, \boldsymbol{v_m}$, and corresponding c_i, to test

$$H_0 : \forall i\, \boldsymbol{v_i}^\top\Sigma^{-1}\boldsymbol{\mu} = c_i \quad \text{versus} \quad H_1 : \exists i\, \boldsymbol{v_i}^\top\Sigma^{-1}\boldsymbol{\mu} \neq c_i,$$

reject at the α level if

$$\sum_{i} \left(\frac{{\boldsymbol{v}_i}^{\top} \hat{\Sigma}^{-1} \hat{\boldsymbol{\mu}} - c_i}{\hat{s}_i} \right)^2 > {\chi^2}_{1-\alpha}(m),$$

where ${\chi^2}_q(\nu)$ is the q quantile of the (central) chi-square distribution with ν degrees of freedom. Note that the set of vectors $\boldsymbol{v_i}$ must have rank m for this test to have (approximately) the nominal type I rate.

The condition of zero squared maximal signal-noise ratio is equivalent to zero portfolio weight. That is

$$\zeta_*^2 = 0 \Leftrightarrow \forall i\, \boldsymbol{e}_i^{\top} \Sigma^{-1} \boldsymbol{\mu} = 0.$$

Thus to test

$$H_0 : \zeta_* = 0 \quad \text{versus} \quad H_1 : \zeta_* > 0,$$

reject at the α level if

$$\sum_{i=1}^{k} \left(\frac{\boldsymbol{e}_i^{\top} \hat{\Sigma}^{-1} \hat{\boldsymbol{\mu}}}{\hat{s}_i} \right)^2 > {\chi^2}_{1-\alpha}(k),$$

where $\hat{s}_i$ is the square root of the estimated variance of $\left[0, -\boldsymbol{e}_i^{\top}\right] \Theta^{-1} \boldsymbol{e}_1$, and where ${\chi^2}_q(\nu)$ is the q quantile of the (central) chi-square distribution with ν degrees of freedom.

1 sample hedged portfolio weight test

To test hypotheses on a single contrast on hedged portfolio weights for contrast vector $\boldsymbol{v}$, estimate Ω and H in Equation 7.42, then use the multivariate delta method to estimate the variance of $\left[0, -\boldsymbol{v}^{\top}\right] \Delta_{\mathsf{I} \setminus \mathsf{G}} \Theta^{-1} \boldsymbol{e}_1$. Let the square root of that estimated variance be $\hat{s}$. Then to test the contrast against a two-sided alternative

$$H_0 : \boldsymbol{v}^{\top} \boldsymbol{\nu}_{*,\mathsf{I} \setminus \mathsf{G}} = c \quad \text{versus} \quad H_1 : \boldsymbol{v}^{\top} \boldsymbol{\nu}_{*,\mathsf{I} \setminus \mathsf{G}} \neq c,$$

for $\boldsymbol{\nu}_{*,\mathsf{I} \setminus \mathsf{G}} = \Sigma^{-1} \boldsymbol{\mu} - \mathsf{G}^{\top} \left(\mathsf{G} \Sigma \mathsf{G}^{\top}\right)^{-1} \mathsf{G} \boldsymbol{\mu}$, reject if $\left|\boldsymbol{v}^{\top} \hat{\boldsymbol{\nu}}_{*,\mathsf{I} \setminus \mathsf{G}} - c\right| > \hat{s} z_{1-\alpha/2}$, where z_q is the q quantile of the standard normal distribution. This test has approximately α type I error rate.

To test against a one-sided alternative,

$$H_0 : \boldsymbol{v}^{\top} \boldsymbol{\nu}_{*,\mathsf{I} \setminus \mathsf{G}} = c \quad \text{versus} \quad H_1 : \boldsymbol{v}^{\top} \boldsymbol{\nu}_{*,\mathsf{I} \setminus \mathsf{G}} > c,$$

reject if $\boldsymbol{v}^{\top} \hat{\boldsymbol{\nu}}_{*,\mathsf{I} \setminus \mathsf{G}} > c + \hat{s} z_{1-\alpha}$.

It is not clear if multiple hypothesese can be approximately tested via a χ^2 test, as it is likely the case that the hedging constraint could reduce the degrees of freedom. In particular, it is not known how to test the hypothesis $\Delta_{\mathsf{I}\backslash\mathsf{G}}\zeta_*^2 = 0$ via testing $\boldsymbol{\nu}_{*,\mathsf{I}\backslash\mathsf{G}} = \mathbf{0}$.

1 sample passthrough weight test

To test hypotheses on a single contrast on the Markowitz passthrough matrix under the conditional expectation model (optionally with conditional heteroskedasticity), estimate the Ω and H in Equation 7.54 as applied to the estimates of Θ_f. Then for conformable matrix V, use the multivariate delta method to estimate the variance of $\operatorname{tr}\left(\mathsf{V}{\Theta_f}^{-1}\right)$. Let the square root of that estimated variance be $\hat{s}$. Then to approximately test the contrast against a two-sided alternative

$$H_0 : \operatorname{tr}\left(\mathsf{V}{\Theta_f}^{-1}\right) = c \quad \text{versus} \quad H_1 : \operatorname{tr}\left(\mathsf{V}{\Theta_f}^{-1}\right) \neq c,$$

reject if $\left|\operatorname{tr}\left(\mathsf{V}{\Theta_f}^{-1}\right) - c\right| > \hat{s}z_{1-\alpha/2}$, where z_q is the q quantile of the standard normal distribution. Note that the contrast matrix V should only have non-zero elements corresponding to the passthrough matrix, as the block diagonal elements of ${\Theta_f}^{-1}$ are only slowly asymptotically normal.

To test against a one-sided alternative,

$$H_0 : \operatorname{tr}\left(\mathsf{V}{\Theta_f}^{-1}\right) = c \quad \text{versus} \quad H_1 : \operatorname{tr}\left(\mathsf{V}{\Theta_f}^{-1}\right) > c,$$

reject if $\operatorname{tr}\left(\mathsf{V}{\Theta_f}^{-1}\right) > c + \hat{s}z_{1-\alpha}$.

To test multiple conditions, say given $\mathsf{V}_1, \mathsf{V}_2, \ldots, \mathsf{V}_m$, and corresponding c_i, to test

$$H_0 : \forall i \operatorname{tr}\left(\mathsf{V}_i{\Theta_f}^{-1}\right) = c_i \quad \text{versus} \quad H_1 : \exists i \operatorname{tr}\left(\mathsf{V}_i{\Theta_f}^{-1}\right) \neq c_i,$$

reject at the α level if

$$\sum_i \left(\frac{\operatorname{tr}\left(\mathsf{V}_i\hat{\Theta}_f^{-1}\right) - c_i}{\hat{s}_i}\right)^2 > {\chi^2}_{1-\alpha}\left(m\right),$$

where ${\chi^2}_q\left(\nu\right)$ is the q quantile of the (central) chi-square distribution with ν degrees of freedom.

Example 7.4.13 (Fama French Four Factors, testing the maximal signal-noise ratio). We perform the same analysis of Example 6.3.8, on the returns of the Market, SMB, HML, and UMD portfolios, but using the asymptotic machinery. We are testing the hypothesis $H_0 : \zeta_* = 0.1 \text{ mo.}^{-1/2}$. Using the heteroskedasticity and autocorrelation consistent covariance estimator we compute $d^2 = 0.001$. The cutoff for rejection at the 0.05 level is computed as 0.03 mo.^{-1}, which is smaller than $\hat{\zeta}_*^2 = 0.095 \text{ mo.}^{-1}$, and we reject the null hypothesis. ⊣

Example 7.4.14 (Fama French Four Factors, testing hedged portfolio maximal signal-noise ratio). We perform the same analysis of Example 6.3.9, on the Fama French four factors with a hedge against the Market, but using the asymptotic machinery. We are testing the hypothesis $H_0 : \Delta_{\mathsf{I} \backslash \mathsf{G}}\zeta_* = 0.25 \text{ mo.}^{-1/2}$. Using the heteroskedasticity and autocorrelation consistent covariance estimator we compute $d^2 = 0.001$. The cutoff for rejection at the 0.05 level is computed as 0.098 mo.^{-1}, which is larger than $\Delta_{\mathsf{I} \backslash \mathsf{G}}\hat{\zeta}_*^2 = 0.063 \text{ mo.}^{-1}$, and we fail to reject the null hypothesis. ⊣

Example 7.4.15 (Portfolio on the Fama French Four Factors, value of SMB). We reconsider the analysis of Example 6.3.11, using the asymptotic machinery. That is, we test the hypothesis that the SMB loading in the Markowitz portfolio on the Market, SMB, HML, and UMD portfolios is zero. We compute $\left|\boldsymbol{v}^\top \hat{\Sigma}^{-1}\hat{\boldsymbol{\mu}} - c\right| = 0.007\%^{-1}$, and estimate $\hat{s} = 0.004\%^{-1}$. Based on that we fail to reject the null hypothesis at the 0.05 level. ⊣

Example 7.4.16 (Conditional Portfolio on the Fama French Four Factors). Continuing Example 6.1.6, we consider the returns of the Fama French four factors under a conditional expectation model with three features: an intercept, a running historical mean return, and the log of a running historical standard deviation of returns. We compute the second moment matrix, then estimate its inverse, $\hat{\Theta}_f^{-1}$ as

	intercept	*mean*	*log_sd*	*Mkt*	*SMB*	*HML*	*UMD*
intercept	12.511	−0.895	−7.511	−0.048	0.068	−0.077	−0.073
mean	−0.895	0.412	0.378	−0.007	0.009	−0.001	−0.020
log_sd	−7.511	0.378	5.070	0.009	−0.057	0.024	0.022
Mkt	−0.048	−0.007	0.009	0.044	−0.020	−0.006	0.014
SMB	0.068	0.009	−0.057	−0.020	0.111	−0.004	0.002
HML	−0.077	−0.001	0.024	−0.006	−0.004	0.099	0.028
UMD	−0.073	−0.020	0.022	0.014	0.002	0.028	0.060

$\%^{-2}$ mo.

One can see the (negative) passthrough matrix $\hat{\mathsf{N}}_*$ from Example 6.1.6 in the corners of this matrix.

We now consider the hypothesis that the passthrough from the log of volatility to momentum (UMD) is zero. By the multivariate delta method, we estimate $\hat{s} = 0.018\%^{-2}$ mo. As the estimated passthrough coefficient is $-0.022\%^{-2}$ mo., we fail to reject the null hypothesis at the 0.05 level. ⊣

7.5 † Likelihoodist inference on the second moment

While this chapter is focused on the distribution of $\hat{\zeta}_*^2$ and $\hat{\boldsymbol{\nu}}_*$ under deviations from *i.i.d.* Gaussian returns, there is not much we can say about the likelihood without assuming a distribution on returns. However, since we have introduced the second moment matrix in this chapter, we will consider inference on it here via likelihood. The density of $\hat{\Theta}$ under *i.i.d.* Gaussian returns is given in Equation 7.6. In this case $\hat{\Theta}$ is the maximum likelihood estimator of Θ. Here we consider inference on Θ^{-1} via likelihood.

Suppose for $i = 1, 2, \ldots, m$, A_i are some $(k+1) \times (k+1)$ symmetric matrices and a_i are given constants. Consider the null hypothesis

$$H_0 : \forall i \operatorname{tr}\left(\mathsf{A}_i \Theta^{-1}\right) = a_i, \quad \text{versus} \quad H_1 : \exists i \operatorname{tr}\left(\mathsf{A}_i \Theta^{-1}\right) \neq a_i. \tag{7.61}$$

The constraints have to be sensible. For example, they cannot violate the positive definiteness of Θ^{-1}, symmetry, *etc.*

The MLE under the null takes the form

$$\Theta_0 = \hat{\Theta} - \sum_i \lambda_i \mathsf{A}_i, \tag{7.62}$$

For some scalars λ_i. The maximum likelihood estimator under the constraints has to be found numerically by solving for the λ_i, subject to the constraints in Equation 7.61. This framework slightly generalizes Dempster's "Covariance Selection," which is a test on the sparsity of Σ^{-1}. [Dempster, 1972; Pav, 2013]

An iterative technique for finding the MLE based on a Newton step would proceed as follow. [Nocedal and Wright, 2006] Let $\boldsymbol{\lambda}^{(0)}$ be some initial estimate of the vector of λ_i, for example the zeros vector. The residual of the kth estimate, $\boldsymbol{\lambda}^{(k)}$ is

$$\epsilon_i^{(k)} := \operatorname{tr}\left(\mathsf{A}_i \left[\hat{\Theta} - \sum_j \lambda_j^{(k)} \mathsf{A}_j\right]^{-1}\right) - a_i. \tag{7.63}$$

The Jacobian of this residual with respect to the lth element of $\boldsymbol{\lambda}_i^{(k)}$ s

$$\frac{\mathrm{d}\epsilon_i^{(k)}}{\mathrm{d}\lambda_l^{(k)}} = \operatorname{vec}\left(\mathsf{A}_i\right)^\top \left(\left[\hat{\Theta} - \sum_j \lambda_j^{(k)} \mathsf{A}_j\right]^{-1} \otimes \left[\hat{\Theta} - \sum_j \lambda_j^{(k)} \mathsf{A}_j\right]^{-1}\right) \operatorname{vec}\left(\mathsf{A}_l\right). \tag{7.64}$$

Newton's method is then the iterative scheme

$$\boldsymbol{\lambda}^{(k+1)} \leftarrow \boldsymbol{\lambda}^{(k)} - \left(\frac{\mathrm{d}\boldsymbol{\epsilon}^{(k)}}{\mathrm{d}\boldsymbol{\lambda}^{(k)}}\right)^{-1} \boldsymbol{\epsilon}^{(k)}_{.} \tag{7.65}$$

When the iterative scheme converges plug $\boldsymbol{\lambda}^{(k)}$ into Equation 7.62 to find the MLE. The likelihood ratio test statistic is

$$-2\log\Lambda = n\left(\log\left|\Theta_0\hat{\Theta}^{-1}\right| + \operatorname{tr}\left({\Theta_0}^{-1}\hat{\Theta}\right) - (k+1)\right). \tag{7.66}$$

By Wilks' Theorem, under the null hypothesis, $-2\log\Lambda$ is, asymptotically in n, distributed as a chi-square with m degrees of freedom. [Wilks, 1938] However, the null hypothesis is likely to be on the boundary of acceptable values of Θ, and thus simulation under the null may be necessary to find the distribution of $-2\log\Lambda$.

Example 7.5.1 (Fama French Four Factors, Likelihood test on maximal signal-noise ratio). We consider the optimal portfolio on the returns of the Market, SMB, HML, and UMD portfolios. We test the null hypothesis $\zeta_* = 0$ which we can test via the four linear constraints $\boldsymbol{\nu}_* = \mathbf{0}$. We start with $\boldsymbol{\lambda}^{(0)} = \mathbf{0}$; after 5 steps, we have converged on $\boldsymbol{\lambda}^{(5)} = \left[5.71 \quad 4.80 \quad -8.00 \quad 0.95\right]^{\top}$. We compute $-2\log\Lambda = 285.423$. Compared to a chi-squared distribution with 4 degrees of freedom, we easily reject the null hypothesis at the 0.05 level. ⊣

Example 7.5.2 (Portfolio on the Fama French Four Factors, value of SMB). We reconsider the analysis of Example 6.3.11, using the likelihood test. That is, we test the hypothesis that the SMB loading in the Markowitz portfolio on the Market, SMB, HML, and UMD portfolios is zero. From $\boldsymbol{\lambda}^{(0)} = 0$, we converge on $\boldsymbol{\lambda}^{(3)} = 0.04$. From this we compute $-2\log\Lambda = 0.222$. Compared to a chi-square distribution with 1 degrees of freedom, we again fail to reject the null hypothesis at the 0.05 level. ⊣

7.6 † Bayesian inference via the second moment

We now consider using the approximate distribution of $\hat{\Sigma}^{-1/2}\hat{\boldsymbol{\mu}}$ to perform Bayesian inference on the squared maximal signal-noise ratio ζ_*^2. Recall that we previously claimed a normal approximation $\hat{\Sigma}^{-1/2}\hat{\boldsymbol{\mu}} \approx \mathcal{N}\left(\Sigma^{-1/2}\boldsymbol{\mu}, d^2\mathsf{I}\right)$, where d^2 could be estimated via the multivariate delta method and heteroskedasticity and autocorrelation consistent covariance estimators. This corresponds to the classical multivariate Gaussian case with unknown mean and known covariance. [Murphy, 2007] These can be stated as

$$\Sigma^{-1/2}\boldsymbol{\mu} \propto \mathcal{N}\left(\boldsymbol{\phi}_0, \Psi_0\right), \tag{7.67}$$

where $\boldsymbol{\phi}_0$ and Ψ_0 are the hyper-parameters. After observing $\hat{\Sigma}^{-1/2}\hat{\boldsymbol{\mu}}$, the posterior is then

$$\Sigma^{-1/2}\boldsymbol{\mu} \propto \mathcal{N}\left(\boldsymbol{\phi}_1, \Psi_1\right), \tag{7.68}$$

where

$$\Psi_1 = \left(\Psi_0{}^{-1} + d^{-2}\mathsf{I}\right)^{-1}, \tag{7.69}$$

$$\phi_1 = \Psi_1\left(\Psi_0{}^{-1}\phi_0 + d^{-2}\hat{\Sigma}^{-1/2}\hat{\mu}\right). \tag{7.70}$$

Both the prior and posterior can be reformulated as beliefs on ζ_*^2 which follow a *generalized chi-squared distribution.* [Imhof, 1961; Duchesne and de Micheaux, 2010]

As the quantity $\Sigma^{-1/2}\mu$ does not have a straightforward interpretation, it seems unlikely that one would have an informative prior on it. A noninformative prior corresponds to $\Psi_0 \to \infty\mathsf{I}$. Under this noninformative prior, the posterior belief on ζ_*^2 is

$$d^{-2}\zeta_*^2 \propto \chi^2\left(k; d^{-2}\hat{\zeta}_*^2\right). \tag{7.71}$$

7.7 Exercises

Ex. 7.1 Sum of Mahalanobis distances For d_t^2 defined in Equation 7.1, compute

$$\sum_{t=1}^{n} d_t^2.$$

Ex. 7.2 Log Mahalanobis distances Compute the Mahalanobis distances of the Fama French factors as in Example 7.0.1 and plot the autocorrelation of the *log* of the Mahalanobis distances.

1. ‽ Why does the log autocorrelogram look different than the plain autocorrelogram of Figure 7.2?
2. ‽ Why does the plain autocorrelogram show a spike at nine months?

Ex. 7.3 Inverting Θ Confirm that Equation 7.4 gives the inverse of Θ given in Equation 7.3.

Ex. 7.4 Approximate var., squared maximal Sharpe ratio Show that the approximate variance of $\hat{\zeta}_*^2$ given by $C_{1,1}/n$ for C of Equation 7.16 is equivalent to the variance from Equation 6.64 up to a term of $\mathcal{O}\left(n^{-2}\right)$.

Ex. 7.5 Covariance of Markowitz portfolio Establish empirically that the asymptotic covariance of $\hat{\boldsymbol{\nu}}_*$ given by Equation 7.16 is correct. Perform simulations drawing returns from the Gaussian distribution with $\hat{\zeta}_*^2 > 0$.

Ex. 7.6 Some grad and Hessian Let X be a symmetric matrix. Define

$$f(\mathsf{X}) = \boldsymbol{e}_t^\top \mathsf{X}^{-1}\boldsymbol{e}_j.$$

1. Prove that
$$\nabla_{\mathrm{vech}(\mathsf{X})} f(\mathsf{X}) = -\left(e_j^\top \hat{\Theta}^{-1} \otimes e_t^\top \hat{\Theta}^{-1}\right) \mathsf{D}.$$

2. Prove that
$$H_{\mathrm{vech}(\mathsf{X})} f(\mathsf{X}) = \mathsf{D}^\top \left[e_j^\top \mathsf{X}^{-1} \otimes \mathsf{X}^{-1} \otimes \mathsf{X}^{-1} e_t^\top + \mathsf{X}^{-1} e_j \otimes e_t^\top \mathsf{X}^{-1} \otimes \mathsf{X}^{-1}\right] \mathsf{D}.$$

3. Prove that
$$H_{\mathrm{vech}(\mathsf{X})} f(\mathsf{X}) = \mathsf{D}^\top 2\mathsf{N}\left[\mathsf{X}^{-1} e_j \otimes e_t^\top \mathsf{X}^{-1} \otimes \mathsf{X}^{-1}\right] \mathsf{D},$$
where $\mathsf{N} = \frac{1}{2}\left(\mathsf{I} + \mathsf{K}\right)$.

Ex. 7.7 Covariance, inverse Cholesky second moment Generate a random symmetric positive definite Σ and a random $\boldsymbol{\mu}$, then confirm that the covariance matrix in Equation 7.57 is diagonal up to machine precision.

Ex. 7.8 Covariance, inverse Cholesky second moment II † Prove that the covariance matrix of Equation 7.57 is diagonal.

Ex. 7.9 Chow test, Markowitz portfolio Express the test for difference in Markowitz portfolio in two time periods as an MGLH test. How would you test this using $\hat{\Theta}$?

1. Perform this test on the Fama French Four factor data to compare the Markowitz portfolio in December and January versus the portfolio the rest of the year.

Ex. 7.10 Two sample test for maximal signal-noise ratio † Suppose you observe $\hat{\Theta}_i$ for $i = 1, 2$, and wish to test the null hypothesis $H_0 : \zeta_{*,1}^2 = \zeta_{*,2}^2$.

1. Supposing that returns are independent, how would you approximately test this via the Cholesky inference trick? Implement your test in code and test it under the null hypothesis. Do you get nearly uniform p-values?
2. What if returns are dependent and measured over the same time period? Implement your test in code and test it under the null hypothesis. Do you get nearly uniform p-values?

Ex. 7.11 Inverse heteroskedasticity coverage Repeat the experiments of Example 7.4.6 but take $\lambda = -1$.

Ex. 7.12 Cholesky trick, conditional expectation † Can you adapt the Cholesky trick to perform hypothesis testing on the Hotelling Lawley trace?

Ex. 7.13 Boring hypothesis testing Perform some hypothesis tests on the Fama French six factor data using $\hat{\Theta}$ and the Cholesky trick.

1. Test $H_0 : \zeta_* = 0.25\,\text{mo.}^{-1/2}$.
2. Hedging out the Market component, test $H_0 : \Delta_{\mathsf{I}\setminus\mathsf{G}}\zeta_* = 0.15\,\text{mo.}^{-1/2}$.

Ex. 7.14 PIs, squared maximal Sharpe ratio Suppose you observe $\hat{\zeta}^2_{*,1}$ on n_1 observations, and then will observe n_2 more observations. You wish to compute a prediction interval on the $\hat{\zeta}^2_{*,2}$ in the out-of-sample data. That is, you wish to compute interval $\left(\zeta^2_{*,l}, \zeta^2_{*,u}\right)$ such that $\hat{\zeta}^2_{*,2}$ is in the interval with a fixed probability $1 - \alpha$, where the probability is on replication of the whole experiment. You may wish to use the Cholesky trick.

1. Test your prediction intervals empirically, letting $n_1 = 2,520$, $n_2 = 42$ and $\zeta^2_* = 0.01\,\text{day}^{-1/2}$.

Ex. 7.15 Momentum of factors Consider using the returns of the previous month as features in a portfolio on the Fama French four factor data. With an intercept term this amounts to five features and four assets.

1. Using the conditional expectation model, compute the squared maximal Sharpe ratio and test the hypothesis that its population analogue is zero, using techniques from Section 6.3.2.
2. Using the asymptotic normality of Θ_f defined in Equation 7.51, find the standard errors of each of the coefficients of the Markowitz coefficient. Which of them are more than two standard errors away from zero? (Essentially test each element of the passthrough individually using the test from Section 7.4.2.)
3. Repeat this analysis using the flattening trick.
4. Using the flattening trick, construct the Markowitz passthrough, hedged against the intercept terms. Test the hypothesis that the squared maximal signal-noise ratio is zero.

Ex. 7.16 Prediction intervals, Fama French Using the daily Fama French four factor data, compute $\hat{\zeta}^2_*$ on odd-numbered years, and use this to compute a prediction interval. Test with $\hat{\zeta}^2_*$ computed on the following year. Do you get nominal coverage?

1. Randomly shuffle the days and repeat this test. Do you get nominal coverage?

Ex. 7.17 Overlapping returns Compute rolling 21 market day returns from the daily returns of the Fama French 4 factor data. Compute $\hat{\boldsymbol{\nu}}_*$ and $\hat{\zeta}^2_*$. Compare these to the estimates from the monthly returns in Example 6.1.1.

1. Compute the expected autocorrelation of returns, then compute the ex-

pected bias in $\hat{\zeta}_*^2$ from Equation 7.33. Does this explain the difference in $\hat{\zeta}_*^2$ that you computed versus the version computed on non-overlapping monthly returns?

Ex. 7.18 Testing for diagonal covariance Express the hypothesis that Σ is a diagonal matrix in the form given in Equation 7.61.

1. Perform this hypothesis test on the Fama French monthly data.

Ex. 7.19 Value and Momentum Suppose you thought that the sample estimate of the weight of value (HML) in the Markowitz portfolio on the Fama French Four factor returns was too low due to random error. Using the correlation matrix of the Markowitz portfolio found in Example 7.4.10, what would you say about the weight of momentum (UMD)?

1. Suppose you thought the weight of value was 2 standard deviations too low. What would this imply about the weight of momentum?
2. The sample correlation of the *returns* of the Fama French Four factor portfolios are estimated as

$$\begin{bmatrix} & Mkt & SMB & HML & UMD \\ \hline Mkt & 1.000 & 0.323 & 0.240 & -0.342 \\ SMB & 0.323 & 1.000 & 0.128 & -0.159 \\ HML & 0.240 & 0.128 & 1.000 & -0.413 \\ UMD & -0.342 & -0.159 & -0.413 & 1.000 \end{bmatrix}.$$

   Explain why the returns of UMD might be negatively correlated with the other factors while the *portfolio weights* of UMD might be positively correlated with those of the other factors (as shown in Example 7.4.10).
3. Consider the Markowitz portfolio on the Fama French four factors, hedged against the Market factor. Estimate the correlation of errors of the portfolio weights as in Example 7.4.10. What happens to the correlation of value and momentum?

Ex. 7.20 Distribution of second moment Confirm that for Gaussian returns, the random variable $n\hat{\Theta}$ has the same density, up to a constant in k and n, as a $k+1$-dimensional Wishart random variable with n degrees of freedom and scale matrix Θ. That is $n\hat{\Theta}$ is a *conditional* Wishart, conditional on $\hat{\Theta}_{1,1} = 1$.

1. For Gaussian features and returns, confirm that $n\hat{\Theta}_f$ follows a non-central Wishart distribution.

Ex. 7.21 Bayesian update, second moment Given the conditional Wishart density for $n\hat{\Theta}$ given in Exercise 7.20, devise a Bayesian update for Θ based on a conditional inverse Wishart prior on Θ. Again the conditioning is on $\Theta_{1,1} = 1$.

Ex. 7.22 Kalman filter, Markowitz portfolio Consider the Kalman Filter on the Markowitz portfolio, using the normal approximation on $\hat{\boldsymbol{\nu}}_*$, *cf.* Example 7.4.2. Start from a normal prior

$$\boldsymbol{\nu}_* \propto \mathcal{N}\left(\boldsymbol{\nu}_{*,0}, \Omega_0\right).$$

Suppose the prediction step is of the form $\boldsymbol{\nu}_* \mapsto \mathsf{A}\boldsymbol{\nu}_* + \boldsymbol{b} + \boldsymbol{z}$, where $\boldsymbol{z} \sim \mathcal{N}\left(\mathbf{0}, \Omega\right)$.

1. From the prior, write the prediction step as

$$\boldsymbol{\nu}_*^{-} \propto \mathcal{N}\left(\mathsf{A}\boldsymbol{\nu}_{*,0} + \boldsymbol{b}, \Omega + \mathsf{A}\Omega_0\mathsf{A}^\top\right).$$

2. Suppose you observe $\hat{\boldsymbol{\nu}}_*$, which is approximately normally distributed: $\hat{\boldsymbol{\nu}}_* \sim \mathcal{N}\left(\boldsymbol{\nu}_*^{-}, \mathsf{\Gamma}\right)$. (Recall how to compute $\mathsf{\Gamma}$ via the central limit theorem and delta method on $\hat{\Theta}$.) Derive the posterior belief

$$\begin{aligned}\boldsymbol{\nu}_* &\propto \mathcal{N}\left(\boldsymbol{\nu}_{*,1}, \Omega_1\right),\\ \Omega_1 &= \left({\Omega_1}^{-1} + \mathsf{\Gamma}^{-1}\right)^{-1},\\ \boldsymbol{\nu}_{*,1} &= \Omega_1\left(\mathsf{\Gamma}^{-1}\hat{\boldsymbol{\nu}}_* + {\Omega_0}^{-1}\boldsymbol{\nu}_{*,0}\right).\end{aligned}$$

Ex. 7.23 Kalman filter, Markowitz, Fama French Implement the Kalman filter of Exercise 7.22. Use it on the Fama French four factor daily data, updating the Markowitz portfolio every three months. Take $\mathsf{A} = \mathsf{I}$, $\boldsymbol{b} = \mathbf{0}$, and let $\Omega = c^2\mathsf{I}$.

1. Run your code on several values of c^2 and plot the results over time.
2. Does your predicted $\boldsymbol{\nu}_*^{-}$ match $\hat{\boldsymbol{\nu}}_*$ very well? For what value of c^2 do you get optimal forecasts?

Ex. 7.24 Asymptotic expectation, matrix normal Derive Equation 7.24 and Equation 7.25. Suppose $\mathsf{X} \sim \mathcal{MN}\left(\mathsf{M}, \Psi, \Sigma\right)$ with $\operatorname{tr}\left(\Psi\right) = n$. Define $\boldsymbol{\mu} = \mathsf{M}^\top\mathbf{1}_n/n$ and $c_\psi = {\mathbf{1}_n}^\top\Psi\mathbf{1}_n$.

1. First show that for matrix R with eigenvalues strictly between 0 and 2,

$$\mathsf{R}^{-1} = \mathsf{I} - \left(\mathsf{R} - \mathsf{I}\right) + \left(\mathsf{R} - \mathsf{I}\right)^2 - \left(\mathsf{R} - \mathsf{I}\right)^3 + \ldots \tag{7.72}$$

   Try to confirm this relationship experimentally where R is the correlation matrix of a matrix whose rows are independent Gaussians.

2. Let b be some positive constant chosen to make the eigenvalues work out. Show that

$$\hat{\zeta}_*^2 \approx \frac{n-1}{n^2}\boldsymbol{z}^\top\left[\mathsf{I}_k - \mathsf{Z} + \mathsf{Z}^2 - \ldots\right]\boldsymbol{z},$$

where we have defined

$$\begin{aligned}
z &= nbc_{\psi}^{-1/2}\Sigma^{-1/2}\hat{\mu},\\
Z &= (n-1)\, b^2 c_{\psi}^{-1}\Sigma^{-1/2}\hat{\Sigma}\Sigma^{-1/2} - I_k,\\
&= b^2 c_{\psi}^{-1}\Sigma^{-1/2}X^{\top}X\Sigma^{-1/2} - \frac{1}{n}zz^{\top} - I_k.
\end{aligned}$$

3. Conclude that, roughly,

$$\begin{aligned}
\mathrm{E}\left[\hat{\zeta}_*^2\right] &\approx \frac{n-1}{n^2}\left(\mathrm{E}\left[z^{\top}z\right] - \mathrm{E}\left[z^{\top}Zz\right]\right),\\
\mathrm{E}\left[\hat{\zeta}_*^4\right] &\approx \left(\frac{n-1}{n^2}\right)^2 \mathrm{E}\left[\left(z^{\top}z\right)^2\right].
\end{aligned}$$

4. Confirm that

$$b^{-1}z \sim \mathcal{N}\left(nc_{\psi}^{-1/2}\Sigma^{-1/2}\mu, I_k\right)$$

and thus $b^{-2}z^{\top}z$ follows a non-central χ^2 distribution with k degrees of freedom and non-centrality parameter

$$\delta = n^2 c_{\psi}^{-1}\mu^{\top}\Sigma^{-1}\mu. \tag{7.73}$$

Show then that

$$\begin{aligned}
\mathrm{E}\left[b^{-2}z^{\top}z\right] &= (k+\delta),\\
\mathrm{E}\left[b^{-4}\left(z^{\top}z\right)^2\right] &= \left((k+\delta)^2 + 2(k+2\delta)\right).
\end{aligned}$$

5. Let $Y := bc_{\psi}^{-1/2}X\Sigma^{-1/2}$. Confirm that

$$\begin{aligned}
Y &\sim \mathcal{MN}\left(bc_{\psi}^{-1/2}M\Sigma^{-1/2}, b^2 c_{\psi}^{-1}\Psi, I_k\right),\\
z &= Y^{\top}1_n,\\
Z &= Y^{\top}Y - \frac{1}{n}zz^{\top} - I_k = Y^{\top}\left(I_n - n^{-1}1_n 1_n^{\top}\right)Y - I_k.
\end{aligned}$$

Using the results of Letac and Massam [2008] on $b^{-2}YY^{\top}$, show that

$$\begin{aligned}
b^{-4}1_n^{\top}\mathrm{E}\left[YY^{\top}YY^{\top}\right]1_n &= (k+\delta)\, c_{\psi}^{-1}n + c_{\psi}^{-1}\operatorname{tr}\left(M\Sigma^{-1}M^{\top}\right)\\
&\quad + k(k+1)\, c_{\psi}^{-2}1_n^{\top}\Psi^2 1_n + c_{\psi}^{-2}1_n^{\top}M\Sigma^{-1}M^{\top}M\Sigma^{-1}M^{\top}1_n\\
&\quad + 2c_{\psi}^{-2}(k+1)\, 1_n^{\top}M\Sigma^{-1}M^{\top}\Psi 1_n.
\end{aligned}$$

6. Conclude that

$$\begin{aligned}
\mathrm{E}\left[\hat{\zeta}_*^2\right] &\approx \frac{n-1}{n^2}\left(2\,\mathrm{E}\left[\boldsymbol{z}^\top \boldsymbol{z}\right] - \mathbf{1}_n^\top \mathrm{E}\left[\mathsf{Y}\mathsf{Y}^\top\mathsf{Y}\mathsf{Y}^\top\right]\mathbf{1}_n + \frac{1}{n}\mathrm{E}\left[\left(\boldsymbol{z}^\top \boldsymbol{z}\right)^2\right]\right),\\
&\approx \frac{n-1}{n^2}b^2\left(\left(2-\frac{nb^2}{c_\psi}\right)(k+\delta) - b^2 c_\psi^{-1}\operatorname{tr}\left(\mathsf{M}\Sigma^{-1}\mathsf{M}^\top\right)\right)\\
&\quad - \frac{n-1}{n^2}b^4 c_\psi^{-2}\left(k(k+1)\mathbf{1}_n^\top \Psi^2 \mathbf{1}_n + \mathbf{1}_n^\top \mathsf{M}\Sigma^{-1}\mathsf{M}^\top\mathsf{M}\Sigma^{-1}\mathsf{M}^\top \mathbf{1}_n\right)\\
&\quad - 2\frac{n-1}{n^2}b^4 c_\psi^{-2}\left((k+1)\mathbf{1}_n^\top \mathsf{M}\Sigma^{-1}\mathsf{M}^\top \Psi \mathbf{1}_n\right),\\
&\approx \frac{n-1}{n^2}\left(2b^2 - \frac{nb^4}{c_\psi}\right)(k+\delta).
\end{aligned}$$

7. Show that

$$\begin{aligned}
\mathrm{Var}\left(\hat{\zeta}_*^2\right) &\approx \left(\frac{n-1}{n^2}\right)^2 b^4\left[(k+\delta)^2 + 2(k+2\delta)\right] - \left(\frac{n-1}{n^2}b^2\left(2-\frac{nb^2}{c_\psi}\right)(k+\delta)\right)^2,\\
&\approx \left(\frac{n-1}{n^2}\right)^2 b^4\left(2(k+2\delta) + \left(1-\left[2-\frac{nb^2}{c_\psi}\right]^2\right)(k+\delta)^2\right).
\end{aligned}$$

8. Letting $b^2 = c_\psi/n$ conclude that

$$\mathrm{E}\left[\hat{\zeta}_*^2\right] \approx \frac{n-1}{n^2}\frac{c_\psi}{n}(k+\delta), \tag{7.74}$$

$$\mathrm{Var}\left(\hat{\zeta}_*^2\right) \approx 2\left(\frac{n-1}{n^2}\right)^2 \frac{c_\psi^2}{n^2}(k+2\delta). \tag{7.75}$$

9. When returns are independent, then $c_\psi = n$. Simplify the expressions for the expected value and variance of $\hat{\zeta}_*^2$ in this case, and compare to the formulæ for the *i.i.d.* Gaussian case, *viz.* Equation 6.57 and Equation 6.58.
10. The choice of $b^2 = c_\psi/n$ seems arbitrary, and indeed we have not quantified the truncated terms. The sufficient condition for convergence of Equation 7.72 is that Z has eigenvalues less than one in absolute value. When $\Psi = \mathsf{I}$ and $\mathsf{M} = \mathbf{1}\boldsymbol{\mu}^\top$ the eigenvalues of $\frac{1}{n-1}\mathsf{Y}^\top\left(\mathsf{I}_n - n^{-1}\mathbf{1}_n\mathbf{1}_n^\top\right)\mathsf{Y}$ asymptotically follow the *Marchenko Pastur* distribution. Under this distribution their support is between $b^2 c_\psi^{-1}\left(1 \pm \sqrt{\lambda}\right)^2$, where $\lambda = k/(n-1)$. Show that the interval containing the eigenvalues of Z is "balanced" (*i.e.*, of the form $[-x, x)$) when

$$b^2 = \frac{c_\psi}{n}\frac{n-1}{n-1+k}.$$

Ex. 7.25 Research Problem: Asymptotic expectation, matrix normal ‽ Derive higher order approximations of $\mathrm{E}\left[\hat{\zeta}_*^2\right]$ and $\mathrm{Var}\left(\hat{\zeta}_*^2\right)$ like Equation 7.24 and Equation 7.25, but with with proper error terms.

Ex. 7.26 Asymptotic Markowitz portfolio, matrix normal Continuing Exercise 7.24, here you will find the approximate distribution of $\hat{\boldsymbol{\nu}}_*$ when X follows a matrix normal distribution. Start from

$$\hat{\boldsymbol{\nu}}_* = (n-1)\, b c_\psi^{-1/2} \Xi^{-1/2} \left[\mathsf{I}_k - \mathsf{Z} + \mathsf{Z}^2 - \ldots\right] \boldsymbol{z}.$$

* 1. Using Stein's lemma, prove that $\boldsymbol{z}$ and Z are uncorrelated.
2. Show that

$$\begin{aligned}
\mathrm{E}\left[\hat{\boldsymbol{\nu}}_*\right] &\approx (n-1)\, b c_\psi^{-1/2} \Xi^{-1/2} \left[\mathrm{E}\left[\boldsymbol{z}\right] - \mathrm{E}\left[\mathsf{Z}\boldsymbol{z}\right] + \ldots\right],\\
&\approx (n-1)\, n b^2 c_\psi^{-1} \Xi^{-1/2} \left[\mathsf{I}_k - \mathrm{E}\left[\mathsf{Z}\right] + \ldots\right] \Xi^{-1/2} \boldsymbol{\mu},\\
&\approx (n-1)\, n b^2 c_\psi^{-1} \left(2 + \frac{b^2}{n} + b^2 c_\psi^{-1} n \left(\boldsymbol{\mu}^\top \Xi^{-1} \boldsymbol{\mu} - 1\right)\right) \Xi^{-1} \boldsymbol{\mu}\\
&- (n-1)\, n b^4 c_\psi^{-2} \Xi^{-1/2} \mathsf{M}^\top \Xi^{-1} \mathsf{M} \Xi^{-1/2} \boldsymbol{\mu} + \ldots
\end{aligned}$$

3. Starting from the approximation

$$\hat{\boldsymbol{\nu}}_* \approx (n-1)\, b c_\psi^{-1/2} \Xi^{-1/2} \boldsymbol{z},$$

show that

$$\mathrm{Var}\left(\hat{\boldsymbol{\nu}}_*\right) \approx (n-1)^2 b^4 c_\psi^{-1} \Xi^{-1}$$

4. Plugging in $b^2 = c_\psi n^{-1}$ show that

$$\begin{aligned}
\mathrm{E}\left[\hat{\boldsymbol{\nu}}_*\right] &\approx (n-1) \left[\left(\frac{c_\psi}{n^2} + \boldsymbol{\mu}^\top \Xi^{-1} \boldsymbol{\mu} + 1\right) \Xi^{-1} \boldsymbol{\mu} - \frac{1}{n} \Xi^{-1/2} \mathsf{M}^\top \Xi^{-1} \mathsf{M} \Xi^{-1/2} \boldsymbol{\mu} + \ldots\right],\\
\mathrm{Var}\left(\hat{\boldsymbol{\nu}}_*\right) &= c_\psi \left(\frac{n-1}{n}\right)^2 \Xi^{-1}.
\end{aligned}$$

Ex. 7.27 Optimal conditional moments In Section 5.3.2, we discussed the optimal conditional strategies when the first two moments are conditional on some observed feature. The solution, given in Equation 5.40 is to hold portfolio

$$\boldsymbol{\nu}_*\left(\boldsymbol{f}_{t-1}\right) = c \mathsf{A}_2^{-1}\left(\boldsymbol{f}_{t-1}\right) \boldsymbol{\mu}\left(\boldsymbol{f}_{t-1}\right),$$

conditional on observing the features $\boldsymbol{f}_{t-1}$. As the moments are unknown in general they have to be estimated from the data. Suppose that conditional on observed features $\boldsymbol{f}_{t-1}$ that returns are normally distributed with second moment matrix $\Theta\left(\boldsymbol{f}_{t-1}; \boldsymbol{\beta}\right)$ for some parameters $\boldsymbol{\beta}$ to be determined. The functional form of $\Theta\left(\boldsymbol{f}_{t-1}; \boldsymbol{\beta}\right)$ is general, though it must be symmetric and positive definite.

1. Having observed n returns $\boldsymbol{x}_t$ and corresponding features $\boldsymbol{f}_{t-1}$, write the log-likelihood.

2. What are the first order necessary conditions for the MLE for $\boldsymbol{\beta}$?
3. The $\boldsymbol{\beta}$ which maximizes the likelihood will in general not have a nice closed form, but rather has to be found numerically. Describe how this would be done in code.
4. Can you design a function $\Theta\left(\boldsymbol{f}_{t-1};\boldsymbol{\beta}\right)$ such that there is a nice closed form for the MLE?
5. Suppose that the first moments take the form

$$\begin{aligned}\boldsymbol{\mu}\left(\boldsymbol{f}_{t-1}\right) &= \boldsymbol{\beta}_1^{\top}\boldsymbol{f}_{t-1},\\ \mathsf{A}_2\left(\boldsymbol{f}_{t-1}\right) &= \mathsf{B}_2^{\top}\boldsymbol{f}_{t-1}\boldsymbol{f}_{t-1}^{\top}\mathsf{B}_2.\end{aligned}$$

   How would you find the MLEs of $\boldsymbol{\beta}_1$ and B_2?
6. How would you perform the Wald test for some hypothesis on $\boldsymbol{\beta}$?
7. † Is the MLE related to the $\boldsymbol{\beta}$ which maximizes the Sharpe ratio of the optimal conditional strategy on the data? Is the Wald test related to the significance tests on $\hat{\zeta}$ of the optimal strategy? This would bridge a gap between statistical practice and iterative hill-climbing the backtested Sharpe ratio.

8

Overoptimism and Overfitting

> Beware of geeks bearing formulas.
>
> WARREN BUFFETT

In Chapter 4, we considered "overoptimism by selection," the gap between the signal-noise ratio and the Sharpe ratio of an asset, when the asset is selected based on its Sharpe ratio. In Chapter 6, we discussed the problem of inference on the squared maximal signal-noise ratio for portfolio optimization problems. We noted there that even if one were confident that ζ_*^2 were reasonably large, one could still have misestimated the Markowitz portfolio, resulting in a small achieved signal-noise ratio. As this problem is exacerbated by the enlarging the investment universe, we claim this is a mechanism by which overfitting happens.

8.1 The "haircut"

Throughout this chapter we will use $q(\cdot)$ to refer to the signal-noise ratio function, defined as

$$q(\hat{\boldsymbol{\nu}}) := \frac{\hat{\boldsymbol{\nu}}^\top \boldsymbol{\mu}}{\sqrt{\hat{\boldsymbol{\nu}}^\top \Sigma \hat{\boldsymbol{\nu}}}}. \tag{8.1}$$

As $\hat{\boldsymbol{\nu}}$ is constructed by observing historical data and is subject to misestimation error, we will sometimes refer $q(\hat{\boldsymbol{\nu}})$ as the *achieved signal-noise ratio* of $\hat{\boldsymbol{\nu}}$. It is "achieved" only in the sense that one has observed $\hat{\boldsymbol{\nu}}$. Assuming that $\boldsymbol{\mu}$ is not all zeros (and thus $\zeta_*^2 > 0$), define the *haircut* of $\hat{\boldsymbol{\nu}}$ as

$$h := 1 - \frac{q(\hat{\boldsymbol{\nu}})}{\zeta_*} = 1 - \left(\frac{\hat{\boldsymbol{\nu}}^\top \boldsymbol{\mu}}{{\boldsymbol{\nu}_*}^\top \boldsymbol{\mu}}\right)\left(\frac{\sqrt{{\boldsymbol{\nu}_*}^\top \Sigma \boldsymbol{\nu}_*}}{\sqrt{\hat{\boldsymbol{\nu}}^\top \Sigma \hat{\boldsymbol{\nu}}}}\right), \tag{8.2}$$

where $\boldsymbol{\nu}_*$ is the population optimal portfolio, positively proportional to $\Sigma^{-1}\boldsymbol{\mu}$. The haircut is a loss: a smaller value means that the sample portfolio achieves a larger proportion of possible SNR, and thus one hopes for a smaller haircut.

DOI: 10.1201/9781003181057-8

The haircut takes values in $[0, 2]$; when it is larger than 1, the portfolio $\hat{\boldsymbol{\nu}}$ has *negative* expected returns. From first principles, one suspects that h is larger for smaller ζ_*, smaller n and larger k.

Modeling the haircut is not straightforward because it is a random quantity which is not observed. That is, it mixes the unknown population parameters Σ and $\boldsymbol{\mu}$ with the sample quantity $\hat{\boldsymbol{\nu}}$, which is random.

8.1.1 Haircut of the Markowitz portfolio

We now turn attention to the haircut of the sample Markowitz portfolio, $\hat{\boldsymbol{\nu}}_*$. A simple approximation to h in this case can be formed by supposing that $\hat{\boldsymbol{\nu}}_* \approx \Sigma^{-1}\hat{\boldsymbol{\mu}}$, that is, by assuming that the covariance matrix is modeled perfectly. Under this approximation and assuming Gaussian *i.i.d.* returns, the distribution of h is

$$\tan\left(\arcsin\left(1-h\right)\right) \sim \frac{1}{\sqrt{k-1}} t\left(\sqrt{n}\zeta_*, k-1\right). \tag{8.3}$$

Here $t\left(\delta, \nu\right)$ is a non-central t-distribution with non-centrality parameter δ and ν degrees of freedom. Note that tangent of arcsine is the TAS function of Equation 1.24.

Because mis-estimation of the covariance matrix should contribute some error, one expects this approximation to be a "stochastic lower bound" on the true haircut. Numerical simulations, however, suggest it is a fairly tight bound for large n/k. Moreover, one expects that the deviation from the assumption of Gaussian *i.i.d.* returns will cause the true haircut to be even worse.

Under Gaussian returns and assuming no error in $\hat{\Sigma}$, we note the following rough approximations, which are supported by Monte Carlo simulations for a fairly wide range of n, k and ζ_*:

$$\begin{aligned}
\operatorname{median}\left(h\right) &\approx 1 - \sin\left(\arctan\left(\frac{\sqrt{n}\zeta_*}{\sqrt{k-1}}\right)\right),\\
\operatorname{E}\left[h\right] &\approx 1 - \sqrt{\frac{n\zeta_*^2}{k + n\zeta_*^2}},\\
\operatorname{Var}\left(h\right) &\approx \frac{k}{\left(k + \left[n\zeta_*^2\right]^{1.08}\right)^2}.
\end{aligned} \tag{8.4}$$

The first approximation follows because the median value of the non-central t-distribution is approximately equal to the non-centrality parameter. [Johnson and Welch, 1940; Kraemer and Paik, 1979] The others were found by a genetic programming search. Note that each of these approximations uses the unknown maximal signal-noise ratio, ζ_*; plugging in the sample estimate $\hat{\zeta}_*$ gives poor approximations because $\hat{\zeta}_*$ is biased and not independent from h.

Example 8.1.1 (Simulated haircut). The quality of Approximation 8.3 is checked by Monte Carlo simulations. We construct the sample Markowitz portfolio of Gaussian returns on $k = 6$ assets, with using $n = 1,008$ (4 years of daily observations), and $\zeta_* = 1.25\,\mathrm{yr}^{-1/2}$. Since the population Markowitz portfolio is known, the haircut can be computed exactly. We repeat this experiment $10,000$ times, then Q-Q plot h against the theoretical distribution from Approximation 8.3 in Figure 8.1. We see that the approximation is very good for this choice of n, k, ζ_*.

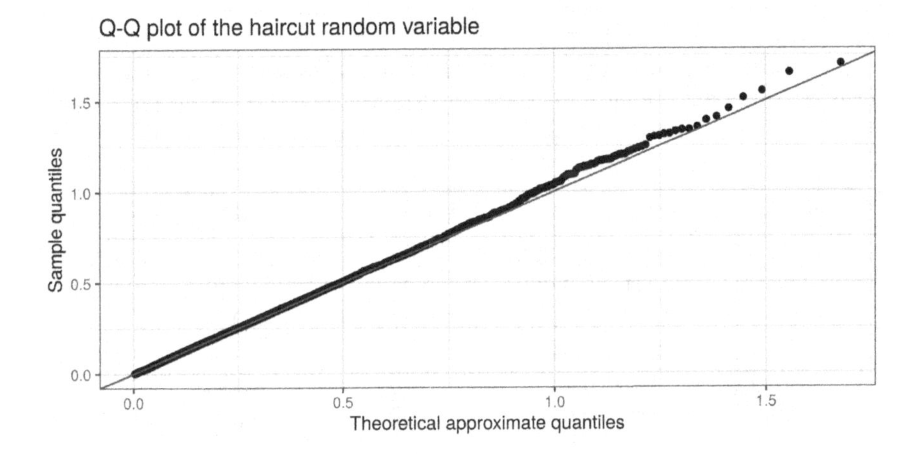

FIGURE 8.1
Q-Q plot of 10,000 simulated haircut values of the Markowitz portfolio versus Approximation 8.3 is shown.

In Table 8.1 we present the empirical median, mean and standard deviation of h under these simulations, along with the approximations from Equation 8.4 The median value of the haircut over the $10,000$ simulations is 0.241, meaning that in more than half the simulations, the sample portfolio has "lost" more than 24% of the optimal population Sharpe ratio. Put another way, with probability around one half, the achieved signal-noise ratio of the sample Markowitz portfolio is less than $0.949\,\mathrm{yr}^{-1/2}$. ⊣

	Empirical	Approximation
Median	0.241	0.255
Mean	0.285	0.286
Standard deviation	0.195	0.185

TABLE 8.1
Empirical approximate values of the median, mean, and standard deviation of the haircut distribution are given for 10,000 Monte Carlo simulations of 1,008 days of Gaussian data for 6 assets with $\zeta_* = 1.25\,\mathrm{yr}^{-1/2}$. The approximations from Equation 8.4 are also reported.

Example 8.1.2 (Simulated haircut, other distributions). Since the Gaussian distribution is a poor model for daily asset returns, we repeat the experiment of Example 8.1.1 using returns drawn from various other distributions. We draw returns from a t-distribution with 4 degrees of freedom, from a Tukey h-distribution with parameter $h = 0.15$, and from a Lambert W $\times$ Gaussian with parameter $\gamma = -0.2$. [Goerg, 2009, 2010] We use these exotic distributions, which have non-trivial kurtosis and skew, to check robustness of Approximation 8.3 to higher order moments. Again, $10,000$ simulations are performed under each of these distributions with the same values of n, k, ζ_* as in the previous example. Some of the empirical quantiles from these simulations are shown in Table 8.2, along with the approximate quantiles from Approximation 8.3. For this choice of n, k, ζ_*, the approximation is very good. ⊣

Quantile	Gaussian	t(4)	Tukey(0.15)	Lam.W(-0.2)	Approx.
10.0	0.088	0.083	0.087	0.086	0.085
50.0	0.241	0.236	0.239	0.240	0.236
75.0	0.375	0.373	0.373	0.375	0.369
95.0	0.655	0.645	0.646	0.664	0.648
97.5	0.782	0.762	0.764	0.775	0.766
99.0	0.949	0.911	0.907	0.920	0.920
99.5	1.092	1.020	0.993	1.053	1.036

TABLE 8.2
Empirical quantiles of the haircut from 10,000 simulations of 1,008 days of 6 assets, with maximal Sharpe ratio of $1.25\,\mathrm{yr}^{-1/2}$ are given, along with the approximate quantiles from Approximation 8.3.

8.1.2 † Attribution of error

Equation 7.8 gives the asymptotic distribution of $\hat{\Theta}^{-1}$, which contains the (negative) Markowitz portfolio and the precision matrix. This allows one to estimate the amount of error in the Markowitz portfolio which is attributable to mis-estimation of the covariance. The remainder one can attribute to mis-estimation of the mean vector, which, is typically implicated as the leading effect. [Chopra and Ziemba, 1993]

The computation is performed as follows: the estimated covariance of $\mathrm{vech}\left(\hat{\Theta}^{-1}\right)$ is turned into a correlation matrix in the usual way[1]. This gives a correlation matrix, call it R, some of the elements of which correspond to the negative Markowitz portfolio, and some to the precision matrix. For a single element of the Markowitz portfolio, let $\boldsymbol{r}$ be a sub-column of R consisting of

[1]That is, by a Hadamard divide of the rank one matrix of the outer product of the diagonal. Or, more practically, by the **R** function `cov2cor`.

the column corresponding to that element of the Markowitz portfolio and the rows of the precision matrix. And let R_{Σ} be the sub-matrix of R corresponding to the precision matrix. The multiple correlation coefficient is then $\boldsymbol{r}^{\top}\mathsf{R}_{\Sigma}^{-1}\boldsymbol{r}$. This is an R^2 number between zero and one, estimating the proportion of variance in that element of the Markowitz portfolio "explained" by error in the precision matrix.

Example 8.1.3 (Fama French Four Factors, attribution of error). We consider the portfolio on the returns of the Market, SMB, HML, and UMD portfolios introduced in Example 1.1.1. For each of these "assets," this squared coefficient of multiple correlation, is expressed as percents in Table 8.3. A HAC estimator for Ω is used. In the "weighted" column, the returns are divided by an 11 month running mean of the median absolute return of the four assets, delayed by a month, assuming a model of "constant maximal Sharpe ratio." We can claim, then, that approximately 55% of the error in the Market position is due to mis-estimation of the precision matrix. ⊣

	Vanilla	Weighted
Mkt	55 %	56 %
SMB	24 %	22 %
HML	42 %	19 %
UMD	71 %	42 %

TABLE 8.3
Estimated multiple correlation coefficients for the three elements of the Markowitz portfolio are presented. These are the percentage of error attributable to mis-estimation of the precision matrix. HAC estimators for Ω are used for both. In the "vanilla" column, no adjustments are made for conditional heteroskedasticity, while in the "weighted" column, returns are divided by the volatility estimate.

8.2 A bound on achieved signal-noise ratio

In practical terms, the haircut presents something of an upper bound on the quality of the Markowitz portfolio constructed from sample estimates, described in normalized terms. In a certain sense, the loss due to estimation error of the portfolio is fundamental and is not limited to just the Markowitz portfolio[2]. Here we will consider a function $\hat{\boldsymbol{\nu}}(\cdot)$ that takes a $n \times k$ matrix of observed historical returns and produces a k-dimensional portfolio allocation.

[2]This section is abridged from the author's paper on the bound on expected achieved signal-noise ratio. [Pav, 2014]

Again, let $q(\cdot)$ be the signal-noise ratio function of Equation 8.1. Under fairly broad conditions it can be shown that [Pav, 2014]

$$\mathrm{E}\left[q\left(\hat{\boldsymbol{\nu}}\left(\mathsf{X}\right)\right)\right] \le \frac{\sqrt{n}\zeta_*^2}{\sqrt{k-1+n\zeta_*^2}}, \tag{8.5}$$

where expectation is over realizations of the $n \times k$ matrix of returns X.

The conditions that must be satisfied by $\hat{\boldsymbol{\nu}}(\cdot)$ for this inequality to hold are to avoid "stopped clocks." For example, if a portfolio construction method always placed all its allocation into $c\boldsymbol{e}_1$, and the first asset happened to have high ζ, then the bound would not be satisfied, in the same way that a stopped clock is right twice a day. The technical condition is implied by *rotational invariance* of the function $\hat{\boldsymbol{\nu}}(\cdot)$, defined as

$$\hat{\boldsymbol{\nu}}\left(\mathsf{XP}\right) = \mathsf{P}^{-1}\hat{\boldsymbol{\nu}}\left(\mathsf{X}\right)$$

for every square non-singular matrix P. The rotational invariance condition is satisfied by the Markowitz portfolio, and thus it is subject to this fundamental bound.

However, not every portfolio construction function satisfies this condition. Certainly the "one over n" portfolio[3], which allocates equal dollars (long) into each asset does not satisfy this condition, but that allocation is not seriously suggested for the long-short framework we are discussing here. [Tu and Zhou, 2011] Moreover, the rotational invariance principle does not seem to be satisfied by the "Risk Parity" portfolio. [Bai *et al.*, 2016] However, rotational invariance implies an invariance in the preferred historical returns of the portfolio. That is, if one transforms the definition of returns by invertible P, then the historical returns of the portfolio are transformed as

$$\mathsf{X}\hat{\boldsymbol{\nu}}\left(\mathsf{X}\right) \mapsto \mathsf{XP}\hat{\boldsymbol{\nu}}\left(\mathsf{XP}\right) = \mathsf{XPP}^{-1}\hat{\boldsymbol{\nu}}\left(\mathsf{X}\right) = \mathsf{X}\hat{\boldsymbol{\nu}}\left(\mathsf{X}\right),$$

which is unchanged. In a sense "only returns matter" for rotationally invariant $\hat{\boldsymbol{\nu}}(\cdot)$. In this light the bound of Inequality 8.5 really does seem fundamental, as the alternative is to embrace an allocation for which something other than just historically modeled returns matters.

Inequality 8.5 is derived from a Cramér-Rao bound on the efficiency of estimators of $\Sigma^{-1}\boldsymbol{\mu}$ on the sphere. One suspects that the Markowitz portfolio, which is built from the Maximum Likelihood Estimators of $\boldsymbol{\mu}$ and Σ, would be asymptotically efficient. This doesn't mean that the bound of Inequality 8.5 can be achieved, since there is some slack in the inequality. But there is not much slack. Ao *et al.* established the asymptotic achieved signal-noise ratio of the Markowitz portfolio. Namely that as $n \to \infty$, assuming k/n converges,

$$q\left(\hat{\boldsymbol{\nu}}\left(\mathsf{X}\right)\right) \xrightarrow{P} \frac{\sqrt{n-k}\zeta_*^2}{\sqrt{k+n\zeta_*^2}}, \tag{8.6}$$

[3] In our nomenclature, "one over k."

where the convergence is in probability. [Ao *et al.*, 2014, Equation (1.1)]

The approximation $\hat{\Sigma} \approx \Sigma$ that leads to Approximation 8.3 can be used to derive the following approximation on the squared quantity

$$q^2\left(\hat{\boldsymbol{\nu}}\left(\mathsf{X}\right)\right) \sim \zeta_*^2 \mathcal{B}\left(n\zeta_*^2, \frac{1}{2}, \frac{k-1}{2}\right), \tag{8.7}$$

where $\mathcal{B}\left(\delta, p, q\right)$ is a non-central Beta distribution with non-centrality δ, and "shape" parameters p and q. [Walck, 1996] However, by describing the distribution of the *square* of $q\left(\hat{\boldsymbol{\nu}}\left(\mathsf{X}\right)\right)$, we cannot easily model the (sometimes significant) probability that it is negative. This form, does, however, give bounds on the variance of $q\left(\hat{\boldsymbol{\nu}}\left(\mathsf{X}\right)\right)$, since the moments of the non-central Beta are known, namely

$$\mathrm{E}\left[q^2\left(\hat{\boldsymbol{\nu}}\left(\mathsf{X}\right)\right)\right] = \zeta_*^2 e^{-\frac{n\zeta_*^2}{2}} \frac{\Gamma\left(\frac{3}{2}\right)}{\Gamma\left(\frac{1}{2}\right)} \frac{\Gamma\left(\frac{k}{2}\right)}{\Gamma\left(\frac{k+2}{2}\right)} {}_2F_2\left(\frac{k}{2}, \frac{3}{2}; \frac{1}{2}, \frac{2+k}{2}; \frac{n\zeta_*^2}{2}\right), \tag{8.8}$$

where ${}_2F_2\left(\cdot, \cdot; \cdot, \cdot; \cdot\right)$ is the Generalized Hypergeometric function. [Olver *et al.*, 2010, sec 16.2] This is a rough upper bound on the variance of $q^2\left(\hat{\boldsymbol{\nu}}\left(\mathsf{X}\right)\right)$; a lower bound can be had using the upper bound on the mean from Inequality 8.5.

Constrained problems

The bound of Inequality 8.5 can be extended to the various constrained portfolio problems we have considered in previous chapters. [Pav, 2014] In each case, the k is replaced with the effective degrees of freedom one has in the portfolio optimization problem. For example, under the subspace constrained optimization problem of Equation 5.18, where one is restricted to a k_g-dimensional subspace, then Inequality 8.5 is replaced by

$$\mathrm{E}\left[q\left(\hat{\boldsymbol{\nu}}\left(\mathsf{X}\right)\right)\right] \le \frac{\sqrt{n}\zeta_*^2}{\sqrt{k_g - 1 + n\zeta_*^2}}. \tag{8.9}$$

Under the hedging constraint of Equation 5.21, which imposes k_g additional constraints on the portfolio, one has

$$\mathrm{E}\left[q\left(\hat{\boldsymbol{\nu}}\left(\mathsf{X}\right)\right)\right] \le \frac{\sqrt{n}\zeta_*^2}{\sqrt{(k - k_g) - 1 + n\zeta_*^2}}. \tag{8.10}$$

The bound can also be extended to the conditional expectation problem of Section 5.3. If one considers the optimization problem of Equation 5.27, but under the further constraint that one is only concerned with the so-called linear strategies, then the bound becomes

$$\mathrm{E}\left[Q\left(\hat{\mathsf{N}}\left(\mathsf{X}, \mathsf{F}\right)\right)\right] \le \frac{\sqrt{n}\zeta_*^2}{\sqrt{fk - 1 + n\zeta_*^2}}, \tag{8.11}$$

where f is the length of the feature vector, and we have defined

$$Q\left(\hat{\mathsf{N}}\right) := \frac{\operatorname{tr}\left(\hat{\mathsf{N}}^{\top}\mathsf{B}\Gamma_f\right)}{\sqrt{\operatorname{tr}\left(\hat{\mathsf{N}}^{\top}\Sigma\hat{\mathsf{N}}\Gamma_f\right)}}. \tag{8.12}$$

This isn't exactly the Hotelling-Lawley trace associated with the passthrough matrix $\hat{\mathsf{N}}$, rather it is the ratio of the expected return to the square root of the expected conditional variance.

8.2.1 Limits to diversification

The bound on achieved signal-noise ratio of Inequality 8.5 has implications for the diversification benefit. Consider the case of $k = 6, n = 1,008\,\text{days}, \zeta_* = 1.25\,\text{yr}^{-1/2}$ versus some superset of this asset universe with $k = 24, n = 1,008\,\text{days}, \zeta_* = 1.6\,\text{yr}^{-1/2}$. Since the optimum cannot decrease over a larger feasible space, we observe that the superset has a higher population signal-noise ratio, ζ_*. One should not, of course, increase the investment universe without *some* concomitant increase in ζ_*. However, in this case the bound on expected achieved signal-noise ratio from Inequality 8.5 for the smaller asset universe is $0.93\,\text{yr}^{-1/2}$, while for the superset it is $0.89\,\text{yr}^{-1/2}$. Diversification has possibly caused a *decrease* in expected achieved signal-noise ratio, even though the opportunity exists to increase achieved signal-noise ratio by a fair amount.

By the fundamental law of active management, one vaguely expects ζ_* to increase as $\sqrt{k}$. If, however, ζ_* scales at a rate slower than $k^{1/4}$, then the derivative of the bound in Inequality 8.5 will be negative for sufficiently large k: adding assets to the universe causes a decrease in expected achieved signal-noise ratio. To see why, note that $\sqrt{n}\zeta_*^2/\sqrt{k-1+n\zeta_*^2}$ has ζ_*^2 in the numerator, and $\sqrt{k}$ in the denominator; if ζ_* grows slower than $k^{1/4}$ the denominator will outpace the numerator, *cf.* Exercise 8.2.

The decreasing upper bound with respect to growing universe size is illustrated in Figure 8.2. Under the assumption $\zeta_* = \zeta_0 k^{\gamma}$, the upper bound of Inequality 8.5 is plotted versus k for different values of γ. The value of ζ_0 is set so that $\zeta_* = 1.25\,\text{yr}^{-1/2}$ when $k = 6$. For $\gamma < \frac{1}{4}$, one sees a local maximum in the upper bound as k increases, a behavior not seen for $\gamma > \frac{1}{4}$, where the bound on achieved signal-noise ratio grows with k.

This loss in value with respect to growing universe size is illustrated in Figure 8.3. Under the assumption $\zeta_* = \zeta_0 k^{\gamma}$, lines of ζ_* and the $0.25, 0.50$, and 0.75 quantiles of $\zeta_*(1-h)$, under Approximation 8.3, are plotted versus k. The facets represent γ values of $0.19, 0.22, 0.25$, and 0.28. The value of ζ_0 is set so that $\zeta_* = 1.25\,\text{yr}^{-1/2}$ when $k = 6$. For $\gamma < 0.25$, one sees a local maximum in achieved Sharpe ratio as k increases, a behavior not seen for $\gamma > 0.25$, where achieved Sharpe ratio grows with k. For the case of "slow growth" of ζ_*, the diversification benefit is not seen by the sample Markowitz

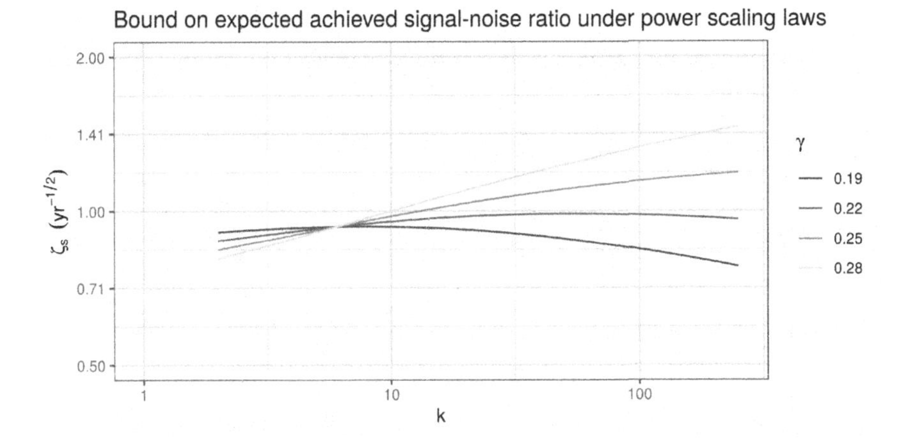

FIGURE 8.2
The upper bound of Inequality 8.5 is plotted versus k for different scaling laws for ζ_*. These scaling laws correspond to $\zeta_* = \zeta_0 k^\gamma$, with γ taking values between 0.19 and 0.28. The constant terms, ζ_0, are adjusted so that $\zeta_* = 1.25\,\mathrm{yr}^{-1/2}$ for $k = 6$ for all the lines. The bound uses $n = 1,008$, corresponding to 4 years of daily observations.

portfolio, rather its practical utility *decreases* because the haircut outpaces the growth of ζ_*.

Example 8.2.1 (Diversification, Industry Portfolios). We consider the monthly returns of five industry portfolios, as introduced in Example 1.2.5. These include 1,104 months of data on five industries from Jan 1927 to Dec 2018. For every combination of the assets, we compute $\hat{\zeta}_*^2$, then estimate ζ_*^2 via the KRS δ_2 estimator of Equation 6.67. We take the square root to estimate ζ_*. In Figure 8.4, we plot $\sqrt{\delta_2}$ versus the universe size, in log-log space. The slope of the best fit line is 0.12, a bit under the 1/4 cutoff for diversification to add benefit. We should note that this method of estimating the power law relationship between ζ_* and k is somewhat questionable since the errors in the δ_2 estimate are not independent. ⊣

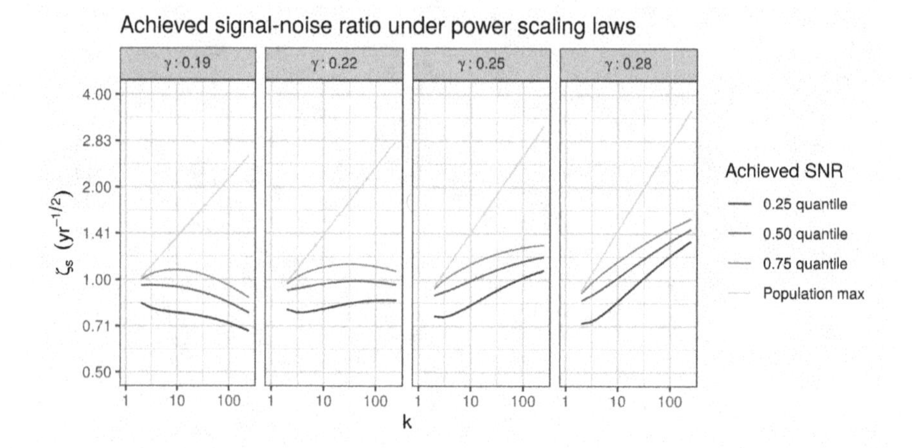

FIGURE 8.3
Some approximate quantiles of the achieved signal-noise ratio of the Markowitz portfolio are plotted versus k for different scaling laws for ζ_*. The facets represent different values of γ. Quantiles are computed via Approximation 8.3.

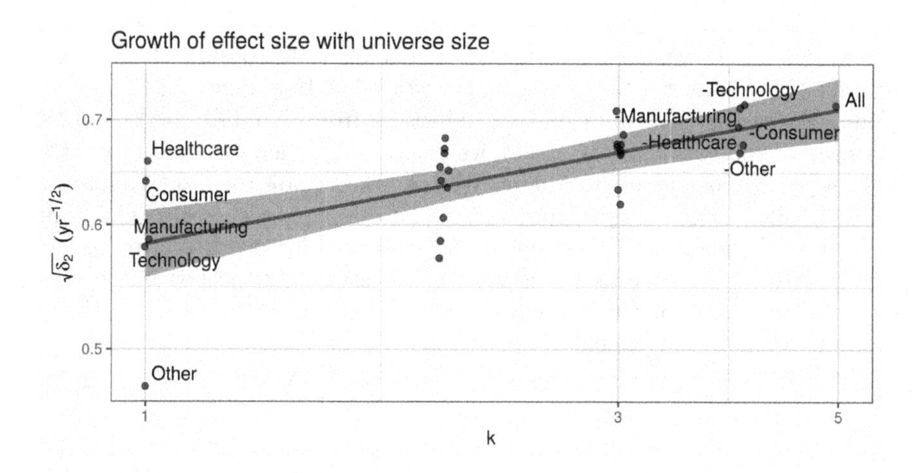

FIGURE 8.4
We plot the square root of the KRS δ_2 estimator of ζ_*^2 against k for the industry portfolios data. The data are in log-log space, and are jittered slightly in the x direction to better show overlapping points. The best fit line has slope 0.12. For $k = 4$ we indicate which industry is *omitted*.

8.2.2 † Choosing a stopped clock

Inequality 8.5 is a fairly harsh bound for the size of ζ_* one expects to encounter in the real world. As an alternative, one could select a "stopped clock" portfolio estimator, one which is biased and so is not subject to Inequality 8.5. However, such a stopped clock may have poor worst-case performance, or may be otherwise incompatible with one's prior beliefs.

This is a topic perhaps best studied from the viewpoint of *statistical decision theory.* [Berger, 1985] For a given portfolio function $\hat{\boldsymbol{\nu}}(\mathsf{X})$, and population parameters $\boldsymbol{\mu}$ and Σ, define the loss function as the negative signal-noise ratio of the portfolio

$$L(\boldsymbol{\mu}, \Sigma, \hat{\boldsymbol{\nu}}(\mathsf{X})) := -q(\hat{\boldsymbol{\nu}}(\mathsf{X})) = -\frac{\boldsymbol{\mu}^{\top}\hat{\boldsymbol{\nu}}(\mathsf{X})}{\sqrt{\hat{\boldsymbol{\nu}}(\mathsf{X})^{\top}\Sigma\hat{\boldsymbol{\nu}}(\mathsf{X})}}. \tag{8.13}$$

One seeks to minimize loss.

The Frequentist risk is the expected loss under realizations of the data. That is

$$R(\boldsymbol{\mu}, \Sigma, \hat{\boldsymbol{\nu}}(\mathsf{X})) := \mathrm{E}\left[L(\boldsymbol{\mu}, \Sigma, \hat{\boldsymbol{\nu}}(\mathsf{X}))\right], \tag{8.14}$$

where expectation is under realizations of X. (In this section, we use the term "risk" in the sense of statistical decision theory, where one risks using the wrong decision rule, not in the sense of volatility of returns.) We wish to measure a portfolio function $\hat{\boldsymbol{\nu}}(\mathsf{X})$ by the highest Frequentist risk it achieves over some set of $\boldsymbol{\mu}$ and Σ, which we call its worst case performance. A portfolio function is then called *minimax* if it has the smallest worst case performance over some set of acceptable portfolio functions. Taking into account our sign flip, this "minimax" portfolio function has the highest worst expected signal-noise ratio. Being minimax is relative to the acceptable portfolio functions and the set of $\boldsymbol{\mu}$ and Σ.

It can be difficult to prove that a given decision rule is indeed the minimax rule; one can sometimes instead prove that the risk of the minimax rule has a certain scaling with sample size, or is bounded from below. [Guntuboyina, 2011] It is also sometimes possible to show that a given decision rule has a maximum higher risk than another rule and thus is not a minimax rule.

Towards that end, we compute the approximate Frequentist risk of the Markowitz portfolio function, which is the negative expected signal-noise ratio. First we consider the Taylor expansion:

$$q(\boldsymbol{\nu} + \boldsymbol{\epsilon}) = q(\boldsymbol{\nu}) + \boldsymbol{\epsilon}^{\top}\nabla_{\boldsymbol{\nu}}q(\boldsymbol{\nu}) + \frac{1}{2}\boldsymbol{\epsilon}^{\top}H_{\boldsymbol{\nu}}q(\boldsymbol{\nu})\boldsymbol{\epsilon} + \ldots,$$

where the gradient and Hessian are evaluated at $\boldsymbol{\nu}$. Now suppose that $\hat{\boldsymbol{\nu}} = \boldsymbol{\nu} + \boldsymbol{\epsilon}$ is random, and $\mathrm{E}[\boldsymbol{\epsilon}] = \mathbf{0}$, which is to say that $\boldsymbol{\nu}$ is the expected value of $\hat{\boldsymbol{\nu}}$. We then have

$$\mathrm{E}\left[q(\hat{\boldsymbol{\nu}})\right] \approx q(\boldsymbol{\nu}) + \frac{1}{2}\operatorname{tr}\left(H_{\boldsymbol{\nu}}q(\boldsymbol{\nu})\operatorname{Var}(\boldsymbol{\epsilon})\right),$$

where the Hessian is evaluated at $\boldsymbol{\nu}$.

Now consider the case where $\hat{\boldsymbol{\nu}} = \hat{\boldsymbol{\nu}}_* = \hat{\Sigma}^{-1}\hat{\boldsymbol{\mu}}$ is the sample Markowitz portfolio. Equations 6.35 and 6.36 give the expectation and covariance terms. In particular we note that $\boldsymbol{\nu} = c\Sigma^{-1}\boldsymbol{\mu}$ for some $c > 0$. Then the Hessian is given by Equation 5.58. Putting them together we have

$$\mathrm{E}\left[q\left(\hat{\boldsymbol{\nu}}_*\right)\right] \approx \zeta_* - \frac{1}{2}\frac{(n-k-2)(k-1)\left(1+\zeta_*^2-\frac{2}{n}\right)}{\zeta_*(n-k-1)(n-k-4)}. \tag{8.15}$$

As this approximation truncates higher order terms, it is only accurate when $n \gg k$. In fact, this approximation is negative when $\zeta_*^2 n/k$ is insufficiently large, while the expected value of $q\left(\hat{\boldsymbol{\nu}}_*\right)$ is non-negative.

In comparison, consider the expected signal-noise ratio of the "one over k," or equal dollar, portfolio. As this portfolio is deterministic, this is simple. If we allow elements of $\boldsymbol{\mu}$ to be negative, the worst case expected value is $-\zeta_*$. If all elements of $\boldsymbol{\mu}$ must be non-negative, the worst case expected value is $\zeta_*/\sqrt{k}$. Thus for sufficiently large n/k and reasonable ζ_*, the Markowitz portfolio has lower risk than the equal dollar portfolio. It should be stressed that the "one over k" portfolio was a strawman introduced for rhetorical purposes, rather than an actionable investment strategy.

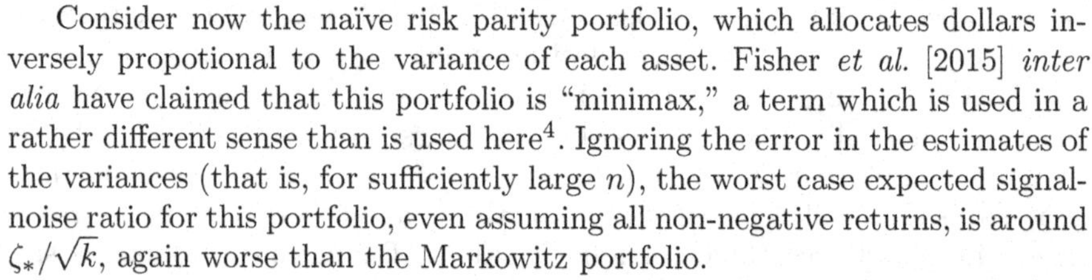

Consider now the naïve risk parity portfolio, which allocates dollars inversely propotional to the variance of each asset. Fisher *et al.* [2015] *inter alia* have claimed that this portfolio is "minimax," a term which is used in a rather different sense than is used here[4]. Ignoring the error in the estimates of the variances (that is, for sufficiently large n), the worst case expected signal-noise ratio for this portfolio, even assuming all non-negative returns, is around $\zeta_*/\sqrt{k}$, again worse than the Markowitz portfolio.

In summary, we have little evidence to reject using the sample Markowitz portfolio when $n \gg k$. The more interesting case when $k > n$ remains an unsolved problem. It is perhaps the case that, depending on one's priors, some stopped clock portfolio yields higher expected achieved signal-noise ratio than $\hat{\boldsymbol{\nu}}_*$. For example, one might have a strong prior conviction that $\boldsymbol{\nu}_*$ has only a few non-zero elements. A portfolio construction techniques that similarly selects only a few non-zero elements in the sample portfolio, perhaps depending on some statistical tests, might have higher expected achieved signal-noise ratio than the sample Markowitz portfolio. However, this prior would seem to be in conflict with the fundamental law of active management, which claims axiomatically that ζ_* grows as $\sqrt{k}$.

Example 8.2.2 (Expected signal-noise ratio, Markowitz portfolio). We simulate returns for $k = 4$ assets over $n = 2,016$ days. We set $\zeta_* = 0.076\,\mathrm{day}^{-\frac{1}{2}}$. We compute $\hat{\boldsymbol{\nu}}_*$ and compute its signal-noise ratio. We repeat this 100,000 times. The mean signal-noise ratio over the simulations is $0.066\,\mathrm{day}^{-\frac{1}{2}}$, the approximate theoretical value from Equation 8.15 is $0.066\,\mathrm{day}^{-\frac{1}{2}}$, and the limiting

[4]In their formulation the asset variances are known, but $\boldsymbol{\mu}$ is completely unknown. Moreover, there is no sample from which one can estimate $\boldsymbol{\mu}$

value from Equation 8.6 is $0.065\,\mathrm{day}^{-\frac{1}{2}}$. The approximation of Equation 8.4 yields an approximate expected value of $0.065\,\mathrm{day}^{-\frac{1}{2}}$. These are all reasonably close. The upper bound from Inequality 8.5 in this case is $0.067\,\mathrm{day}^{-\frac{1}{2}}$. ⊣

8.3 Inference on achieved signal-noise ratio

In Chapter 6, we considered inference on the unknown quantity ζ_*^2. Certainly if ζ_*^2 is small, one should not be terribly interested in investing in the assets. However, now we see that even if ζ_*^2 is large, the haircut effect can cause one to have a small achieved signal-noise ratio. It seems much more important to perform inference on the achieved signal-noise ratio, $q\left(\hat{\boldsymbol{\nu}}_*\right)$. This does not fit in well with the classical statistical paradigm wherein one is performing inference on a fixed unknown population quantity; instead we are now seeking to perform inference on some combination of the unknown population quantities and the observed Markowitz portfolio.

Because this problem is so unusual there are few known results. The main result is the *Sharpe Ratio Information Coefficient* of Paulsen and Söhl [2016] defined as

$$SRIC := \hat{\zeta}_* - \frac{k-1}{n\hat{\zeta}_*}. \tag{8.16}$$

For the case of Gaussian *i.i.d.* returns, and under the assumption $\hat{\Sigma} = \Sigma$, the SRIC is an unbiased estimate of the achieved signal-noise ratio:

$$\mathrm{E}\left[SRIC\right] = \mathrm{E}\left[q\left(\hat{\boldsymbol{\nu}}_*\right)\right]. \tag{8.17}$$

This equation can be viewed as stating that two random variables, one observable, and one of interest, happen to have the same expected value. They both happen to be estimators of some unknown population parameter which is of no interest for our purposes.

We note that Equation 8.16 holds only for $k > 1$. For the $k = 1$ case, again under the approximation $\hat{\Sigma} = \Sigma$, we have

$$\begin{aligned} &\mathrm{E}\left[SRIC\right] = \zeta_*, \quad \text{but} \\ &\mathrm{E}\left[q\left(\hat{\boldsymbol{\nu}}_*\right)\right] = \zeta_*\left(2\Phi\left(\sqrt{n}\zeta_*\right) - 1\right), \end{aligned} \tag{8.18}$$

where here $\Phi\left(x\right)$ is the cumulative distribution of the standard normal.

Example 8.3.1 (Fama French Four Factors, SRIC). We consider the returns of the Market, SMB, HML, and UMD portfolios introduced in Example 1.1.1. Using 1,128 months of monthly data from Jan 1927 to Dec 2020, we compute $\hat{\zeta}_* = 0.308\,\mathrm{mo.}^{-1/2}$. From this we compute $SRIC = 0.299\,\mathrm{mo.}^{-1/2}$. ⊣

Distribution of SRIC

In Example 7.4.1 we found that $\hat{\zeta}_*$ is nearly normal in distribution, but not unbiased for ζ_*. Via Approximation 6.59 and the multivariate delta method we claim that

$$\sqrt{n}\left(SRIC - q\left(\hat{\boldsymbol{\nu}}_*\right)\right) \rightsquigarrow \mathcal{N}\left(0, 1 + \frac{1}{2}\zeta_*^2 + \frac{k}{2n\zeta_*^2}\right). \tag{8.19}$$

Note that the variance in Equation 8.19 is looks like the approximate variance of the Sharpe ratio (*cf.* Equation 2.24), but with an extra term in k. However, the distributional "shape" of $SRIC$ is arguably less normal than that of ζ_*, as shown in the following example.

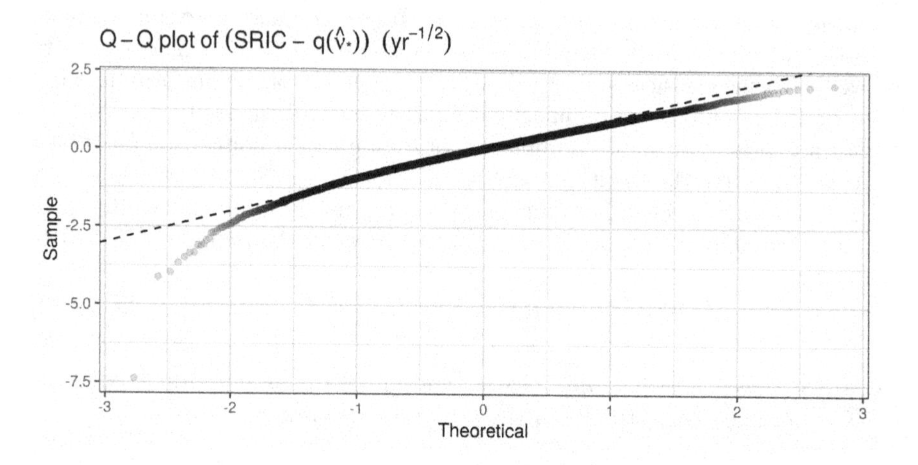

FIGURE 8.5
Q-Q plot of $SRIC - q\left(\hat{\boldsymbol{\nu}}_*\right)$ are shown over 10,000 simulations of *i.i.d.* Gaussian data is given against a theoretical Gaussian law. The variance is implied by Equation 8.19. We also plot the $y = x$ line.

Example 8.3.2 (Asymptotic non-Normality, SRIC). We draw 10,000 simulations of $n = 1,000$ days of data on $k = 8$ assets. Returns are *i.i.d.* Gaussian with $\zeta_* = 1\ \mathrm{yr}^{-1/2}$, assuming 252 days per year. We compute $SRIC$ and $q\left(\hat{\boldsymbol{\nu}}_*\right)$, then Q-Q plot them against a normal law with variance given by Equation 8.19 in Figure 8.5. The empirical standard error of $SRIC - q\left(\hat{\boldsymbol{\nu}}_*\right)$ is around $0.667\ \mathrm{yr}^{-1/2}$, while the theoretical value is $0.712\ \mathrm{yr}^{-1/2}$. The empirical differences are somewhat normal in shape, but with some potentially large deviations in the tails. ⊣

8.3.1 Confidence intervals

The SRIC gives us the expected achieved signal-noise ratio of the Markowitz portfolio. For purposes of inference we seek a confidence interval on $q\left(\hat{\boldsymbol{\nu}}_*\right)$.

A number of confidence intervals based on the asymptotic distribution of Θ have been derived, but these are of limited practical use since they require unreasonably large n/k. [Pav, 2013] A simple confidence bound similar to the SRIC can be derived under the approximation $\hat{\Sigma} \approx \Sigma$, namely [Pav, 2020b]

$$\Pr\left\{\hat{\zeta}_* - \frac{{\chi^2}_{1-\alpha}\left(k, \frac{n\zeta_*^2}{4}\right) - \frac{n\zeta_*^2}{4}}{n\hat{\zeta}_*} \geq q\left(\hat{\boldsymbol{\nu}}_*\right)\right\} = \alpha. \tag{8.20}$$

Here ${\chi^2}_{\alpha}\left(v, \delta\right)$ is the α quantile of the non-central chi-square distribution with v degrees of freedom and non-centrality parameter δ.

Example 8.3.3 (CI Coverage). We test the confidence bound of Equation 8.20 along with three asymptotic bounds of dubious quality. [Pav, 2013] We draw returns from the multivariate normal distribution. We fix the number of days of data, n, the number of assets, k, and the optimal signal-noise ratio, ζ_*, and perform 100,000 simulations. We then let n vary from 50 to 102,400 days; we let k vary from 2 to 16; we let ζ_* vary from 0.5 $\mathrm{yr}^{-1/2}$ to 2 $\mathrm{yr}^{-1/2}$, where we assume 252 days per year. We compute the lower confidence limits on $q\left(\hat{\boldsymbol{\nu}}_*\right)$ *very* optimistically, by plugging in the *actual* ζ_* in the computations when needed. For practical inference, ζ_* would need to be estimated; here we are testing the quality of the confidence bounds with this "clairvoyance" to check the effects of the underlying approximations. For the bound of Equation 8.20, this means were are testing the effects of the approximation $\hat{\Sigma} \approx \Sigma$.

In Figure 8.6 we plot the empirical type I rate at the nominal 0.05 level of the four confidence bounds. The "difference" bound takes the form $\hat{\zeta}_* - c$, while the "product" form is $c\hat{\zeta}_*$. The "TAS" bound is built on the $f_{\text{tas}}\left(x\right)$ function of Equation 1.24. The "SRIC" bound is the one given in Equation 8.20. We restrict the y-axis to an area around the nominal rate to show detail, which causes some of the lines to entirely fall off the plot.

The main takeaways from this experiment are that the SRIC bound performs much better than the asymptotic bounds, giving near-nominal coverage for much of the parameter space considered here, and that the approximation required for the SRIC bound does not seem to cause issues except when $n \leq 100k$ or so. ⊣

Unfortunately the constant $c = {\chi^2}_{1-\alpha}\left(k, \frac{n\zeta_*^2}{4}\right) - \frac{n\zeta_*^2}{4}$ depends on the unknown ζ_*. Moreover, for α small, like 0.05, the constant c seems to be increasing with ζ_*. Thus one cannot assume $\zeta_* = 0$ and construct conservative confidence bounds. Instead it seems one should construct simultaneous confidence bounds on ζ_* and $q\left(\hat{\boldsymbol{\nu}}_*\right)$. To do this, for some $q \in \left[0, 1\right)$ find an upper $q\alpha$ confidence bound on ζ_*, call it $\zeta_{*,u}$, then plug this in to construct the constant c to build a $\left(1-q\right)\alpha$ bound. This approach is not guaranteed to give the nominal coverage to $q\left(\hat{\boldsymbol{\nu}}_*\right)$, since it provides *simultaneous* bounds. In practice, it tends to be somewhat conservative. An alternative approach, also not guaranteed to give nominal coverage, is to plug in δ_2 as an estimate of ζ_*^2 into the computation of the confidence bound, where δ_2 is one of the

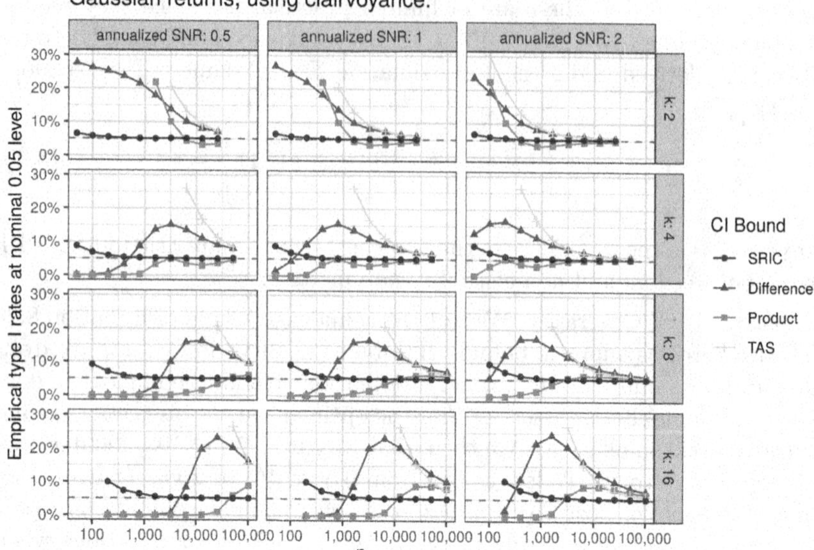

FIGURE 8.6
The empirical type I rate, over 100,000 simulations, of four one-sided confidence bounds for $q(\hat{\boldsymbol{\nu}}_*)$ are shown for a nominal type I rate of 0.05. The daily returns are drawn from multivariate normal distribution with varying ζ_*, n, and k.

KRS estimators of Equation 6.67. This approach also tends to be somewhat conservative, but gives closer to nominal coverage, except when n/k is not large enough.

To be concrete, to find simultaneous confidence intervals on ζ_* and $q(\hat{\boldsymbol{\nu}}_*)$ with coverage $1-\alpha$, first compute $[\zeta_l, \zeta_u)$ interval on ζ_* with coverage $1-q\alpha$ for some $q \in (0,1)$, as in Section 6.3.1. Then for some $s \in (0,1)$ compute

$$c_l = \min\left\{ {\chi^2}_{1-s(1-q)\alpha}\left(k, \frac{n\zeta^2}{4}\right) - \frac{n\zeta^2}{4} \middle| \ \zeta_l \le \zeta \le \zeta_u \right\},$$
$$c_u = \min\left\{ {\chi^2}_{1-(1-s)(1-q)\alpha}\left(k, \frac{n\zeta^2}{4}\right) - \frac{n\zeta^2}{4} \middle| \ \zeta_l \le \zeta \le \zeta_u \right\},$$

then compute interval

$$\left(\hat{\zeta}_* - \frac{c_l}{\hat{\zeta}_*}, \hat{\zeta}_* - \frac{c_u}{\hat{\zeta}_*}\right)$$

for $q(\hat{\boldsymbol{\nu}}_*)$.

For approximate an confidence interval only on $q(\hat{\boldsymbol{\nu}}_*)$, compute the KRS estimator δ_2 of Equation 6.67, then compute

$$c_l = {\chi^2}_{1-s\alpha}\left(k, \frac{n\delta_2}{4}\right) - \frac{n\delta_2}{4},$$
$$c_u = {\chi^2}_{1-(1-s)\alpha}\left(k, \frac{n\delta_2}{4}\right) - \frac{n\delta_2}{4},$$

for $s \in (0, 1)$, then compute interval

$$\left(\hat{\zeta}_* - \frac{c_l}{\hat{\zeta}_*}, \hat{\zeta}_* - \frac{c_u}{\hat{\zeta}_*}\right)$$

for $q\left(\hat{\boldsymbol{\nu}}_*\right)$. This interval will have approximate coverage of $1-\alpha$, but is typically somewhat larger.

The typical use case for confidence intervals is that one wishes to falsify the hypothesis that $q\left(\hat{\boldsymbol{\nu}}_*\right)$ is small. Thus, one typically constructs a one-sided confidence interval. Paradoxically, for the simultaneous bounds this is typically achieved by building a one-sided interval on ζ_* of the form $[0, \zeta_u)$. Again this is because ${\chi^2}_{1-\alpha}\left(k, \frac{n\zeta_*^2}{4}\right) - \frac{n\zeta_*^2}{4}$ appears to grow in ζ_* for small α.

Example 8.3.4 (Feasible CI Coverage). We reconsider the experiments of Example 8.3.3, but compute feasible confidence bounds. We use both the simultaneous CI approach with $q = 0.25$; and plug in δ_2 to construct the bound. In Figure 8.7, we plot the empirical type I rate for both of these bounds versus n, with facets for ζ_* and k. We see that the δ_2 plug-in estimator has coverage closer to the nominal 0.05 rate. Both bounds have issues when n/k is not sufficiently large, a problem stemming from the poor quality of the approximation $\hat{\Sigma} \approx \hat{\hat{\Sigma}}$, and which was seen in the previous example. ⊣

Example 8.3.5 (Fama French Four Factors, Confidence Intervals). We consider a portfolio on the returns of the Market, SMB, HML, and UMD portfolios introduced in Example 1.1.1. We compute $\delta_2 = 0.091\text{ mo.}^{-1}$. Plugging in $\sqrt{\delta_2}$ for ζ_* in Equation 8.20, we compute a two-sided 95% confidence bound on $q\left(\hat{\boldsymbol{\nu}}_*\right)$ as $[0.23, 0.347)\text{ mo.}^{-1/2}$. Note that this is slightly lower than the confidence interval on ζ_* we found for these returns in Example 7.4.8 and Example 6.3.1. ⊣

Constrained problems

The confidence bounds can easily be translated into the subspace-constrained problem by replacing k with the number of "latent" assets in the subspace. The hedge constraint is slightly more involved. Consider the portfolio problem of Equation 6.9. Recall that the sample optimal portfolio is given by

$$\hat{\boldsymbol{\nu}}_{*,\mathsf{I}\backslash\mathsf{G}} = c\left(\hat{\Sigma}^{-1}\hat{\boldsymbol{\mu}} - \mathsf{G}^{\top}\left(\mathsf{G}\hat{\Sigma}\mathsf{G}^{\top}\right)^{-1}\mathsf{G}\hat{\boldsymbol{\mu}}\right).$$

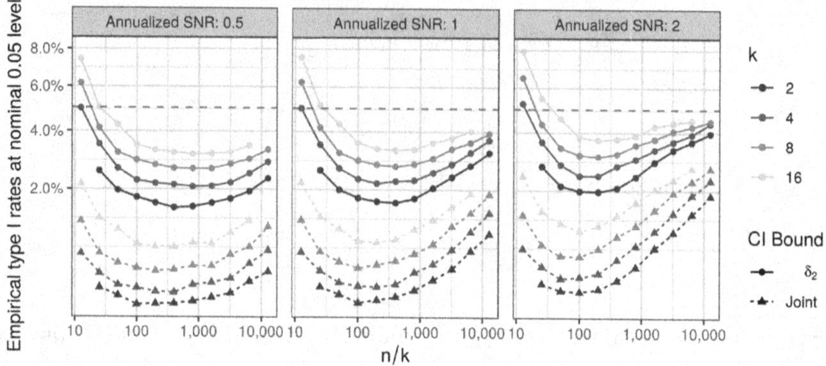

FIGURE 8.7
The empirical type I rate, over 100,000 simulations, of two feasible one-sided confidence bounds for $q(\hat{\boldsymbol{\nu}}_*)$ are shown for a nominal type I rate of 0.05. The daily returns are drawn from multivariate normal distribution with varying ζ_*, n, and p. The y-axis is drawn in square root scale to show detail.

The achieved signal-noise ratio of this portfolio is

$$q\left(\hat{\boldsymbol{\nu}}_{*,\mathsf{I}\setminus\mathsf{G}}\right) = \frac{\boldsymbol{\mu}^{\top}\hat{\boldsymbol{\nu}}_{*,\mathsf{I}\setminus\mathsf{G}}}{\sqrt{\hat{\boldsymbol{\nu}}_{*,\mathsf{I}\setminus\mathsf{G}}{}^{\top}\boldsymbol{\Sigma}\hat{\boldsymbol{\nu}}_{*,\mathsf{I}\setminus\mathsf{G}}}}. \tag{8.21}$$

Then, as in the unhedged case, we have

$$\Pr\left\{q\left(\hat{\boldsymbol{\nu}}_{*,\mathsf{I}\setminus\mathsf{G}}\right) \leq \sqrt{\Delta_{\mathsf{I}\setminus\mathsf{G}}\hat{\zeta}_*^2} - \frac{\chi^2{}_{1-\alpha}\left(k - k_g, \frac{n\Delta_{\mathsf{I}\setminus\mathsf{G}}\zeta_*^2}{4}\right) - \frac{n\Delta_{\mathsf{I}\setminus\mathsf{G}}\zeta_*^2}{4}}{n\sqrt{\Delta_{\mathsf{I}\setminus\mathsf{G}}\hat{\zeta}_*^2}}\right\} = \alpha. \tag{8.22}$$

To perform feasible inference one will need to estimate $\Delta_{\mathsf{I}\setminus\mathsf{G}}\,\zeta_*^2$. Again there are two ways to do this: either via a joint confidence intervals on $q\left(\hat{\boldsymbol{\nu}}_{*,\mathsf{I}\setminus\mathsf{G}}\right)$ and $\Delta_{\mathsf{I}\setminus\mathsf{G}}\,\zeta_*^2$, essentially using Equation 6.29; or by using the KRS-type estimators of Equation 6.68. So, as above, to find simultaneous confidence intervals on $\Delta_{\mathsf{I}\setminus\mathsf{G}}\,\zeta_*^2$ and $q\left(\hat{\boldsymbol{\nu}}_{*,\mathsf{I}\setminus\mathsf{G}}\right)$ with coverage $1-\alpha$, first compute $[\zeta_l, \zeta_u)$ interval on $\sqrt{\Delta_{\mathsf{I}\setminus\mathsf{G}}\zeta_*^2}$ with coverage $1 - q\alpha$ for some $q \in (0,1)$, as in Section 6.3.1. Then for some $s \in (0,1)$ compute

$$c_l = \min\left\{\chi^2{}_{1-s(1-q)\alpha}\left(k - k_g, \frac{n\zeta^2}{4}\right) - \frac{n\zeta^2}{4}\,\middle|\, \zeta_l \leq \zeta \leq \zeta_u\right\},$$
$$c_u = \min\left\{\chi^2{}_{1-(1-s)(1-q)\alpha}\left(k - k_g, \frac{n\zeta^2}{4}\right) - \frac{n\zeta^2}{4}\,\middle|\, \zeta_l \leq \zeta \leq \zeta_u\right\},$$

then compute interval

$$\left(\hat{\zeta}_* - \frac{c_l}{\hat{\zeta}_*}, \hat{\zeta}_* - \frac{c_h}{\hat{\zeta}_*}\right)$$

for $q\left(\hat{\boldsymbol{\nu}}_{*,\mathsf{I}\backslash\mathsf{G}}\right)$.

For an approximate confidence interval only on $q\left(\hat{\boldsymbol{\nu}}_{*,\mathsf{I}\backslash\mathsf{G}}\right)$, compute the KRS estimator δ_2 of Equation 6.68, then compute

$$c_l = {\chi^2}_{1-s\alpha}\left(k - k_g, \frac{n\delta_2}{4}\right) - \frac{n\delta_2}{4},$$
$$c_u = {\chi^2}_{1-(1-s)\alpha}\left(k - k_g, \frac{n\delta_2}{4}\right) - \frac{n\delta_2}{4},$$

for $s \in (0, 1)$, then compute interval

$$\left(\hat{\zeta}_* - \frac{c_l}{\hat{\zeta}_*}, \hat{\zeta}_* - \frac{c_h}{\hat{\zeta}_*}\right)$$

for $q\left(\hat{\boldsymbol{\nu}}_{*,\mathsf{I}\backslash\mathsf{G}}\right)$. Again this interval only has approximate coverage, and tends to be conservative.

Example 8.3.6 (Feasible CI Coverage, hedged portfolios). As in the unhedged case, we first perform simulations where the population parameter $\Delta_{\mathsf{I}\backslash\mathsf{G}}\ \zeta_*^2$ is known, to assess the effects of the approximation $\hat{\Sigma} \approx \Sigma$. In our simulations, we set $k_g = k/2$, and let G be the first k_g rows of the identity matrix. We set $\boldsymbol{\mu} = c\mathbf{1}$ and $\Sigma = \mathsf{I}$. We perform 100,000 simulations for different values of $\Delta_{\mathsf{I}\backslash\mathsf{G}}\zeta_*^2$, k and n, computing $q\left(\hat{\boldsymbol{\nu}}_{*,\mathsf{I}\backslash\mathsf{G}}\right)$ for the hedged portfolio, as well as $\Delta_{\mathsf{I}\backslash\mathsf{G}}\ \hat{\zeta}_*^2$ and $\hat{\zeta}_{*,\mathsf{G}}^2$. We compute the lower 0.05 bound using knowledge of $\Delta_{\mathsf{I}\backslash\mathsf{G}}\ \zeta_*^2$ and compute the empirical type I rate over the 100,000 simulations, which we plot versus n/k in Figure 8.8. Again we see that the nominal type I rate is nearly achieved when $n > 100k$ or so.

As above we analyze the data from the hedged experiments, but compute feasible confidence bounds. We use both the simultaneous CI approach with $q = 0.25$; and plug in $\Delta_{\mathsf{I}\backslash\mathsf{G}}\zeta_*^2 = \sqrt{\delta_2}$ to construct the bound. In Figure 8.9, we plot the empirical type I rate for both of these bounds versus n, with facets for ζ_* and k. Once again, the δ_2 plug-in estimator has coverage closer to the nominal 0.05 rate, and both bounds are anti-conservative when n/k is not sufficiently large. ⊣

Example 8.3.7 (Fama French Four Factors, confidence intervals hedged portfolio). We continue Example 6.3.2, where we found a one-sided confidence interval on $\Delta_{\mathsf{I}\backslash\mathsf{G}}\ \zeta_*^2$ for the problem of finding an optimal portfolio on the Fama French four factors with a hedge against the Market. In Example 6.2.6, we found the KRS estimator in this case to be $\delta_2 = 0.06\,\text{mo.}^{-1}$. From this we compute a one sided 95% confidence bound on $q\left(\hat{\boldsymbol{\nu}}_{*,\mathsf{I}\backslash\mathsf{G}}\right)$ to be $[0.186, \infty)\ \text{mo.}^{-1/2}$. ⊣

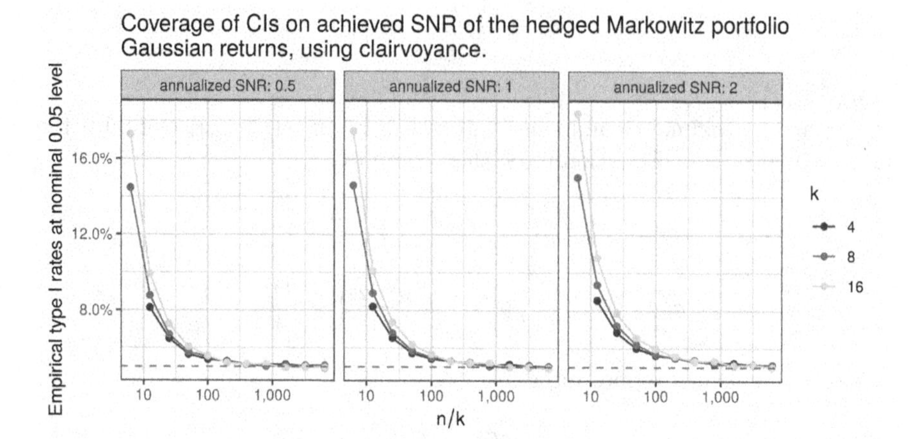

FIGURE 8.8
The empirical type I rate, over 100,000 simulations, of a one-sided confidence bound for $q\left(\hat{\boldsymbol{\nu}}_{*,\mathsf{I}\setminus\mathsf{G}}\right)$ are shown for the hedged portfolio problem. We set $k_g = k/2$ and take $\boldsymbol{\mu} \propto \mathbf{1}$. The SNR in the facet titles refers to $\sqrt{\Delta_{\mathsf{I}\setminus\mathsf{G}}\zeta_*^2}$.

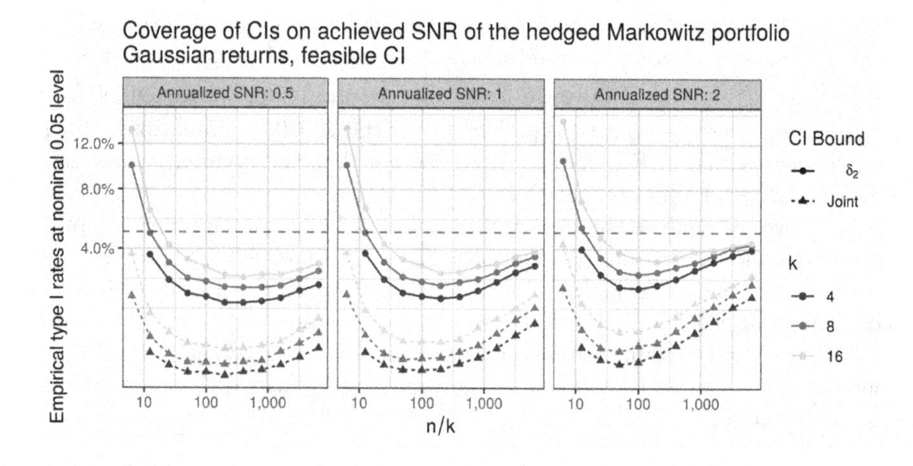

FIGURE 8.9
The empirical type I rate, over 100,000 simulations, of two feasible one-sided confidence bounds for $q\left(\hat{\boldsymbol{\nu}}_{*,\mathsf{I}\setminus\mathsf{G}}\right)$ are shown for a nominal type I rate of 0.05. The daily returns are drawn from multivariate normal distribution with varying ζ_*, n, and p. The y-axis is drawn in square root scale to show detail.

8.3.2 Hypothesis testing

As usual, the confidence intervals on the achieved signal-noise ratio can be re-expressed as hypothesis tests.

one-sided test on achieved signal-noise ratio

To test

$$H_0 : q\left(\hat{\boldsymbol{\nu}}_*\right) = \zeta_0 \quad \text{versus} \quad H_1 : q\left(\hat{\boldsymbol{\nu}}_*\right) \geq \zeta_0,$$

reject approximately at the α level if

$$\hat{\zeta}_* - \frac{{\chi^2}_{1-\alpha}\left(k, \frac{n\delta_2}{4}\right) - \frac{n\delta_2}{4}}{\hat{\zeta}_*} \geq \zeta_0,$$

Here ${\chi^2}_{\alpha}\left(v, \delta\right)$ is the α quantile of the non-central chi-square distribution with v degrees of freedom and non-centrality parameter δ, and δ_2 is computed as in Equation 6.67.

one-sided test on hedged achieved signal-noise ratio

To test

$$H_0 : q\left(\hat{\boldsymbol{\nu}}_{*,\mathsf{I}\backslash\mathsf{G}}\right) = \zeta_0 \quad \text{versus} \quad H_1 : q\left(\hat{\boldsymbol{\nu}}_{*,\mathsf{I}\backslash\mathsf{G}}\right) \geq \zeta_0,$$

reject approximately at the α level if

$$\sqrt{\Delta_{\mathsf{I}\backslash\mathsf{G}}\hat{\zeta}_*^2} - \frac{{\chi^2}_{1-\alpha}\left(k - k_g, \frac{n\delta_2}{4}\right) - \frac{n\delta_2}{4}}{\sqrt{\Delta_{\mathsf{I}\backslash\mathsf{G}}\hat{\zeta}_*^2}} \geq \zeta_0,$$

Here ${\chi^2}_{\alpha}\left(v, \delta\right)$ is the α quantile of the non-central chi-square distribution with v degrees of freedom and non-centrality parameter δ, and δ_2 is computed as in Equation 6.68.

Example 8.3.8 (Fama French Four Factors, Hypothesis Testing). As a companion to Example 8.3.5, we test the hypothesis $H_0 : q\left(\hat{\boldsymbol{\nu}}_*\right) = 0.2\ \text{mo.}^{-\frac{1}{2}}$. for the Fama French four factor data. Again we compute $\delta_2 = 0.091\ \text{mo.}^{-1}$, and a lower 95% confidence bound on $q\left(\hat{\boldsymbol{\nu}}_*\right)$ of $0.242\ \text{mo.}^{-1/2}$. As this is bigger than ζ_0 we reject at the approximately 0.05 level. ⊣

Example 8.3.9 (Fama French Four Factors, Hedged Hypothesis Testing). As dual to Example 8.3.7, we test the hypothesis $H_0 : q\left(\hat{\boldsymbol{\nu}}_{*,\mathsf{I}\backslash\mathsf{G}}\right) = 0.25\ \text{mo.}^{-\frac{1}{2}}$. for the Fama French four factor data with a hedge against the Market. Again we compute $\delta_2 = 0.06\ \text{mo.}^{-1}$, and a lower 95% confidence bound on $q\left(\hat{\boldsymbol{\nu}}_{*,\mathsf{I}\backslash\mathsf{G}}\right)$ of $0.186\ \text{mo.}^{-1/2}$. As this is smaller than ζ_0 we fail to reject at the approximately 0.05 level. ⊣

8.3.3 Strategy selection

The Sharpe Ratio Information Criterion was constructed to satisfy Equation 8.17, but it was also shown to be similar to the Akaike Information Criterion, though flipped in sign so that one prefers a higher $SRIC$ instead of a lower one. [Pawitan, 2001; Paulsen and Söhl, 2016] Because of the connection to the AIC, SRIC is also advertised as a means of selecting one among many strategies. Since for a fixed k, $SRIC$ is monotonically increasing in $\hat{\zeta}_*^2$, it is only useful in cases where one is selecting from among strategies with different k.

However, since the likelihood is known for the case of *i.i.d.* Gaussian returns, one could compute what appears to be the information criteria exactly under this assumption, without requiring the approximation $\hat{\Sigma} \approx \Sigma$. There is an error in the logic here, however, and we will present a putative AIC, and its cousin BIC, defined as

$$\text{fAIC} = 2k - 2 \log \mathcal{L}_{SR,*}\left(\zeta_{\text{MLE}}^2 \middle| \hat{\zeta}_*^2, n, k\right), \tag{8.23}$$

$$\text{fBIC} = k \log n - 2 \log \mathcal{L}_{SR,*}\left(\zeta_{\text{MLE}}^2 \middle| \hat{\zeta}_*^2, n, k\right). \tag{8.24}$$

Again, the "f" here is for "fake." From the density of $\hat{\zeta}_*^2$ given in Equation 6.46, we can express the log likelihood as

$$\log \mathcal{L}_{SR,*}\left(x \middle| \hat{\zeta}_*^2, n, k\right) = \log \frac{n\,(n-k)}{(n-1)\,k} + \log f_f\left(\frac{n}{n-1}\frac{n-k}{k}\hat{\zeta}_*^2; k, n-k, nx\right).$$

Here $f_f\left(x; \nu_1, \nu_2, \delta\right)$ is the density of the non-central F-distribution with non-centrality parameter δ and ν_1, ν_2 degrees of freedom. The MLE of ζ_*^2, ζ_{MLE}^2 has to be found numerically, but otherwise the computation is straightforward.

There is something quite odd about using these two (fake) information criteria for strategy selection. In Figure 8.10, we plot contours of the (negative) fAIC, (negative) fBIC, the SRIC and the δ_0 of Equation 6.67. The latter is included because, beyond some minor differences in the constants, $SRIC$ is approximately equal to $\delta_0/\hat{\zeta}_*$. While $SRIC$ and δ_0 clearly are increasing in $\hat{\zeta}_*$ and decreasing in k, the plot seems to indicate that fAIC can sometimes prefer smaller $\hat{\zeta}_*$ for a fixed k.

Despite the advertised use of AIC and BIC for "selection among models," they are actually prescribed for selection among models *on the same data.* Which is to say, given multiple competing models for the underlying process that generates the matrix of returns, X, one maximizes the likelihood under each of them, then selects the model with lowest information criterion. This is rather different from the advertised use of SRIC, which is a comparison of the inferred achieved signal-noise ratio of the Markowitz portfolio built on different, possibly disjoint, subsets of some universe of assets. For this purpose, the fAIC and fBIC are poorly equipped, and will often select subsets with very poor achieved signal-noise ratio. Do *not* use these fAIC and fBIC for model comparison. On the other hand, use of δ_0 as selection criterion yields

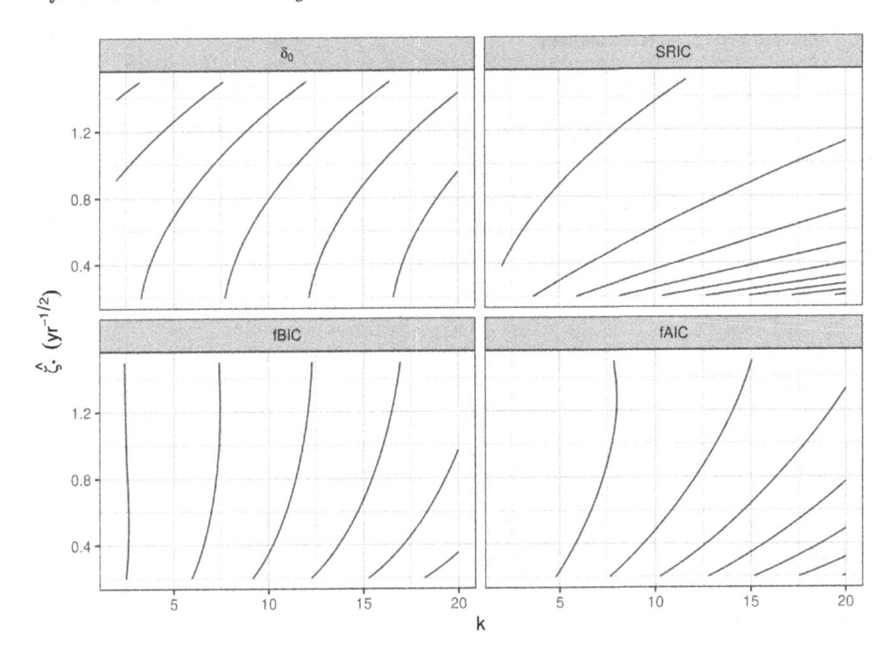

FIGURE 8.10
Contours of δ_0, SRIC, negative fBIC and negative fAIC are plotted versus k and $\hat{\zeta}_*$, based on sample size of 4 years of data at a rate of 252 days per year.

results very similar to using $SRIC$. The advantage to the former is that its distribution is well understood, and we can model the effects of overoptimism by selection.

Example 8.3.10 (Strategy selection simulations). We simulate the selection of strategies based on observed information, using either δ_0, $SRIC$ or the fAIC to select among strategies. We create a population where ζ_* follows a power law $\zeta_* = \zeta_0 k^{\gamma}$. For a fixed γ, we let k vary uniformly from 2 to 16, and let ζ_0 vary uniformly from 0 $\mathrm{yr}^{-1/2}$ to 0.6 $\mathrm{yr}^{-1/2}$. We then draw $n = 1,008$ days of *i.i.d.* Gaussian returns returns, and compute $\hat{\Sigma}^{-1}\hat{\boldsymbol{\mu}}$, $\hat{\zeta}_*^2$, and $q\left(\hat{\boldsymbol{\nu}}_*\right)$. We compute δ_0, $SRIC$ and the fAIC. We perform this experiment m times, then select the strategy with the highest selection metric and note its $q\left(\hat{\boldsymbol{\nu}}_*\right)$. Returns from different strategies tested are independent. We repeat this many times to find the empirical expected $q\left(\hat{\boldsymbol{\nu}}_*\right)$ as a function of m, the number of strategies considered. We let m vary.

We consider "high" and "low" values of γ as compared to $\frac{1}{4}$, which is the cutoff for diversification to contribute value. In Figure 8.11 we plot $q\left(\hat{\boldsymbol{\nu}}_*\right)$ versus m for the different selection methods. The empirical mean values of $q\left(\hat{\boldsymbol{\nu}}_*\right)$ are computed over 20,000 simulations. We generally find that selection via δ_0 and $SRIC$ succeed in finding better strategies as m increases, while fAIC selects worse strategies, and can not be recommended for this use. We see that

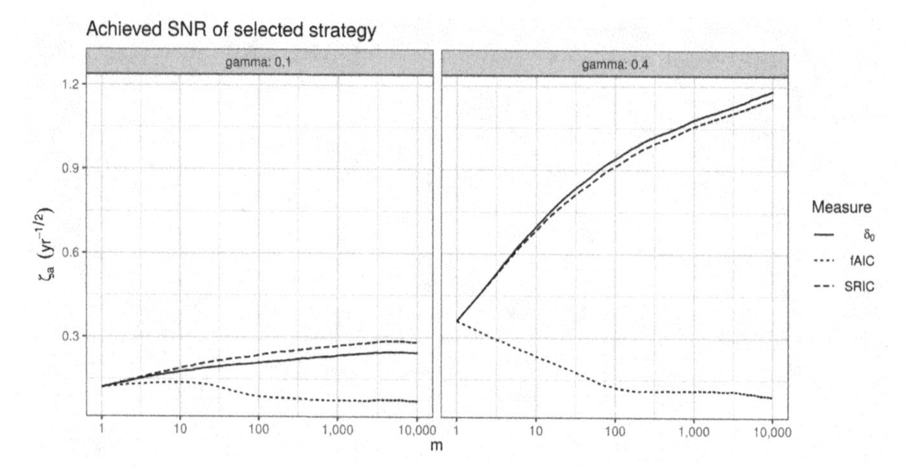

FIGURE 8.11
We plot the empirical expected value of $q(\hat{\boldsymbol{\nu}}_*)$ versus m, the number of strategies over which selection is performed. Empirical means are computed over 20,000 simulations. Selection via δ_0 and $SRIC$ give similar results, and greatly outperform fAIC.

δ_0 and $SRIC$ give similar performance, with $SRIC$ perhaps performing better for low γ, but the roles reversed for high γ. While not shown here, experiments with fBIC indicate it gives performance identical to fAIC. $\dashv$

8.3.4 † Cross validation

Among practicing quants, *cross validation* is a folk remedy prescribed to cure all ills, probably beneficial but of unknown efficacy. Cross validation consists of fitting a model (in our case a portfolio, say) with some of the data held aside. That data held aside is then used to estimate the performance of the portfolio. This procedure is then repeated over multiple such divisions, or *splits*, of the data into *training* and *validation* sets. A model is built on the training set of one split, then tested on the corresponding validation set. The performance on validation sets over multiple splits is then somehow averaged together.

Two kinds of cross validation are commonly used in testing strategies, namely *m-fold* cross validation, and *walk-forward* (or "rolling") cross validation. To illustrate these, assume the sample size n is a multiple of m. In m-fold cross validation, the validation set for the jth split consists of $\boldsymbol{x}_t$ for $(j-1)\,n/m < t \le jn/m$, and the training data consists of the rest of the data. In walk-forward cross validation, the training set for the jth split consists of $\boldsymbol{x}_t$ for $t \le (j-1)\,n/m$, and the validation set are $\boldsymbol{x}_t$ for $(j-1)\,n/m < t \le jn/m$. These are illustrated in Figure 8.12. We note that walk-forward cross validation actually loses a split due to availability of training data. Moreover in some splits the training data is rather short, and thus the trained models are perhaps much noisier than one fit on the entire data set.

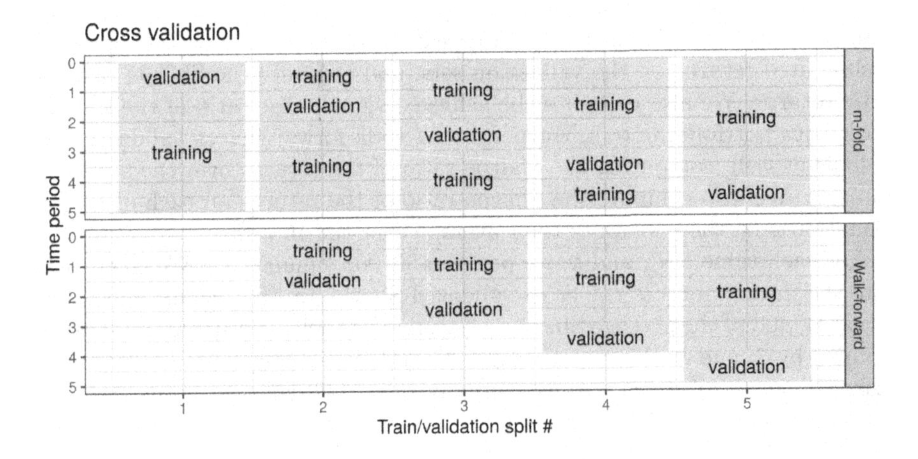

FIGURE 8.12
Two kinds of cross validation, m-fold and walk-forward are shown for the case of $m = 5$. A walk-forward cross validation takes into account the time ordering of the data, and only tests a model on validation data which comes after it in time.

For simplicity, consider the unconditional means case where we wish to estimate the the achieved signal-noise ratio of the portfolio $\hat{\boldsymbol{\nu}}_*$ fit on all of X, although this can easily be generalized to accomodate linear passthrough and more complex strategies. That sample Markowitz portfolio is constructed on training data then evaluated on the validation data, presumably by simulating the returns of the portfolio on that validation data. In common use, one would assemble the simulated returns on the validation data sets of the splits into one series, then compute the Sharpe ratio of these returns. It appears, however, that the Sharpe ratio computed in this way is biased, as shown in the following example. Moreoever, walk-forward cross validations has an apparently worse bias problem for larger ζ_*.

Instead, we suggest an average of Sharpe ratios approach, in which the Sharpe ratio is computed on the simulatated returns of each of the m validation sets, then these m values are averaged together. Empirically, this seems to be a less biased estimator of $q\left(\hat{\boldsymbol{\nu}}_*\right)$, although the presence of bias is probably affected by the sample size and number of folds.

We stress that this issue has not been completely described theoretically, and the general theory of cross validation is still area of active research. [Bates *et al.*, 2021; Bayle *et al.*, 2020; Austern and Zhou, 2020; Benkeser *et al.*, 2020] Moreover, the author is not aware of research specifically on the distribution of Sharpe ratio on cross validated returns as described above. We encourage the reader to check a particular cross validation scheme via simulations before any critical use.

Example 8.3.11 (Cross Validation). We perform 10-fold cross validation on $n = 1,260$ days. In our simulations we take $k = 5, 10$, and let ζ_* vary from $0.125\,\mathrm{yr}^{-1/2}$ to $1.875\,\mathrm{yr}^{-1/2}$. In each simulation we construct a series of simulated returns on the validation sets, and compute the Sharpe ratio of those returns. We also compute the achieved signal-noise ratio of the sample Markowitz portfolio, $q\left(\hat{\boldsymbol{\nu}}_*\right)$. We perform a walk-forward cross validation on 10 folds as well, computing the Sharpe ratio of the returns over the 9 validation sets. We also compute the Sharpe ratio of the returns on each of the 10 validation sets, then compute their average over the 10 sets.

For each value of ζ_* and k, we perform 50,000 simulations. We then compute the mean values of the cross validated Sharpe ratios (regular and walk-forward), and of the average Sharpe ratios, and the achieved signal-noise ratio, grouped by ζ_* and k. We view the three feasible measures as estimators of the achieved signal-noise ratio, so we compute their difference. In Figure 8.13 we plot these versus ζ_*. We see a persistent negative bias in the Sharpe ratio of the cross validated returns: in expectation they take smaller value than $q\left(\hat{\boldsymbol{\nu}}_*\right)$, which is, of course, smaller than ζ_*. The walk-forward measure is even worse for large ζ_*. ⊣

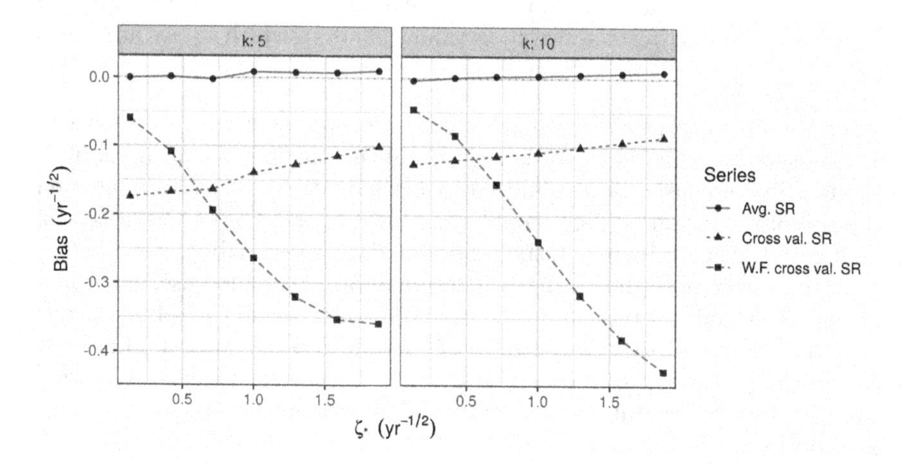

FIGURE 8.13
The bias of 10-fold cross validated Sharpe ratio is plotted versus ζ_*, along with the walk-forward Sharpe ratio and that of the average of fold Sharpe ratios. Bias is defined as the measure minus the achieved signal-noise ratio, $q\left(\hat{\boldsymbol{\nu}}_*\right)$.

8.4 Exercises

Ex. 8.1 Haircut of the Markowitz portfolio Consider the distribution of the haircut of the Markowitz portfolio under Gaussian returns

1. Show that the distribution of the haircut is invariant to rotations. That is, suppose that instead of observing returns $\boldsymbol{x}$, one had observed $\mathsf{P}^{\top}\boldsymbol{x}$ for invertible square matrix P.
2. If $\boldsymbol{x} \sim \mathcal{N}(\boldsymbol{\mu}, \Sigma)$, then $\mathsf{P}^{\top}\boldsymbol{x} \sim \mathcal{N}\left(\mathsf{P}^{\top}\boldsymbol{\mu}, \mathsf{P}^{\top}\Sigma\mathsf{P}\right)$. Conclude that, without loss of generality, when studying the haircut one can assume that $\boldsymbol{x} \sim \mathcal{N}(\zeta_{*}\boldsymbol{e}_1, \mathsf{I})$.
3. Assuming $\boldsymbol{x} \sim \mathcal{N}(\zeta_{*}\boldsymbol{e}_1, \mathsf{I})$, and under the approximation $\Sigma = \hat{\Sigma}$ show that
$$h = 1 - \frac{(\zeta_{*} + z_1)\,\zeta_{*}}{\zeta_{*}\sqrt{(\zeta_{*} + z_1)^2 + z_2^2 + \ldots z_k^2}}.$$
From there, derive the stochastic representation of Approximation 8.3.

Ex. 8.2 Diversification benefit Let B be the bound on achieved signal-noise ratio from Inequality 8.5:

$$B := \frac{\sqrt{n}\zeta_{*}^2}{\sqrt{k - 1 + n\zeta_{*}^2}}.$$

1. Show that
$$\frac{\mathrm{d}\log B}{\mathrm{d}k} \geq 0 \Leftrightarrow \frac{1}{2n\zeta_{*}^2 + 4(k-1)} \leq \frac{\mathrm{d}\log \zeta_{*}}{\mathrm{d}k},$$
2. Show that
$$\frac{1}{4(k-1)} \leq \frac{\mathrm{d}\log \zeta_{*}}{\mathrm{d}k} \Rightarrow \frac{1}{2n\zeta_{*}^2 + 4(k-1)} \leq \frac{\mathrm{d}\log \zeta_{*}}{\mathrm{d}k}.$$
3. Find $\frac{\mathrm{d}\log B}{\mathrm{d}k}$ when $\zeta_{*} = c\,(k-1)^{1/4}$.

Ex. 8.3 Gamma, Fama French Repeat Example 8.2.1 but on the Fama French 6 factor data to estimate γ.

Ex. 8.4 Contours of equal bounds Plot contours, in the space of k and ζ_{*}, of the function $\sqrt{n}\zeta_{*}^2\left(k - 1 + n\zeta_{*}^2\right)^{-1/2}$, the upper bound of Inequality 8.5. Use the same sample size as in Figure 8.10. Compare to the sample contours of equal *SRIC* in Figure 8.10.

Ex. 8.5 Bounds on achieved signal-noise ratio Confirm empirically that the upper bound of Inequality 8.5 is respected for Gaussian returns. Use $n = 2,000$ days and $k = 8$; vary ζ_*.

1. How much lower is the empirical expected value of $\mathrm{E}\left[q\left(\hat{\boldsymbol{\nu}}\left(\mathsf{X}\right)\right)\right]$ than the upper bound?
2. Do you see convergence to the asymptotic value of Equation 8.6?

Ex. 8.6 Asymptotic bounds Suppose that $k/n \to c_a$ as $n \to \infty$. Compute

$$\lim_{n\to\infty} \frac{\sqrt{n}\zeta_*^2}{\sqrt{k-1+n\zeta_*^2}} \frac{\sqrt{k+n\zeta_*^2}}{\sqrt{n-k\zeta_*^2}}.$$

This is the ratio of the upper bound of Inequality 8.5 and the asymptotic value of Equation 8.6. What does this say about the (finite n) upper bound and the asymptotic value of $q\left(\hat{\boldsymbol{\nu}}_*\right)$?

Ex. 8.7 Squared achieved signal-noise ratio Confirm the relation of Approximation 8.7 empirically by Q-Q plotting empirical quantiles of $q^2\left(\hat{\boldsymbol{\nu}}_*\right)$ against the non-central Beta law.

1. By the approximation as a non-central Beta distribution find an upper bound on the variance of $q^2\left(\hat{\boldsymbol{\nu}}_*\right)$.
2. Find a lower bound on the variance of $q^2\left(\hat{\boldsymbol{\nu}}_*\right)$.
3. Similarly, find upper and lower bounds on the variance of $q\left(\hat{\boldsymbol{\nu}}\left(\mathsf{X}\right)\right)$. What does this say about the achieved signal-noise ratio of $\hat{\boldsymbol{\nu}}_*$ as $n \to \infty$?

Ex. 8.8 Expected achieved signal-noise ratio For $\zeta_* > 0$, prove that

$$\mathrm{E}\left[q\left(\hat{\boldsymbol{\nu}}_*\right)\right] > 0.$$

Hint: use Equation 6.35.

Ex. 8.9 Risk of other portfolios ‽ Define the naïve risk parity portfolio as $\hat{\boldsymbol{\nu}} = c\boldsymbol{\sigma}^{-2}$, where the power is element-wise.

1. Find the expected value and variance of this portfolio.
2. Using a Taylor expansion, compute the approximate expected value $\mathrm{E}\left[q\left(\hat{\boldsymbol{\nu}}\right)\right]$. Confirm this is smaller than that of the Markowitz portfolio for large n.

Ex. 8.10 SRIC units Confirm that the units of the SRIC of Equation 8.16 make sense.

Ex. 8.11 SRIC, one asset Confirm empirically that Equation 8.17 does *not* hold when $k = 1$.

1. Derive Equation 8.18.

Ex. 8.12 SRIC, MGLH Derive the *SRIC* for the Hotelling-Lawley trace using one of the F distribution approximations.

1. Check empirically whether the expectation equation analogous to Equation 8.17 is met.

Ex. 8.13 Fama French achieved signal-noise ratio CI Compute a 95% confidence interval on the achieved signal-noise ratio of $\hat{\boldsymbol{\nu}}_*$ for the Fama French six factors using monthly data.

1. Repeat for the case where you impose a hedging constraint against the Market.

Ex. 8.14 Fama French achieved signal-noise ratio testing For the Fama French six factor monthly data test the null hypothesis $H_0 : q\left(\hat{\boldsymbol{\nu}}_*\right) = 0.2\,\text{mo.}^{-1/2}$.

1. Impose a hedging constraint against the Market, and test $H_0 : q\left(\hat{\boldsymbol{\nu}}_{*,\mathsf{I}\backslash\mathsf{G}}\right) = 0.25\,\text{mo.}^{-1/2}$.

Ex. 8.15 True AIC Generate a population $\boldsymbol{\mu}$ and Σ, with $\boldsymbol{\mu} \neq \mathbf{0}$, and spawn a matrix of n *i.i.d.* vectors of returns in X. Grab two random possibly overlapping subsets of the columns and compute $\hat{\zeta}^2_{*,i}$ on them. Compute the fake AIC based on the likelihood of the maximum likelihood estimators of $\zeta^2_{*,i}$. Compute the actual AIC based on the analysis of Section 6.4.2. Do the two AIC measures prefer the same model? Repeat with different random seeds.

1. Repeat this experiment with m random subsets of the columns. How do the orderings compare under the different metrics?

Ex. 8.16 Overfitting the SRIC If *SRIC* is used to choose from among multiple strategies, and then used to estimate the achieved signal-noise ratio of the selected strategy, it too will be subject to the overoptimism of Chapter 4. Use the normal approximation of Equation 8.19 along with the overoptimism corrections studied earlier.

1. Empirically draw independent returns from m distinct populations with $\zeta^2_* = 0$, and the same aspect ratio k/n but let n and k vary. Select the backtest with the highest *SRIC*. Repeat this hundreds of times. Is the empirical distribution of the maximum of many *SRIC* consistent with the maximum of normals? (Take the variance from Equation 8.19.)
2. Do your results vary considerably with m?
3. Repeat that experiment but let the aspect ratio vary as well. Is there appreciable degradation to the results?
4. Repeat this experiment, but take $\zeta^2_* > 0$. Record the achieved signal-noise ratio of your selected backtest each time. Then compute $z = SRIC - q\left(\hat{\boldsymbol{\nu}}_*\right)$, and find the backtest which maximizes z, call it z_*.

Record the z_* and repeat hundreds of times. Is the empirical distribution of the z_* consistent with the maximum of normals?

5. Generate correlated $SRIC$: generate a single $n \times k$ matrix of returns X with rows *i.i.d.* Gaussian having mean $\boldsymbol{\mu}$ and the structured covariance $\Sigma = (1-\rho)\,\mathsf{I} + \rho\left(\mathbf{1}\mathbf{1}^{\top}\right)$. Then randomly select subsets of the columns of X of a fixed size, compute $\hat{\boldsymbol{\nu}}_*$ on those columns, generate $\hat{\zeta}_*^2$ and $SRIC$. Compute $q\left(\hat{\boldsymbol{\nu}}_*\right)$ and then compute $z = SRIC - q\left(\hat{\boldsymbol{\nu}}_*\right)$. Are the z correlated? How is the correlation of z affected by ρ? Find the maximum z over your randomly selected subsets. Record the z_* and repeat hundreds of times. Is the empirical distribution of the z_* consistent with the maximum of correlated normals?

Ex. 8.17 Overfitting the SRIC, II Generate correlated $SRIC$: generate a single $n \times k$ matrix of returns X with rows *i.i.d.* Gaussian having mean $\boldsymbol{\mu}$ and a diagonal covariance matrix Σ. Let the signal-noise ratio of each column vary. Then randomly select subsets of the columns of X, compute $\hat{\boldsymbol{\nu}}_*$ on those columns, and compute $\hat{\zeta}_*^2$, $SRIC$ and $q\left(\hat{\boldsymbol{\nu}}_*\right)$. Compute m such backtests, and select the one with the highest $SRIC$. Perform the hypothesis test

$$H_0 : \forall i, j \; q\left(\hat{\boldsymbol{\nu}}_{*,i}\right) = q\left(\hat{\boldsymbol{\nu}}_{*,j}\right)$$

via the MHT, the one-sided test, and Follman's test. Assume the normal approximation for $SRIC$ from Equation 8.19.

1. Perform it under the null where the mean of X is zero: $\boldsymbol{\mu} = \mathbf{0}$. Repeat the experiment hundreds of times. Do you get the nominal type I rate?

2. Perform it under the alternative, where $\boldsymbol{\mu} \neq \mathbf{0}$. Repeat the experiment hundreds of times, and compute the empirical power. Because the "effect size" here is random (it depends on $\hat{\boldsymbol{\nu}}_*$), it is hard to estimate power as a function of the effect size. Compute and plot the power versus buckets of the $q\left(\hat{\boldsymbol{\nu}}_*\right)$ of the selected backtest. Which techniques have the highest power for this hypothesis test? Do you see the same general pattern as in Example 4.1.13?

Ex. 8.18 Expectation of SRIC Fix ζ_*^2, n, and k. Using Taylor's theorem expand $SRIC$ as a quadratic function around $\mathrm{E}\left[\hat{\zeta}_*^2\right]$, which is given in Equation 6.56.

1. Derive the approximate expected value of $SRIC$.
2. Check your approximation via simulations.

Ex. 8.19 Standard error of SRIC Derive the approximate standard error of $SRIC$. Confirm your approximation through simulations. For identical n and ζ_*, do you see wider standard error for larger k? What implications might this have for strategy selection via $SRIC$?

1. Check this empirically. Create two populations with different k and ζ_*, but for which the expected value of $SRIC$ is the same. (*cf.* Exercise 8.18) Draw n returns from both populations and compute $SRIC$ on both, check which you select, and record $q(\hat{\boldsymbol{\nu}}_*)$. Repeat this hundreds of times. Do you see significantly different probabilities of selecting one of the populations?

Ex. 8.20 **Achieved hedging** Just as the sample portfolio will not achieve a signal-noise ratio of ζ_*, one fears that a hedged portfolio built on sample data will not meet the population hedging constraint.

1. Under the approximation $\hat{\Sigma} \approx \Sigma$, show that the sample portfolio satisfies $\hat{\boldsymbol{\nu}}_{*,\mathsf{I}\backslash\mathsf{G}}{}^{\top}\Sigma\mathsf{G} = \mathbf{0}$.
2. Test this relationship empirically. For what values of k/n do you see the hedging relationship achieved in expectation? What happens for smaller values of k/n?

Ex. 8.21 **Choosing a stopped clock** For a given universe of assets, one has to choose between the bound of Inequality 8.5, or a "stopped clock" (or "biased") portfolio. Make an argument in favor of a stopped clock portfolio.

1. Devise a stopped clock portfolio estimator. (This should be easy: any estimator that subselects the assets is a stopped clock.) Construct some prior belief that would justify your stopped clock.

Ex. 8.22 **Random subset portfolios** Consider the *random subset portfolio* selection technique: given k assets, it randomly selects a subset of the assets of fixed size k_s, computes $\hat{\zeta}_*^2$ on those assets, and repeats, selecting the subset with highest $\hat{\zeta}_*^2$.

1. Do you think the achieved signal-noise ratio of this portfolio selection technique could exceed the bound of Inequality 8.5?
2. Suppose that $\boldsymbol{\mu} = c\mathbf{1}$ and $\Sigma = c^2\mathsf{I}$. Then ζ_* grows as universe size to the $\frac{1}{2}$ power. So what happens, then, to the bound of Inequality 8.5 as the subset size increases?
3. Suppose that $\boldsymbol{\mu}$ is not evenly distributed among the assets (but keep $\Sigma = c^2\mathsf{I}$). What do you expect to happen now? Of course, it depends on how much $\boldsymbol{\mu}$ deviates from $c\mathbf{1}$.
4. Check this empirically. Set $k = 100$, $n = 120\,\text{mo.}$ and let $\zeta_*^2 = 1\,\text{mo.}^{-1/2}$. (Note that this is very high. What is upper bound of Inequality 8.5 for the achieved signal-noise ratio of $\hat{\boldsymbol{\nu}}_*$?) Select random subsets of size $k_s = 10$, and select the subset with the highest $\hat{\zeta}_*^2$, and record $q(\hat{\boldsymbol{\nu}}_*)$. Repeat this hundreds of times. Can you get empirically expected $q(\hat{\boldsymbol{\nu}}_*)$ greater than the upper bound of Inequality 8.5?
5. Repeat that experiment, but choose different values of k_s. Also check

the sensitivity of the results to the arrangement of $\boldsymbol{\mu}$. How would you quantify the sensitivity to the "uniformity" of $\boldsymbol{\mu}$?

Ex. 8.23 **PCA portfolios** Consider a variation of the random subset portfolio selector given in Exercise 8.22 that performs *Principal Components Analysis* (PCA) on the matrix of returns. Effectively this consists of Z-scoring each column, then performing a Singular Value Decomposition to find the left and right singular vectors and the singular values. One then selects the optimal portfolio in the subspace of the top k_s right singular vectors.

1. Is this a "stopped clock" portfolio, or is it rotationally invariant?
2. Implement this portfolio selection technique and check it empirically. Set $k = 100$, $n = 120\,\text{mo.}$ and let $\zeta_*^2 = 1\,\text{mo.}^{-1/2}$. Select the top $k_s = 10$ singular vectors as the subspace G, and record $q\left(\hat{\nu}_{*,\mathsf{G}}\right)$. Repeat this hundreds of times.
3. Repeat that experiment, but choose different values of k_s. Also check the sensitivity of the results to the arrangement of $\boldsymbol{\mu}$. How would you quantify the sensitivity to the "uniformity" of $\boldsymbol{\mu}$?

Ex. 8.24 **Distribution of backtested returns** † Suppose that n/k is sufficiently large we can make the approximation $\hat{\Sigma} \approx \Sigma$. What is the approximate distribution of the vector $\mathsf{X}\hat{\nu}_*$?

Ex. 8.25 **Exchangeability** We say that a portfolio construction function is *exchangeable* with respect to a class of linear transformations $\mathcal{P}$ if

$$\forall \mathsf{P} \in \mathcal{P} : \hat{\nu}\left(\mathsf{X}\mathsf{P}\right) = \mathsf{P}^{-1}\hat{\nu}\left(\mathsf{X}\right).$$

1. Let $\mathcal{P}$ be the set of all square diagonal matrices with non-negative diagonals. Show that the equal-dollar allocation is not exchangeable with respect to $\mathcal{P}$.
2. Consider the minimum variance portfolio which solves

$$\max_{\hat{\nu}:\hat{\nu}^{\top}\hat{\Sigma}\hat{\nu}\le R^2} \mathbf{1}^{\top}\hat{\nu}.$$

   Show that the minimum variance portfolio is exchangeable with respect to the set of all square diagonal matrices with non-negative diagonals. Show that it is not exchangeable with respect to the set of all rotations.

3. Consider the equal risk contribution portfolio which solves [Forseth and Tricker, 2019]

$$\max_{\hat{\nu}:\hat{\nu}^{\top}\hat{\Sigma}\hat{\nu}\le R^2} \sum_i \log \hat{\nu}_i.$$

   Show that the this portfolio is not exchangeable with respect to the set of all square diagonal matrices with non-negative diagonals.

Ex. 8.26 Random populations Fix Σ and k. Suppose that a *Jinn* will grant you a set of assets whose returns have covariance Σ, and whose expectation the Jinn draws randomly via $\boldsymbol{\mu} \sim \mathcal{N}\left(\mathbf{0}, c^2\Sigma\right)$.

1. What is the expectation, under draws of the Jinn, of ζ_*^2?
2. Suppose that the Jinn hands you a universe of assets and you hold the equal dollar (long) portfolio on those assets. Show that the $\mathrm{E}\left[q\left(\hat{\boldsymbol{\nu}}\right)\right] = 0$, where expectation is under draws of the Jinn.
3. † Suppose instead the Jinn shows you n rows of returns, X, drawn from this population, and you compute the sample Markowitz portfolio on returns. What is $\mathrm{E}\left[q\left(\hat{\boldsymbol{\nu}}_*\right)\right]$?
4. Simulate draws from the Jinn, generation of returns, computation of $\hat{\boldsymbol{\nu}}_*$, to estimate $\mathrm{E}\left[q\left(\hat{\boldsymbol{\nu}}_*\right)\right]$.
5. Using simulation, estimate $\mathrm{E}\left[q\left(\hat{\boldsymbol{\nu}}\right)\right]$ where $\hat{\boldsymbol{\nu}}$ is the equal risk contribution portfolio. (*cf.* Exercise 8.25.)

Ex. 8.27 Proper Bayesian portfolios † Given Σ, consider the "loss" function associated with a portfolio $\hat{\boldsymbol{\nu}}$ as

$$\mathcal{L}\left(\hat{\boldsymbol{\nu}}; \Sigma\right) = 1 - \frac{\hat{\boldsymbol{\nu}}^\top \Sigma \left(\Sigma^{-1}\boldsymbol{\mu}\right)}{\sqrt{\hat{\boldsymbol{\nu}}^\top \Sigma \hat{\boldsymbol{\nu}}}\sqrt{\left(\Sigma^{-1}\boldsymbol{\mu}\right)^\top \Sigma \left(\Sigma^{-1}\boldsymbol{\mu}\right)}}.$$

1. Confirm that $\mathcal{L}\left(\hat{\boldsymbol{\nu}}; \Sigma\right) = 1 - \frac{q(\hat{\boldsymbol{\nu}})}{\zeta_*}$.
2. Why might one describe $\mathcal{L}\left(\hat{\boldsymbol{\nu}}; \Sigma\right)$ as a "cosine loss" function?
3. * Consider the Bayesian posterior of Equation 6.90, namely

$$\Sigma \propto \mathcal{IW}\left(m_1\Omega_1, m_1\right), \quad \boldsymbol{\mu}\,|\Sigma \propto \mathcal{N}\left(\boldsymbol{\mu}_1, \frac{1}{n_1}\Sigma\right).$$

   Conditional on Σ, show how the expected loss of a portfolio $\hat{\boldsymbol{\nu}}$ could be computed from the expected value of a Projected Normal Distribution. [Hernandez-Stumpfhauser *et al.*, 2017]
4. * What is the portfolio that minimizes the Bayesian risk under this loss function? Compare to the Bayesian portfolio of Section 6.5.

Ex. 8.28 Boring CI computations

1. Compute a 95% confidence interval on $q\left(\hat{\boldsymbol{\nu}}_*\right)$ for $\hat{\boldsymbol{\nu}}_*$ the Markowitz portfolio on the Fama French six factor data. Use both the simultaneous bound and δ_2 approaches.
2. Repeat that but hedging out exposure to the Market factor.

Ex. 8.29 CI monotonicity Letting $\alpha = 0.05$, plot $c = {\chi^2}_{1-\alpha}\left(k, \frac{n\zeta_*^2}{4}\right) - \frac{n\zeta_*^2}{4}$ versus ζ_* for reasonable values. Confirm that c is increasing in ζ_*.

1. For what values of α does this appear to hold?

Ex. 8.30 Cross validation Replicate the simulations of Example 8.3.11, but with some variations:

1. Compute the mean and the standard deviation of validated period returns. Repeat the experiment thousands of times, and scatter values against each other. Does this explain the bias of the validated Sharpe ratio? (*cf.* Exercise 3.37.)
2. Repeat the experiment but consider $n = 1,000$ days, and use 100 folds. Does the bias of the cross validated returns Sharpe ratio get better or worse?

Ex. 8.31 Cross validation for inference ☨ How would you use cross validation to perform inference on $q(\hat{\boldsymbol{\nu}}_*)$? Assume you will use the average of Sharpe ratios over validation folds to estimate $q(\hat{\boldsymbol{\nu}}_*)$.

Ex. 8.32 Cross validation sign Sharpe ratio Replicate the simulations of Example 8.3.11, but compute the sign Sharpe ratio of Equation 2.73 on validation set returns. Show that this gives a nearly unbiased estimate of $q(\hat{\boldsymbol{\nu}}_*)$.

1. Does this seem like a promising way to test for achieved signal-noise ratio? What can go wrong?

* **Ex. 8.33 Research problem: minimax portfolio** ☨ Define the Frequentist risk as in Equation 8.14.

1. Find a lower bound on the minimax risk. To use the standard machinery, you will likely have to redefine the loss function to be non-negative. Consider using $\zeta_* - q(\hat{\boldsymbol{\nu}}(\mathsf{X}))$ or the haircut loss, $1 - q(\hat{\boldsymbol{\nu}}(\mathsf{X}))/\zeta_*$.
2. Is the Markowitz portfolio a minimax portfolio rule?

* **Ex. 8.34 Research problem: Bayes portfolio** ☨ Define the Frequentist risk as in Equation 8.14. Given a prior distribution on $\boldsymbol{\mu}$ and Σ, call it π, the *Bayes risk* is the expected value under the prior of the Frequentist risk. That is, it is defined as

$$r(\boldsymbol{\mu}, \Sigma, \hat{\boldsymbol{\nu}}(\mathsf{X})) := \mathrm{E}_{\pi}\left[R(\boldsymbol{\mu}, \Sigma, \hat{\boldsymbol{\nu}}(\mathsf{X}))\right].$$

A Bayes portfolio is one that minimizes the Bayes risk. Is there a analytical way to express the Bayes portfolio? Find the Bayes portfolio for the prior of Equation 6.88.

* **Ex. 8.35 Research problem: online Markowitz** ☨ Does online Markowitz portfolio selection have bounded log regret? [Luo *et al.*, 2018]

** **Ex. 8.36 Research problem: overfitting bounds** ☨ There are traditional methods for bounding the out-of-sample performance of an estimator in terms of the in-sample performance, and some inherent "complexity" of the data itself, for example VC-dimension or Rademacher complexity. Adapt one of these methods to the problem of selecting a portfolio to maximize signal-noise ratio. You may find it easier to use the sign Sharpe ratio of Equation 2.73. More generally, apply it to the problem of selecting some parameters of a parametrized trading system based on backtested values. [Efron, 2004; Anguita *et al.*, 2014]

** **Ex. 8.37 Research problem: achieved SNR CIs** ☦ Improve the confidence intervals on $q(\hat{\nu}_*)$ of Section 8.3.1:

1. Find a bound that gives exact coverage conditional on ζ_*, without having to rely on the approximation $\Sigma \approx \hat{\Sigma}$.
2. Find a bound that gives near nominal coverage by using a better estimate of ζ_*^2 or taking into account the variation in the estimate of ζ_*^2 and the deviation of $q(\hat{\nu}_*)$ from its expected value.

** **Ex. 8.38 Research problem: SR prediction intervals** ☦ Construct prediction intervals for the out-of-sample Sharpe ratio of the sample Markowitz portfolio, in analogue to the confidence intervals of Section 8.3.1. That is, let $\hat{\nu}_*$ be the Markowitz portfolio built on a sample of size n_1: $\hat{\nu}_* = \hat{\Sigma}_1^{-1}\hat{\mu}_1$. Suppose that you will observe a future sample of size n_2, and construct $\hat{\mu}_2$, and $\hat{\Sigma}_2$ based on that sample. Find prediction interval on

$$\frac{\hat{\nu}_*^{\top}\hat{\mu}_2}{\sqrt{\hat{\nu}_*^{\top}\hat{\Sigma}_2\hat{\nu}_*}}.$$

Recall that a prediction interval is constructed conditional on observing $\hat{\mu}_1$ and $\hat{\Sigma}_1$, and contains the future Sharpe ratio with probability $1-\alpha$, where the error rate is over replications of the entire experiment (*i.e.*, observations of both samples). You should probably follow the strategy of performing the analysis under the approximation $\Sigma_i \approx \hat{\Sigma}$, $i = 1, 2$, then find when that approximation is satisfactory.

** **Ex. 8.39 Research problem: achieved SNR paired test** ☦ Suppose you observe n paired samples of the returns of two universe of assets, with possibly correlated returns, then construct the Markowitz portfolio on both. Construct a test for the hypothesis that they have the same achieved signal-noise ratio.

9

Market Timing

If money can't buy you love,
What are you saving it for?

MERYN CADELL
Steam Clean Express

In the first part of this book we considered inference on the signal-noise ratio and the ex-factor signal-noise ratio. The use case for that analysis is, roughly, when one can choose to invest, or not, a predefined positive amount of capital in an asset: a binary decision on a single asset, or choosing one asset from many. In the second part, we considered the portfolio construction problem, wherein one has to select a vector of arbitrary allocations to a number of assets, sometimes with the aid of time-varying state variables or features to inform our decision. This section, on *market timing*, by which we mean allocating a continuous (possibly negative) amount to a single predefined asset with use of conditioning information, was originally intended to sit between those sections as a kind of segue. However the analysis we want to perform on market timing is largely a special case of the portfolio inference problems. At times this can be viewed as *folding time into space* in the sense that we can describe multi-period optimization problems on a single asset as a single-period optimization problem on multiple assets[1].

Sharpe himself analyzed a market timing model, using actual market data, and was skeptical of the gains. [Sharpe, 1975] Sharpe was fairly conservative in his assumptions: the market participant traded once a year, incurring 2% transaction costs, and could not short the market. Under these assumptions, an active manager would need to be very skillful to beat a passive ("buy-and-hold") manager. We will not impose the restrictions that Sharpe did, allowing our manager to short at will, and mostly ignoring transaction costs. Ignoring these details is convenient, but sacrifices fidelity of our model.

[1]Plus this just sounds really cool.

DOI: 10.1201/9781003181057-9

9.1 Market timing with a single binary feature

This is perhaps the simplest model of market timing and arguably the least realistic. Suppose you observe $f_{t-1} \in \{-1, +1\}$ prior to the time required to capture the returns x_t. Moreover, independent of x_t, we have

$$\Pr\{\text{sign}(f_{t-1}) = \text{sign}(x_t)\} = \frac{1+\rho}{2},$$

where $-1 \leq \rho \leq 1$ is the *timing ability* of the manager[2].

One should pause and consider whether this is a realistic model of a market timing manager. In particular, it seems odd that the ability of the manager is independent of the *magnitude* of x_t, since presumably if $|x_t|$ is small compared to the volatility of the market, it could have "more easily gone the other way."

The manager who holds the asset long or short according to $\text{sign}(f_{t-1})$, incurs no transaction costs and can short freely, will capture returns of

$$+|x_t|\text{, with probability } \frac{1+\rho}{2},$$
$$-|x_t|\text{, with probability } \frac{1-\rho}{2},$$

assuming x_t are *relative returns*. (*cf.* Section 1.1.1 on why these should be relative and not log returns.) Then the expected first and second moments of the manager's returns are, respectively, $\rho\,\mathrm{E}[|x|]$ and $\mathrm{E}[x^2]$. Thus the signal-noise ratio of the manager is

$$\zeta = \frac{\rho\,\mathrm{E}[|x|]}{\sqrt{\mathrm{E}[x^2] - \rho^2\,\mathrm{E}[|x|]^2}} = f_{\text{tas}}\left(\frac{\rho\,\mathrm{E}[|x|]}{\sqrt{\mathrm{E}[x^2]}}\right), \tag{9.1}$$

where $f_{\text{tas}}(\cdot)$ is defined as in Equation 1.24.

The quantity

$$\kappa := \frac{\mathrm{E}[|x|]}{\sqrt{\mathrm{E}[x^2]}} \tag{9.2}$$

limits the optimal signal-noise ratio that a timing strategy can achieve. It reflects the inherent uncertainty in the absolute returns of the market asset. For normally distributed x, it takes value $\kappa = \sqrt{\frac{2}{\pi}} \approx 0.798$, where we omit the units[3]. Using the Fama French "Market" monthly relative returns data from

[2]Throughout this section we will assume that returns take value zero with zero probability. While this is a fair approximation for longer time scales, it may be inadequate on shorter time scales because assets are denominated in discrete units.

[3]Normally distributed log returns should be normal at every time scale, but normally distributed relative returns are not! The former follows from Cramér's decomposition theorem, assuming independent sub-returns.

Jan 1927 to Dec 2020, we estimate $\kappa \approx 0.719$. For daily "Market" returns from 1926-11-03 to 2020-12-31, we estimate $\kappa \approx 0.64$. Thus, a manager with perfect market timing ability on monthly returns would have a signal-noise ratio of around $3.6\,\mathrm{yr}^{-1/2}$. This is unusually large, but not infinite.

By this analysis we have split the problem in two: we have to estimate κ from historical data (though presumably we need not perform inference on it), and we have to estimate ρ, and probably we have to perform inference on it.

First consider ρ. Collecting n observations of f_{t-1} and corresponding x_t, we estimate ρ via

$$\hat{\rho} := \frac{1}{n} \sum_{1 \le i \le n} \operatorname{sign}\left(f_{t-1}\right) \operatorname{sign}\left(x_t\right), \tag{9.3}$$

where the sign function takes values ± 1.

Note that we can treat $\operatorname{sign}\left(f_{t-1}\right) \operatorname{sign}\left(x_t\right)$ like a coin flip, in which case $\left(1 + \hat{\rho}\right)/2$ is the estimated probability of landing a "heads." Standard theory regarding inference on the Binomial distribution can then be employed. [Bain and Engelhardt, 1992] That is, $n\left(1 + \hat{\rho}\right)/2 \sim B\left(n, \left(1 + \rho\right)/2\right)$ follows a Binomial distribution. Thus to test

$$H_0 : \rho = \rho_0 \quad \text{versus} \quad H_1 : \rho > \rho_0,$$

compute $\hat{\rho}$, and reject the null hypothesis if

$$\hat{\rho} \ge \frac{2}{n} B_{1-\alpha}\left(n, \left(1 + \rho_0\right)/2\right) - 1,$$

where $B_{1-\alpha}\left(n, p\right)$ is the $1 - \alpha$ quantile of the Binomial distribution of n observations and probability p.

We also know that, by the central limit theorem, as $n \to \infty$,

$$\frac{\hat{\rho} - \rho}{\sqrt{\left(1 - \rho^2\right)/n}} \to \mathcal{N}\left(0, 1\right). \tag{9.4}$$

Thus $\sqrt{\left(1 - \rho^2\right)/n}$ is an approximate standard error for $\hat{\rho}$, which one would estimate by plugging in $\hat{\rho}$ for ρ.

Example 9.1.1 (Timing the Market with Volatility). Consider the daily relative returns of the Market, introduced in Example 1.1.1. We compute the running standard deviation of the daily relative returns over a 120 market day window, then delayed by 5 days. We then let the feature f_{t-1} be equal to the sign of $0.01\,\mathrm{day}^{-1/2}$ minus this volatility. We then consider daily returns x_t from 1950-01-03 to 2020-12-31, a total of $n = 17,953$ days.

From these we compute the estimate $\hat{\rho} = 0.071$. To check the null hypothesis $H_0 : \rho = 0.06$ versus the hypothesis that ρ is larger, we compute the 0.95

quantile of the Binomial distribution to arrive at the cutpoint of 0.072. Since $\hat{\rho}$ is larger than this critical value, we reject at the $\alpha = 0.05$ level.

We estimate the standard error of $\hat{\rho}$ as $\sqrt{\left(1-\hat{\rho}^2\right)/n} = 0.007$. This implies a 90% confidence interval for ρ is $[0.058, 0.083)$, and we reject the null hypothesis using the Normal approximation. ⊣

Example 9.1.1 raises a number of questions:

1. Why did we choose $H_0 : \rho = 0.06$ for the null hypothesis? Perhaps a more natural benchmark would be the the returns of the "buy-and-hold" portfolio simply holds the Market asset long. This is essentially Sharpe's lament. [Sharpe, 1975]
2. The feature we chose is defined over a 120 day window, and so should be fairly autocorrelated. However, the daily Market returns considered here should not exhibit such autocorrelation for long periods. Moreover, we delayed the signal by a few days, which seems odd for a daily trading strategy. Have we done something wrong?
3. Why did we perform inference on $\hat{\rho}$ but not on $\hat{\kappa}$?
4. Why did we analyze $\hat{\rho}$ at all instead of performing a "backtest" by computing the Sharpe ratio of the simulated strategy returns (*i.e.*, $f_{t-1}x_t$) and performing inference on that? That is, why bother with $\hat{\rho}$ when we can directly compute $\hat{\zeta}$?
5. Does this market timer satisfy the independence assumptions?

The first question is a bit annoying: on the one hand it is tinged by a survivorship bias[4], but also because the explicit goal of the manager might be to deliver returns which are somehow orthogonal to the Market asset. We will consider it nonetheless. The second question is superseded by a more general form which we will consider in Section 9.3.1.

The last three questions are related to each other. By assuming that the probability of correctly guessing sign is independent of $|x_t|$, we ought to gain statistical power, and thus can discriminate finer effects by directly performing inference on ρ.

Example 9.1.2 (Timing the Market with Volatility, II). We return to the market timing setup of Example 9.1.1, where we considered daily relative returns of the Fama French Market term, and a market timer using a delayed running volatility estimate. For that market timer, we estimated $\hat{\rho} = 0.071$, with approximate standard error of 0.007. Over the same period, however, 1950-01-03 to 2020-12-31, we estimate the timing "ability" of the buy-and-hold timer as $\hat{\rho}_h = 0.11$, with approximate standard error of 0.007. While this isn't a proper statistical test, it appears the ability of buy and hold timer is about 4.7 standard errors higher than that of our volatility timer. We certainly have no evidence that the timing strategy beats buy-and-hold.

In Figure 9.1, we show the time series of f_t, which confirms visually that it

---

[4]As we wish to consider timing of any asset, we must accept that some assets have negative expected returns.

is fairly autocorrelated, as one would suspect from its construction. We might be better off analyzing a monthly version of this signal, see Exercise 9.1.

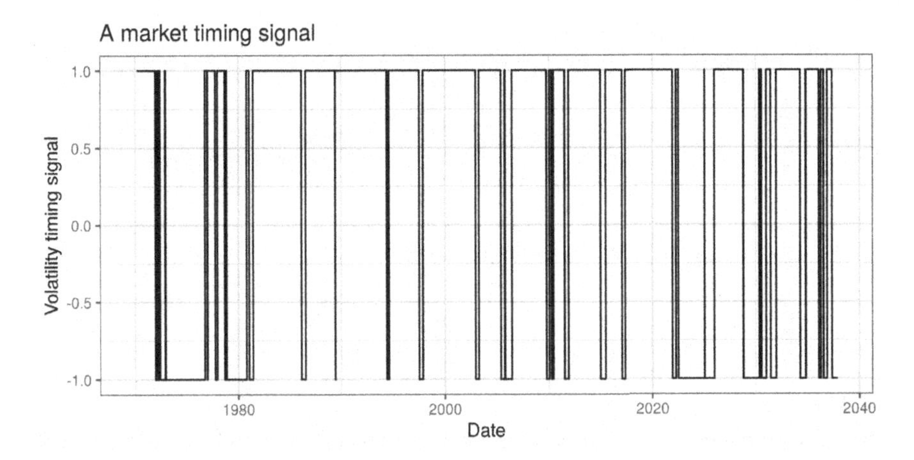

FIGURE 9.1
The market timing signal is shown over time.

From the data we estimate $\hat{\kappa} = 0.66$. Via the delta method, using the heteroskedasticity and autocorrelation consistent covariance estimator, we estimate the standard error of $\hat{\kappa}$ to be around 0.013. [Pav, 2016b; Zeileis, 2004] Performing that computation with the "vanilla" covariance estimator results in a somewhat smaller standard error estimate of around 0.008. Thus there is relatively little error in our estimate of κ, due to the relatively large sample size.

For the observed values of $\hat{\rho}$ and $\hat{\kappa}$, we use Equation 9.1 to estimate the signal-noise ratio as $0.75\,\mathrm{yr}^{-1/2}$. We also perform a "backtest," estimating the returns of the market timing manager to be $\operatorname{sign}(x_t f_{t-1})\,|x_t|$. We then compute the Sharpe ratio of that series as $0.26\,\mathrm{yr}^{-1/2}$ with a standard error of 0.12. There is a very large discrepancy between these numbers. Recall that we have assumed that accuracy is independent of $|x_t|$. If that is not the case, then Equation 9.1 does not give the signal-noise ratio of the timing strategy. In Figure 9.2, we plot a non-parametric estimate of ρ as a function of $|x_t|$ along with a horizontal line at the overall estimate of 0.071. The accuracy appears to be non-constant across $|x_t|$, which could explain the mismatch between the two estimates of the signal-noise ratio of the timing strategy. Worse still the non-parametric $\hat{\rho}$ appears negative for large values of $|x_t|$, while the overall average $\hat{\rho}$ is positive.

To further test this possible dependence, we separate the returns into two samples, based on whether the returns have the same sign as the signal. We take the absolute values of these, then compare the two samples using the Baumgartner-Weiß-Schindler test, a non-parametric two-sample test of equality of distributions. [Baumgartner *et al.*, 1998; Pav, 2016a] This test yields a

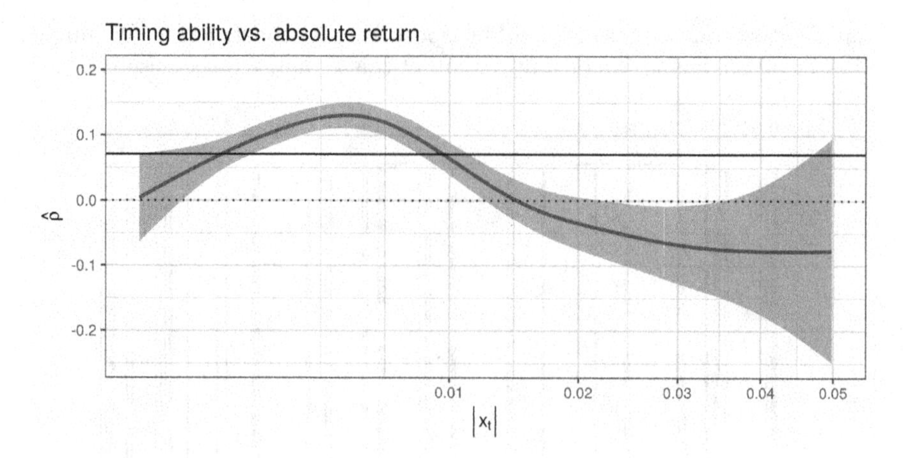

FIGURE 9.2
A non-parametric estimate of ρ is shown for corresponding values of $|x_t|$, along with the long term average value of 0.071. The x-axis is in square root space. The assumption of independence is clearly violated.

p-value of 2×10^{-7}, and we reject the null of equal distributions. The classical Kolmogorov-Smirnov test also rejects the null, with a p-value underflowing far below 0.01. ⊣

9.1.1 Gain in power

The timing model in this section is simple to work with. By assuming away some degrees of freedom, one gains statistical power. Consider the power of the hypothesis test outlined above. To test $H_0 : \rho = \rho_0$ versus $H_1 : \rho > \rho_0$, we reject the null if

$$\hat{\rho} \geq \frac{2}{n} B_{1-\alpha} \left(n, \left(1 + \rho_0\right)/2\right) - 1.$$

Suppose that $\rho = \rho_1 > \rho_0$. Then the power of this test, the probability of (correctly) rejecting the null hypothesis in favor of the alternative, is

$$1 - \beta = 1 - F_B \left(B_{1-\alpha} \left(n, \left(1 + \rho_0\right)/2\right); n, \left(1 + \rho_1\right)/2\right), \tag{9.5}$$

where $F_B \left(x; n, p\right)$ is the (discrete) cumulative distribution function Binomial distribution of n observations and probability p.

Let us compare this power to the power of a one-sided test of significant signal-noise ratio as outlined in Section 2.5.3 for the case of daily returns. Consider the case where we are testing $\rho > \rho_0 = 0.075$ at the 0.05 rate. Assume $\kappa = 0.7$ and assume we have observed 1,260 days of returns and the signal. In Figure 9.3, we compare the power of the two tests for ρ_1 ranging from 0.075 to 0.15. These correspond to the signal-noise ratio ranging from $0.835\,\mathrm{yr}^{-1/2}$ to $1.676\,\mathrm{yr}^{-1/2}$. The timing test has higher power.

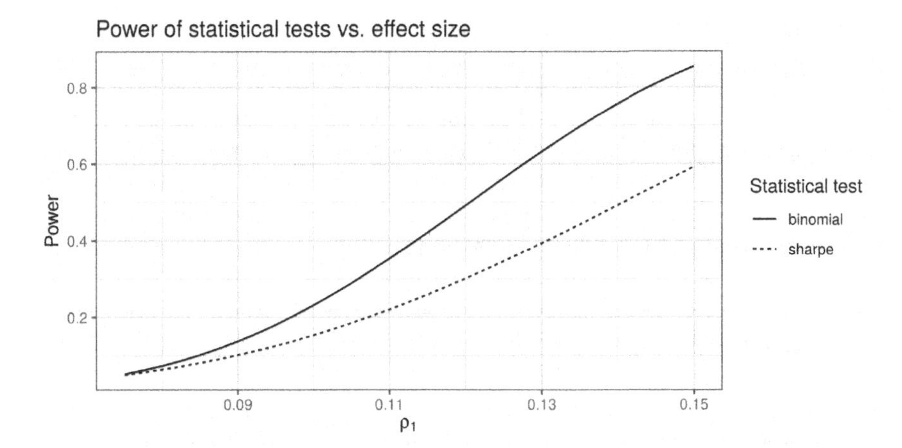

FIGURE 9.3
The power, the probability of correctly rejecting the null, is compared for two tests. One is the binomial-based one-sided test for significant ρ, the other is the one-sided test for significant signal-noise ratio based on normal returns. Power is plotted versus the true timing ability ρ_1, assuming 1,260 days of daily returns, with $\kappa = 0.7$, testing against $\rho_0 = 0.075$ and a type I rate of 0.05. The timing test shows higher power than the test on the Sharpe ratio of backtested returns.

9.1.2 Testing assumptions

In our development of this timing model we assumed that the timing ability of the manager was independent of the magnitude of returns. As it happens, this is a *sufficient* condition for the analysis that followed, but not a *necessary* condition. We used independence when we claimed that the expected return of the timing manager was $\rho\,\mathrm{E}\left[|x|\right]$. The more general form, without assumption, is

$$\begin{aligned}\mu &= \mathrm{E}\left[\mathrm{sign}\left(f_{t-1}\right) x_t\right],\\ &= \mathrm{E}\left[\mathrm{sign}\left(f_{t-1}\right) \mathrm{sign}\left(x_t\right) |x_t|\right],\\ &= \mathrm{Cov}\left(\mathrm{sign}\left(f_{t-1}\right) \mathrm{sign}\left(x_t\right), |x_t|\right) + \mathrm{E}\left[\mathrm{sign}\left(f_{t-1}\right) \mathrm{sign}\left(x_t\right)\right] \mathrm{E}\left[|x_t|\right],\\ &= \mathrm{Cov}\left(\mathrm{sign}\left(f_{t-1}\right) \mathrm{sign}\left(x_t\right), |x_t|\right) + \rho\,\mathrm{E}\left[|x_t|\right].\end{aligned} \tag{9.6}$$

If the win or loss of the market timer (*i.e.*, the product $\mathrm{sign}\left(f_{t-1}\right) \mathrm{sign}\left(x_t\right)$) is independent of the magnitude of returns, $|x_t|$, then their covariance $\mathrm{Cov}\left(\mathrm{sign}\left(f_{t-1}\right) \mathrm{sign}\left(x_t\right), |x_t|\right)$ is equal to zero and we have correctly computed the signal-noise ratio of our timing manager.

It is this zero correlation condition that is necessary for our analysis, not the assumption of independence. To use this model of market timing, we ought to test this condition. Turning the above equations around, the zero correlation condition is equivalent to the following:

$$\text{Cov}\left(\text{sign}\left(f_{t-1}\right)\text{sign}\left(x_t\right), \left|x_t\right|\right) = 0 \Leftrightarrow \text{E}\left[\text{sign}\left(f_{t-1}\right)x_t\right] = \rho\,\text{E}\left[\left|x_t\right|\right].$$

Under the dubious assumption that both ρ and $\text{E}\left[\left|x_t\right|\right]$ are known, testing the equation on the right is equivalent to a test of the mean of the series $\text{sign}\left(f_{t-1}\right)x_t$, which has unknown variance. This would be performed by a t-test (see Section 2.5.3), and the apparent gain in power has evaporated. This is another instance of the following vague principle:

Assume Only True Things!

If you can assume away something true, you gain inferential power; conversely if an assumption of your analysis is false, you lose power.

Not all is lost, however. The analysis in this section can useful in some situations: assessing the putative signal-noise ratio of a market timer when the predictive ability is quoted in a paper, say, or by a data vendor, when the series of returns and timing signal are not provided.

9.2 Market timing with a discrete feature

We now consider market timing with a single discrete feature. That is, suppose that prior to the investment decision required to capture returns x_t, you observe a scalar feature, f_{t-1} that can take one of a finite number of values or "states," $\mathcal{S}_1, \mathcal{S}_2, \ldots, \mathcal{S}_J$. Note that this subsumes the case where one observes a vector of discrete features, since their Cartesian product could be converted into a single discrete feature, unless one wished to impose a hierarchical structure of some kind[5].

We furthermore assume the feature, f_{t-1} is not within control of the investor. For example, the following fit into our idea of a discrete feature: 1) a ±1 indicator for whether the previous day's return was positive or negative, 2) the day of the week or month of the year, 3) an indicator of market volatility taking values `Low`, `Medium` and `High`, 4) a running average of the high temperature in New York City, in Celsius, rounded to 1 significant digit, and so on. We note that some of these examples have an apparent order, but we will not typically assume that the order is important.

[5]For example, one might posit that one discrete feature affects expected returns, while another affects volatility. Keeping the two features separate would simplify the bookkeeping in some hypothesis testing.

We let π_j be the probability the feature takes the jth state:

$$\pi_j = \Pr\left\{f_{t-1} = \mathcal{S}_j\right\}.$$

We assume that probability is independent of t. For deterministic features, like the day of the week, let π_j be the long term average proportion of f_t that equal $\mathcal{S}_j$.

Let μ_j and σ_j^2 be the mean and variance of x conditional on observing the jth state:

$$\begin{aligned}\mu_j &= \mathrm{E}\left[x_t \,|\, f_{t-1} = \mathcal{S}_j\right],\\ \sigma_j^2 &= \mathrm{Var}\left(x_t \,|\, f_{t-1} = \mathcal{S}_j\right).\end{aligned}$$

We note that in some cases it may be desireable to impose restrictions on this model, *e.g.*, homoskedasticity where $\sigma_j^2 = \sigma^2$ for all j, or equality of the signal-noise ratio of each state, *etc.* The assumption of homoskedasticity corresponds to maximizing the single-period signal-noise ratio, while allowing σ_j^2 to vary with j corresponds to maximizing the multi-period signal-noise ratio. If returns truly are homoskedastic, then one will benefit from improved estimation by pooling across states.

We assume a strategy of allocating ν_j proportion of your wealth in the asset upon observing $f_t = \mathcal{S}_j$. We will follow our usual plan of describing the optimal allocation assuming the population parameters are known; then discuss how to estimate the population parameters; then consider inference on the unknowns. We note that the problem can be transformed into a traditional linear portfolio problem.

Let μ and σ^2 be the mean and variance of our strategy's returns. It is simple to show that

$$\begin{aligned}\mu &= \sum_j \pi_j \nu_j \mu_j, &(9.7)\\ \sigma^2 &= \sum_j \pi_j {\nu_j}^2\left(\sigma_j^2 + {\mu_j}^2\right) - \left(\sum_j \pi_j \nu_j \mu_j\right)^2.\end{aligned}$$

This can be written more compactly as

$$\begin{aligned}\mu &= \boldsymbol{\nu}^{\top}\left(\boldsymbol{\pi} \odot \boldsymbol{\mu}\right), &(9.8)\\ \sigma^2 &= \boldsymbol{\nu}^{\top}\left(\mathrm{Diag}\left(\boldsymbol{\pi} \odot \left(\boldsymbol{\mu}^2 + \boldsymbol{\sigma}^2\right)\right) - \left(\boldsymbol{\pi} \odot \boldsymbol{\mu}\right)\left(\boldsymbol{\pi} \odot \boldsymbol{\mu}\right)^{\top}\right)\boldsymbol{\nu},\end{aligned}$$

where $\boldsymbol{\pi}$ is the J-vector of the π, $\boldsymbol{\mu}$ and $\boldsymbol{\sigma}^2$ are vectors of the μ and σ^2, $\odot$ is the Hadamard (or "element-wise") multiplication operator, *cf.* Exercise 9.5.

Per the general theory of portfolio optimization (*cf.* Section 5.2), the multi-period signal-noise ratio is maximized when

$$\begin{aligned}\boldsymbol{\nu} &= c\Big(\mathrm{Diag}\left(\boldsymbol{\pi} \odot \left(\boldsymbol{\mu}^2 + \boldsymbol{\sigma}^2\right)\right) - \left(\boldsymbol{\pi} \odot \boldsymbol{\mu}\right)\left(\boldsymbol{\pi} \odot \boldsymbol{\mu}\right)^{\top}\Big)^{-1}\left(\boldsymbol{\pi} \odot \boldsymbol{\mu}\right),\\ &= c'\left(\mathrm{Diag}\left(\boldsymbol{\pi} \odot \left(\boldsymbol{\mu}^2 + \boldsymbol{\sigma}^2\right)\right)\right)^{-1}\left(\boldsymbol{\pi} \odot \boldsymbol{\mu}\right). &(9.9)\end{aligned}$$

The latter equation follows from the Sherman-Morrison-Woodbury identity, Equation A.3. With ζ_* the signal-noise ratio of this strategy, again by the general theory we have

$$\zeta_*^2 = \frac{q^2}{1-q^2}, \text{ where } q^2 = \sum_j \pi_j \frac{{\mu_j}^2}{\sigma_j^2 + {\mu_j}^2}. \tag{9.10}$$

We can rewrite this equation in terms of the $f_{\text{tas}}(\cdot)$ function of Equation 1.24:

$$\zeta_*^2 = \left(f_{\text{tas}}(q)\right)^2,$$
$$q^2 = \sum_j \pi_j \left(f_{\text{tas}}^{-1}(\zeta_j)\right)^2,$$

where $\zeta_j = \mu_j/\sigma_j$ is the Hansen ratio for the jth state. [Černý, 2020] Thus ζ_*^2 is some kind of transformed average of the ζ_j^2.

9.2.1 Estimation and inference

Suppose we observe n observations of the feature f_{t-1} and corresponding x_t. Let $1_{\{\cdot\}}$ be the indicator function:

$$1_{\{f_{t-1}=\mathcal{S}_j\}} = \begin{cases} 1 & \text{if } f_{t-1} = \mathcal{S}_j, \\ 0 & \text{otherwise.} \end{cases}$$

The usual sample estimates of the probabilities, mean and second moment of returns in each state is given by

$$\hat{\pi}_j = \frac{1}{n}\sum_t 1_{\{f_{t-1}=\mathcal{S}_j\}}, \tag{9.11}$$

$$\hat{\mu}_j = \frac{\sum_t 1_{\{f_{t-1}=\mathcal{S}_j\}} x_t}{\sum_t 1_{\{f_{t-1}=\mathcal{S}_j\}}} = \frac{1}{n\hat{\pi}_j}\sum_t 1_{\{f_{t-1}=\mathcal{S}_j\}} x_t, \tag{9.12}$$

$$\hat{\alpha}_{2,j} = \frac{\sum_t 1_{\{f_{t-1}=\mathcal{S}_j\}} {x_t}^2}{\sum_t 1_{\{f_{t-1}=\mathcal{S}_j\}}} = \frac{1}{n\hat{\pi}_j}\sum_t 1_{\{f_{t-1}=\mathcal{S}_j\}} {x_t}^2. \tag{9.13}$$

Let F be the $n \times J$ matrix whose rows are the indicator vectors

$$\mathsf{F}_{t,j} = 1_{\{f_{t-1}=\mathcal{S}_j\}}.$$

We let $\boldsymbol{x}$ be the n-vector of returns. The apparent mismatch in indexing from f_{t-1} to rows of $\mathsf{F}_{t,:}$ allows us to "line up" the matrix F and the vector $\boldsymbol{x}$.

Homoskedastic errors

As noted above, the homoskedastic model where $\sigma_j^2 = \sigma^2$ for all j, corresponds to the conditional expectation model wherein one is maximizing the single period signal-noise ratio. Under this assumption we have:

$$\begin{aligned} \operatorname{E}\left[x_t \,\middle|\, \boldsymbol{f}_{t-1}\right] &= \beta^\top \boldsymbol{f}_{t-1}, \\ \operatorname{Var}\left(x_t \,\middle|\, \boldsymbol{f}_{t-1}\right) &= \sigma^2, \end{aligned} \tag{9.14}$$

where $\boldsymbol{f}_{t-1}$ is $\mathsf{F}_{t,:}{}^\top$. The estimates from regression are

$$\begin{aligned} \hat{\beta} &= \left(\mathsf{F}^\top \mathsf{F}\right)^{-1} \mathsf{F}^\top \boldsymbol{x}, \\ \hat{\sigma}^2 &= \left(\hat{\boldsymbol{e}}^\top \hat{\boldsymbol{e}}\right) / (n - J), \quad \text{where } \hat{\boldsymbol{e}} = \boldsymbol{x} - \mathsf{F}\hat{\beta}. \end{aligned} \tag{9.15}$$

Note that $\mathsf{F}^\top\mathsf{F}$ is a diagonal matrix because each row of F has exactly one non-zero element, and $\hat{\boldsymbol{\pi}} = \left(\mathsf{F}^\top\mathsf{F}\right)^{-1}\mathsf{F}^\top\mathbf{1}$ is a vector of the $\hat{\pi}_j$. Moreover

$$\hat{\beta}_j = \frac{\sum_t x_t \mathsf{F}_{t,j}}{\sum_t \mathsf{F}_{t,j}^2} = \hat{\mu}_j.$$

Again, $\hat{\sigma}^2$ is the pooled estimate, which takes the value

$$\hat{\sigma}^2 = \frac{n}{n-J} \sum_{j=1}^{J} \hat{\pi}_j \left(\hat{\alpha}_{2,j} - \hat{\mu}_j^2\right).$$

To maximize the single-period signal-noise ratio conditional on observing $\boldsymbol{f}_{t-1}$, one should allocate proportional to $\hat{\beta}^\top \boldsymbol{f}_{t-1}/\hat{\sigma}^2$. However, this just means one should hold proportional to $\operatorname{sign}(\hat{\mu}_j)$ when $f_{t-1} = \mathcal{S}_j$, up to one's risk budget. The single-period Sharpe ratio of this portfolio is $\left|\hat{\mu}_j\right|/\hat{\sigma}$. The average of the squared single-period Sharpe ratio of this strategy is

$$\hat{T} = \sum_{j=1}^{J} \hat{\pi}_j \frac{\hat{\mu}_j^2}{\hat{\sigma}^2}. \tag{9.16}$$

This is the sample Hotelling Lawley statistic. Inference on T, the sample analogue, is via the MGLH tests of Section 6.3.2, *i.e.*, via an approximation as a non-central F-distribution.

Example 9.2.1 (Timing the Market). We consider the case of market timing on the monthly Market returns from the Fama French four factors data. We compute three separate binary "features":

1. The 3-month running sum of the absolute value of all four Fama French factors is computed, and lagged a month. The binary feature is whether this lagged sum exceeds 39%.
2. A binary feature is computed representing whether the previous period SMB returns exceeded the previous period's HML returns.

3. A binary feature is computed representing whether the previous period UMD returns exceeded 1%.

We then define f_{t-1} to be one plus the first binary feature plus two times the second plus four times the third binary feature. Thus f_{t-1} can take one of eight possible values.

As we have lost a few months for computation of the features, we have $n = 1,127$ mo. of returns, ranging from Feb 1927 to Dec 2020. We estimate $\hat{\boldsymbol{\pi}}$, $\hat{\boldsymbol{\mu}}$ and $\hat{\boldsymbol{\alpha}}_2$ as

$$\begin{aligned}
\hat{\boldsymbol{\pi}}^{\top} &= \begin{bmatrix} 0.22 & 0.06 & 0.19 & 0.06 & 0.17 & 0.06 & 0.17 & 0.07 \end{bmatrix}, \\
\hat{\boldsymbol{\mu}}^{\top} &= \begin{bmatrix} 1.09 & 0.99 & 0.86 & 0.82 & 0.60 & 2.33 & 1.12 & -0.03 \end{bmatrix} \%\,\text{mo.}^{-1}, \\
\hat{\boldsymbol{\alpha}}_2{}^{\top} &= \begin{bmatrix} 18.08 & 71.81 & 20.32 & 23.63 & 19.12 & 95.45 & 19.05 & 50.10 \end{bmatrix} \%^2\,\text{mo.}^{-1}.
\end{aligned}$$

From this we compute $\hat{T} = 0.039$ mo.$^{-1}$. Note that the squared Sharpe ratio of the (buy and hold) Market returns in this case is 0.031 mo.$^{-1}$, which is to say there is no great increase in Sharpe ratio from timing via this discrete feature. Note that our discrete features are somewhat oddly defined: we chose them by trial and error such that at least one of the $\hat{\mu}_j$ was negative. As the Market returns are largely positive mean, if you select discrete features haphazardly, all $\hat{\mu}_j$ might be positive, and then the conditional portfolio is indistinguishable from the buy-and-hold portfolio!

We perform the test for the hypothesis that $T = 0$ using McKeon's approximation of the HLT to an F distribution, as in Section 6.3.2. We compute the cutoff using McKeon's approximation to be 0.014 mo.$^{-1}$, and we reject at the 0.05 level, since $\hat{T}$ exceeds this value. ⊣

We note that homoskedasticity of errors is an assumption, one that may not be defensible for a given set of features and returns. For example, in the previous example, there is great variability in the $\hat{\boldsymbol{\alpha}}_2$, so much so that the assumption of equal variances in each class is in some doubt. The classical hypothesis test for equality of variances in independent groups is *Levene's test*. Levene's test is a kind of ANOVA on the absolute errors. Defining σ_j^2 as the variance of x_t conditional on $f_{t-1} = \mathcal{S}_j$, to test

$$H_0 : \forall j\, \sigma_j^2 = \sigma_1^2 \quad \text{versus} \quad H_1 : \exists i, j\, \sigma_i^2 \neq \sigma_j^2,$$

compute

$$\begin{aligned}
Z_j &= \frac{1}{n\hat{\pi}_j} \sum_t 1_{\{f_{t-1}=\mathcal{S}_j\}} \left|x_t - \hat{\mu}_j\right|, \\
\bar{Z} &= \sum_j \hat{\pi}_j Z_j, \\
W &= \frac{n-J}{J-1} \frac{n \sum_j \hat{\pi}_j \left(Z_j - \bar{Z}\right)^2}{\sum_j \sum_t 1_{\{f_{t-1}=\mathcal{S}_j\}} \left(\left|x_t - \hat{\mu}_j\right| - Z_j\right)^2},
\end{aligned}$$

and reject at the α level if $W > f_{1-\alpha}(J-1, n-J)$, where $f_q(v_1, v_2)$ is the q quantile of the (central) F-distribution with v_1 and v_2 degrees of freedom.

Example 9.2.2 (Timing the Market, Levene's Test). We return to the setup of Example 9.2.1 to examine the assumption of homoskedasticity. We perform Levene's test, computing $W = 15$. As this is larger than the cutoff 2.02 we reject the null hypothesis of homoskedasticity at the 0.05 level. ⊣

Heteroskedastic errors, multi-period optimization

The heteroskedastic errors case corresponds to the multi-period signal-noise ratio optimization procedure. In Section 6.1.1, we described how to perform inference in this case via the flattening trick. So we create some J "pseudoassets" with returns vector

$$\boldsymbol{x}_t = x_t \boldsymbol{f}_{t-1}.$$

Note that at every time period, $J-1$ of the elements of $\boldsymbol{x}_t$ are zero, and thus the Gaussian distribution is a terrible model for $\boldsymbol{x}$. Then as in Equation 6.1, we estimate the mean and covariance of $\boldsymbol{x}$ as

$$\begin{aligned} \mathrm{E}\left[\boldsymbol{x}\right] &= \frac{1}{n}\sum_t \boldsymbol{x}_t = \frac{1}{n}\mathsf{F}^\top \boldsymbol{x} = \hat{\boldsymbol{\pi}} \odot \hat{\boldsymbol{\mu}}, \\ \mathrm{Var}\left(\boldsymbol{x}\right) &= \frac{1}{n-1}\sum_t \left(\boldsymbol{x}_t - \mathrm{E}\left[\boldsymbol{x}\right]\right)\left(\boldsymbol{x}_t - \mathrm{E}\left[\boldsymbol{x}\right]\right)^\top, \\ &= \frac{n}{n-1}\left(\mathrm{Diag}\left(\hat{\boldsymbol{\pi}} \odot \hat{\boldsymbol{\alpha}}_2\right) - \left(\hat{\boldsymbol{\pi}} \odot \hat{\boldsymbol{\mu}}\right)\left(\hat{\boldsymbol{\pi}} \odot \hat{\boldsymbol{\mu}}\right)^\top\right). \end{aligned} \tag{9.17}$$

Again, the multi-period Sharpe ratio is maximized when one takes allocation

$$\hat{\boldsymbol{\nu}} = c\left(\mathrm{Diag}\left(\hat{\boldsymbol{\pi}} \odot \hat{\boldsymbol{\alpha}}_2\right)\right)^{-1}\left(\hat{\boldsymbol{\pi}} \odot \hat{\boldsymbol{\mu}}\right) = c\hat{\boldsymbol{\mu}} \oslash \hat{\boldsymbol{\alpha}}_2. \tag{9.18}$$

The squared maximal Sharpe ratio of this portfolio is

$$\begin{aligned} \hat{\zeta}_*^2 &= \frac{n-1}{n}\left(f_{\mathrm{tas}}\left(q\right)\right)^2, \\ q^2 &= \left(\hat{\boldsymbol{\pi}} \odot \hat{\boldsymbol{\mu}}\right)^\top \mathrm{Diag}\left(\hat{\boldsymbol{\pi}} \odot \hat{\boldsymbol{\alpha}}_2\right)^{-1}\left(\hat{\boldsymbol{\pi}} \odot \hat{\boldsymbol{\mu}}\right) = \sum_j \hat{\pi}_j \frac{\hat{\mu}_j^2}{\hat{\alpha}_{2,j}}. \end{aligned}$$

Example 9.2.3 (Timing the Market, II). We consider the discrete feature market timing from Example 9.2.1, but let the variance vary among groups. We find the Markowitz portfolio is proportional to

$$\hat{\boldsymbol{\nu}}_*^\top \propto \begin{bmatrix} 0.60 & 0.14 & 0.42 & 0.35 & 0.31 & 0.24 & 0.59 & -0.01 \end{bmatrix}.$$

We compute $\hat{\zeta}_*^2 = 0.042\ \mathrm{mo.}^{-1}$. We test the null hypothesis that $\zeta_* = 0$ via the connection to Hotelling's T^2 as in Section 6.3.2. We compute the cutoff as $0.003\ \mathrm{mo.}^{-1}$ which is smaller than $\hat{\zeta}_*^2$ and we reject at the 0.05 level. Additionally we compute a 95% confidence interval on ζ_*^2, using the technique outlined in Section 6.3.1, as $[0.021, 0.07)\ \mathrm{mo.}^{-1}$. ⊣

One advantage of flattening is that we can easily impose portfolio constraints. For example, one can impose the hedging constraint that the flattened portfolio should have zero covariance with the rows of the $k_g \times k$ matrix G. One standard form of this analysis is hedging against the buy-and-hold portfolio, which corresponds to $\mathsf{G} = \mathbf{1}^\top$. The optimal portfolio is

$$\hat{\nu}_{*,\mathsf{I}\setminus\mathsf{G}} = c\left(\operatorname{Diag}\left(\hat{\pi} \odot \hat{\alpha}_2\right) - \left(\hat{\pi} \odot \hat{\mu}\right)\left(\hat{\pi} \odot \hat{\mu}\right)^\top\right)^{-1}\left(\hat{\pi} \odot \hat{\mu}\right)$$
$$- c\mathsf{G}^\top\left(\mathsf{G}\left(\operatorname{Diag}\left(\hat{\pi} \odot \hat{\alpha}_2\right) - \left(\hat{\pi} \odot \hat{\mu}\right)\left(\hat{\pi} \odot \hat{\mu}\right)^\top\right)\mathsf{G}^\top\right)^{-1}\mathsf{G}\left(\hat{\pi} \odot \hat{\mu}\right).$$

The squared Sharpe ratio of the optimal portfolio, $\Delta_{\mathsf{I}\setminus\mathsf{G}}\,\hat{\zeta}_*^2$, can be found by plugging in the expected mean and covariance into Equation 6.11. Inference is via the standard techniques applied to the flattened pseudoassets.

Example 9.2.4 (Timing the Market, Hedged). We continue Example 9.2.3, but impose a hedge against the buy-and-hold portfolio, corresponding to $\mathsf{G} = \mathbf{1}^\top$. We find the hedged Markowitz portfolio is proportional to

$$\hat{\nu}_{*,\mathsf{I}\setminus\mathsf{G}}^\top \propto \left[0.30 \quad -0.19 \quad 0.11 \quad 0.03 \quad -0.00 \quad -0.08 \quad 0.29 \quad -0.34\right].$$

We compute $\Delta_{\mathsf{I}\setminus\mathsf{G}}\hat{\zeta}_*^2 = 0.012\ \text{mo.}^{-1}$. Using the standard techniques from Section 6.3.1, we compute a one-sided 95% confidence interval on $\Delta_{\mathsf{I}\setminus\mathsf{G}}\ \zeta_*^2$ as $[0, \infty)\ \text{mo.}^{-1}$. ⊣

9.3 Market timing with a continuous feature

We now consider market timing model with a single continuous feature. Suppose that we observe some continuous random variable $f_{t-1} \in \mathbb{R}^n$ prior to the time required to capture the returns x_t. We assume that

$$x_t = \mu + \beta f_{t-1} + \epsilon_t,$$

where the error ϵ_t has zero mean, variance σ^2, and is independent of f_{t-1}. The sample estimates of β and σ^2, the optimal portfolio and linear passthrough, and inference on these quantities follows exactly by the techniques of Chapter 6 and Chapter 7. There is very little that stands out in this $k = 1$ case, and we mostly refer the reader to the previous chapters.

9.3.1 Reasonable timing of signals

We have entirely ignored the issue of *integration* wherein a time series is not stationary unless one performs some number of differencing operations. While we do not expect our returns series to be stationary *pe se*, as the volatility may change, we generally rely on some kind of long term average mean and volatility of returns. When one performs a linear regression on non-stationary time series, one may find that the relationship appears statistically significant even when the time series are independent, a phenomenon known as *spurious correlation*. While equity returns are generally not integrated, you should take care not to try to use features series which are integrated in market timing, as there is a "stationarity mismatch."

Here we consider a weaker form of this mismatch in the form of an autocorrelation analysis of the signal and returns. First, we outline a number of hypothetical market timing feature and returns pairs, and ask you to consider whether such a relationship seems plausible, conditional on momentum being a plausible strategy:

1. The previous trading day's returns of a stock used as a signal to predict today's returns.
2. The returns of a stock in the last five minutes of the previous trading day used to predict the next year's returns.
3. The returns of a stock over the past year used to predict today's daily returns of the stock.
4. The returns of a stock on this day one year ago used to predict today's daily returns of the stock.

There are at least three moving parts in each of these hypotheticals: the autocorrelation of the feature, the autocorrelation of the return, and the gap between them. There is little that we can say mathematically about the gap. Common sense would suggest that more recent events are more likely to be significant, and that high frequency random events from far in the past are unlikely to have any effect on future price movement. However, we can link the autocorrelation of returns, the autocorrelation of features, and the effect size. Under the model

$$x_t = \beta f_{t-1} + \epsilon_t,$$

where the ϵ are independent, we have

$$\frac{\rho_f}{\rho_x} = 1 + \frac{1}{T}, \tag{9.19}$$

where ρ_f is the autocorrelation of f_t, ρ_x is the autocorrelation of returns x_t, and T is the Hotelling Lawley trace of the strategy which allocates proportional to βf_{t-1}.

Thus we see that $\rho_f > \rho_x$, which is not surprising: the addition of the independent uncorrelated noise ϵ_t to βf_{t-1} should only decrease its autocorrelation. This means, however, that a high frequency feature cannot be used to predict a low frequency return. The model of item 2 above is not just implausible, it is impossible unless the returns over the last five minutes of a trading day are highly autocorrelated from one day to the next.

The other implication is that if f_{t-1} is a low frequency signal and thus highly autocorrelated, but x_t is not, then T should be very low. The model of item 3 above may be plausible, but we should expect $\rho_f \approx 1$, ρ_x should be very small, and thus $T \approx \rho_x$, which is to say the feature has not added much beyond simply using yesterday's returns instead.

9.4 Nonparametric market timing

One interesting case that we can consider here that is not a specific case of the more general problems we studied elsewhere is the problem of nonparametric market timing. In this case one observes scalar feature f_{t-1} prior to the time required to invest and capture returns x_t. Conditional on f_{t-1} one holds a certain amount long or short in the asset. We considered the optimal allocation in Section 5.3.2 when the population parameters are known, namely

$$\boldsymbol{\nu}_* \left(f_{t-1}\right) = c\mathsf{A}_2^{-1} \left(f_{t-1}\right) \boldsymbol{\mu} \left(f_{t-1}\right), \tag{9.20}$$

for some c, *cf.* Equation 5.40. The sample analogue for our scalar case is

$$\hat{\nu}_* \left(f_{t-1}\right) = c \frac{\hat{\mu} \left(f_{t-1}\right)}{\hat{\alpha}_2 \left(f_{t-1}\right)}. \tag{9.21}$$

Since the forms of the mean and second moment functions are unknown, the $\hat{\mu}(f)$ and $\hat{\alpha}_2(f)$ must be estimated somehow. The field of nonparametric (or semiparametric) regression is too large to be summarized in the space we have here. [Ruppert *et al.*, 2003] Essentially the functions $\hat{\mu}(f)$ and $\hat{\alpha}_2(f)$ are fit via splines or kernels or some other statistical wizardry. For this application we stress that care must be taken to ensure that the estimated $\hat{\alpha}_2(f)$ be strictly positive.

One situation in which this nonparametric problem arises is in the analysis of sizing of a fund manager with floating leverage. In this case, the observed x_t are the returns of the manager's holdings *adjusted to leverage 1*, and f_{t-1} is the manager's chosen leverage. If the fit $\hat{\mu}(f)/\hat{\alpha}_2(f)$ is close to $y = x$, then the manager has chosen their leverage optimally. If it is not, then as overseer of the manager you might choose to double down (or halve down!) on their decisions in order to optimize signal-noise ratio.

Example 9.4.1 (Nonparametric Timing the Market with Volatility). We continue Example 9.1.1, taking the daily relative returns of the Market as the returns. For a feature f_{t-1} we use the raw signal of the running standard deviation of the daily relative returns over a 120 market day window, then delayed by 5 days. We then estimate $\hat{\mu}(f)$ and $\hat{\alpha}_2(f)$ via a general additive model using the **mgcv** package. [Wood, 2017] In Figure 9.4 we plot the fit $\hat{\mu}(f)/\hat{\alpha}_2(f)$ versus f. We see a general decrease, by a factor of around 10, in the optimal allocation across the values of f tested: higher volatility suggests lower, though still positive, allocation into the Market. ⊣

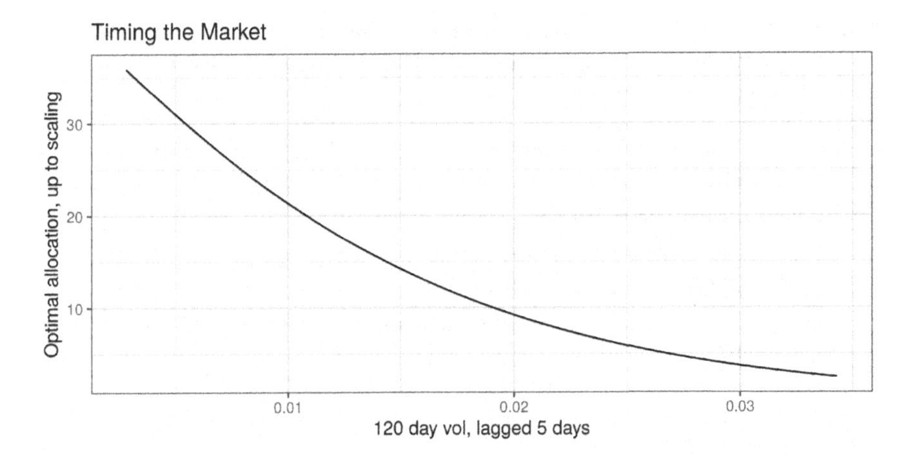

FIGURE 9.4
We plot the optimal allocation, $\hat{\mu}(f)/\hat{\alpha}_2(f)$ versus the feature f, defined to be the 120 day running volatility of the Market returns, lagged by 5 days. The first two moment functions were fit by a GAM.

9.5 Exercises

Ex. 9.1 Monthly timing signal Repeat the analysis of Example 9.1.1, but use a monthly signal. For simplicity, define the signal f_{t-1} based on the average absolute monthly returns of the Market over a 10 month window, then delayed by 2 months. That is, if x_t are the returns which correspond, say, to the month December, 2014, then compute the average absolute returns of January, 2014 through October, 2014, inclusive. Then let f_{t-1} be negative if that average is greater than 4% and positive otherwise.

1. Compute $\hat{\rho}$, and an approximate standard error on it. Estimate ζ from $\hat{\rho}$.
2. Compute $\hat{\kappa}$, and an approximate standard error on it.
3. Directly compute $\hat{\zeta}$ via a backtest.

Ex. 9.2 Range of κ The domain of $f_{\text{tas}}(\cdot)$ is $(-1, 1)$. For the signal-noise ratio of Equation 9.1 to be well defined, we should bound κ (trivially, $|\rho| \leq 1$).

1. Via Jensen's Inequality, show that $\kappa \leq 1$. Find a distribution such that equality holds.
2. Suppose x_t are drawn from a discrete distribution taking two values, one of which is zero. Find κ for these returns. Show that arbitrarily small κ are possible.

* 3. Suppose that we rescaled returns so that $\mathrm{E}\left[|x|\right] = 1$. Furthermore, suppose that the fourth moment of x_t is bounded, say less that M. Find a lower bound on κ.

Ex. 9.3 Timer covariance In Example 9.1.2, we found that the "backtested" Sharpe ratio of our toy market timing model was much lower than implied by Equation 9.1.

1. From this information, guess the *sign* of the covariance $\mathrm{Cov}\left(\mathrm{sign}\left(f_{t-1}\right)\mathrm{sign}\left(x_t\right), |x_t|\right)$.
2. Using the same data, compute a sample estimate of the covariance $\mathrm{Cov}\left(\mathrm{sign}\left(f_{t-1}\right)\mathrm{sign}\left(x_t\right), |x_t|\right)$.

Ex. 9.4 Insane abilities In Example 9.1.2, we estimated the $\hat{\kappa} = 0.66$ for daily returns of the Market.

1. Plot the implied signal-noise ratio versus ρ for ρ ranging from 0 to 1.
2. A *very* silly paper [Lachanski and Pav, 2017] claims that a market timer using several days' stale public information achieves $\rho = 13/15$. What is the implied signal-noise ratio of that timer, in annualized units? Does this seem very high? Does it seem plausible? (*cf.* Section 1.2.2 for plausible values of achieved Sharpe ratio.)

Ex. 9.5 Discrete state moments Derive the forms of moments of returns under discrete states.

1. Prove Equation 9.7.
2. Prove Equation 9.8.

Ex. 9.6 Discrete state optimal portfolio Derive the form of the optimal portfolio under discrete states.

1. Prove Equation 9.9.
2. Prove that for $q \geq 0$, the ζ_* of Equation 9.10 is increasing in q.

Ex. 9.7 Likelihood test for heteroskedasticity Under the discrete state model, suppose that conditional on the state, returns are normally distributed with mean μ_j and variance σ_j^2. That is,

$$x_t \,|\, f_{t-1} = \mathcal{S}_j \sim \mathcal{N}\left(\mu_j, \sigma_j^2\right).$$

Derive the likelihood based test for testing the hypothesis

$$H_0 : \forall j\, \sigma_j^2 = \sigma_1^2 \quad \text{versus} \quad H_1 : \exists i, j\, \sigma_i^2 \neq \sigma_j^2,$$

as an alternative to Levene's test.

1. Apply your test to the states and returns from Example 9.2.2.

Ex. 9.8 Calendar effects on the market Let the state indicators be the 12 months of the year. Using the monthly Market data from the Fama French factors, compute μ_j and σ_j^2.

1. Perform Levene's (or the likelihood) test for equality of the σ_j^2.
2. Compute the optimal allocation per month to maximize the multi-period Sharpe ratio assuming heteroskedasticity. Perform inference on ζ_*, and compare against the buy-and-hold signal-noise ratio.
3. Repeat that process but with a hedge against the buy-and-hold allocation. Perform inference on $\Delta_{\mathsf{I}\setminus\mathsf{G}}\,\zeta_*$.

* 4. Given approximate normality of the elements of $\hat{\boldsymbol{\nu}}_*$, how would you test the hypothesis that elements of $\boldsymbol{\nu}_*$ are all equal? Apply that test to this problem.

Ex. 9.9 Meta momentum Using the nonparametric timing approach, check on the optimal allocation into the Fama French UMD portfolio using the previous period's UMD returns as a feature. See also Exercise 6.10.

Ex. 9.10 Crisis market timing As in Exercise 6.39, define "post-crisis" as the Market level being 10% or more below its highest historical value. Using this binary state, compute the $\hat{\mu}_j$ and $\hat{\sigma}_j^2$, compute the portfolio to maximize multi-period $\hat{\zeta}_*^2$.

Ex. 9.11 Reasonable timing Derive Equation 9.19 assuming the ϵ are independent.

Ex. 9.12 Unreasonable timing Suppose you have a feature observable each day that is highly predictive of the returns over (only) the next 20 trading days. Describe how to turn this signal, and historical returns, into a signal highly predictive of (only) the 19th trading day from now.

1. Confirm your signal empirically.
2. Does this seem sensible to you?

Ex. 9.13 Single-period signal-noise ratio testing ☥ Consider market timing with continuous feature, $x_t = \mu + \beta f_{t-1} + \epsilon_t$. Describe how you would attempt to maximize the single period signal-noise ratio given some historical observations of returns and feature.

1. Using the techniques from Chapter 2, how might you test

$$H_0 : \left[1, f_{t-1}\right]^\top [\mu, \beta] = \sigma c \quad \text{versus} \quad H_1 : \left[1, f_{t-1}\right]^\top [\mu, \beta] > \sigma c$$

for fixed c?

* 2. That hypothesis test does not capture the fact that you have opportunistically chosen the direction of your trade. Instead, how would you

test the hypothesis

$$H_0 : \text{sign}\left(\hat{\mu} + \hat{\beta} f_{t-1}\right) [1, f_{t-1}]^{\top} [\mu, \beta] = \sigma c \quad \text{versus}$$
$$H_1 : \text{sign}\left(\hat{\mu} + \hat{\beta} f_{t-1}\right) [1, f_{t-1}]^{\top} [\mu, \beta] > \sigma c$$

for fixed c?

* **Ex. 9.14 Nonparametric calendar effects** Let the feature be a kind of normalized Julian date, defined as

$$2\pi \frac{\text{Julian date} - \text{Julian date for July 1 of that year}}{\text{Julian date for December 31 of that year}}.$$

This weird looking normalization takes care of leap years, maps July 1 to zero, New Year's Day to $-\pi$ and New Year's Eve to π.

1. Using the daily Market data from the Fama French factors, compute the mean and second moment of returns as some non-parametric function of the normalized Julian date. You will have to take into account the periodicity of the x variable. Is there a January effect?
2. Find the optimal allocation, to maximize the multi-period signal-noise ratio as a function of date. Plot it in polar coordinates, or some other way that takes into account the periodicity of the x.
3. What is the $\hat{\zeta}_*^2$ of this allocation?
4. How would you perform inference on the ζ_*^2 of this allocation? How would you perform inference on $q\left(\hat{\boldsymbol{\nu}}_*\right)$?
5. How would you adapt this procedure to hedge out the buy-and-hold allocation? How would you perform inference on the hedged allocation?

Ex. 9.15 Discrete state portfolio, multiple assets † Adapt the discrete state market timing model of Section 9.2 to the case where you have k assets.

1. Assuming heteroskedastic errors, how would you maximize the multi-period signal-noise ratio? How would you perform inference on ζ_*^2? How would you test the hypothesis that the optimal portfolio is the same in each state? Is this any different than just using the flattening trick?
2. Assuming homoskedastic errors, how would you maximize the single-period signal-noise ratio? How would you perform inference on ζ_*^2? Is this any different than the conditional expectation model?

10

† *Backtesting*

> Writing code means always having to say you're sorry.
>
> Paul Ford

10.1 On backtesting

In the previous chapters we analyzed the returns of assets, but made no mention of where returns come from. For liquid assets traded on an exchange, computation of returns from prices is straightforward. To analyze baskets or portfolios of assets, for example the Fama French factors, or complex trading strategies, one must estimate their returns. The goal of *backtesting* is to produce an unbiased estimate of historical returns of a basket, portfolio, or trading strategy. You will almost certainly backtest a quantitative strategy before allocating money to it[1].

Creating a suite of signal generation and backtesting routines that are fast, flexible and easy to use, and time-safe is a serious software engineering challenge. In this, the least technical chapter of this book, we offer our opinions on backtesting:

- One should recognize that backtested returns are indeed only estimates, as there is uncertainty in the price at which you could have bought or sold the real assets on the market, an uncertainty which depends on the simulated traded volume. Moreover, there is uncertainty in second-order effects which are much harder to quantify. For example, other market participants might notice your trading activity and react to it: front-running your orders, copying your trades if you are profitable, and so on.
- As backtests are estimates, they include error. Some of that error is natural, arising from the uncertainty of achieved price of your trades: if you

[1] Though it is said that Chuck Norris does not backtest his strategies. [Wikipedia contributors, 2020]

DOI: 10.1201/9781003181057-10

had submitted a limit order of a given size at a given time, it is hard to say exactly how many shares would have been filled, and how your order would have affected bid and ask prices immediately following your trades. One hopes that by careful design and engineering, this uncertainty in your historical simulations is unbiased and small in magnitude. Some errors, however, are due to code bugs or logical fallacies. These are more pernicious as they are usually not unbiased. A strategy that appears profitable only because of bugs in your backtester is jocularly said to be exploiting *backtest arbitrage.* Because most trading ideas that you conjure up are not actually profitable, unless your backtesting system is free from such bugs, strategies which look profitable are more likely to be the result of type I errors.

- If you are a casual quantitative trader, or just exploring quantitative trading, you will likely mingle your predictive code and trade simulation code together[2]. This is a natural pattern for quick studies, but is likely to be buggy. You should move to a design pattern where there is separate dedicated code for performing the predictive part (or generation of signals) and separate standalone code for performing simulations on those predictions. The latter we might call a "backtester" or "backtesting harness." As there are not many freely available standalone backtesting programs, you will likely have to write these yourself.

- We would argue that one might want several such backtesting routines, for different purposes. We envision them as a pyramid, as illustrated in Figure 10.1: at the bottom are the commonly-used backtesters which are quick but low-fidelity; moving up the period gives one increasingly higher fidelity simulations but at greater computational cost and complexity. The lower fidelity backtesters will have a higher type I rate, in that their simplified model of trading costs will likely be too optimistic. But these can have high enough power to pass profitable strategies up to the next level, while rejecting obviously unprofitable strategies. One will likely spend more time working at the base of the pyramid, as profitable strategies are quite rare.

    1. At the bottom of the pyramid we have placed "thought experiments." While we recommend *actually thinking about* the market inefficiency you wish to capture, this backtester is almost useless due to its extremely high type I rate: most quants could convince themselves that any given strategy could be profitable, and then do the same for the other side of the trade. However, thinking about one's theory for market inefficiency can potentially reveal side effects other than trading profits that can be analyzed separately. For example, a purely technical trading rule could potentially be applicable in markets other than the one you are backtesting in; a strategy on volatility futures

[2]Typically in a 3,000 line script.

FIGURE 10.1
Different forms of backtesters are illustrated. At the bottom of the pyramid are low fidelity simulations, which should be run frequently. As we move up the pyramid, the computational burden increases, while the type I rate decreases.

might embed a forecast of volatility that can be analyzed by itself; a pattern in one kind of agricultural futures should likely appear in others, but perhaps not in precious metals futures. Moreover, while this may seem surprising, potential investors are often reassured by the existence of a justification for a trading strategy, and will shy away from inexplicable money-printing schemes that seem to arise purely from the alchemy of computation.

2. Next up the pyramid we include "correlation studies." By this we mean one construct a time-by-assets matrix of signals, line these up against a corresponding matrix of *adjusted returns*, and measure correlation. This is justified by Example 5.3.4, where we found that $\zeta_* \approx |\rho|$ where ρ is the correlation between signal and return. There are at least two ways to define "correlation" here: for each asset, one can consider the correlation of signal and returns over time for that asset, which you might consider the "longitudinal" correlation; or for each prediction time, one can consider the correlation of signal and returns across assets, the "cross-sectional" correlation. Moreover, while the relation $\zeta_* \approx |\rho|$ refers to the Pearson correlation, one should be able to turn significant Spearman or Kendall correlation into a profitable trading strategy.

   Besides correlation studies, we would here add backtests on forecasts of quantities other than returns. Suppose, for example, that you thought you could forecast the beta of stocks to the market better than the consensus forecast. You might then backtest your beta forecasting method, perhaps comparing to some strawman forecasting method, evaluating your method based on known statistical properties. It is rarely clear how to turn these "other forecasts" into a profitable trading strategy, however. For example, having a good forecast of beta to the market doesn't obviously translate into a trading strategy. Before performing a backtest of an other forecast, you should devise the trading strategy that follows from it and backtest it with some amount of clairvoyance of the actual other quantity. In our example, you would cheat and use the actual out-of-sample beta to the market of stocks to first test your trading strategy. Obviously if having perfect forecasts of out-of-sample beta is not profitable, you need not improve your beta forecasts. This advice applies to the correlation studies as well: you should confirm that you can convert a significant Pearson (or Spearman or Kendall) correlation, either longitudinal or cross-sectional, into a profitable strategy by backtesting simulated signals with a backtester from higher up in the pyramid.

3. Next we include what might properly be called a "backtester": a proportional allocation simulator. This takes a time-by-asset matrix of signals[3] and simulates trading to achieve allocation under some very simple rules. One such rule would be to target a dollarwise allocation that is an affine shift of the signal for the given point in time, with the two free parameters fixed by a zero dollar mean constraint and a leverage constraint. That is, if vector $\boldsymbol{f}_t$ is a k-vector of features at time t, one allocate dollars proportional to $\hat{\boldsymbol{\nu}}_t = c_0 + c_1\boldsymbol{f}_t$ for some $c_1 > 0$ such that $\mathbf{1}^\top \hat{\boldsymbol{\nu}}_t = 0$ and $\mathbf{1}^\top \left|\hat{\boldsymbol{\nu}}_t\right| = 1$.

   One can entertain other methods for translating a signal into an allocation, each of which might require a different backtester for reasons of efficiency and coder sanity. These methods are described by the equivalence classes of signals they implicitly define. For example, under the proportional allocation model we describe above, two signals are equivalent if they are a positive affine shift of each other for every time row, since they result in the same $\hat{\boldsymbol{\nu}}_t$. Another, coarser, equivalence class would be to call two signals equivalent if they have the same sign for every asset for every time row. Under this model, only the sign of the signal matters, and one imagines the corresponding allocation is $\hat{\boldsymbol{\nu}}_t = c_0 + c_1 \operatorname{sign}\left(\boldsymbol{f}_t\right)$. Another method is implied by considering two signals equivalent if they order the assets the same for every time row. In this case the allocation is some function of $rank\left(\boldsymbol{f}_t\right)$, the vector of ranks of $\boldsymbol{f}_t$. A good allocation in this case is known under a long-only constraint. [Chriss and Almgren, 2005]

4. The simple fill backtester might take only *adjusted prices* (of which, more below), but then cannot properly account for per-share com-

[3]Or potentially a higher dimensional object, time-by-asset-by-some-knob-you-want-to-test.

missions. If the backtester also accepts actual prices, then per-share commissions can be properly accounted for. Typically one models a fixed rate of *slippage* as a proportion of dollar turnover. That is, for every dollar traded, assume you lose one tenth of a cent, or ten bps, to trade impact[4]. This is a very rough model of impact, but is perhaps less harmful than assuming no impact at all. In reality your achieved price and volume will depend on how you trade, but also on the volatility and liquidity of the asset. So next up the pyramid we envision backtesters with the same inflexible rules for allocation, but a more refined model for impact, which require more data and impose certain assumptions about impact. The impact model of Almgren *et al.* [2005] can be implemented with a few extra pieces of data, namely the volatility, typical volume and liquidity of each asset. This model expresses the temporary impact J, defined as a relative return to the arrival price, as

$$J = \sigma \left(\eta \left| \frac{X}{VT} \right|^{\beta} + \frac{\gamma T}{2} \left| \frac{X}{VT} \right|^{\alpha} \left(\frac{\Theta}{V} \right)^{\delta} \right) + \text{noise}, \tag{10.1}$$

where X is the number of shares traded, VT is the typical volume traded over the same time window, Θ/V is the inverse turnover of the stock, and $\alpha, \beta, \gamma, \delta$ and η are fit by a regression.

One can write a backtester that accepts this information and outputs not just a time series of simulated net asset value, but rather a general object that can produce simulated NAV conditional on a certain amount invested at each time step. The trick is to express the shares traded in a stock, X_i as some proportion of your total simulated NAV: $X_i = w_i A$, where A is the simulated NAV. Then the expected temporary impact relative to the simulated NAV, as a loss, is

$$A^{\beta} \sum_i w_i S_{0,i} \sigma_i \eta \left| \frac{w_i}{V_i T_i} \right|^{\beta} + \frac{A^{\alpha}}{2} \sum_i w_i S_{0,i} \sigma_i T_i \left| \frac{w_i}{V_i T_i} \right|^{\alpha} \left(\frac{\Theta_i}{V_i} \right)^{\delta}, \tag{10.2}$$

where $S_{0,i}$ is the arrival price of stock i (we drop time indexing to avoid confusion), and all other i indices refer to the attributes of stock i. This temporary impact for a given trading period can be re-expressed as $c_1 A^{\alpha} + c_2 A^{\beta}$ with two constants for each time period. Thus one could create a backtester that produces the raw impact-free returns as well as two time series of the constants c_1, c_2. From this one could then easily reconstruct the simulated NAV for different starting NAV values. This gives you the ability to quickly judge the *capacity* of a trading strategy, by tweaking the starting NAV.

5. As previously noted the backtesters at the base of the pyramid require

[4]We do not mean to suggest this exact value, this is only an example.

less data and run quickly. A correlation between signal and adjusted returns, once both are computed, can typically be computed in much less than a second. The proportional allocation backtester can easily be implemented to take two price matrices and signal and produce simulated NAV values in around a second for simulated daily trading of hundreds of stocks over around ten years. The impact backtester requires two more matrices and takes perhaps twice as long. However, these fast backtesters assume very simple allocation rules which are not typically appropriate for actual use. For example, one may wish to impose additional hedging constraints to one's portfolio, or place limit orders in a certain manner that cannot easily fit into the simplistic API of one of the fast backtesters.

So next up the pyramid we find what would traditionally be considered a backtester: a fairly general but probably slow testing harness that accepts an object which can perform fill simulation, an object with a method to turn data and current allocations into requested trades. This backtester: wakes up on a specified schedule; simulates the fill of any pending orders; presents market data, time-safe signal and current simulated allocation to the trade-making method; passes the requested trades to the fill simulator; and marks the value of the portfolio. This backtester does very little other than orchestrate the passing of data between the various components, and provide a guarantee of time safety, which is in contrast with the fast backtesters which did everything in one piece of code.

If written with sufficient generality the trade-making method can, and perhaps should, be used in real trading. This "protoduction" development model avoids having to re-implement trading code for actual trading, which speeds time to trading, but can also inject "bugs" into your actual net asset value.

The allocation method is often written to use an optimizer to create desired trades. While this is a very good way to ensure that certain portfolio constraints are met (*e.g.*, zero net exposure, zero beta, sector gross leverage constraints, *etc.*), using such an allocation can slow down your backtests dramatically. Moreover, your objective function, whether it is forecast signal-noise ratio or some other quantity, will include copious amounts of noise in the estimation of returns. You may find that an optimizer that works well when population quantities are plugged into the objective performs rather poorly when forecast quantities are used.

Optimizers are notoriously ill-behaved in this domain. You may find, for example, that when the objective is maximizing the forecast signal-noise ratio, given zero net exposure and maximum position sizing constraints, that an optimizer returns a target position which is c proportion of your wealth long in $k/2$ of the assets and c proportion short in the remaining assets. The further imposition of a zero beta constraint, say, might only change the allocation in one asset. Optimized allocations lack nuance. You may spend a fair amount of time and energy tuning your optimizer; recognize, however, that if you are changing your constraints and objective to match some preconceived notion of what a portfolio should look like, then the optimizer is no longer really doing anything other than confirming your priors.

6. Next we have paper trading, which is the real-time use of strategy code to generate and submit orders to some test endpoint. Paper trading is too slow[5] to establish statistical significance of the Sharpe ratio of your strategy, rather it is used to find any bugs in your model of when data are available, and to ensure that your trading code doesn't submit insane orders. This is more important if you plan to submit orders directly to the market, rather than have a human in the loop.
7. At the apex of the pyramid we have actual trading of actual money. Like paper trading, live trading is a slow way to collect data, but actual returns have a much lower type I rate than backtested returns.

   We would argue that, despite widespread use of the terms, there is no in-sample and out-of-sample data in quantitative trading, rather there is in-sample data and live trading with real money. After all, if you cross-validated a trading strategy and found it performed poorly in the "out-of-sample" data, you wouldn't actually deploy the strategy, so how is it out-of-sample?

- In the lower half of the pyramid computational efficiency is key, as you will find yourself conducting hundreds or thousands of such simple backtests. Computational efficiency is aided by the use of adjusted prices or adjusted returns. Adjusted prices are prices that have been adjusted for splits and dividends (via some assumed reinvestment scheme). While this is relatively straightforward, adjusting for mergers, acquisitions, spin-offs and other corporate actions is less clear.

  Use of adjusted returns is a leading source of *time traveling*, a form of information leakage that is only available in simulation. [Dama, 2011; David, 2019] For example, suppose you created backwards-adjusted relative prices of stocks, in the sense that they match actual prices at the last time of the simulation, but not earlier due to splits and dividends. A strategy that simply invests long in lower (adjusted) price stocks will tend to catch stocks that have been successful over the investment period.

  This example illustrates the importance of point-in-time universe construction. Your experiments at the lower end of the pyramid must be restricted to only those assets which would have passed some filtering conditions at

[5] You collect only one day's data per day.

that point in simulated time. Reconciling this with mergers and spinoffs and the rise and demise of companies is complicated. The idea that a company is a fixed quantity throughout time and that its shares reflect ownership in some stationary business activities is highly susceptible to Ship of Theseus arguments. For example, Citigroup came to exist through the merger of Citicorp and Travelers Group in 1998, only to see Travelers spun off in two pieces in the early 2000's, then Citigroup was partly nationalized and nearly de-listed following the Subprime Mortgage Crisis, then denationalized. The story of "Citigroup" over time does not easily reconcile to a single linear time series, yet if you query, say, Google Finance for ticker symbol `C` it will happily produce a price series going back to 1986. It is not clear, however, what stock (or stocks) you could have bought in 1986 to result in the holding of one share of `C` today. Your price-adjustment scheme, universe membership mask, time series operations and fast backtesters have to make some sense of this mess, producing a coherent time series for stocks that would have been available at a given point in time without leaking information from the future.

Universe composition issues will also complicate use of data from third party vendors. Vendors will often backfill stocks into data on the request of customers. You should assume that data you purchase from vendors requires special handling to remove backfill and survivorship biases. In this context, backfill bias is the retroactive inclusion of the historical record of a stock into a database because it passes for inclusion at the current point in time. Survivorship bias here means that stocks which are acquired or de-listed do not have any record in the database.

- Simulation of price impact, or slippage, is a major contributor of uncertainty in backtests. While there are several plausible models of slippage, they each require calibration of parameters. Calibration is difficult since it would seem to require one to estimate the difference between observed and unobserved alternatives, namely what the fill price would have been had one submitted an order when one actually had not. Instead impact models are estimated via observed price movement and achieved average fill price relative to the arrival price, which is the midpoint price when the order is submitted. For example, the impact model of Almgren *et al.* [2005] requires estimation of parameters in two equations

$$\frac{S_{post} - S_0}{\sigma S_0} \sim \gamma T \left|\frac{X}{VT}\right|^{\alpha} \left(\frac{\Theta}{V}\right)^{\delta} + \text{noise},$$

$$\frac{\bar{S} - S_0}{\sigma S_0} - \frac{1}{2}\frac{S_{post} - S_0}{\sigma S_0} \sim \eta \left|\frac{X}{VT}\right|^{\beta} + \text{noise},$$

where S_0 is the arrival price, S_{post} is the price some time after trading has completed, $\bar{S}$ is the volume-weighted average price achieved, X is the share size, σ is the volatility of the log price, T is the length of time over which the trade ocurred, VT is the average shares traded over the same time period historically and Θ/V is the illuiqidity.

We would argue that this model is actually underspecified. In particular the first equation which represents "permanent impact" (and perhaps the second) should be expanded to include a factor model of some kind. When the author fit the first equation with an added term for the contemporaneous returns of the S&P 500 index, the fit coefficients changed radically, and the market term was by far the most significant one in the regression[6]. This opens the gates for adding more terms to the model to represent other factor exposures.

Additionally, if you chose to calibrate such a model to your own trading, the perceived trading impact will likely vary with your strategy. Suppose, for example, that you trade a strategy that accurately forecasts a relatively slow movement in a stock's price, but your strategy triggers halfway through the price movement. If you measure the difference in arrival price and fill price, you should see apparent slippage whether you have actually traded or not. In fact the *entire point of your strategy* is to capture such a price movement. You might simply be "riding a wave," rather than "causing a wake." A momentum strategy should then appear to have different slippage than a mean-reversion strategy. This is why the coefficients published by Almgren *et al.* should be viewed with a fair amount of suspicion, since the client trades analyzed (trades executed by Citigroup Equity Trading desk for 19 months in 2001 through 2003) arise from some unknown process and might be exposed to movements of the market as a whole or have other factor exposures.

- The approximate backtesters towards the bottom of the pyramid are useful for establishing relationships in your data. These relationships may be enlightening even if the putative effect would not survive realistic impact modeling. For example, a common exercise is the *lead-lag profiling* of a signal, where the same signal is quickly backtested with artificial leads and lags added to the observable time of the signal. The result is a plot of backtested signal-noise ratio versus Δt, where one has added Δt to the effective time at which one could observe the signal. This will allow you to broadly classify the signal as "trend following," "mean reverting" or neither by looking at the signal-noise ratio for slightly negative Δt. Also it allows one to determine the alpha decay profile of the signal: how quickly must it be traded, how much would it help to compute it earlier if possible, *etc.* A relatively flat profile across Δt is cause for suspicion if the signal is nominally high-frequency, as this would indicate that the predictive value is not achieved from what the signal computes but from

[6] Given the size of the trades analyzed, it would be hard to argue that the S&P 500 index was impacted by the analyzed trades.

the background universe construction or your backtesting system. A flat profile is an indicator of time leakage.

- The output from a backtest is often a simulated NAV in the form of a scalar time series. More generally, however, the backtesters near the bottom of the pyramid should probably be designed to accept multidimensional signals and output multidimensional time series, for purposes of efficiency and for improved bookkeeping. One should be able to simultaneously test multiple variants of a signal. But also a backtester could output multiple results based on assumptions about fill.

  More generally it may help to think of the types of input and output, and whether they should support certain algebraic operations. The most obvious would be the ability to glue together non-overlapping time series, and have that operation commute with the backtester. For example, the "append" operator is a well-defined non-commutative operator for time series: if you have time series data for multiple stocks through March, you can easily append the data for April to the end, and the order matters. The same append operator makes sense for the output of a backtester, obviously so if it is a single time series of simulated NAV values. A clever algebraic trick, which trivializes common walk-forward cross validation analyses is for the backtesting operation and the append operation to commute with each other. [Izbicki, 2013] This makes backtesting embarrassingly parallel, because you could backtest each year on a separate thread.

  Taking this idea further, certain other operations on the signals can be made to commute with the backtesting operation. For the proportional allocation backtester, if you add a single constant to the time series, or scale it by a positive constant, the backtest output is unchanged, so these operations can trivially be made to commute with backtesting. By careful design one could enlargen the output of the backtester so that certain modifications to the signal which would appear to be *preprocessing* can be achieved in *post-processing*. This requires careful design of the objects and the supported operations.

- A backtest simulates a kind of counterfactual: that you had traded something that you in fact did not. There are different kinds of counterfactual that you could test, and you should be very clear which you are testing. For example, you might wish to backtest the returns of a static dollarwise portfolio, which should be very simple. However if that static portfolio is the Markowitz portfolio of the assets in your universe, as computed using all the data through 2019, then you could not have known that information prior to 2019. This is perfectly reasonable if your goal is to evaluate the returns of the 2019 Markowitz portfolio, though the techniques of Chapter 8 should be used to dampen your overoptimism. However, if you want to estimate the profitability of the strategy that recomputes the Markowitz portfolio monthly and allocates to it, you would likely simulate the returns of that historically. This has a greater degree of "time hygiene" than backtesting the 2019 Markowitz portfolio, but in the early period of your simulations, the Markowitz portfolio is built with less data and so has greater error than you might expect going forward, adding to the noise of your simulations. The counterfactual now is that you would have chosen to use the Markowitz portfolio to select portfolios. This seems reasonable as the Markowitz portfolio is a well-established technique, but if you are developing a more elaborate method for portfolio selection, you might not have known to do that in the past. The results from backtests always contain some amount of information in the future of the period simulated.

- Whatever you feel about machine learning and automatic strategy generation, they are excellent techniques for exposing bugs in your simulation code and metrics for choosing from among strategies. Humans find bugs at a slower rate.
- Backtesting routines have biases, mostly in their models of fill simulation. If you trade multiple strategies based on backtests, they will have correlated errors due to the use of common fill simulation models. The natural way to diversify these errors away is to trade strategies at different frequencies, or in different markets. There are obvious operational difficulties with this.

Given the complexities of producing a time-safe backtest, the uncertainty of trade impact, the low incidence rate of profitable trading strategies, and the temptation to overfit, Bayes' rule suggests that a strategy which looks profitable in a backtest is likely instead to be the product of some error. In brief:

A bitter truth

It's probably a type I error.

10.2 Exercises

Ex. 10.1 Defining correlation Given multiple observations of aligned returns and feature, how would you compute their "correlation?" Should you compute the correlation on each asset and take an average? Or compute the correlation laterally on each observation and take an average?

1. How is this affected by the covariance of returns and the covariance of the feature vector?

Ex. 10.2 Detecting backfill How would you detect backfill bias in a vendor-supplied dataset? Pose the problem as a statistical test.

Ex. 10.3 Detecting time travel Suppose that your backtesting code is actually leaking information about future splits and dividends. How might you detect this effect in backtested returns?

Ex. 10.4 Testing anomalies A caricature of quant strategy development is that the quant creates a trading rule parametrized by vector $\boldsymbol{\theta}$, then backtests hundreds of values of $\boldsymbol{\theta}$. For example, the elements of $\boldsymbol{\theta}$ in a technical trading strategy might encode the lookback window to some moving Z-score of returns as well as a threshold for making a trade, and so on.

1. Suppose that the signal-noise ratio of this family of trading strategies is linear in some transformation of the parameters: $\zeta_*(\boldsymbol{\theta}) = \boldsymbol{\beta}^\top f(\boldsymbol{\theta})$ for some vector-valued $f(\cdot)$. Does this model make sense? What kinds of $f(\cdot)$ seem sensical to you? Can you think of potential problems with unbounded functions?
2. Stealing an idea from Gaussian processes, one could model the correlation of returns of two strategies, $\boldsymbol{\theta_1}$ and $\boldsymbol{\theta_2}$ as $K(g(\boldsymbol{\theta_1}), g(\boldsymbol{\theta_2}))$, for some vector-valued transformation $\boldsymbol{g}\cdot$ and some "Kernel" function $K(\cdot,\cdot)$. How would this affect the correlation of the $\hat{\zeta}_*^2$ of backtested returns?
3. Suppose for given $f(\cdot)$, $g(\cdot)$, and $K(\cdot,\cdot)$ you wished to estimate the $\boldsymbol{\beta}$ that determines the signal-noise ratio. You have backtested hundreds of $\boldsymbol{\theta_i}$. How would you estimate $\boldsymbol{\beta}$ taking into account the correlation of backtested returns?

Ex. 10.5 Research problem: estimating market impact How would you design an experiment to estimate the impact of your own trading? What confounding variables should you control for?

1. How can you be sure you are measuring the "wake" of your trading and not the "wave" your strategy is riding?
2. How much would it cost, in real dollars, to perform your experiment?

A

Prerequisites

> The trouble with being educated is that it takes a long time; it uses up the better part of your life and when you are finished what you know is that you would have benefited more by going into banking.
>
> PHILLIP K. DICK
> *The Transmigration of Timothy Archer*

This chapter contains some preliminary material assumed in the rest of the text: a hodgepodge of definitions, notational declarations, *etc.* This chapter cannot include all the necessary prerequisites, and we generally assume working familiarity with linear algebra and some foundation in probability and statistics.

A.1 Linear algebra

As much as possible, we will denote vectors by bold lower case Roman letters: $\boldsymbol{v}$, $\boldsymbol{x}$, $\boldsymbol{y}$, *etc.* Matrices will typically be denoted in Roman or Greek bold upper case letters: A, X, Σ. The transpose of $\boldsymbol{x}$ is denoted $\boldsymbol{x}^\top$. For non-square or rank-deficient A, we will sometimes write A^C for a matrix whose rows span the null space of the rows of A; that is $\mathsf{A}^C\mathsf{A}^\top = \mathsf{0}$. A vector is typically considered a column vector so that $\mathsf{A}\boldsymbol{x}$ and $\boldsymbol{x}^\top\mathsf{Q}\boldsymbol{x}$ are well-formed.

We will denote a vector of all ones by $\mathbf{1}$; all zeros by $\mathbf{0}$. The identity matrix is denoted I; the zero matrix $\mathsf{0}$. Let $\boldsymbol{e}_i$ be the ith column of the identity matrix, where the size of the vector should be implicit from context. Similarly, define the "single entry" matrix, J^{ij} as the matrix of all zeros but for a single 1 in the i, j location. Equivalently, we can write $\mathsf{J}^{ij} = \boldsymbol{e}_i{\boldsymbol{e}_j}^\top$.

Definition A.1.1 (Matrix operations). We will write A^{-1} for the inverse of non-singular matrix A. For square symmetric non-singular matrix A, let $\mathsf{A}^{1/2}$ be the lower triangular *Cholesky factor* of A, defined as a lower triangular

DOI: 10.1201/9781003181057-A

matrix such that $\mathsf{A}^{1/2}\mathsf{A}^{\top/2} = \mathsf{A}$. Let the inverse of $\mathsf{A}^{1/2}$ be denoted by $\mathsf{A}^{-1/2}$; the inverse of $\mathsf{A}^{\top/2}$ is $\mathsf{A}^{-\top/2}$.

Definition A.1.2 (Matrix functionals)**.** For square matrix A, the *trace* of A, denoted $\operatorname{tr}(\mathsf{A})$ is the sum of the diagonal elements:

$$\operatorname{tr}(\mathsf{A}) = \sum_i A_{ii}.$$

Let $|\mathsf{A}|$ be the determinant of A.

The trace has a convenient property that products inside the trace can be rotated: $\operatorname{tr}(\mathsf{AB}) = \operatorname{tr}(\mathsf{BA})$. So in particular one has the identity $\boldsymbol{x}^\top\boldsymbol{x} = \operatorname{tr}\left(\boldsymbol{x}\boldsymbol{x}^\top\right)$.

Definition A.1.3 (Converting between matrices and vectors)**.** For vector $\boldsymbol{x}$, define $\operatorname{Diag}(\boldsymbol{x})$ as the diagonal matrix whose diagonal equals $\boldsymbol{x}$. For matrix A, define $\operatorname{diag}(\mathsf{A})$ as the vector of the diagonal of A. For matrix A, let $\operatorname{vec}(\mathsf{A})$, and $\operatorname{vech}(\mathsf{A})$ be the vector and half-space vector operators. The former turns an $p \times p$ matrix into an p^2 vector of its columns stacked on top of each other; the latter vectorizes a symmetric (or lower triangular) matrix into a vector of the non-redundant elements. Let L be the "Elimination matrix," a matrix of zeros and ones with the property that $\operatorname{vech}(\mathsf{A}) = \mathsf{L}\operatorname{vec}(\mathsf{A})$. The "Duplication matrix," D, is the matrix of zeros and ones that reverses this operation: $\mathsf{D}\operatorname{vech}(\mathsf{A}) = \operatorname{vec}(\mathsf{A})$ for symmetric A. [Magnus and Neudecker, 1980] Note that this implies that

$$\mathsf{LD} = \mathsf{I}\,(\neq \mathsf{DL}).$$

Let K be the "commutation matrix," the matrix whose rows are a permutation of the rows of the identity matrix such that $\mathsf{K}\operatorname{vec}(\mathsf{A}) = \operatorname{vec}\left(\mathsf{A}^\top\right)$ for square matrix A. [Magnus and Neudecker, 2007, 1979] Let N be the "symmetric idempotent matrix" $\mathsf{N} := \frac{1}{2}(\mathsf{I} + \mathsf{K})$. (Here we use ":=" to mean "defined as." The equation above should be read as "N is defined as $\frac{1}{2}(\mathsf{I} + \mathsf{K})$.") This matrix has many interesting properties, for which we refer the reader to Magnus and Neudecker [1986].

Definition A.1.4 (Matrix products)**.** For conformable matrices A, B, the matrix product is denoted by string concatenation: AB is the product. The *Kronecker product* will be denoted by $\mathsf{A} \otimes \mathsf{B}$. This is defined blockwise as

$$\mathsf{A} \otimes \mathsf{B} := \begin{bmatrix} A_{11}\mathsf{B} & A_{12}\mathsf{B} & \dots & A_{1n}\mathsf{B} \\ A_{21}\mathsf{B} & A_{22}\mathsf{B} & \dots & A_{2n}\mathsf{B} \\ \vdots & \vdots & \ddots & \vdots \\ A_{m1}\mathsf{B} & A_{m2}\mathsf{B} & \dots & A_{mn}\mathsf{B} \end{bmatrix}. \tag{A.1}$$

A handy identity connecting the Kronecker product and vectorization function is that

$$\operatorname{vec}(\mathsf{AXB}) = \left(\mathsf{B}^\top \otimes \mathsf{A}\right)\operatorname{vec}(\mathsf{X}). \tag{A.2}$$

The *Hadamard product*, or elementwise product, is denoted by $\mathsf{A} \odot \mathsf{B}$. This is defined only for matrices of the same size, and is defined elementwise: $(\mathsf{A} \odot \mathsf{B})_{ij} = A_{ij}B_{ij}$. Similarly we may define Hadamard ratios: $\mathsf{A} \oslash \mathsf{B}$ is the elementwise ratio $(\mathsf{A} \oslash \mathsf{B})_{ij} = A_{ij}/B_{ij}$. At times we may write the Hadamard power (*i.e.*, elementwise) of a vector or matrix as $\boldsymbol{v}^k$ or A^k.

Definition A.1.5 (Sherman-Morrison-Woodbury identity)**.** The "Woodbury identity" is useful for computing the inverse of a matrix which is a low-rank update to a matrix with an easily, or previously computed inverse:

$$(\mathsf{A} + \mathsf{UCV})^{-1} = \mathsf{A}^{-1} - \mathsf{A}^{-1}\mathsf{U}\left(\mathsf{C} + \mathsf{VA}^{-1}\mathsf{U}\right)^{-1}\mathsf{VA}^{-1}. \tag{A.3}$$

When U and V are vectors, this identity is often referred to as the "Sherman-Morrison" identity.

A.2 Probability distributions

A comprehensive treatment of probability distributions is beyond the scope of this book, or any one single book. The reader is directed to the texts of: Walck [1996], a good free overview of univariate distributions; Krishnamoorthy [2006], for a more theoretical, but similar overview; Press [2012], for multivariate distributions.

We assume a familiarity with the basics of probability distributions: the *density* of a distribution gives the local likelihood of a realization of a random variable. We may also refer to this as the probability density function or PDF. For a univariate distribution, the integral of the density gives the *(cumulative) distribution* function, or CDF. The CDF of random variable X is the function $f(x) = \Pr\{X \leq x\}$. The *quantile function* is the inverse of the cumulative distribution function.

A.2.1 Univariate probability distributions

Here we list the univariate distributions we may need in this book, along with our notational preferences.

normal distribution

Also known as the *Gaussian distribution*, this is the granddaddy of continuous distributions, and tends to arise whenever you sum independent random variables, a consequence of the central limit theorem. We write $x \sim \mathcal{N}(\mu, \sigma^2)$ to denote that x is normally distributed with mean μ and variance σ^2, which means that x has the PDF

$$\phi(x) = \frac{1}{\sqrt{2\pi\sigma^2}} e^{-\frac{(x-\mu)^2}{2\sigma^2}}.$$

The CDF of x has no nice closed form, but we write it as $\Phi(x)$. We often write Gaussian variables as z. The symbol for the pth quantile of the normal distribution is z_p.

chi-square

This distribution has the stochastic representation as the sum of the squares of independent zero-mean normal random variables. These can also arise approximately as the sums of squares of independent random variables, due to a kind of central limit theorem for squared variables. [Hall, 1983] When one sums v independent zero-mean normals, we say that the resultant sum has v *degrees of freedom.* We write $x \sim \chi^2(v)$ to mean that x follows a chi-square law with v degrees of freedom. We write $f_{\chi^2}(x; v)$ for the density, and $F_{\chi^2}(x; v)$ for the cumulative distribution of the chi-square distribution with v degrees of freedom. We write ${\chi^2}_p(v)$ for the pth quantile. To be precise, we have

$$f_{\chi^2}(x; v) = \frac{1}{2^{v/2}\Gamma(v/2)} x^{v/2-1} e^{-x/2}.$$

The chi-square is a specific realization of the more general *Gamma distribution.*

non-central chi-square

This is a chi-square where the summed normal variates do not have zero-mean. That is, if $z_i \sim \mathcal{N}(\mu_i, 1)$ are independent normal random variables then $x = \sum_{1 \le i \le v} z_i^2$ follows a non-central chi-square with v degrees of freedom and *non-centrality parameter* $\delta = \sum_i \mu_i^2$. We write this as $x \sim \chi^2(v, \delta)$. We write the density, distribution, and quantile functions as $f_{\chi^2}(x; v, \delta)$, $F_{\chi^2}(x; v, \delta)$, and ${\chi^2}_p(v, \delta)$, respectively. When $\delta = 0$ we recover the usual, or "central" chi-square distribution. In general, the non-central variant of a distribution describes the case where some variable in the stochastic representation does not have zero-mean.

chi

This is the positive square root of a chi-square. That is, if $x \sim \chi^2(v)$, then $\sqrt{x}$ follows a chi distribution.

t distribution

This arises as the ratio of a normal to an independent chi variate. Typically this occurs when dividing a sample average by some estimate of the standard deviation of the process. That is, if $z \sim \mathcal{N}(0, 1)$ and $x \sim \chi^2(v)$ are independent then

$$t = \frac{z}{\sqrt{x}}$$

follows a t distribution with v degrees of freedom. We write $t \sim t(v)$. We write the density, distribution and quantile of this distribution as $f_t(x; v)$, $F_t(x; v)$ and $t_p(v)$, respectively. For large v this distribution is very close to normal.

non-central t distribution

The ratio of a normal with non-zero-mean to an independent chi. That is

$$t = \frac{z}{\sqrt{x}}$$

when $z \sim \mathcal{N}(\delta, 1)$ and $x \sim \chi^2(v)$ are independent follows a non-central t distribution with v degrees of freedom and non-centrality parameter δ, written as $t \sim t(\delta, v)$. We write the density, distribution and quantile of this distribution as $f_t(x; \delta, v)$, $F_t(x; \delta, v)$ and $t_p(\delta, v)$, respectively.

F distribution

This is the (rescaled) ratio of independent chi-square variates. That is if $x_i \sim \chi^2(v_i)$ for $i = 1, 2$ are independent, then

$$f = \frac{x_1/v_1}{x_2/v_2}$$

follows an F distribution with v_1 and v_2 degrees of freedom, which we write as $f \sim F(v_1, v_2)$. We write the density, distribution and quantile of this distribution as $f_f(x; v_1, v_2)$, $F_f(x; v_1, v_2)$ and $f_p(v_1, v_2)$, respectively. The square of a t variable is a F variable, up to scaling. This distribution typically arises as the ratio of the square of some averages to the estimated variance of the process, and is used to test several hypothetical equalities simultaneously. For large sample sizes, the variation in the denominator is often ignored, and a chi-square approximation is made.

non-central F distribution

This is a F distribution with a non-central chi-square in the numerator. That is if $x_1 \sim \chi^2(v_1, \delta)$ and $x_2 \sim \chi^2(v_2)$ are independent, then

$$f = \frac{x_1/v_1}{x_2/v_2}$$

follows a non-central F distribution with v_1 and v_2 degrees of freedom, and non-centrality parameter δ, which we write as $f \sim F(v_1, v_2, \delta)$. We write the density, distribution and quantile of this distribution as $f_f(x; v_1, v_2, \delta)$, $F_f(x; v_1, v_2, \delta)$ and $f_p(v_1, v_2, \delta)$, respectively.

beta distribution

A beta random variable arises as the ratio of a chi-square to itself plus another independent chi-square. That is if $x_i \sim \chi^2(v_i)$ for $i = 1, 2$ are independent, then

$$b = \frac{x_1}{x_1 + x_2}$$

follows an beta distribution with $v_1/2$ and $v_2/2$ shape parameters. We write this as $b \sim \mathcal{B}(v_1/2, v_2/2)$. We write the density, distribution and quantile of this distribution as $f_\beta(x; v_1/2, v_2/2)$, $F_\beta(x; v_1/2, v_2/2)$ and $\beta_p(v_1/2, v_2/2)$, respectively. A trigonometric transform (Equation 1.24) relates the beta to the (rescaled) F distribution.

non-central beta distribution

This generalizes the beta distribution to the case where the numerator chi-square variate is noncentral. Thus if $x_1 \sim \chi^2(v_1, \delta)$ and $x_2 \sim \chi^2(v_2)$ are independent, then

$$b = \frac{x_1}{x_1 + x_2}$$

follows an beta distribution with $v_1/2$ and $v_2/2$ shape parameters and non-centrality parameter δ. We write this as $b \sim \mathcal{B}(v_1/2, v_2/2, \delta)$. We write the density, distribution and quantile of this distribution as $f_\mathcal{B}(x; v_1/2, v_2/2, \delta)$, $F_\mathcal{B}(x; v_1/2, v_2/2, \delta)$ and $\mathcal{B}_p(v_1/2, v_2/2, \delta)$, respectively.

uniform distribution

This distribution does not naturally arise, except perhaps via transformations of other variates. For $\alpha < \beta$, the uniform distribution on $[\alpha, \beta)$ is the one with constant density on that interval and zero elsewhere:

$$f_\mathcal{U}(x; \alpha, \beta) = \frac{1_{\{\alpha \le x \le \beta\}}}{\beta - \alpha}.$$

We write $x \sim \mathcal{U}(\alpha, \beta)$ to denote that x is so distributed. We write the density, distribution and quantile of this distribution as $f_\mathcal{U}(x; \alpha, \beta)$, $F_\mathcal{U}(x; \alpha, \beta)$, and $\mathcal{U}_p(\alpha, \beta)$, respectively.

A.2.2 Multivariate probability distributions

Many of the univariate distributions have multivariate analogues which we will encounter.

multivariate normal distribution

We say that k-vector $\boldsymbol{x}$ follows a *multivariate normal* (or *Gaussian*) distribution with mean $\boldsymbol{\mu}$ and covariance Σ, if the density of $\boldsymbol{x}$ is

$$\phi(\boldsymbol{x}) = \frac{1}{\sqrt{(2\pi)^k |\Sigma|}} \exp\left(-\frac{1}{2}(\boldsymbol{x} - \boldsymbol{\mu})^\top \Sigma^{-1} (\boldsymbol{x} - \boldsymbol{\mu})\right). \tag{A.4}$$

We write this as $\boldsymbol{x} \sim \mathcal{N}(\boldsymbol{\mu}, \Sigma)$.

The family of multivariate normal random variables is "closed" under affine shifts: let B be a $p \times k$ matrix whose row space has rank p, for $p \le k$, and let $\boldsymbol{a}$ be a p-vector. Then $\mathsf{B}\boldsymbol{x} + \boldsymbol{a} \sim \mathcal{N}(\mathsf{B}\boldsymbol{\mu} + \boldsymbol{a}, \mathsf{B}\Sigma\mathsf{B}^\top)$. By taking B to be rows of the identity matrix, one can then easily see that the multivariate distribution is closed under taking *marginals*. That is, any sub-vector of normally distributed vector $\boldsymbol{x}$ is normally distributed. Furthermore, when Σ is diagonal the individual elements of the vector $\boldsymbol{x}$, the x_i, are independent. Thus you could express vector $\boldsymbol{x} \sim \mathcal{N}(\boldsymbol{\mu}, \Sigma)$ as $\boldsymbol{x} = \boldsymbol{\mu} + \Sigma^{1/2}\boldsymbol{z}$ for $\boldsymbol{z} \sim \mathcal{N}(\boldsymbol{0}, \mathsf{I})$.

The density of the normal distribution can be expressed on a matrix parameter that combines the first two moments. [Pav, 2013] Suppose $\boldsymbol{x} \sim \mathcal{N}(\boldsymbol{\mu}, \Sigma)$. Define the matrix Θ as

$$\Theta := \begin{bmatrix} 1 & \boldsymbol{\mu}^\top \\ \boldsymbol{\mu} & \Sigma + \boldsymbol{\mu}\boldsymbol{\mu}^\top \end{bmatrix}. \tag{A.5}$$

Then the density of $\boldsymbol{x}$ is

$$\phi\left(\boldsymbol{x}\right) = \frac{e^{\frac{1}{2}}}{\sqrt{\left(2\pi\right)^{k}\left|\Theta\right|}} \exp\left(-\frac{1}{2}\operatorname{tr}\left(\Theta^{-1}\tilde{\boldsymbol{x}}\tilde{\boldsymbol{x}}^{\top}\right)\right), \tag{A.6}$$

where $\tilde{\boldsymbol{x}} := \left[1, \boldsymbol{x}^{\top}\right]^{\top}$.

elliptical distributions

We can generalize the multivariate normal distribution to a fatter tailed distribution by considering *elliptical distributions*. The idea is that the norm of the vector-valued random variable is determined separately from its direction, and the direction follows a simple covariance-like structure. One simple way to define an elliptical distribution is to let $\boldsymbol{z}$ follow an n-dimensional normal distribution with mean $\mathbf{0}$ and covariance I. Then let

$$\boldsymbol{x} = \boldsymbol{\mu} + a\Lambda^{1/2}\frac{\boldsymbol{z}}{\left\|z\right\|_2},$$

where a is some random variable, $\boldsymbol{\mu}$ is the mean of $\boldsymbol{x}$, and Λ is the covariance of $\boldsymbol{x}$ up to scaling. When a takes a chi distribution (not chi-square) with n degrees of freedom, we recover the multivariate normal.

Moments of products of elements of $\boldsymbol{x}$ are given, for the normal case, by *Isserlis' Theorem*; an extension of this theorem to elliptical distributions gives the following moment relations. [Isserlis, 1918; Vignat and Bhatnagar, 2007] By this theorem, we have

$$\operatorname{E}\left[x_i x_j\right] = \mu_i\mu_j + \frac{\operatorname{E}\left[a^2\right]}{n}\lambda_{i,j}, \tag{A.7}$$

and thus the covariance of $\boldsymbol{x}$ is

$$\Sigma = \frac{\operatorname{E}\left[a^2\right]}{n}\Lambda.$$

The third moment is

$$\operatorname{E}\left[x_i x_j x_k\right] = \mu_i\mu_j\mu_k + \mu_i\Sigma_{j,k} + \mu_j\Sigma_{i,k} + \mu_k\Sigma_{i,l}. \tag{A.8}$$

When considering a centered version of $\boldsymbol{x}$, we can treat $\boldsymbol{\mu}$ as the zero vector, in which case the third moment is zero. elliptically distributed variables have no skew, which may make them a poor choice for modeling asset returns[1].

The fourth moment is

$$\begin{aligned}\operatorname{E}\left[x_i x_j x_k x_l\right] = \mu_i\mu_j\mu_k\mu_l &+ \frac{n}{n+2}\frac{\operatorname{E}\left[a^4\right]}{\operatorname{E}\left[a^2\right]^2}\left(\Sigma_{i,j}\Sigma_{k,l} + \Sigma_{i,k}\Sigma_{j,l} + \Sigma_{i,l}\Sigma_{j,k}\right)\\ &+ \mu_i\mu_j\Sigma_{k,l} + \mu_i\mu_k\Sigma_{j,l} + \mu_i\mu_l\Sigma_{j,k}\\ &+ \mu_j\mu_k\Sigma_{i,l} + \mu_j\mu_l\Sigma_{i,k} + \mu_k\mu_l\Sigma_{i,j}.\end{aligned} \tag{A.9}$$

[1]Although slightly better than Gaussian, which have too modest kurtosis.

The kurtosis (not the *excess* kurtosis) of the ith element of $\boldsymbol{x}$ is then

$$\frac{3n}{n+2}\frac{\mathrm{E}\left[a^4\right]}{\mathrm{E}\left[a^2\right]^2}.$$

Often we need only consider the first four moments of an elliptical distribution. In this case, we can think of the elliptical distribution as parametrized by $\boldsymbol{\mu}$, Σ, and the *kurtosis factor* defined as

$$\kappa := \frac{n}{n+2}\frac{\mathrm{E}\left[a^4\right]}{\mathrm{E}\left[a^2\right]^2}. \tag{A.10}$$

The multivariate normal is an elliptical distribution with $\kappa = 1$. Distributions with larger kurtosis factor are considered "fat tailed."

An alternative characterization is as follows: random vector $\boldsymbol{x}$ has an elliptical distribution if and only if the density of $\boldsymbol{x}$ is some function of the quadratic form $(\boldsymbol{x} - \boldsymbol{\mu})^{\top} \Lambda^{-1} (\boldsymbol{x} - \boldsymbol{\mu})$.

lambda-prime and lambda-square distributions

For fixed t, if $ty^2 \sim \chi^2(t)$ and $\boldsymbol{x}\,|y^2 \sim \mathcal{N}\left(\sqrt{y^2}\boldsymbol{a}, \mathsf{I}_m\right)$, then we say that x follows a (multivariate) *lambda-prime* distribution with t degrees of freedom and parameter $\boldsymbol{a}$, which we write as $\boldsymbol{x} \sim \lambda'(t, \boldsymbol{a})$. The lambda-prime distribution is a special case of the more general *K-prime* distribution. [Lecoutre, 1999] In the case of scalar a and x, we write the density, cumulative distribution, and quantile functions of this distribution as, respectively, $f_{\lambda'}(x; t, a)$, $F_{\lambda'}(x; t, a)$, and $\lambda'_p(t, a)$.

If $ty^2 \sim \chi^2(t)$ and $x\,|y^2 \sim \chi^2\left(m; a^2y^2\right)$ then we say that x follows a *Lambda square* distribution with m and t degrees of freedom and "eccentricity" a^2. We write this as $x \sim \Lambda^2_{m,t}\left(a^2\right)$. The density of x is [Lecoutre, 1999; Rouanet and Lecoutre, 1983]

$$\frac{1}{2}\frac{1}{\Gamma\left(\frac{t}{2}\right)}\left(\frac{t}{t+a^2}\right)^{t/2}\left(\frac{x}{2}\right)^{m/2-1}e^{-\frac{x}{2}}\sum_{j=0}^{\infty}\frac{1}{j!}\frac{\Gamma\left(\frac{t+2j}{2}\right)}{\Gamma\left(\frac{m+2j}{2}\right)}\left(\frac{a^2x}{2\left(t+a^2\right)}\right)^j. \tag{A.11}$$

In this case x has expected value $m + a^2$ and variance $2\left(m + 2a^2 + a^4/t\right)$. The lambda-square distribution is a special case of the more general *K-square* distribution. [Lecoutre, 1999] We note that if $\boldsymbol{x} \sim \lambda'(t, \boldsymbol{a})$ for m-vector $\boldsymbol{a}$ then $\boldsymbol{x}^{\top}\boldsymbol{x} \sim \Lambda^2_{m,t}\left(\boldsymbol{a}^{\top}\boldsymbol{a}\right)$.

Wishart distribution

Let the rows of the $n \times p$ matrix X be *i.i.d.* multivariate normals with zero-mean and covariance $\boldsymbol{\Sigma}$. Then $\mathsf{W} = \mathsf{X}^{\top}\mathsf{X}$ follows a Wishart distribution with parameter $\boldsymbol{\Sigma}$ and n degrees of freedom, written as $\mathsf{W} \sim \mathcal{W}(\Sigma, n)$, or sometimes as $\mathsf{W} \sim \mathcal{W}_p(\Sigma, n)$ to emphasize that W is $p \times p$. [Press, 2012; Anderson, 2003;

Timm, 2002] The Wishart distribution generalizes the chi-square distribution, or rather the gamma distribution, of which the chi-square is an example.

The density of the Wishart is

$$f_{\mathcal{W}}(\mathsf{W}; \Sigma, n) = \frac{|\mathsf{W}|^{n-p-1/2}}{2^{np/2} |\Sigma|^{n/2} \Gamma(n/2)} e^{-\frac{1}{2}\operatorname{tr}(\Sigma^{-1}\mathsf{W})}. \tag{A.12}$$

This density is defined for non-integral n, a case not covered by the stochastic representation above.

The Wishart enjoys a "closure property": informally, a projection of a Wishart is also a Wishart. If A is a $p \times k$ matrix of rank k where $k \leq p$, then

$$\mathsf{W} \sim \mathcal{W}_p(\Sigma, n) \Rightarrow \mathsf{A}^\top \mathsf{W} \mathsf{A} \sim \mathcal{W}_k\left(\mathsf{A}^\top \Sigma \mathsf{A}, n\right). \tag{A.13}$$

non-central Wishart distribution

As with typical non-central distributions, the non-central Wishart distribution arises when non-centered variates are used in place of centered variates. So imagine that the rows of the $n \times p$ matrix X are *i.i.d.* multivariate normals with mean $\boldsymbol{\mu}$ and covariance Σ. Then $\mathsf{W} = \mathsf{X}^\top \mathsf{X}$ follows a non-central Wishart distribution with parameter Σ, n degrees of freedom, and non-centrality paramater $\boldsymbol{\mu}$, written as $\mathsf{W} \sim \mathcal{W}_p(\Sigma, \boldsymbol{\mu}, n)$. [Letac and Massam, 2004] The non-central Wishart also satisfies a closure property.

inverse Wishart distribution

Let $\mathsf{Y}^{-1} \sim \mathcal{W}\left(\Psi^{-1}, n\right)$. Then Y follows an Inverse Wishart Distribution with parameter Ψ and n degrees of freedom, written as $\mathsf{Y} \sim \mathcal{IW}(\Psi, n)$.

The density of the inverse Wishart is

$$f_{\mathcal{IW}}(\mathsf{Y}; \Psi, n) = \frac{|\mathsf{Y}|^{n+p+1/2} |\Psi|^{n/2}}{2^{np/2} \Gamma(n/2)} e^{-\frac{1}{2}\operatorname{tr}(\Psi \mathsf{Y})}. \tag{A.14}$$

multivariate t distribution

The *multivariate t distribution* generalizes the t-distribution to the multivariate setting. [Kotz and Nadarajah, 2004] There are two equivalent stochastic representations of this distribution. Suppose that

$$Y \sim \mathcal{N}(\mathbf{0}_p, \Sigma),$$
$$S^2 \sim \chi^2(n),$$

with Y and S^2 independent. Then $X = \frac{Y}{\sqrt{S^2/n}} + \boldsymbol{\mu}$ follows a multivariate t distribution with n degrees of freedom, matrix Σ, and location parameter $\boldsymbol{\mu}$, written as

$$X \sim \mathcal{T}(n, \Sigma, \boldsymbol{\mu}). \tag{A.15}$$

X has the density [Dickey, 1967]

$$f_{\mathcal{T}}(X; n, \Sigma, \mu) = \frac{\Gamma\left(\frac{1}{2}(n+p)\right) n^{n/2}}{\Gamma\left(\frac{1}{2}n\right) \pi^{p/2} |\Sigma|^{\frac{1}{2}}} \left(n + (X-\mu)^{\top} \Sigma^{-1} (X-\mu)\right)^{-n+p/2}. \quad \text{(A.16)}$$

The multivariate t distribution admits an alternative stochastic representation. [Dickey, 1967; Ando and Kaufman, 1965; Kotz and Nadarajah, 2004] Let

$$Y \sim \mathcal{N}(\mathbf{0}_p, n\mathsf{I}_p),$$
$$U \sim \mathcal{W}\left(\Sigma^{-1}, n+p-1\right),$$

with Y and U independent. Then

$$X = \left(U^{1/2}\right)^{-1} Y + \mu \sim \mathcal{T}(n, \Sigma, \mu),$$

where here $\mathsf{A}^{1/2}$ represents the symmetric square root of A. Another way of stating this is in terms of an inverse Wishart, as follows:

$$\begin{aligned} W &\sim \mathcal{IW}(\Sigma, n+p-1), \\ X \mid W &\sim \mathcal{N}(\mu, nW). \end{aligned} \quad \text{(A.17)}$$

This form will appear in the context of posterior marginal distributions in Bayesian analysis.

Kotz and Nadarajah [2004, page 1] claim that the distribution is "said to be central if $\mu = \mathbf{0}$; otherwise, it is said to be noncentral." However, the $\mu \neq \mathbf{0}_p$ case should emphatically *not* be called a "non-central" multivariate t, since it does *not* reduce to the scalar non-central t in the $p = 1$ case. Instead one should consider the μ as a locational shift. The distribution of X is spherically symmetric around μ, a property not shared by the scalar non-central t.

The Multivariate t satisfies a multiplicative closure property. If A is a $p \times k$ matrix of rank k where $k \leq p$, then

$$X \sim \mathcal{T}(n, \Sigma, \mu) \Rightarrow \mathsf{A}^{\top} X \sim \mathcal{T}\left(n, \mathsf{A}^{\top} \Sigma \mathsf{A}, \mathsf{A}^{\top} \mu\right). \quad \text{(A.18)}$$

As a consequence, marginals of the Multivariate t distribution follow the (scaled, shifted) scalar t distribution.

The Multivariate t distribution is an elliptical distribution, with mean μ, covariance $\frac{n}{n-2}\Sigma$ and kurtosis factor $\kappa = \frac{n-2}{n-4}$.

non-central multivariate t distribution

Kshirsagar [1961] described a truly non-central multivariate t distribution that generalizes the scalar non-central t distribution in the $p = 1$ case. Let

$$Y \sim \mathcal{N}\left(\mu, \sigma^2 R\right),$$
$$S^2 \sim \chi^2(n),$$

where R is a correlation matrix, *i.e.*, symmetric positive definite with ones along the diagonal. Then

$$X = \frac{Y}{\sqrt{\sigma^2 S^2/n}}$$

follows the non-central multivariate t distribution with covariance $\Sigma = \sigma^2 R$, degrees of freedom n and non-centrality parameter $\boldsymbol{\mu}$. Kshirsagar gives the density of this distribution and notes that its marginals follow the scalar non-central t distribution. [Kotz and Nadarajah, 2004; Kshirsagar, 1961] This distribution is *not* an elliptical distribution.

A.2.3 Matrixvariate probability distributions

Finally, we consider a matrix-variate distribution.

matrix normal distribution

We say that an $n \times k$ matrix X follows a *matrix normal* (or *Gaussian*) distribution with mean M and covariance matrices U and V if $\operatorname{vec}(\mathsf{X}) \sim \mathcal{N}(\operatorname{vec}(\mathsf{M}), \mathsf{V} \otimes \mathsf{U})$. [Gupta *et al.*, 2013; Barratt, 2018] Here M is an $n \times k$ matrix and U and V are symmetric positive definite definite matrices of size $n \times n$ and $k \times k$, respectively. We write this as

$$\mathsf{X} \sim \mathcal{MN}_{n,k}(\mathsf{M}, \mathsf{U}, \mathsf{V}).$$

The density of X is

$$f_{\mathcal{MN}}(\mathsf{X}; \mathsf{M}, \mathsf{U}, \mathsf{V}) = \frac{\exp\left(-\frac{1}{2}\operatorname{tr}\left(\mathsf{V}^{-1}(\mathsf{X}-\mathsf{M})^\top \mathsf{U}^{-1}(\mathsf{X}-\mathsf{M})\right)\right)}{\sqrt{(2\pi)^{nk}|\mathsf{V}|^n|\mathsf{U}|^k}}. \tag{A.19}$$

We note that the matrix normal distribution has a closure property. If $\mathsf{X} \sim \mathcal{MN}(\mathsf{M}, \mathsf{U}, \mathsf{V})$ and A and B are conformable matrices then $\mathsf{AX} \sim \mathcal{MN}(\mathsf{AM}, \mathsf{AUA}^\top, \mathsf{V})$ and $\mathsf{XB} \sim \mathcal{MN}(\mathsf{MB}, \mathsf{U}, \mathsf{B}^\top\mathsf{VB})$.

A.3 Statistical practice

We assume the reader has some familiarity with common statistical practices, which we summarize here. In this book we will consider performing inference on some unknown population quantities. Typographically we follow the convention of denoting most population quantities with Greek letters, and denoting some sample estimate of that quantity with a "hat." For example, we will use μ to stand for the expected value of a random variable, and $\hat{\mu}$ for the mean of a sample of realizations of that random variable. Noting that $\hat{\mu}$

is itself a random variable, since it is the mean of random variables, it has a distribution. An estimator is said to be *unbiased* when its expected value is the population quantity of interest. For example, the usual sample mean, $\hat{\mu}$, is an unbiased estimate of μ. This does not mean that any particular realization of $\hat{\mu}$ equals μ, but rather that the process that generates $\hat{\mu}$ has expected value equal to μ. The *standard error* refers to the standard deviation of some sample estimator of a population quantity.

Some statistical practice focuses on analyzing estimators of population quantities and trying to understand and reduce the standard error. However, the majority of statistical theory deals more broadly with inference on the unknown population quantities. If you have taken a basic course in statistics, you have likely encountered the *Frequentist* paradigm. Inference as practiced by Frequentist statisticians consists of conjuring up a hypothetical fact about population parameters, referred to as a *null hypothesis.* Usually the null hypothesis is a strawman of sorts, a fact that represents an uninteresting or inconvenient fact that one then wishes to falsify. For example, one might wish to disprove that the returns of a given investment are negative, which we will write as $H_0 : \mu \leq 0$. Sometimes one is interested in falsifying the null in favor of another restriction on the population parameters; we refer to this other restriction as the *alternative hypothesis.* Continuing the example above, we might express the null that the mean of some variable is negative against the alternative hypothesis that it is positive as

$$H_0 : \mu \leq 0 \quad \text{versus} \quad H_1 : \mu > 0.$$

The Frequentist then computes some statistic, the distribution of which is known conditional on the null hypothesis being true. If the statistic takes an unusually large or small value, then the Frequentist takes confidence in *rejecting the null.* Often the statistic is converted to a *p-value*, which should be uniformly distributed under the null hypothesis. If the p-value is smaller than some critical value α, usually chosen arbitrarily to be 0.05, then the null hypothesis is "rejected at the α level." The α is the *type I rate*, and refers to the rate of "type I" errors. A type I error is an error where the null hypothesis is actually true, but the p-value is small just because that happens sometimes, and one falsely rejects the null hypothesis. Conditional on the null hypothesis being true, the Frequentist expects to make type I errors at the rate α, over the long term. This is why they are called "Frequentist," because the entire process is calibrated to have a certain frequency, over many replications of such experiments, of type I errors. [Wasserman, 2004]

The Frequentist is also concerned with *type II errors*, which occur when the null hypothesis is false, but the statistician fails to reject the null hypothesis, because the weight of evidence is not strong enough. We denote the type II rate by β. The *power* of a statistical test is $1 - \beta$, which is the probability of correctly rejecting the null when it is false. Of course, the β rate depends on how far wrong the null hypothesis is, but it also depends on the statistical test in use. For two statistical tests with the same type I rate, one should prefer the test with higher power, since it is more likely to correctly reject when the null is false. However, it is often the case that some statistical tests perform better than others under certain alternative hypotheses than other.

The Frequentist's *confidence interval* is a complement to the hypothesis test. One can view a confidence interval as the region of hypothetical values for the population parameter for which one would not reject the null hypothesis at a given type I rate. It is important to realize that the confidence interval, constructed as it is from the observed data, is itself a random quantity. Usually the confidence interval contains the population parameter of interest. The probability that it does, under replications of the experiment, is the *coverage* of the confidence interval, and is the complement of the type I rate. Just as Frequentists typically perform hypothesis tests at the 0.05 level, these translate into 95% confidence intervals. Again it is important to note that the probability guarantee refers to replications of the experiment: generating data from the process, constructing the confidence interval, and observing whether it contains the population parameter. For one particular realization, we cannot claim to know the "probability" that the population parameter is inside it: since the Frequentist views the population parameter as fixed, once the confidence interval is observed, there is no randomness left, and either the parameter is inside the interval or not. A "probability" can be assigned to the process itself, not any realization.

Inference in the *likelihoodist* paradigm focuses on the *likelihood* function. The likelihood function of a random variable is simply the probability density function, but viewed as a function of the unknown population parameter with the observed variable fixed. For example, if one observes n independent observations z_i of a normal random variable with unknown mean μ, but known variance of 1. The sample mean

$$\hat{\mu} = \frac{1}{n}\sum_i z_i$$

is normally distributed: $\hat{\mu} \sim \mathcal{N}\left(\mu, \frac{1}{n}\right)$. The likelihood function for μ is then

$$\mathcal{L}_{\mathcal{N}}\left(\hat{\mu}\,|\mu\right) = \frac{1}{\sqrt{2\pi/n}} e^{-\frac{(\hat{\mu}-\mu)^2}{2/n}}.$$

Within the likelihood framework, one often estimates the unknown population parameter with the *Maximum Likelihood Estimator* (MLE), which is a hypothetical value of μ that maximizes the likelihood function. In our running example, $\hat{\mu}$ is the MLE. The likelihood function is also used in the likelihood ratio test, which is used in Frequentist inference. [Pawitan, 2001]

The Frequentist conceives of a population parameter as a fixed non-random value, and views their life's work as a sequence of hypothesis tests over which they should commit some small fixed proportion of type I errors. The Bayesian, on the other hand, represents uncertainty in their "belief" of the population parameter by supposing that the population parameter is a random quantity, updating their beliefs upon observing relevant data. The central relation underlying Bayesian inference is Bayes' theorem,

$$\Pr\{B \mid A\} = \frac{\Pr\{A \mid B\} \Pr\{A\}}{\Pr\{B\}}, \tag{A.20}$$

where A and B are events. By $\Pr\{B \mid A\}$ we mean the probability of B conditional on A being true or observed. Bayes' theorem is easily derived from the definition of conditional probability, namely

$$\Pr\{A \mid B\} := \frac{\Pr\{A \cap B\}}{\Pr\{B\}}, \quad \text{if} \quad \Pr\{B\} \neq 0. \tag{A.21}$$

Bayes' theorem is employed in *Bayesian inference* to compute the probability of some hypothesis about the population parameter conditional on some observed evidence. One starts with a *prior belief* about the hypothesis, call it $\Pr\{H\}$, observes some evidence E, then computes $\Pr\{H \mid E\}$, which is called the *posterior belief.* Strictly speaking the computation is

$$\Pr\{H \mid E\} = \frac{\Pr\{E \mid H\} \Pr\{H\}}{\Pr\{E\}}.$$

In practice the denominator $\Pr\{E\}$ is unneeded, as it is the same for every hypothesis which one might hold regarding the population parameter. Instead one computes the posterior up to some unknown constant, which we write as

$$\Pr\{H \mid E\} \propto \Pr\{E \mid H\} \Pr\{H\}, \tag{A.22}$$

where $\propto$ is read as "proportional to." Equation A.22 is the Bayesian update formula, which one can read as "posterior belief is likelihood times prior belief." As one will compute the posterior only up to a constant, often the prior is also expressed only as a proportion.

In this book we will mostly focus on *conjugate priors*, as these are the most convenient to deal with. A conjugate prior is one where the prior and posterior take the same distributional form, but with different parameters. Under conjugate priors, one could almost view the prior as "pseudo observations," which is to say your prior belief has come to you in the form of some imaginary sample of the random variable at hand, which you will then combine with the actual observed data to arrive at the posterior. The use of conjugate priors also simplifies sequential collection of samples, since the distributional form is unchanged at each step.

The Bayesian may also compute a *credible interval,* which is a counterpart to the Frequentist confidence interval. The credible interval views the population parameter as a random quantity, distributed according to the posterior distribution. Then one only need to compute quantiles of the posterior to produce the credible interval. [Gelman *et al.*, 2013; Press, 2012]

There is a fair amount of squabbling among statisticians regarding the merits of the various approaches. There is even disagreement on what qualifies an approach as being Frequentist or Bayesian[2]. We would counsel you to avoid such philosophical debates and familiarize yourself with both paradigms. In our exploration of the Frequentist and Bayesian approaches we will find that they usually produce near-identical results computationally, but via different praxis and with different interpretations. One strength of the Bayesian approach is that it is simple to use complex models; however these are beyond the scope of this book. The Frequentist approach requires somewhat less theoretical machinery, and so we weight our attention more toward Frequentist inference.

A.4 † Matrix derivatives

In the more advanced parts of this text, we will need to compute the derivatives of matrices and the derivatives of quantities with respect to matrices. This is somewhat complicated by considerations of notation (the derivative of an $m \times n$ matrix with respect to a $p \times q$ matrix consists of $mnpq$ distinct elements) as well as symmetry (*e.g.*, we typically want to consider the derivative of a quantity with respect to a symmetric matrix, only considering changes which respect symmetry). For more details, Magnus and Neudecker [1986, 2007] give a very good introduction. Petersen and Pedersen [2012] provides a good cheat-sheet of results.

Definition A.4.1 (Derivatives)**.** For m-vector $\boldsymbol{x}$, and n-vector $\boldsymbol{y}$, let the derivative $\frac{\mathrm{d}\boldsymbol{y}}{\mathrm{d}\boldsymbol{x}}$ be the $n \times m$ matrix whose jth column is the partial derivative of $\boldsymbol{y}$ with respect to x_j. This follows the so-called "numerator layout" convention. For matrices Y and X, define

$$\frac{\mathrm{d}\mathsf{Y}}{\mathrm{d}\mathsf{X}} := \frac{\mathrm{d}\operatorname{vec}(\mathsf{Y})}{\mathrm{d}\operatorname{vec}(\mathsf{X})}.$$

When $n = 1$, the derivative is commonly called the "gradient," however, by the numerator layout convention we express it as a row vector, rather than a column vector, which is unusual.

We list several matrix derivative identities which may be of use:

- For symmetric matrices Y and X,

$$\frac{\mathrm{d}\operatorname{vech}(\mathsf{Y})}{\mathrm{d}\operatorname{vec}(\mathsf{X})} = \mathsf{L}\frac{\mathrm{d}\mathsf{Y}}{\mathrm{d}\mathsf{X}}, \quad \frac{\mathrm{d}\operatorname{vec}(\mathsf{Y})}{\mathrm{d}\operatorname{vech}(\mathsf{X})} = \frac{\mathrm{d}\mathsf{Y}}{\mathrm{d}\mathsf{X}}\mathsf{D}, \quad \frac{\mathrm{d}\operatorname{vech}(\mathsf{Y})}{\mathrm{d}\operatorname{vech}(\mathsf{X})} = \mathsf{L}\frac{\mathrm{d}\mathsf{Y}}{\mathrm{d}\mathsf{X}}\mathsf{D}. \quad \text{(A.23)}$$

[2]Frequentists certainly use Bayes' theorem, and Bayesians care about the probability of accepting erroneous hypotheses.

- For invertible matrix A,

$$\frac{d\mathsf{A}^{-1}}{d\mathsf{A}} = -\left(\mathsf{A}^{-\top} \otimes \mathsf{A}^{-1}\right) = -\left(\mathsf{A}^{\top} \otimes \mathsf{A}\right)^{-1}. \tag{A.24}$$

Note how this result generalizes the scalar derivative: $\frac{dx^{-1}}{dx} = -\left(x^{-1}x^{-1}\right)$. For *symmetric* A, the derivative with respect to the non-redundant part is

$$\frac{d\,\mathrm{vech}\left(\mathsf{A}^{-1}\right)}{d\,\mathrm{vech}\left(\mathsf{A}\right)} = -\mathsf{L}\left(\mathsf{A}^{-1} \otimes \mathsf{A}^{-1}\right)\mathsf{D}. \tag{A.25}$$

- Given conformable, symmetric, matrices X, Y, Z, and constant matrix J,

$$\frac{d\mathsf{XY}}{d\mathsf{Z}} = (\mathsf{I} \otimes \mathsf{X})\frac{d\mathsf{Y}}{d\mathsf{Z}} + \left(\mathsf{Y}^{\top} \otimes \mathsf{I}\right)\frac{d\mathsf{X}}{d\mathsf{Z}}. \tag{A.26}$$

$$\frac{d\mathsf{XX}^{\top}}{d\mathsf{Z}} = (\mathsf{I} + \mathsf{K})(\mathsf{X} \otimes \mathsf{I})\frac{d\mathsf{X}}{d\mathsf{Z}}. \tag{A.27}$$

$$\frac{d\,\mathrm{tr}\left(\mathsf{XY}\right)}{d\mathsf{Z}} = \mathrm{vec}\left(\mathsf{X}^{\top}\right)^{\top}\frac{d\mathsf{Y}}{d\mathsf{Z}} + \mathrm{vec}\left(\mathsf{Y}\right)^{\top}\frac{d\mathsf{X}^{\top}}{d\mathsf{Z}}. \tag{A.28}$$

$$\frac{d\left|\mathsf{X}\right|}{d\mathsf{Z}} = \left|\mathsf{X}\right|\mathrm{vec}\left(\mathsf{X}^{-\top}\right)^{\top}\frac{d\mathsf{X}}{d\mathsf{Z}}. \tag{A.29}$$

$$\frac{d\left|\mathsf{XY}\right|}{d\mathsf{Z}} = \left|\mathsf{XY}\right|\left(\mathrm{vec}\left(\mathsf{X}^{-\top}\right)^{\top}\frac{d\mathsf{X}}{d\mathsf{Z}} + \mathrm{vec}\left(\mathsf{Y}^{-\top}\right)^{\top}\frac{d\mathsf{Y}}{d\mathsf{Z}}\right). \tag{A.30}$$

$$\frac{d\left|(\mathsf{XY})^{-1}\right|}{d\mathsf{Z}} = \left|\mathsf{XY}\right|^{-1}\left(\mathrm{vec}\left(\mathsf{X}^{-\top}\right)^{\top}\frac{d\mathsf{X}}{d\mathsf{Z}} + \mathrm{vec}\left(\mathsf{Y}^{-\top}\right)^{\top}\frac{d\mathsf{Y}}{d\mathsf{Z}}\right). \tag{A.31}$$

where K is the commutation matrix.

- Let X be a symmetric positive definite matrix. Let Y be its lower triangular Cholesky factor. That is, Y is the lower triangular matrix such that $\mathsf{YY}^{\top} = \mathsf{X}$. Then

$$\frac{d\,\mathrm{vech}\left(\mathsf{Y}\right)}{d\,\mathrm{vech}\left(\mathsf{X}\right)} = \left(\mathsf{L}\left(\mathsf{I} + \mathsf{K}\right)\left(\mathsf{Y} \otimes \mathsf{I}\right)\mathsf{L}^{\top}\right)^{-1}, \tag{A.32}$$

where K is the "commutation matrix."

- Let X be a symmetric positive definite matrix. Let Y be the lower triangular Cholesky factor of the *inverse* of X. That is, Y is the lower triangular matrix such that $\mathsf{YY}^{\top} = \mathsf{X}^{-1}$, sometimes written as $\mathsf{Y} = \mathsf{X}^{-1/2}$. Then

$$\frac{d\,\mathrm{vech}\left(\mathsf{Y}\right)}{d\,\mathrm{vech}\left(\mathsf{X}\right)} = -\left(\mathsf{L}\left(\mathsf{I} + \mathsf{K}\right)\left(\left(\mathsf{Y}^{\top}\right)^{-1} \otimes \left(\mathsf{YY}^{\top}\right)^{-1}\right)\mathsf{L}^{\top}\right)^{-1}. \tag{A.33}$$

Note how this result generalizes the scalar derivative: $\frac{dx^{-1/2}}{dx} = -\left(2x^{1/2}x\right)^{-1}$.

A.5 The central limit theorem and delta method

The *central limit theorem* is widely used in statistical practice to establish the asymptotic distribution of sample statistics. Roughly speaking this theorem tells us that the sample mean is approximately normally distributed. Formally, the Lindberg-Lévy form of the theorem states that if $x_1, x_2, \ldots, x_n$ are independent and identically distributed random variables with $\mathrm{E}\left[x_i\right] = \mu$ and $\mathrm{Var}\left(x_i\right) = \sigma^2 < \infty$, then as $n \rightarrow \infty$, $\sqrt{n}\left(\hat{\mu} - \mu\right)$ converges in distribution to a normal distribution with mean zero and variance σ^2, where $\hat{\mu} = \frac{1}{n}\sum_i x_i$ is the sample mean[3]. We will write the conclusion of the central limit theorem as

$$\sqrt{n}\left(\hat{\mu} - \mu\right) \rightsquigarrow \mathcal{N}\left(0, \sigma^2\right). \tag{A.34}$$

There are variants of the central limit theorem which loosen the restriction of *i.i.d.* variables in various ways, which is to say the result is fairly robust to this assumption. [Adams, 2009; Wasserman, 2004] Under somewhat stronger assumptions, the *Berry-Esseen theorem* gives the rate of convergence of the cumulative distribution of $\sqrt{n}\hat{\mu}/\sigma$ to the normal distribution. [Feller, 1968]

There is a multivariate version of the central limit theorem, which we will also encounter. It states that if $\boldsymbol{x}_1, \boldsymbol{x}_2, \ldots, \boldsymbol{x}_n$ are p-variate vectors, independent and identically distributed with $\mathrm{E}\left[\boldsymbol{x}_i\right] = \boldsymbol{\mu}$ and $\mathrm{Var}\left(\boldsymbol{x}_i\right) = \Sigma$, then

$$\sqrt{n}\left(\hat{\boldsymbol{\mu}} - \boldsymbol{\mu}\right) \rightsquigarrow \mathcal{N}\left(0, \Sigma\right) \tag{A.35}$$

as $n \rightarrow \infty$, where $\hat{\boldsymbol{\mu}} = \frac{1}{n}\sum_i \boldsymbol{x}_i$ is the sample mean.

While the central limit theorem is useful for sample averages, we often consider transformations of averages to produce more elaborate sample statistics. For example, in this text we compute the Sharpe ratio, which is the ratio of the sample mean to the sample standard deviation, which together can be computed from the mean of observed x_i and the mean of x_i^2. The *delta method* is a technique for finding the asymptotic distribution of such transformed statistics. The delta method tells us that for function $f\left(\cdot\right)$ if

$$\sqrt{n}\left(\hat{\mu} - \mu\right) \rightsquigarrow \mathcal{N}\left(0, \sigma^2\right),$$

then

$$\sqrt{n}\left(f\left(\hat{\mu}\right) - f\left(\mu\right)\right) \rightsquigarrow \mathcal{N}\left(0, \sigma^2\left(f'\left(\mu\right)\right)^2\right),$$

when the derivative $f'\left(\cdot\right)$ exists and is non-zero at μ.

There is also a multivariate delta method. Suppose that $g\left(\cdot\right)$ is a function which takes a p-vector to an m-vector. Furthermore, suppose that the first derivative of $g\left(\cdot\right)$ is not the zero vector at $\boldsymbol{\mu}$. Then if

$$\sqrt{n}\left(\hat{\boldsymbol{\mu}} - \boldsymbol{\mu}\right) \rightsquigarrow \mathcal{N}\left(0, \Sigma\right),$$

[3] "Convergence in distribution" means that the probability distribution converges to the normal distribution for every value in its domain, although the rate of convergence can be non-uniform.

then

$$\sqrt{n}\left(g\left(\hat{\boldsymbol{\mu}}\right)-g\left(\boldsymbol{\mu}\right)\right) \rightsquigarrow \mathcal{N}\left(0, \frac{\mathrm{d}g\left(\boldsymbol{\mu}\right)}{\mathrm{d}\boldsymbol{\mu}}\Sigma\frac{\mathrm{d}g\left(\boldsymbol{\mu}\right)}{\mathrm{d}\boldsymbol{\mu}}^{\top}\right),$$

where the derivative follows the "numerator layout" of Section A.4.

Note that the delta method is trivial to prove, by Slutsky's Theorem, when the function f is an affine function, since it simply corresponds to an offset and rescaling of an asymptotically normal distribution. The delta method essentially allows us to replace the function f by its affine approximation around the point $(\mu, f(\mu))$. That is, the convergence of $\hat{\mu}$ to μ guarantees we can treat the function f as if it were linear, for purposes of the asymptotic convergence of $f(\hat{\mu})$. This can feel counterintuitive, especially when f is badly behaved around μ. However, one should keep in mind that this is a purely asymptotic result. When applied to finite sample sizes, one can be led astray by the delta method.

A.6 Exercises

Ex. A.1 **Trace** Let A be a square matrix.

1. Show that if A is diagonal, then $\operatorname{tr}(\mathsf{A}) = \mathbf{1}^{\top}\mathsf{A}\mathbf{1}$.
2. Show that $\operatorname{tr}\left(\mathsf{A}^2\right) = \operatorname{vec}(\mathsf{A})^{\top}\operatorname{vec}(\mathsf{A})$ for symmetric matrix A.
3. Show tha the trace of a matrix is equal to the sum of its eigenvalues. Do you recall a similar property of the matrix determinant?

Ex. A.2 **Elimination and Duplication** Write out the 3×4 Elimination matrix. Write out the 4×3 Duplication matrix.

Ex. A.3 **Commutation** Let $\mathsf{K}_{n,n}$ be the commutation matrix for $n \times n$ matrices, *i.e.*, $\mathsf{K}_{n,n}\operatorname{vec}(\mathsf{A}) = \operatorname{vec}\left(\mathsf{A}^{\top}\right)$ for $n \times n$ matrices A. Let A and B be $n \times n$ matrices.

1. Show that
$$\mathsf{K}_{n,n}\left(\mathsf{A} \otimes \mathsf{B}\right) = \left(\mathsf{B} \otimes \mathsf{A}\right)\mathsf{K}_{n,n}.$$
2. Letting $\mathsf{N} = \frac{1}{2}\left(\mathsf{I} + \mathsf{K}\right)$, show that $\mathsf{DLN} = \mathsf{N}$.

Ex. A.4 **Matrix calculus** Assume X is symmetric.

1. What is $\frac{\mathrm{d}\operatorname{tr}(\mathsf{X})}{\mathrm{d}\mathsf{X}}$?
2. Compute $\frac{\mathrm{d}\operatorname{tr}\left(\mathsf{X}^2\right)}{\mathrm{d}\mathsf{X}}$.

Ex. A.5 **Cholesky factorization** Is the Cholesky factorization unique?

Ex. A.6 **Sherman Morrison Woodbury** Show that

$$\left(\mathsf{A} + \boldsymbol{u}\boldsymbol{v}^{\top}\right)^{-1}\boldsymbol{u} = \frac{1}{1 + \boldsymbol{v}^{\top}\mathsf{A}^{-1}\boldsymbol{u}}\mathsf{A}^{-1}\boldsymbol{u}.$$

Ex. A.7 **Inverse Kronecker** Suppose A and B are square non-singular matrices. Prove that

$$(\mathsf{A} \otimes \mathsf{B})^{-1} = \mathsf{A}^{-1} \otimes \mathsf{B}^{-1}.$$

Ex. A.8 **Distribution connections** Consider the connections between univariate distributions:

1. Let $t \sim t(v)$. Show that $vt^2 \sim F(1, v)$. What if t follows a non-central t distribution?
2. Let $f \sim F(v_1, v_2)$. Show that

$$\frac{f}{(v_2/v_1) + f} \sim \mathcal{B}(v_1/2, v_2/2).$$

   What if f follows a non-central F distribution?

Ex. A.9 **Numerical derivatives** Use a differencing scheme to approximate the derivative of matrix-valued operations of matrix input. That is, letting ϵ be a small quantity, compute, for some input matrix X and function f

$$\frac{f\left(\mathsf{X} + \epsilon \mathsf{J}^{ij}\right) - f(\mathsf{X})}{\epsilon},$$

vectorize it, and set this equal to the i, jth column of the approximate derivative.

1. Using the numerical differencing scheme, confirm the result of Equation A.32. Note that some implementations of the Cholesky operation will assume the input is symmetric and thus ignore the lower or upper triangular part of the input. Take care to deal with this in your numerical approximation. (Hint: apply the symmetric idempotent matrix to your input.)
2. Using the numerical differencing scheme, confirm the result of Equation A.33.

Ex. A.10 **Wishart facts** 1. Prove the closure property of the Wishart distribution, Equation A.13.

References

Good artists borrow. Great artists steal.

VARIOUSLY ATTRIBUTED

Milton Abramowitz and Irene A. Stegun. *Handbook of Mathematical Functions with Formulas, Graphs, and Mathematical Tables.* Dover, New York, ninth Dover printing, tenth GPO printing edition, 1964. URL `http://people.math.sfu.ca/~cbm/aands/toc.htm`.

W. J. Adams. *The Life and Times of the Central Limit Theorem.* History of mathematics. American Mathematical Society, second edition, 2009. ISBN 9780821890547. doi: 10.1090/hmath/035. URL `https://books.google.com/books?id=6-vLk8VuNbQC`.

Masafumi Akahira. A higher order approximation to a percentage point of the non-central t-distribution. *Communications in Statistics – Simulation and Computation*, 24(3):595–605, 1995. doi: 10.1080/03610919508813261. URL `http://www.tandfonline.com/doi/abs/10.1080/03610919508813261`.

Masafumi Akahira, Michikazu Sato, and Norio Torigoe. On the new approximation to non-central t-distributions. *Journal of the Japan Statistical Society*, 25(1):1–18, 1995. doi: 10.14490/jjss1995.25.1. URL `https://www.jstage.jst.go.jp/article/jjss1995/25/1/25_1_1/_pdf`.

Robert Almgren, Chee Thum, Emmanuel Hauptmann, and Hong Li. Direct estimation of equity market impact. *Risk*, 18(7):58–62, 2005. URL `https://media.gradebuddy.com/documents/968401/2bd40052-60f6-470e-92b2-b3165ec8e336.pdf`.

T. W. Anderson. *An Introduction to Multivariate Statistical Analysis.* Wiley Series in Probability and Statistics. Wiley, 2003. ISBN 9780471360919. URL `http://books.google.com/books?id=Cmm9QgAACAAJ`.

Albert Ando and G. M. Kaufman. Bayesian analysis of the independent multinormal process; Neither mean nor precision known. *Journal of the American Statistical Association*, 60(309):347–358, 1965. ISSN 01621459. doi: 10.1080/01621459.1965.10480797. URL `http://www.jstor.org/stable/2283159`.

Donald W. K. Andrews. Testing when a parameter is on the boundary of the maintained hypothesis. *Econometrica*, 69(3):683–734, 2001. URL http://citeseerx.ist.psu.edu/viewdoc/download?doi=10.1.1.222.2587&rep=rep1&type=pdf.

Davide Anguita, Alessandro Ghio, Luca Oneto, and Sandro Ridella. A deep connection between the Vapnik–Chervonenkis entropy and the Rademacher complexity. *IEEE Transactions on Neural Networks and Learning Systems*, 25(12):2202–2211, 2014. doi: 10.1109/TNNLS.2014.2307359. URL https://www.academia.edu/download/45236822/A_Deep_Connection_Between_the_VapnikCher20160430-742-arn1gi.pdf.

Mengmeng Ao, Yingying Li, and Xinghua Zheng. Solving the Markowitz optimization problem for large portfolios, 2014. URL http://fmaconferences.org/SanDiego/Papers/MAXSER_MengmengAo.pdf.

David R. Aronson. *Evidence-Based Technical Analysis: Applying the Scientific Method and Statistical Inference to Trading Signals.* Wiley, Hoboken, NJ, November 2006. URL http://www.evidencebasedta.com/.

Clifford S. Asness, Andrea Frazzini, and Lasse Heje Pedersen. Quality minus junk. Privately Published, October 2013. doi: 10.2139/ssrn.2312432. URL http://ssrn.com/abstract=2312432.

Attaboy and Annie Owens. *Hi-Fructose: New Contemporary Fashion.* Cernunnos, 2019.

Morgane Austern and Wenda Zhou. Asymptotics of cross-validation, 2020. URL https://arxiv.org/abs/2001.11111.

Xi Bai, Katya Scheinberg, and Reha Tütüncü. Least-squares approach to risk parity in portfolio selection. *Quantitative Finance*, 16(3):357–376, 2016. doi: 10.1080/14697688.2015.1031815. URL https://doi.org/10.1080/14697688.2015.1031815.

L. J. Bain and M. Engelhardt. *Introduction to Probability and Mathematical Statistics.* Classic Series. Cengage Learning, 1992. ISBN 9780534380205. URL http://books.google.com/books?id=MkFRIAAACAAJ.

Bruno Baldessari. The distribution of a quadratic form of normal random variables. *The Annals of Mathematical Statistics*, 38(6):1700–1704, 1967. doi: 10.1214/aoms/1177698604. URL https://repository.lib.ncsu.edu/bitstream/handle/1840.4/2157/ISMS_1966_476.pdf.

Yong Bao. Estimation risk-adjusted Sharpe ratio and fund performance ranking under a general return distribution. *Journal of Financial Econometrics*, 7(2):152–173, 2009. doi: 10.1093/jjfinec/nbn022. URL https://doi.org/10.1093/jjfinec/nbn022.

Yong Bao and Aman Ullah. Moments of the estimated Sharpe ratio when the observations are not IID. *Finance Research Letters*, 3(1), 2006. doi: 10.1016/j.frl.2005.11.001. URL `https://papers.ssrn.com/sol3/papers.cfm?abstract_id=2738158#`.

Shane Barratt. A matrix Gaussian distribution. *arXiv preprint arXiv:1804.11010*, 2018. URL `https://arxiv.org/abs/1804.11010`.

Stephen Bates, Trevor Hastie, and Robert Tibshirani. Cross-validation: what does it estimate and how well does it do it?, 2021. URL `https://arxiv.org/abs/2104.00673`.

W. Baumgartner, P. Weiss, and H. Schindler. A nonparametric test for the general two-sample problem. *Biometrics*, 54(3):1129–1135, 1998. doi: 10.2307/2533862. URL `https://www.jstor.org/stable/2533862`.

Pierre Bayle, Alexandre Bayle, Lucas Janson, and Lester Mackey. Cross-validation confidence intervals for test error, 2020. URL `https://arxiv.org/abs/2007.12671`.

Martin Becker. Exact simulation of final, minimal and maximal values of Brownian motion and jump-diffusions with applications to option pricing. *Computational Management Science*, 7(1):1–17, 2010. doi: 10.1007/s10287-007-0065-9. URL `http://www.oekonometrie.uni-saarland.de/papers/SimBrownWP.pdf`.

David Benkeser, Maya Petersen, and Mark J. van der Laan. Improved small-sample estimation of nonlinear cross-validated prediction metrics. *Journal of the American Statistical Association*, 115(532):1917–1932, 2020. doi: 10.1080/01621459.2019.1668794. URL `https://doi.org/10.1080/01621459.2019.1668794`.

James O. Berger. *Statistical decision theory and Bayesian analysis*. Springer Series in Statistics. Springer, second edition, 1985.

Roger L. Berger. Likelihood ratio tests and intersection-union tests. Technical Report 2288, Department of Statistics, North Carolina State University, September 1996. URL `http://www.stat.ncsu.edu/information/library/mimeo.archive/ISMS_1996_2288.pdf`. Institute of Statistics, Mimeo Series.

Dimitris Bertsimas, Vishal Gupta, and Ioannis Ch. Paschalidis. Inverse optimization: A new perspective on the Black-Litterman model. *Operations research*, 60(6):1389–1403, 2012. doi: 10.1287/opre.1120.1115. URL `https://www.mit.edu/~dbertsim/papers/Finance/Inverse%20Optimization%20-%20A%20New%20Perspective%20on%20the%20Black-Litterman%20Model.pdf`.

Michael J. Best. *Portfolio Optimization*. CRC Press, 2010.

Martin Bilodeau and David Brenner. *Theory of Multivariate Statistics (Springer Texts in Statistics)*. Springer, 1 edition, August 1999. ISBN 0387987398. doi: 10.1.1.172.3290. URL http://citeseerx.ist.psu.edu/viewdoc/summary?doi=10.1.1.172.3290.

Fisher Black, M. C. Jensen, and Myron Scholes. The capital asset pricing model: Some empirical tests. *Studies in the theory of capital markets*, 81 (3):79–121, 1972. doi: 10.1.1.665.9887. URL https://papers.ssrn.com/sol3/papers.cfm?abstract_id=908569.

Taras Bodnar and Yarema Okhrin. On the product of inverse Wishart and normal distributions with applications to discriminant analysis and portfolio theory. *Scandinavian Journal of Statistics*, 38(2):311–331, 2011. ISSN 1467-9469. doi: 10.1111/j.1467-9469.2011.00729.x. URL http://dx.doi.org/10.1111/j.1467-9469.2011.00729.x.

Jacob Boudoukh, Ronen Israel, and Matthew P. Richardson. Long Horizon Predictability: A Cautionary Tale. *SSRN eLibrary*, 2018. doi: 10.2139/ssrn.3142575. URL https://ssrn.com/paper=3142575.

Kris Boudt, Brian Peterson, and Christophe Croux. Estimation and decomposition of downside risk for portfolios with non-normal returns, 2008. doi: 10.2139/ssrn.1024151. URL http://ssrn.com/abstract=1024151.

Michael W. Brandt. Portfolio choice problems. *Handbook of financial econometrics*, 1:269–336, 2009. doi: 10.1.1.119.8765. URL https://faculty.fuqua.duke.edu/~mbrandt/papers/published/portreview.pdf.

Michael W. Brandt and Pedro Santa-Clara. Dynamic portfolio selection by augmenting the asset space. *The Journal of Finance*, 61(5):2187–2217, 2006. ISSN 1540-6261. doi: 10.1111/j.1540-6261.2006.01055.x. URL http://faculty.fuqua.duke.edu/~mbrandt/papers/published/condport.pdf.

Richard P. Brent. *Algorithms for Minimization without Derivatives*. Prentice-Hall, Englewood Cliffs, N.J., 1973. URL http://books.google.com/books?id=FR_RgSsC42EC.

Frank Bretz, Torsten Hothorn, and Peter Westfall. *Multiple comparisons using R*. Chapman and Hall/CRC, 2016. URL http://www.ievbras.ru/ecostat/Kiril/R/Biblio_N/R_Eng/Bretz2011.pdf.

Mark Britten-Jones. The sampling error in estimates of mean-variance efficient portfolio weights. *The Journal of Finance*, 54(2):655–671, 1999. doi: 10.1111/0022-1082.00120. URL http://www.jstor.org/stable/2697722.

Mark Britten-Jones and Anthony Neuberger. Improved inference and estimation in regression with overlapping observations, 2007. doi: 10.1111/j.1468-5957.2011.02244.x. URL http://www2.warwick.ac.uk/fac/soc/wbs/subjects/finance/faculty1/anthony_neuberger/improved.pdf.

Peter J. Brockwell and Richard A. Davis. *Introduction to Time Series and Forecasting*. Springer, 2nd edition, March 2002. ISBN 9780387953519. URL `http://books.google.com/books?vid=ISBN0387953515`.

Warren E. Buffett. Berkshire Hathaway shareholder letter 2016. Privately Published, February 2017. URL `http://www.berkshirehathaway.com/letters/2016ltr.pdf`.

Warren E. Buffett and Ted Seides. Bet 362. Privately Published, 2007. URL `http://longbets.org/362/`.

John Y. Campbell, Andrew W. Lo, and A. Craig MacKinlay. *The econometrics of financial markets*. Princeton University press, 2012.

Mark M. Carhart. On persistence in mutual fund performance. *Journal of Finance*, 52(1):57, 1997. doi: 10.2307/2329556. URL `http://www.jstor.org/stable/2329556`.

CBOE. The CBOE volatility index - VIX. Privately Published, 2009a. URL `http://www.cboe.com/micro/vix/vixwhite.pdf`.

CBOE. CBOE VIX data. Privately Published, 2017b. URL `http://www.cboe.com/products/vix-index-volatility/vix-options-and-futures/vix-index/vix-historical-data`.

Aleš Černý. The Hansen ratio in mean–variance portfolio theory, 2020. URL `https://arxiv.org/abs/2007.15980`.

Damien Challet. Sharper asset ranking from total drawdown durations, 2015. URL `http://arxiv.org/abs/1505.01333`.

R. Chattamvelli and R. Shanmugam. An enhanced algorithm for noncentral t-distribution. *Journal of Statistical Computation and Simulation*, 49 (1-2):77–83, 1994. doi: 10.1080/00949659408811561. URL `http://www.tandfonline.com/doi/abs/10.1080/00949659408811561`.

Long Chen, Robert Novy-Marx, and Lu Zhang. An alternative three-factor model. Privately Published, April 2011. doi: 10.2139/ssrn.1418117. URL `http://ssrn.com/abstract=1418117`.

Victor Chernozhukov, Iván Fernández-Val, and Alfred Galichon. Rearranging Edgeworth-Cornish-Fisher expansions. Privately Published, 2007. URL `http://arxiv.org/abs/0708.1627`.

Ludwig B. Chincarini and Daehwan Kim. *Quantitative equity portfolio management*. McGraw-Hill, 2010. ISBN 9780071459396.

Vijay Kumar Chopra and William T. Ziemba. The effect of errors in means, variances, and covariances on optimal portfolio choice. *The Journal of Portfolio Management*, 19(2):6–11, 1993. doi: 10.3905/jpm.1993.

409440. URL http://faculty.fuqua.duke.edu/~charvey/Teaching/BA453_2006/Chopra_The_effect_of_1993.pdf.

Gregory C. Chow. Tests of equality between sets of coefficients in two linear regressions. *Econometrica*, 28(3):591–605, 1960. ISSN 00129682. doi: 10.2307/1910133. URL http://www.jstor.org/stable/1910133.

Neil A Chriss and Robert Almgren. Portfolios from sorts, 2005. doi: 10.2139/ssrn.720041. URL https://dx.doi.org/10.2139/ssrn.720041.

John Howland Cochrane. *Asset pricing.* Princeton Univ. Press, Princeton [u.a.], 2001. ISBN 0691074984. URL http://press.princeton.edu/titles/7836.html.

Gregory Connor. Sensible return forecasting for portfolio management. *Financial Analysts Journal*, 53(5):44–51, 1997. ISSN 0015198X. doi: 10.2469/faj.v53.n5.2116. URL https://faculty.fuqua.duke.edu/~charvey/Teaching/BA453_2006/Connor_Sensible_Return_Forecasting_1997.pdf.

Rama Cont. Empirical properties of asset returns: stylized facts and statistical issues. *Quantitative Finance*, 1(2):223–236, 2001. doi: 10.1080/713665670. URL http://personal.fmipa.itb.ac.id/khreshna/files/2011/02/cont2001.pdf.

Gerard Cornuejols and Reha Tütüncü. *Optimization Methods in Finance.* Mathematics, Finance and Risk. Cambridge University Press, 2006. ISBN 9781139460569. URL http://web.math.ku.dk/~rolf/CT_FinOpt.pdf.

Max Dama. Max Dama on automated trading. Privately Published, 2011. URL http://isomorphisms.sdf.org/maxdama.pdf.

Zachary David. Information leakage in financial machine learning research. *Algorithmic Finance*, 8(1-2):1–4, 2019. doi: 10.3233/AF-190900. URL https://content.iospress.com/articles/algorithmic-finance/af190900.

A. P. Dempster. Covariance selection. *Biometrics*, 28(1):157–175, 1972. ISSN 0006341X. doi: 10.2307/2528966. URL http://www.jstor.org/stable/2528966.

James M. Dickey. Matricvariate generalizations of the multivariate t distribution and the inverted multivariate t distribution. *The Annals of Mathematical Statistics*, 38(2):511–518, 4 1967. doi: 10.1214/aoms/1177698967. URL http://dx.doi.org/10.1214/aoms/1177698967.

Matthew F. Dixon, Igor Halperin, and Paul Bilokon. *Machine Learning in Finance.* Springer, 2020.

P. Duchesne and P. Lafaye de Micheaux. Computing the distribution of quadratic forms: Further comparisons between the Liu-Tang-Zhang approximation and exact methods. *Computational Statistics and Data Analysis*, 54:858–862, 2010.

D. Duffie. *Dynamic Asset Pricing Theory*. Princeton Series in Finance. Princeton University Press, 3rd edition, 2010. ISBN 9781400829200. URL https://books.google.com/books?id=f2Wv-LDpsoUC.

Pierre Dutilleul. The MLE algorithm for the matrix normal distribution. *Journal of Statistical Computation and Simulation*, 64(2):105–123, 1999. doi: 10.1080/00949659908811970. URL https://doi.org/10.1080/00949659908811970.

Dirk Eddelbuettel and Romain François. Rcpp: Seamless R and C++ integration. *Journal of Statistical Software*, 40(8):1–18, 2011. doi: 10.18637/jss.v040.i08. URL http://www.jstatsoft.org/v40/i08/.

Bradley Efron. Student's t-test under non-normal conditions. Technical Report 21, Department of Statistics, Harvard University, May 1968. URL https://apps.dtic.mil/sti/pdfs/AD0670743.pdf.

Bradley Efron. The estimation of prediction error: covariance penalties and cross-validation. *Journal of the American Statistical Association*, 99(467):619–632, 2004. doi: 10.1198/016214504000000692. URL http://pages.stat.wisc.edu/~wahba/stat860public/pdf1/efron.covariance.04.pdf.

Edwin J. Elton, Martin J. Gruber, and Christopher R. Blake. Fundamental economic variables, expected returns, and bond fund performance. *The Journal of Finance*, 50(4):1229–1256, 1995. ISSN 00221082. doi: 10.1111/j.1540-6261.1995.tb04056.x. URL http://www.jstor.org/stable/2329350.

Eugene F. Fama and Kenneth R. French. The cross-section of expected stock returns. *Journal of Finance*, 47(2):427, 1992. doi: 10.1111/j.1540-6261.1992.tb04398.x. URL http://www.jstor.org/stable/2329112.

William Feller. On the Berry-Esseen theorem. *Zeitschrift für Wahrscheinlichkeitstheorie und Verwandte Gebiete*, 10(3):261–268, 1968. doi: 10.1007/BF00536279. URL http://static.stevereads.com/papers_to_read/feller--on_the_berry-esseen_theorem_.pdf.

Mark Finkelstein and Edward O. Thorp. Nontransitive dice with equal means. *Optimal Play: Mathematical Studies in Games and Gambling*, pages 293–310, 2000. URL http://www.math.uci.edu/~mfinkels/dice9.pdf.

Gregg S. Fisher, Philip Z. Maymin, and Zakhar G. Maymin. Risk parity optimality. *The Journal of Portfolio Management*, 41(2):42–56,

2015. URL https://gersteinfisher.com/wp-content/uploads/2013/05/SSRN_Risk_Parity_Optimality.pdf.

Ronald Aylmer Fisher. The sampling error of estimated deviates, together with other illustrations of the properties and applications of the integrals and derivatives of the normal error function. In *Introduction to British A.A.S. Math. Tables, I.* 1931a. URL http://hdl.handle.net/2440/15209.

Ronald Aylmer Fisher. *Properties and applications of* Hh *functions*, pages xxvi–xxxv. 1931b. URL http://scholar.google.com/scholar?cites=8550436429828449531\&\#38;as_sdt=4005\&\#38;sciodt=4000\&\#38;hl=en.

Dean Follmann. A simple multivariate test for one-sided alternatives. *Journal of the American Statistical Association*, 91(434):854–861, 1996. ISSN 01621459. doi: 10.2307/2291680. URL http://www.jstor.org/stable/2291680.

Paul E. Ford. What is code? *Bloomberg Businessweek*, 11, 2015. URL https://www.bloomberg.com/graphics/2015-paul-ford-what-is-code/.

Erik Forseth and Ed Tricker. Equal risk contribution portfolios. Technical report, Graham Capital Management, 3 2019. URL https://www.grahamcapital.com/Equal%20Risk%20Contribution%20April%202019.pdf.

Andrea Frazzini, David Kabiller, and Lasse Heje Pedersen. Buffett's alpha. Privately Published, 2012. URL http://www.econ.yale.edu/~af227/pdf/Buffett%27s%20Alpha%20-%20Frazzini,%20Kabiller%20and%20Pedersen.pdf.

Kenneth French. Data library. Privately Published, 2017. URL http://mba.tuck.dartmouth.edu/pages/faculty/ken.french/data_library.html.

Kenneth French. 5 industry portfolios. Privately Published, 2019. URL http://mba.tuck.dartmouth.edu/pages/faculty/ken.french/Data_Library/det_5_ind_port.html.

Claudio Fuentes and Vik Gopal. A constrained conditional likelihood approach for estimating the means of selected populations, 2017. URL http://arxiv.org/abs/1702.07804. cite arxiv:1702.07804.

Claudio Fuentes, George Casella, and Martin T. Wells. Confidence intervals for the means of the selected populations. *Electron. J. Statist.*, 12(1): 58–79, 2018. doi: 10.1214/17-EJS1374. URL https://doi.org/10.1214/17-EJS1374.

Martin Gardner. *The Colossal Book of Mathematics: Classic Puzzles, Paradoxes, and Problems : Number Theory, Algebra, Geometry, Probability, Topology, Game Theory, Infinity, and Other Topics of Recreational Mathematics.* Norton, 2001. ISBN 9780393020236. URL `https://books.google.com/books?id=orz0SDEakpYC`.

A. Gelman, J. B. Carlin, H. S. Stern, D. B. Dunson, A. Vehtari, and D. B. Rubin. *Bayesian Data Analysis, Third Edition.* Chapman & Hall/CRC Texts in Statistical Science. Taylor & Francis, 2013. ISBN 9781439840955. URL `http://books.google.com/books?id=ZXL6AQAAQBAJ`.

David C. Gerard and Peter D. Hoff. A higher-order LQ decomposition for separable covariance models, 2014. URL `https://arxiv.org/abs/1410.1094`.

Michael R. Gibbons, Stephen A. Ross, and Jay Shanken. A test of the efficiency of a given portfolio. *Econometrica: Journal of the Econometric Society*, pages 1121–1152, 1989. doi: 10.2307/1913625. URL `https://teach.business.uq.edu.au/courses/FINM6905/files/module-1/readings/Gibbons%20Ross%20Shanken.pdf`.

Narayan C. Giri. On the likelihood ratio test of a normal multivariate testing problem. *The Annals of Mathematical Statistics*, 35(1):181–189, 1964. doi: 10.1214/aoms/1177703740. URL `http://projecteuclid.org/euclid.aoms/1177703740`.

Georg M. Goerg. Lambert W Random Variables – A New Family of Generalized Skewed Distributions with Applications to Risk Estimation. *ArXiv e-prints*, December 2009. URL `http://arxiv.org/abs/0912.4554`.

Georg M. Goerg. The Lambert Way to Gaussianize skewed, heavy tailed data with the inverse of Tukey's h transformation as a special case. *ArXiv e-prints*, October 2010. URL `http://arxiv.org/abs/1010.2265`.

Georg M. Goerg. *LambertW: Probabilistic Models to Analyze and Gaussianize Heavy-Tailed, Skewed Data*, 2016. R package version 0.6.4.

William Sealy Gosset. The probable error of a mean. *Biometrika*, 6(1):1–25, March 1908. doi: 10.2307/2331554. URL `http://dx.doi.org/10.2307/2331554`. Originally published under the pseudonym "Student".

Franklin A. Graybill and George Marsaglia. Idempotent matrices and quadratic forms in the general linear hypothesis. *Ann. Math. Statist.*, 28(3):678–686, 09 1957. doi: 10.1214/aoms/1177706879. URL `https://doi.org/10.1214/aoms/1177706879`.

R. Grinold and R. Kahn. *Active Portfolio Management: A Quantitative Approach for Producing Superior Returns and Selecting Superior Returns and Controlling Risk.* McGraw-Hill Library of Investment and Finance.

McGraw-Hill Education, 1999. ISBN 9780070248823. URL http://books.google.com/books?id=a1yB8LTQn0EC.

William C. Guenther. Evaluation of probabilities for the noncentral distributions and the difference of two t-variables with a desk calculator. *Journal of Statistical Computation and Simulation*, 6(3-4):199–206, 1978. doi: 10.1080/00949657808810188. URL http://dx.doi.org/10.1080/00949657808810188.

Adityanand Guntuboyina. *Minimax lower bounds*. PhD thesis, Yale University, 2011. URL https://www.stat.berkeley.edu/~aditya/resources/Thesis.pdf.

A. K. Gupta, T. Varga, and T. Bodnar. *Elliptically Contoured Models in Statistics and Portfolio Theory*. SpringerLink : Bücher. Springer New York, 2nd edition, 2013. ISBN 9781461481546. URL https://books.google.com/books?id=C_66BAAAQBAJ.

Peter Hall. Chi squared approximations to the distribution of a sum of independent random variables. *The Annals of Probability*, 11(4):1028–1036, 1983. ISSN 00911798. URL https://projecteuclid.org/euclid.aop/1176993451.

Peter Hall and Qiying Wang. Exact convergence rate and leading term in central limit theorem for Student's t statistic, 2004. doi: 10.1214/009117904000000252. URL http://arxiv.org/abs/math/0410148.

Peter Reinhard Hansen. Asymptotic tests of composite hypotheses. Working Paper 2003-09, Brown University, Department of Economics, Providence, RI, 2003. URL http://hdl.handle.net/10419/80222.

Peter Reinhard Hansen. A test for superior predictive ability. *Journal of Business and Economic Statistics*, 23(4), 2005. doi: 10.1198/073500105000000063. URL http://pubs.amstat.org/doi/abs/10.1198/073500105000000063.

Johanna Hardin and David M. Rocke. The distribution of robust distances. *Journal of Computational and Graphical Statistics*, 14(4):928–946, 2005. doi: 10.1198/106186005X77685. URL http://dmrocke.ucdavis.edu/Robdist5.pdf.

Campbell R. Harvey and Yan Liu. Backtesting. *The Journal of Portfolio Management*, 42(1):13–28, 2015. doi: 10.3905/jpm.2015.42.1.013. URL https://www.cmegroup.com/education/files/backtesting.pdf.

Joel Hasbrouck. *Empirical Market Microstructure: The Institutions, Economics, and Econometrics of Securities Trading*. Oxford University Press, USA, 2007. ISBN 9780195301649. URL http://books.google.com/books?id=KIEHmQEACAAJ.

Roy D. Henriksson and Robert C. Merton. On market timing and investment performance. II. statistical procedures for evaluating forecasting skills. *The Journal of Business*, 54(4):513–533, 1981. URL `http://www.jstor.org/stable/2352722`.

Daniel Hernandez-Stumpfhauser, F. Jay Breidt, and Mark J. van der Woerd. The general projected normal distribution of arbitrary dimension: Modeling and Bayesian inference. *Bayesian Analysis*, 12(1):113–133, 2017. doi: 10.1214/15-BA989. URL `https://projecteuclid.org/download/pdfview_1/euclid.ba/1453211962`.

Ulf Herold and Raimond Maurer. Tactical asset allocation and estimation risk. *Financial Markets and Portfolio Management*, 18(1):39–57, 2004. ISSN 1555-4961. doi: 10.1007/s11408-004-0104-2. URL `http://dx.doi.org/10.1007/s11408-004-0104-2`.

G. W. Hill. Algorithm 396: Student's t-quantiles. *Commun. ACM*, 13(10):619–620, October 1970. ISSN 0001-0782. doi: 10.1145/355598.355600. URL `http://doi.acm.org/10.1145/355598.355600`.

G. W. Hill. Remark on "Algorithm 396: Student's t-quantiles [s14]". *ACM Trans. Math. Softw.*, 7(2):250–251, June 1981. ISSN 0098-3500. doi: 10.1145/355945.355956. URL `http://doi.acm.org/10.1145/355945.355956`.

Stewart Hodges. *A Generalization of the Sharpe Ratio and Its Applications to Valuation Bounds and Risk Measures.* FORC preprint: Financial Options Research Centre. Financial Options Research Centre, Warwick Business School, University of Warwick, 1998. URL `http://www2.warwick.ac.uk/fac/soc/wbs/subjects/finance/research/wpaperseries/1998/98-88.pdf`.

D. Hogben, R. S. Pinkham, and M. B. Wilk. The moments of the non-central t-distribution. *Biometrika*, 48(3/4):465–468, 1961. ISSN 00063444. doi: 10.2307/2332772. URL `http://www.jstor.org/stable/2332772`.

William Huber. What is the distribution of a bivariate normal component conditional on the max of the other component? Cross Validated, 2019. URL `https://stats.stackexchange.com/q/416791`. [Online; accessed 2019-07-10].

Gur Huberman and Shmuel Kandel. Mean-variance spanning. *The Journal of Finance*, 42(4):873–888, 1987. ISSN 00221082. doi: 10.2307/2328296. URL `http://www.jstor.org/stable/2328296`.

Jiunn T. Hwang. Empirical Bayes estimation for the means of the selected populations. *Sankkhyâ: The Indian Journal of Statistics*, 55:285–304, 1993. URL `https://www.jstor.org/stable/25050934`.

J. P. Imhof. Computing the distribution of quadratic forms in normal variables. *Biometrika*, 48(3-4):419–426, 12 1961. ISSN 0006-3444. doi: 10.1093/biomet/48.3-4.419. URL https://doi.org/10.1093/biomet/48.3-4.419.

L. Isserlis. On a formula for the product-moment coefficient of any order of a normal frequency distribution in any number of variables. *Biometrika*, 12 (1/2):134–139, 1918. ISSN 00063444. doi: 10.2307/2331932. URL http://www.jstor.org/stable/2331932.

Toshiya Iwashita. Asymptotic null and nonnull distribution of Hotelling's T^2-statistic under the elliptical distribution. *Journal of Statistical Planning and Inference*, 61(1):85 – 104, 1997. ISSN 0378-3758. doi: 10.1016/S0378-3758(96)00153-X. URL http://www.sciencedirect.com/science/article/pii/S037837589600153X.

Michael Izbicki. Algebraic classifiers: a generic approach to fast cross-validation, online training, and parallel training. In *International Conference on Machine Learning*, pages 648–656, 2013. URL http://proceedings.mlr.press/v28/izbicki13.html.

Stefan R. Jaschke. The Cornish-Fisher-expansion in the context of Delta - Gamma - Normal approximations. Technical Report 2001,54, Humboldt University of Berlin, Interdisciplinary Research Project 373: Quantification and Simulation of Economic Processes, 2001. URL http://www.jaschke-net.de/papers/CoFi.pdf.

D. R. Jensen. The joint distribution of traces of Wishart matrices and some applications. *The Annals of Mathematical Statistics*, 41(1):133–145, 1970. ISSN 00034851. doi: doi:10.1214/aoms/1177697194. URL https://projecteuclid.org/download/pdf_1/euclid.aoms/1177697194.

J. D. Jobson and Bob M. Korkie. Estimation for Markowitz efficient portfolios. *Journal of the American Statistical Association*, 75(371):544–554, 1980. doi: 10.1080/01621459.1980.10477507. URL https://www.tandfonline.com/doi/abs/10.1080/01621459.1980.10477507.

J. D. Jobson and Bob M. Korkie. Performance hypothesis testing with the Sharpe and Treynor measures. *The Journal of Finance*, 36(4):889–908, 1981. ISSN 00221082. doi: 10.2307/2327554. URL http://www.jstor.org/stable/2327554.

N. L. Johnson and B. L. Welch. Applications of the non-central t-distribution. *Biometrika*, 31(3-4):362–389, March 1940. doi: 10.1093/biomet/31.3-4.362. URL http://dx.doi.org/10.1093/biomet/31.3-4.362.

Zura Kakushadze and Willie Yu. Notes on Fano ratio and portfolio optimization, 2017. URL http://arxiv.org/abs/1711.10640.

Raymond Kan and Guofu Zhou. Tests of mean-variance spanning. *Annals of Economics and Finance*, 13(1), 2012. doi: 10.2139/ssrn.231522. URL http://www.aeconf.net/Articles/May2012/aef130105.pdf.

Yutaka Kano. An asymptotic expansion of the distribution of Hotelling's T^2-statistic under general distributions. *American Journal of Mathematical and Management Sciences*, 15(3-4):317–341, 1995. doi: 10.1080/01966324.1995.10737403. URL https://doi.org/10.1080/01966324.1995.10737403.

Robert E. Kass and Larry Wasserman. The selection of prior distributions by formal rules. *Journal of the American statistical Association*, 91(435):1343–1370, 1996. doi: 10.1080/01621459.1996.10477003. URL http://stat.cmu.edu/~kass/papers/rules.pdf.

Peter Kennedy. *A guide to econometrics.* Blackwell Pub., 6th edition, 2008. ISBN 9781405182577. URL http://books.google.com/books?id=ax1QcAAACAAJ.

Jongphil Kim and Anthony J. Hayter. Efficient confidence interval methodologies for the non centrality parameter of a non central t distribution. *Communications in Statistics - Simulation and Computation*, 37(4):660–678, 2008. doi: 10.1080/03610910701739605. URL http://www.tandfonline.com/doi/abs/10.1080/03610910701739605.

Igor Kotsiuba and Stepan Mazur. On the asymptotic and approximate distributions of the product of an inverse Wishart matrix and a Gaussian vector. Technical Report 2015:11, Department of Statistics, School of Economics and Management, Lund University, 2015. URL http://journals.lub.lu.se/index.php/stat/article/view/15167/13723.

S. Kotz and S. Nadarajah. *Multivariate t-Distributions and Their Applications.* Cambridge University Press, 2004. ISBN 9780521826549. URL http://drsmorey.org/bibtex/upload/Kotz:Nadarajah:2004.pdf.

Helena Chmura Kraemer and Minja Paik. A central t approximation to the noncentral t distribution. *Technometrics*, 21(3):357–360, 1979. ISSN 00401706. doi: 10.2307/1267759. URL http://www.jstor.org/stable/1267759.

Andreas Krause. An overview of asset pricing models. Privately Published, 2001. URL http://people.bath.ac.uk/mnsak/Research/Asset_pricing.pdf.

K. Krishnamoorthy. *Handbook of Statistical Distributions with Applications.* Statistics: A Series of Textbooks and Monographs. CRC Press, 2006. ISBN 9781420011371. URL http://www.crcnetbase.com/isbn/9781420011371.

William Kruskal. The monotonicity of the ratio of two noncentral t density functions. *The Annals of Mathematical Statistics*, 25(1):162–165, 1954. ISSN 00034851. URL http://www.jstor.org/stable/2236523.

A. M. Kshirsagar. Some extensions of the multivariate t-distribution and the multivariate generalization of the distribution of the regression coefficient. *Mathematical Proceedings of the Cambridge Philosophical Society*, 57(01): 80–85, 1 1961. ISSN 1469-8064. doi: 10.1017/S0305004100034885. URL http://journals.cambridge.org/article_S0305004100034885.

Tatsuya Kubokawa, C. P. Robert, and A. K. Md. E. Saleh. Estimation of noncentrality parameters. *Canadian Journal of Statistics*, 21(1):45–57, Mar 1993. doi: 10.2307/3315657. URL http://www.jstor.org/stable/3315657.

Michael Lachanski and Steven E. Pav. Shy of the character limit: "Twitter mood predicts the stock market" revisited. *Econ Journal Watch*, 14(3): 302–345, September 2017. URL https://econjwatch.org/articles/shy-of-the-character-limit-twitter-mood-predicts-the-stock-market-revisited.

Daniel Lakens. Equivalence tests: A practical primer for t-tests, correlations, and meta-analyses. 2016. doi: 10.1177/1948550617697177. URL https://osf.io/preprints/psyarxiv/97gpc/.

Nico F. Laubscher. Normalizing the noncentral t and F distributions. *Ann. Math. Statist.*, 31(4):1105–1112, 12 1960. doi: 10.1214/aoms/1177705682. URL https://doi.org/10.1214/aoms/1177705682.

Bruno Lecoutre. Two useful distributions for Bayesian predictive procedures under normal models. *Journal of Statistical Planning and Inference*, 79:93–105, 1999. doi: 10.1016/S0378-3758(98)00231-6. URL http://www.researchgate.net/publication/242997315_Two_useful_distributions_for_Bayesian_predictive_procedures_under_normal_models/file/5046352b9cba661c83.pdf.

Bruno Lecoutre. Another look at confidence intervals for the noncentral T distribution. *Journal of Modern Applied Statistical Methods*, 6(1):107–116, 2007. doi: 10.22237/jmasm/1177992600. URL http://www.univ-rouen.fr/LMRS/Persopage/Lecoutre/telechargements/Lecoutre_Another_look-JMSAM2007_6(1).pdf.

Olivier Ledoit and Michael Wolf. Robust performance hypothesis testing with the Sharpe ratio. *Journal of Empirical Finance*, 15(5):850–859, December 2008. ISSN 0927-5398. doi: 10.1016/j.jempfin.2008.03.002. URL http://www.ledoit.net/jef2008_abstract.htm.

Jason D. Lee, Dennis L. Sun, Yuekai Sun, and Jonathan E. Taylor. Exact post-selection inference, with application to the lasso, 2013. doi: 10.1214/15-AOS1371. URL `http://arxiv.org/abs/1311.6238`. cite arxiv:1311.6238 Comment: Published at http://dx.doi.org/10.1214/15-AOS1371 in the Annals of Statistics (http://www.imstat.org/aos/) by the Institute of Mathematical Statistics (http://www.imstat.org).

Yoong-Sin Lee and Ting-Kwong Lin. Algorithm AS 269: High order Cornish-Fisher expansion. *Journal of the Royal Statistical Society. Series C (Applied Statistics)*, 41(1):233–240, 1992. ISSN 00359254. doi: 10.2307/2347649. URL `http://www.jstor.org/stable/2347649`.

Robert Lehr. Sixteen S-squared over D-squared: a relation for crude sample size estimates. *Statistics in Medicine*, 11(8):1099–102, 1992. ISSN 0277-6715. doi: 10.1002/sim.4780110811. URL `http://www.biomedsearch.com/nih/Sixteen-S-squared-over-D/1496197.html`.

Russell V. Lenth. Statistical algorithms: Algorithm AS 243: Cumulative distribution function of the non-central t distribution. *Journal of Applied Statistics*, 38(1):185–189, March 1989. ISSN 0035-9254 (print), 1467-9876 (electronic). doi: 10.2307/2347693. URL `http://lib.stat.cmu.edu/apstat/243`.

Gérard Letac and Hélène Massam. A tutorial on the non central Wishart distribution. Privately Published, 2004. URL `http://www.math.univ-toulouse.fr/~letac/Wishartnoncentrales.pdf`.

Gérard Letac and Hélène Massam. The noncentral Wishart as an exponential family, and its moments. *Journal of Multivariate Analysis*, 99(7):1393 – 1417, 2008. ISSN 0047-259X. doi: 10.1016/j.jmva.2008.04.006. URL `http://www.sciencedirect.com/science/article/pii/S0047259X08001139`. Special Issue: Multivariate Distributions, Inference and Applications in Memory of Norman L. Johnson.

Pui-Lam Leung and Wing-Keung Wong. On testing the equality of multiple Sharpe ratios, with application on the evaluation of iShares. *Journal of Risk*, 10(3):15–30, 2008. doi: 10.2139/ssrn.907270. URL `http://www.risk.net/digital_assets/4760/v10n3a2.pdf`.

Wenbo V. Li and Qi-Man Shao. A normal comparison inequality and its applications. *Probability Theory and Related Fields*, 122(4):494–508, 2002. doi: 10.1.1.147.1513. URL `http://citeseerx.ist.psu.edu/viewdoc/download?doi=10.1.1.147.1513&rep=rep1&type=pdf`.

Andrew W. Lo. The statistics of Sharpe ratios. *Financial Analysts Journal*, 58 (4), July/August 2002. doi: 10.2469/faj.v58.n4.2453. URL `http://ssrn.com/paper=377260`.

Haipeng Luo, Chen-Yu Wei, and Kai Zheng. Efficient online portfolio with logarithmic regret. In S. Bengio, H. Wallach, H. Larochelle, K. Grauman, N. Cesa-Bianchi, and R. Garnett, editors, *Advances in Neural Information Processing Systems 31*, pages 8235–8245. Curran Associates, Inc., 2018. URL `http://papers.nips.cc/paper/8045-efficient-online-portfolio-with-logarithmic-regret.pdf`.

Hal Lux. The secret world of Jim Simons. *Institutional Investor*, 34(11), November 2000.

C. R. MacCluer. *Calculus of Variations: Mechanics, Control, and Other Applications*. Pearson/Prentice Hall, 2005. ISBN 9780131423831. URL `https://books.google.com/books?id=rqQrAAAAYAAJ`.

Malik Magdon-Ismail, Amir F. Atiya, Amrit Pratap, and Yaser S. Abu-Mostafa. The Sharpe ratio, range, and maximal drawdown of a Brownian motion. Technical report, Technical Report TR 02-13, RPI Computer Science, 2002. URL `http://intelligenthedgefundinvesting.com/pubs/rb-mapa1.pdf`.

Jan R. Magnus and H. Neudecker. The commutation matrix: Some properties and applications. *Ann. Statist.*, 7(2):381–394, 03 1979. doi: 10.1214/aos/1176344621. URL `https://doi.org/10.1214/aos/1176344621`.

Jan R. Magnus and H. Neudecker. The elimination matrix: Some lemmas and applications. *SIAM Journal on Algebraic Discrete Methods*, 1(4):422–449, December 1980. doi: 10.1137/0601049. URL `http://www.janmagnus.nl/papers/JRM008.pdf`.

Jan R. Magnus and H. Neudecker. Symmetry, 0-1 matrices and Jacobians. *Econometric Theory*, 2:157–190, 1986. doi: 10.1017/S0266466600011476. URL `http://www.janmagnus.nl/papers/JRM014.pdf`.

Jan R. Magnus and H. Neudecker. *Matrix Differential Calculus with Applications in Statistics and Econometrics*. Wiley Series in Probability and Statistics: Texts and References Section. Wiley, 3rd edition, 2007. ISBN 9780471986331. URL `http://www.janmagnus.nl/misc/mdc2007-3rdedition`.

Burton G. Malkiel. The efficient market hypothesis and its critics. *Journal of Economic Perspectives*, 17(1):59–82, 2003. doi: 10.1257/089533003321164958. URL `http://eml.berkeley.edu/~craine/EconH195/Fall_14/webpage/Malkiel_Efficient%20Mkts.pdf`.

K. V. Mardia. 9 tests of unvariate and multivariate normality. In *Analysis of Variance*, volume 1 of *Handbook of Statistics*, pages 279 – 320. Elsevier, 1980. doi: 10.1016/S0169-7161(80)01011-5. URL `http://www.sciencedirect.com/science/article/pii/S0169716180010115`.

Stepan Mazur. personal communication, 2020.

James J. McKeon. F approximations to the distribution of Hotelling's T^2. *Biometrika*, 61(2):381–383, 1974. doi: 10.2307/2334369. URL https://www.jstor.org/stable/2334369.

Elmar Mertens. Comments on variance of the IID estimator in Lo (2002). Technical report, Working Paper University of Basel, Wirtschaftswissenschaftliches Zentrum, Department of Finance, 2002. URL http://www.elmarmertens.com/research/discussion/soprano01.pdf.

Robert E. Miller and Adam K. Gehr. Sample size bias and Sharpe's performance measure: A note. *Journal of Financial and Quantitative Analysis*, 13(05):943–946, 1978. doi: 10.2307/2330636. URL https://doi.org/10.2307/2330636.

Matthew R. Morey and Hrishikesh D. Vinod. A double Sharpe ratio. *Advances in Investment Analysis and Portfolio Management*, 8:57–65, 2001. URL https://elib.umkendari.ac.id/eb_el/Cheng-Few_Lee-Advances_in_Investment_Analysis_and_Portfolio_Management,_Volume_8,_Volume_8-JAI_Pr.pdf#page=72.

Keith E. Muller and Bercedis L. Peterson. Practical methods for computing power in testing the multivariate general linear hypothesis. *Computational Statistics & Data Analysis*, 2(2):143–158, 1984. ISSN 0167-9473. doi: 10.1016/0167-9473(84)90002-1. URL http://www.sciencedirect.com/science/article/pii/0167947384900021.

C. Munk. *Financial Asset Pricing Theory*. OUP Oxford, 2nd edition, 2013. ISBN 9780199585496. URL https://books.google.com/books?id=3WVs9pFkimIC.

Kevin P. Murphy. Conjugate Bayesian analysis of the Gaussian distribution. Privately Published, 2007. URL http://www.cs.ubc.ca/~murphyk/Papers/bayesGauss.pdf.

Daniel B. Nelson. Conditional heteroskedasticity in asset returns: A new approach. *Econometrica*, 59(2):347–370, 1991. ISSN 00129682. doi: 10.2307/2938260. URL http://finance.martinsewell.com/stylized-facts/distribution/Nelson1991.pdf.

NIST. Digital library of mathematical functions, 2011. URL http://dlmf.nist.gov/5.11#iii. F. W. J. Olver, A. B. Olde Daalhuis, D. W. Lozier, B. I. Schneider, R. F. Boisvert, C. W. Clark, B. R. Miller and B. V. Saunders, eds.

J. Nocedal and S. J. Wright. *Numerical Optimization*. Springer series in operations research and financial engineering. Springer, 2006. ISBN 9780387400655. URL http://books.google.com/books?id=VbHYoSyelFcC.

Ralph G. O'Brien and Gwowen Shieh. Pragmatic, unifying algorithm gives power probabilities for common F tests of the multivariate general linear hypothesis, 1999. URL http://www.bio.ri.ccf.org/Power/.

R. E. Odeh and J. O. Evans. Algorithm AS 70: The percentage points of the normal distribution. *Journal of the Royal Statistical Society. Series C (Applied Statistics)*, 23(1):96–97, 1974. ISSN 00359254, 14679876. doi: 10.2307/2347061. URL http://www.jstor.org/stable/2347061.

Frank W. Olver, Daniel W. Lozier, Ronald F. Boisvert, and Charles W. Clark. *NIST Handbook of Mathematical Functions*. Cambridge University Press, New York, NY, USA, 1st edition, 2010. ISBN 0521140633, 9780521140638. URL http://dlmf.nist.gov/.

J. D. Opdyke. Comparing Sharpe ratios: So where are the p-values? *Journal of Asset Management*, 8(5), 2007. doi: 10.1057/palgrave.jam.2250084. URL http://ssrn.com/abstract=886728.

D. B. Owen. Survey of properties and applications of the noncentral t-distribution. *Technometrics*, 10(3):445–478, 1968. doi: 10.1080/00401706.1968.10490594. URL http://dx.doi.org/10.1080/00401706.1968.10490594.

Dirk Paulsen and Jakob Söhl. Noise fit, estimation error and a Sharpe information criterion, 2016. URL http://arxiv.org/abs/1602.06186.

Steven E. Pav. Asymptotic distribution of the Markowitz portfolio. Privately Published, 2013. URL http://arxiv.org/abs/1312.0557.

Steven E. Pav. Bounds on portfolio quality. Privately Published, 2014. URL http://arxiv.org/abs/1409.5936.

Steven E. Pav. Safety third: Roy's criterion and higher order moments. Privately Published, 2015. URL http://arxiv.org/abs/1506.04227.

Steven E. Pav. *BWStest: Baumgartner Weiss Schindler Test of Equal Distributions*, 2016a. URL https://CRAN.R-project.org/package=BWStest. R package version 0.2.0.

Steven E. Pav. *madness: Automatic Differentiation of Multivariate Operations*, 2016b. URL https://github.com/shabbychef/madness. R package version 0.2.0.

Steven E. Pav. *MarkowitzR: Statistical Significance of the Markowitz Portfolio*, 2018. URL https://github.com/shabbychef/MarkowitzR. R package version 1.0.1.

Steven E. Pav. *aqfb.data: Data for the AQFB Book*, 2019a. URL https://github.com/shabbychef/aqfb_data. R package version 0.1.6.

Steven E. Pav. Conditional inference on the asset with maximum Sharpe ratio. Privately Published, 2019b. URL `http://arxiv.org/abs/1906.00573`.

Steven E. Pav. A *post hoc* test on the Sharpe ratio. Privately Published, 2019c. URL `http://arxiv.org/abs/1911.04090`. code available from `https://github.com/shabbychef/posthoc\_sr`.

Steven E. Pav. *SharpeR: Statistical Significance of the Sharpe Ratio*, 2020a. URL `https://github.com/shabbychef/SharpeR`. R package version 1.2.1.

Steven E. Pav. Inference on achieved signal noise ratio, 2020b. URL `http://arxiv.org/abs/2005.06171`. cite arxiv:2005.06171.

Steven E. Pav. *tsrsa: Code and Data for The Sharpe Ratio: Statistics and Applications Book*, 2021. URL `https://github.com/shabbychef/tsrsa`. R package version 0.1.1.

Yudi Pawitan. *In all likelihood: statistical modelling and inference using likelihood.* Oxford science publications. Clarendon press, Oxford, 2001. ISBN 978-0-19-850765-9. URL `http://books.google.com/books?id=8T8fAQAAQBAJ`.

Michael D. Perlman. One-sided testing problems in multivariate analysis. *The Annals of Mathematical Statistics*, 40(2):549–567, 1969. doi: 10.1214/aoms/1177697723. URL `https://projecteuclid.org/euclid.aoms/1177697723`.

Michael D. Perlman and Lang Wu. The emperor's new tests. *Statistical Science*, 14(4):355–369, 1999. URL `https://projecteuclid.org/euclid.ss/1009212517`.

Kaare Brandt Petersen and Michael Syskind Pedersen. The matrix cookbook, November 2012. URL `http://www2.imm.dtu.dk/pubdb/p.php?3274`. Version 20121115.

Brian G. Peterson and Peter Carl. *PerformanceAnalytics: Econometric Tools for Performance and Risk Analysis*, 2018. URL `https://CRAN.R-project.org/package=PerformanceAnalytics`. R package version 1.5.2.

Hans-Peter Piepho. An algorithm for a letter-based representation of all-pairwise comparisons. *Journal of Computational and Graphical Statistics*, 13(2):456–466, 2004. doi: 10.1198/1061860043515. URL `https://doi.org/10.1198/1061860043515`.

K. C. Sreedharan Pillai. On the distribution of the largest or the smallest root of a matrix in multivariate analysis. *Biometrika*, 43(1/2):122–127, 1956. doi: 10.2307/2333585. URL `https://www.jstor.org/stable/2333585`.

K. C. Sreedharan Pillai and Pablo Samson, Jr. On Hotelling's generalization of T^2. *Biometrika*, pages 160–168, 1959. doi: 10.2307/2332818. URL `https://www.jstor.org/stable/2332818`.

Jacques Poitevineau and Bruno Lecoutre. Implementing Bayesian predictive procedures: The K-prime and K-square distributions. *Computational Statistics and Data Analysis*, 54(3):724–731, 2010. doi: 10.1016/j.csda.2008.11.004. URL `http://arxiv.org/abs/1003.4890v1`.

S. J. Press. *Applied Multivariate Analysis: Using Bayesian and Frequentist Methods of Inference.* Dover Publications, Incorporated, 2012. ISBN 9780486139388. URL `http://books.google.com/books?id=WneJJEHYHLYC`.

Edward E. Qian, Ronald H. Hua, and Eric H. Sorensen. *Quantitative equity portfolio management: modern techniques and applications.* CRC Press, 2007.

R Core Team. *R: A Language and Environment for Statistical Computing.* R Foundation for Statistical Computing, Vienna, Austria, 2015. URL `http://www.R-project.org/`. ISBN 3-900051-07-0.

C. Radhakrishna Rao. *Advanced Statistical Methods in Biometric Research.* John Wiley and Sons, 1952. URL `http://books.google.com/books?id=HvFLAAAAMAAJ`.

C. Radhakrishna Rao, Helge Toutenburg, Shalabh, Christian Heumann, and Michael Schomaker. *Linear Models and Generalizations: Least Squares and Alternatives.* Springer Series in Statistics Series. Springer, 3rd edition, 2010. ISBN 9783642093531. URL `http://books.google.com/books?id=seedcQAACAAJ`.

G. C. Reinsel and R. P. Velu. *Multivariate Reduced-Rank Regression: Theory and Applications.* Lecture Notes in Statistics. Springer, 1998. ISBN 9780387986012. URL `http://books.google.com/books?id=QZJ1QgAACAAJ`.

Adam Rej, Philip Seager, and Jean-Philippe Bouchaud. You are in a drawdown. when should you start worrying?, 2017. URL `http://arxiv.org/abs/1707.01457`.

Alvin C. Rencher. *Methods of Multivariate Analysis.* Wiley series in probability and mathematical statistics. Probability and mathematical statistics. J. Wiley, 2002. ISBN 9780471418894. URL `http://books.google.com/books?id=SpvBd7IUCxkC`.

Igor Rivin. What is the Sharpe ratio, and how can everyone get it wrong?, 2018. URL `http://arxiv.org/abs/1802.04413`.

Dale W. R. Rosenthal. *A Quantitative Primer on Investments with R.* Q36 LLC, 2018. ISBN 9781732235601. URL https://books.google.com/books?id=ffBCtwEACAAJ.

Stephen A. Ross. The arbitrage theory of capital asset pricing. *Journal of Economic Theory*, 13(3):341–360, 1976. doi: 10.1016/0022-0531(76)90046-6. URL http://linkinghub.elsevier.com/retrieve/pii/0022053176900466.

Henry Rouanet and Bruno Lecoutre. Specific inference in ANOVA: From significance tests to Bayesian procedures. *British Journal of Mathematical and Statistical Psychology*, 36(2):252–268, 1983. doi: 10.1111/j.2044-8317.1983.tb01131.x. URL https://onlinelibrary.wiley.com/doi/abs/10.1111/j.2044-8317.1983.tb01131.x.

A. D. Roy. Safety first and the holding of assets. *Econometrica*, 20(3):431–449, 1952. ISSN 00129682. doi: 10.2307/1907413. URL http://www.jstor.org/stable/1907413.

D. Ruppert, M. P. Wand, and R. J. Carroll. *Semiparametric Regression.* Cambridge Series in Statistical and Probabilistic Mathematics. Cambridge University Press, 2003. ISBN 9780521785167. URL https://books.google.com/books?id=Y4uEvXFP2voC.

Jeffrey A. Ryan and Joshua M. Ulrich. *xts: eXtensible Time Series*, 2020. URL https://CRAN.R-project.org/package=xts. R package version 0.12-0.

Stephen Satchell and Alan Scowcroft. A demystification of the Black-Litterman model: Managing quantitative and traditional portfolio construction. *Journal of Asset Management*, 1(2):138–150, 2000. doi: 10.1057/palgrave.jam.2240011. URL https://link.springer.com/content/pdf/10.1057/palgrave.jam.2240011.pdf.

Fritz Scholz. Applications of the noncentral t-distribution. Privately Published, April 2007. URL http://www.stat.washington.edu/fritz/DATAFILES498B2008/NoncentralT.pdf.

Pranab Kumar Sen. Union-intersection principle and constrained statistical inference. *Journal of Statistical Planning and Inference*, 137(11):3741–3752, 2007. ISSN 0378-3758. doi: 10.1016/j.jspi.2007.03.046. URL http://www.sciencedirect.com/science/article/pii/S0378375807001310. Special Issue: In Celebration of the Centennial of The Birth of Samarendra Nath Roy (1906-1964).

William F. Sharpe. Mutual fund performance. *Journal of Business*, 39:119, 1965. doi: 10.1086/294846. URL http://ideas.repec.org/a/ucp/jnlbus/v39y1965p119.html.

William F. Sharpe. Likely gains from market timing. *Financial Analysts Journal*, 31(2):60–69, 1975. ISSN 0015198X. URL http://www.jstor.org/stable/4477805.

Gwowen Shieh. A comparative study of power and sample size calculations for multivariate general linear models. *Multivariate Behavioral Research*, 38(3):285–307, 2003. doi: 10.1207/S15327906MBR3803_01. URL http://www.tandfonline.com/doi/abs/10.1207/S15327906MBR3803_01.

Gwowen Shieh. Power and sample size calculations for multivariate linear models with random explanatory variables. *Psychometrika*, 70(2):347–358, 2005. doi: 10.1007/s11336-003-1094-0. URL https://ir.nctu.edu.tw/bitstream/11536/13599/1/000235235100007.pdf.

Mervyn J. Silvapulle and Pranab Kumar Sen. *Constrained statistical inference : inequality, order, and shape restrictions.* Wiley-Interscience, Hoboken, N.J., 2005. ISBN 0471208272. URL http://books.google.com/books?isbn=0471208272.

Kesar Singh, Minge Xie, and William E. Strawderman. Confidence distribution (CD) – distribution estimator of a parameter. In Regina Liu, William Strawderman, and Cun-Hui Zhang, editors, *Complex Datasets and Inverse Problems*, volume Volume 54 of *Lecture Notes–Monograph Series*, pages 132–150. Institute of Mathematical Statistics, Beachwood, Ohio, USA, 2007. doi: 10.1214/074921707000000102. URL http://dx.doi.org/10.1214/074921707000000102.

David Slepian. The one-sided barrier problem for Gaussian noise. *Bell System Technical Journal*, 41(2):463–501, 1962. doi: 10.1002/j.1538-7305.1962.tb02419.x. URL https://onlinelibrary.wiley.com/doi/abs/10.1002/j.1538-7305.1962.tb02419.x.

Kent Smetters and Xingtan Zhang. A sharper ratio: A general measure for correctly ranking non-normal investment risks. Working Paper 19500, National Bureau of Economic Research, October 2013. doi: 10.3386/w19500. URL http://www.nber.org/papers/w19500.

Thomas Smith. The Sharpe ratio ratio, 2019. URL https://riskyfinance.com/wp-content/uploads/2019/07/The_Sharpe_Ratio_Ratio_ThomasSmith-FINAL.pdf.

M. R. Spiegel and L. J. Stephens. *Schaum's Outline of Statistics.* Schaum's Outline Series. Mcgraw-hill, 2007. ISBN 9780071594462. URL http://books.google.com/books?id=qdcBmgs3N3AC.

M. C. Spruill. Computation of the maximum likelihood estimate of a noncentrality parameter. *Journal of Multivariate Analysis*, 18(2):216–224, 1986. ISSN 0047-259X. doi: 10.1016/0047-259X(86)90070-9. URL http://www.sciencedirect.com/science/article/pii/0047259X86900709.

Stephen M. Stigler. Stigler's law of eponymy. *Transactions of the New York Academy of Sciences*, 39(1 Series II):147–157, 1980. ISSN 2164-0947. doi: 10.1111/j.2164-0947.1980.tb02775.x. URL http://dx.doi.org/10.1111/j.2164-0947.1980.tb02775.x.

Edward J. Sullivan. A.D. Roy: The forgotten father of portfolio theory. *Research in the History of Economic Thought and Methodology*, 29:73–82, 2011. doi: 10.1108/S0743-4154(2011)000029A008. URL http://www.emeraldinsight.com/books.htm?chapterid=1943472.

W. Y. Tan. On the distribution of quadratic forms in normal random variables. *Canadian Journal of Statistics*, 5(2):241–250, 1977. doi: 10.2307/3314784. URL https://onlinelibrary.wiley.com/doi/abs/10.2307/3314784.

Dei-In Tang. Uniformly more powerful tests in a one-sided multivariate problem. *Journal of the American Statistical Association*, 89(427):1006–1011, 1994. ISSN 01621459. doi: 10.2307/2290927. URL http://www.jstor.org/stable/2290927.

N. H. Timm. *Applied multivariate analysis: methods and case studies.* Springer Texts in Statistics. Physica-Verlag, 2002. ISBN 9780387227719. URL http://amzn.to/TMdgaE.

Jun Tu and Guofu Zhou. Markowitz meets Talmud: A combination of sophisticated and naive diversification strategies. *Journal of Financial Economics*, 99(1):204–215, 2011. doi: 10.1016/j.jfineco.2010.08.013. URL http://ink.library.smu.edu.sg/cgi/viewcontent.cgi?article=2104&context=lkcsb_research.

Vladimir V. Ulyanov. Cornish-Fisher expansions. In Miodrag Lovric, editor, *International Encyclopedia of Statistical Science*, pages 312–315. Springer Berlin Heidelberg, 2014. ISBN 978-3-642-04897-5. doi: 10.1007/978-3-642-04898-2_193. URL http://dx.doi.org/10.1007/978-3-642-04898-2_193.

Rossen Valkanov. Long-horizon regressions: theoretical results and applications. *Journal of Financial Economics*, 68(2):201 – 232, 2003. ISSN 0304-405X. doi: 10.1016/S0304-405X(03)00065-5. URL http://www.sciencedirect.com/science/article/pii/S0304405X03000655.

Gerald van Belle. *Statistical Rules of Thumb (Wiley Series in Probability and Statistics)*. Wiley-Interscience, 1 edition, March 2002. ISBN 0471402273. URL http://www.vanbelle.org/.

C. Vignat and S. Bhatnagar. An extension of Wick's theorem, 2007. URL http://arxiv.org/abs/0709.1999.

Michael Vock. *"One-sided" Statistical Inference for a Multivariate Location Parameter.* PhD thesis, University of Bern, 2007. URL http://www.imsv.unibe.ch/e19451/e226662/e309213/e309217/e309218/files309221/Diss_Vock_3rd_eng.pdf.

Christian Walck. *Hand-book on STATISTICAL DISTRIBUTIONS for experimentalists.* Privately Published, December 1996. URL http://www.stat.rice.edu/~dobelman/textfiles/DistributionsHandbook.pdf.

Qiying Wang and Peter Hall. Relative errors in central limit theorems for Student's t statistic, with applications. *Statistica Sinica*, pages 343–354, 2009. URL https://pdfs.semanticscholar.org/75b0/d953ebcd84c493c71f89e6903c42571c15d1.pdf.

Larry Wasserman. *All of Statistics: A Concise Course in Statistical Inference.* Springer Texts in Statistics. Springer, 2004. ISBN 9780387402727. URL http://books.google.com/books?id=th3fbFI1DaMC.

Halbert White. A reality check for data snooping. *Econometrica*, 68:1097–1127, 2000. doi: 10.1111/1468-0262.00152. URL https://www.ssc.wisc.edu/~bhansen/718/White2000.pdf.

Hadley Wickham. *ggplot2: Elegant Graphics for Data Analysis.* Springer-Verlag New York, 2016. ISBN 978-3-319-24277-4. URL https://ggplot2.tidyverse.org.

Hadley Wickham, Romain François, Lionel Henry, and Kirill Müller. *dplyr: A Grammar of Data Manipulation*, 2020. URL https://CRAN.R-project.org/package=dplyr. R package version 1.0.1.

Wikipedia contributors. Noncentral t-distribution — wikipedia, the free encyclopedia, 2011. URL http://en.wikipedia.org/w/index.php?title=Noncentral_t-distribution&oldid=423475834. [Online; accessed 18-April-2011].

Wikipedia contributors. Chuck Norris facts — Wikipedia, the free encyclopedia, 2020. URL https://en.wikipedia.org/w/index.php?title=Chuck_Norris_facts&oldid=953622972. [Online; accessed 17-May-2020].

Samuel S. Wilks. The large-sample distribution of the likelihood ratio for testing composite hypotheses. *The Annals of Mathematical Statistics*, 9 (1):60–62, 1938. ISSN 00034851. doi: 10.1214/aoms/1177732360. URL https://projecteuclid.org/euclid.aoms/1177732360.

Samuel S. Wilks. *Mathematical Statistics.* Wiley publication in mathematical statistics. Wiley, 1962. URL https://books.google.com/books?id=zREJAQAAIAAJ.

Viktor Witkovsky. A note on computing extreme tail probabilities of the noncentral t distribution with large noncentrality parameter. Privately Published, 2013. URL http://arxiv.org/abs/1306.5294.

Frank A Wolak. An exact test for multiple inequality and equality constraints in the linear regression model. *Journal of the American Statistical Association*, 82(399):782–793, 1987. doi: 10.1080/01621459.1987.10478499. URL https://web.stanford.edu/group/fwolak/cgi-bin/sites/default/files/files/An%20Exact%20Test%20for%20Multiple%20Equality%20and%20Inequality%20Constraints%20in%20the%20Linear%20Regression%20Model_Wolak(1).pdf.

S. N. Wood. *Generalized Additive Models: An Introduction with R*. Chapman and Hall/CRC, 2 edition, 2017.

John Alexander Wright, Sheung Chi Phillip Yam, and Siu Pang Yung. A test for the equality of multiple Sharpe ratios. *Journal of Risk*, 16(4), 2014. doi: 10.21314/JOR.2014.289. URL http://www.risk.net/journal-of-risk/journal/2340044/latest-issue-of-the-journal-of-risk-volume-16-number-4-2014.

Min-ge Xie and Kesar Singh. Confidence distribution, the frequentist distribution estimator of a parameter: A review. *International Statistical Review*, 81(1):3–39, 2013. ISSN 1751-5823. doi: 10.1111/insr.12000. URL http://dx.doi.org/10.1111/insr.12000.

Yihui Xie. *knitr: A General-Purpose Package for Dynamic Report Generation in R*, 2018. URL https://yihui.name/knitr/. R package version 1.19.

Hirokazu Yanagihara. Asymptotic expansions of the null distributions of three test statistics in a nonnormal GMANOVA model. *Hiroshima Mathematical Journal*, 31(2):213–262, 07 2001. doi: 10.32917/hmj/1151105700. URL http://projecteuclid.org/euclid.hmj/1151105700.

Ruoyong Yang and James O Berger. *A catalog of noninformative priors*. Institute of Statistics and Decision Sciences, Duke University, 1996. URL https://www.stat.duke.edu/~berger/papers/catalog.ps.

Chuancun Yin. Stochastic orderings of multivariate elliptical distributions, 2019. URL https://arxiv.org/abs/1910.07158.

Valeri Zakamouline and Steen Koekebakker. Portfolio Performance Evaluation with Generalized Sharpe Ratios: Beyond the Mean and Variance. *SSRN eLibrary*, 2008. doi: 10.2139/ssrn.1028715. URL http://ssrn.com/paper=1028715.

Achim Zeileis. Econometric computing with HC and HAC covariance matrix estimators. *Journal of Statistical Software*, 11(10):1–17, 2004. URL http://www.jstatsoft.org/v11/i10/.

Ofer Zeitouni. GAUSSIAN FIELDS notes for lectures. unpublished notes, 2017. URL http://www.wisdom.weizmann.ac.il/~zeitouni/notesGauss.pdf.

Jian Zhang. *Path-Dependence Properties of Leveraged Exchange-Traded Funds: Compounding, Volatility and Option Pricing.* PhD thesis, New York University, 2010. URL http://math.cims.nyu.edu/faculty/avellane/thesis_Zhang.pdf.

Ming Zhou and Yongzhao Shao. A powerful test for multivariate normality. *Journal of Applied Statistics*, 41(2):351–363, 2014. doi: 10.1080/02664763.2013.839637. URL https://www.ncbi.nlm.nih.gov/pmc/articles/PMC3927875/.

Gregory Zuckerman. *The Man Who Solved the Market: How Jim Simons Launched the Quant Revolution.* Penguin Publishing Group, 2019. ISBN 9780735217980. URL https://books.google.com/books?id=2vmODwAAQBAJ.

Index

> Index I copy from old
> Vladivostok telephone directory.
>
> TOM LEHRER
> *Lobachevsky*

Printed in the United States
by Baker & Taylor Publisher Services